Metal-Ion Separation
and Preconcentration

ACS SYMPOSIUM SERIES **716**

Metal-Ion Separation and Preconcentration

Progress and Opportunities

Andrew H. Bond, EDITOR
Argonne National Laboratory

Mark L. Dietz, EDITOR
Argonne National Laboratory

Robin D. Rogers, EDITOR
The University of Alabama

American Chemical Society, Washington, DC

The paper used in this publication meets the minimum requirements of American National Standard for Information Sciences—Permanence of Paper for Printed Library Materials, ANSI Z39.48–1984.

Copyright © 1999 American Chemical Society

ISBN 0–8412–3594–5

Distributed by Oxford University Press

All Rights Reserved. Reprographic copying beyond that permitted by Sections 107 or 108 of the U.S. Copyright Act is allowed for internal use only, provided that a per-chapter fee of $20.00 plus $0.25 per page is paid to the Copyright Clearance Center, Inc., 222 Rosewood Drive, Danvers, MA 01923, USA. Republication or reproduction for sale of pages in this book is permitted only under license from ACS. Direct these and other permissions requests to ACS Copyright Office, Publications Division, 1155 16th Street, N.W., Washington, DC 20036.

The citation of trade names and/or names of manufacturers in this publication is not to be construed as an endorsement or as approval by ACS of the commercial products or services referenced herein; nor should the mere reference herein to any drawing, specification, chemical process, or other data be regarded as a license or as a conveyance of any right or permission to the holder, reader, or any other person or corporation, to manufacture, reproduce, use, or sell any patented invention or copyrighted work that may in any way be related thereto. Registered names, trademarks, etc., used in this publication, even without specific indication thereof, are not to be considered unprotected by law.

PRINTED IN THE UNITED STATES OF AMERICA

Advisory Board

ACS Symposium Series

Mary E. Castellion
ChemEdit Company

Arthur B. Ellis
University of Wisconsin at Madison

Jeffrey S. Gaffney
Argonne National Laboratory

Gunda I. Georg
University of Kansas

Lawrence P. Klemann
Nabisco Foods Group

Richard N. Loeppky
University of Missouri

Cynthia A. Maryanoff
R. W. Johnson Pharmaceutical
  Research Institute

Roger A. Minear
University of Illinois
  at Urbana–Champaign

Omkaram Nalamasu
AT&T Bell Laboratories

Kinam Park
Purdue University

Katherine R. Porter
Duke University

Douglas A. Smith
The DAS Group, Inc.

Martin R. Tant
Eastman Chemical Co.

Michael D. Taylor
Parke-Davis Pharmaceutical
  Research

Leroy B. Townsend
University of Michigan

William C. Walker
DuPont Company

Foreword

THE ACS SYMPOSIUM SERIES was first published in 1974 to provide a mechanism for publishing symposia quickly in book form. The purpose of the series is to publish timely, comprehensive books developed from ACS-sponsored symposia based on current scientific research. Occasionally, books are developed from symposia sponsored by other organizations when the topic is of keen interest to the chemistry audience.

Before agreeing to publish a book, the proposed table of contents is reviewed for appropriate and comprehensive coverage and for interest to the audience. Some papers may be excluded in order to better focus the book; others may be added to provide comprehensiveness. When appropriate, overview or introductory chapters are added. Drafts of chapters are peer-reviewed prior to final acceptance or rejection, and manuscripts are prepared in camera-ready format.

As a rule, only original research papers and original review papers are included in the volumes. Verbatim reproductions of previously published papers are not accepted.

ACS BOOKS DEPARTMENT

Contents

Preface...xi

Overviews

1. Progress in Metal Ion Separation and Preconcentration:
   An Overview...2
   Andrew H. Bond, Mark L. Dietz, and Robin D. Rogers

2. Future of Separation Science in U.S. Department of Energy
   Processing and Remediation...13
   Gregory R. Choppin

3. Solvent Extraction in the Treatment of Acidic High-Level Liquid
   Waste: Where Do We Stand?..20
   E. Philip Horwitz and Wallace W. Schulz

Aqueous Systems

4. Aqueous Complexes in f-Element Separation Science.........................52
   Kenneth L. Nash

5. Metal Ion Separations in Aqueous Biphasic Systems and Using
   Aqueous Biphasic Extraction Chromatography....................................79
   Jonathan G. Huddleston, Scott T. Griffin, Jinhua Zhang,
   Heather D. Willauer, and Robin D. Rogers

Extractant Design and Synthesis

6. A Molecular Mechanics Method for Predicting the Influence
   of Ligand Structure on Metal Ion Binding Affinity..............................102
   Benjamin P. Hay

7. Ligand Design for Small Cations: The Li^+/14-Crown-4 System.............114
   Richard A. Sachleben and Bruce A. Moyer

8. Synthesis of Novel Azamacrocyclic Metal Ion Receptors Using a
   Modified Mannich Aminomethylation Reaction 133
   Jerald S. Bradshaw, Andrei V. Bordunov, Xian Xin Zhang,
   Victor N. Pastushok, and Reed M. Izatt

Separations Using Liquid–Liquid Systems

9. Metal Ion Separations with Proton-Ionizable Lariat Ethers 146
   Richard A. Bartsch

10. Redox-Recyclable Extraction and Recovery of Heavy Metal Ions
    and Radionuclides from Aqueous 156
    Steven H. Strauss

Separations Using Solid–Liquid Systems

11. Structural Basis of Selectivity in Tunnel Type Inorganic
    Ion Exchangers..168
    Abraham Clearfield, Damodara M. Poojary,
    Elizabeth A. Behrens, Roy A. Cahill, Anatoly I. Bortun,
    and Lyudmila N. Bortun

12. Metal Ion Separations with Lariat Ether Ion-Exchange Resins............183
    Richard A. Bartsch and Takashi Hayashita

13. Design of Novel Polymer-Supported Reagents
    for Metal Ion Separations..194
    Spiro D. Alexandratos and Latiff A. Hussain

14. Recent Advances in the Chemistry and Applications
    of the Diphonix Resins...206
    E. Philip Horwitz, Renato Chiarizia, Spiro D. Alexandratos,
    and Michael Gula

15. Reillex-HPQ Anion Exchange Column Chromatography:
    Removal of Pertechnetate Ion from DSSF-5 Simulant.........................219
    Norman C. Schroeder, Susan D. Radzinski, Jason R. Ball,
    Kenneth R. Ashley, and Glenn D. Whitener

16. Extraction Chromatography: Progress and Opportunities....................234
    Mark L. Dietz, E. Philip Horwitz, and Andrew H. Bond

17. Metal-Ion Separations Using SuperLig or AnaLig Materials
    Encased in Empore Cartridges and Disks......................................251
        Garold L. Goken, Ronald L. Bruening,
        Krzysztof E. Krakowiak, and Reed M Izatt

18. Biologically Generated Materials for Metal-Ion Binding:
    Answers to Some Fundamental Chemical Questions........................260
        Gary D. Rayson, Lawrence R. Drake, Hongying Xia,
        Shan Lin, and Paul J. Jackson

Separations Using Membranes

19. Use of Ligand-Modified Micellar-Enhanced Ultrafiltration
    to Selectively Remove Copper from Water......................................280
        Susan B. Shadizadeh, Richard W. Taylor, John F. Scamehorn,
        Annette L. Schovanec, and Sherril D. Christian

20. Water-Soluble Metal-Binding Polymers with Ultrafiltration................294
        Barbara F. Smith, Thomas W. Robison, and Gordon D. Jarvinen

Separations Using Chromatographic and Supercritical Fluid Extraction Systems

21. Different Two-Phase Liquid Systems for Inorganic Separations
    by Countercurrent Chromatography...333
        Boris Ya. Spivakov, Tatiana A. Maryutina, Petr S. Fedotov,
        and Svetlana N. Ignatova

22. Fundamental Aspects of Metal-Ion Separations by Centrifugal
    Partition Chromatography...347
        Subramaniam Muralidharan and Henry Freiser

21. Extraction and Separation of Uranium and Lanthanides
    with Supercritical Fluids...390
        Chien M. Wai, Yuehe Lin, Min Ji, Karen L. Toews, and
        Neil G. Smart

Indexes

Author Index...403

Subject Index..405

Preface

In many respects, separations science can be regarded as a mature field. Numerous separations methods (e.g., ion exchange and precipitation) are very well established and, indeed, have considerable economic impact on a variety of industrial and other processes. In recent years, however, the need for new and improved separations technologies, particularly for metal ions, has grown substantially. In part, this growth is a result of increased awareness of the problems associated with the introduction of toxic metals into the environment and the resultant need for processes for the treatment of industrial effluents. At the same time, there has been increasing need to separate and preconcentrate various metals for subsequent determination, a need driven by both environmental regulation and the demand for high purity materials.

In 1987, the National Academy of Sciences published a report on the field of separations and outlined high priority research needs. Now, a decade later, we believe that it is an appropriate time to consider the progress toward meeting those needs that has been made recently in the area of metal ion separations. The twenty-three chapters of this book cover both novel separations media and methodology and recent developments in established separation techniques. Because of the enormous amount of research activity in this field, this volume is not intended to be all-inclusive. Rather the objective is to provide the reader with a general indication of the state-of-the-art in the field, in terms of both fundamental conceptual and technical advances and applications. We expect this book to be of use to both the novice and the experienced practitioner, to basic researchers who work in metal ion complexation and separation, and to those interested in the application of metal ion separations methods to problems in waste treatment, environmental remediation, and chemical analysis.

Acknowledgments

We gratefully acknowledge the financial support of the Office of Basic Energy Sciences of the United States Department of Energy, the Separation Science and Technology Subdivision of the ACS Division of Industrial and Engineering Chemistry, Eichrom Industries, Inc., and Reilly Industries, Inc. for the Symposium on Recent Advances in Metal Ion Separation and Preconcentration, held at the 214th ACS National Meeting in Las Vegas, Nevada, September 7-11, 1997.

We also acknowledge the members of the ACS Books Department for their assistance in assembling this volume. Finally we thank all those who participated in the symposium, the authors for preparing their manuscripts, and the referees for reviewing each chapter.

MARK L. DIETZ
Chemistry Division
Argonne National Laboratory
Argonne, IL 60439

ANDREW H. BOND
Chemistry Division
Argonne National Laboratory
Argonne, IL 60439

ROBIN D. ROGERS
Department of Chemistry
The University of Alabama
Tuscaloosa, AL 35487

OVERVIEWS

Chapter 1

Progress in Metal Ion Separation and Preconcentration: An Overview

Andrew H. Bond[1], Mark L. Dietz[1], and Robin D. Rogers[2]

[1]Chemistry Division, Argonne National Laboratory, 9700 South Cass Avenue, Argonne, IL 60439–4831
[2]Department of Chemistry, The University of Alabama, Tuscaloosa, AL 35487

> A brief historical perspective covering the most mature chemically-based metal ion separation methods is presented, as is a summary of the recommendations made in the 1987 National Research Council (NRC) report entitled "Separation and Purification: Critical Needs and Opportunities". An examination of *Metal Ion Separation and Preconcentration: Progress and Opportunities* shows that advances are occurring in each area of need cited by the NRC. Following an explanation of the objectives and general organization of this book, the contents of each chapter are briefly summarized and some future research opportunities in metal ion separations are presented.

Physical and pyrochemical processes for the separation and purification of metals have been employed for thousands of years; however, the ability to selectively partition metal ions using chemical means is a more recent development. Selective purification and concentration of metals in solution has primarily been achieved using liquid-liquid extraction and ion exchange. These techniques have achieved a high level of success on the analytical, preparative, and process-scales and, consequently, hold an important place in the hierarchy of metal ion separations methods.

More than a century has passed since liquid-liquid extraction, more commonly known as solvent extraction, was first used on an analytical-scale to partition mercuric chloride from an aqueous solution into ether (*1*). In 1942, after the potential of nuclear weapons and energy was recognized, solvent extraction was employed on a preparative-scale for the separation of plutonium from uranium targets for studies of its chemical and nuclear properties (*2-4*). Shortly thereafter, ether extraction for the purification of uranium was transferred to the industrial-scale for weapons development (*1,5*). The effectiveness of tri-*n*-butyl phosphate as an extractant for uranium and plutonium was recognized in 1949 (*6*), and the first PUREX (Plutonium-Uranium Recovery and EXtraction) processing canyon was constructed at the United States' Savannah River site in 1951 (*7*). The success of solvent extraction in the production of high purity metals for the nuclear defense and energy programs, together with the development of a variety of selective metal ion extractants, has led to the widespread application of solvent extraction in hydrometallurgical separations (*8-10*) and in nuclear reprocessing and waste treatment applications (*5,6,11,12*).

Nearly a century has also passed since Tswett's initial use of $CaCO_3$ in a chromatographic mode to separate biological pigments, work which paved the way for the preparation of the first ion-exchange resins in 1935 (13) and their subsequent characterization and application. Ion-exchange chromatography was so well developed by the 1950s that it was employed as the sole method of chemical confirmation in the initial report of the discovery of mendelevium (4,14).

By the 1970s, the design and synthesis of ligands showing metal ion and/or molecular recognition had begun to attract considerable attention. This work was led by the pioneering efforts of Pederson, Cram, and Lehn and has had an immeasurable impact on separation science.

Membrane-based (15-17) and supercritical fluid (18-20) processes for metal ion extraction are newer developments and are the subjects of intense research interest. Numerous applications for these techniques have been proposed, and many are likely to receive considerable attention in the future.

Despite these advances, it has long been apparent that certain research needs in separation science have remained unmet. Ten years ago, the National Research Council (NRC) published a report entitled "Separation and Purification: Critical Needs and Opportunities" (21) in which the current status and future directions of separations science were discussed. Various shortcomings of existing methodologies were noted, and several broadly applicable generic research needs were identified (21,22):

(1) Improved Selectivity
(2) Selective Concentration from Dilute Solution
(3) Study of Interfacial Phenomena
(4) Improvements to Processing Rates and Capacity of Separation Systems to Minimize Capital Equipment Costs
(5) Development of New Process Configurations and Integration
(6) Improvement of Energy Efficiency

Many of these proposed needs are interdependent, and advances in one area often effect improvements in others. The principal advances in metal ion separations over the last 10-20 years have occurred in the realm of improved selectivity, cited by the NRC committee as a vital need. Indeed, enhanced selectivity in virtually any separations process can favorably affect process operation, efficiency, and economics.

One of the objectives of this book is to present an overview of the state-of-the-art in metal ion separations, thereby permitting a comparison of the past decade's progress with the perceived needs presented in the NRC report. A review of *Metal Ion Separation and Preconcentration: Progress and Opportunities* shows that many of the proposed needs are being addressed and that significant advances in each of the generic research areas are occurring. The development of ion-selective extractants and separations processes is well represented in this book, as are separations from dilute solutions, studies of interfacial phenomena and extraction kinetics, and the development of new separation systems that promise minimal adverse environmental impacts and improved energy efficiencies.

Separation science has been described as a mature field of study (23), and the ubiquitous scientific and industrial needs for purity and, hence, chemical separations have long been recognized. As separation science matures, however, so to do technology and societal priorities; thus, new needs are continually being created. For example, the significant current research thrusts in environmental remediation, pollution prevention, and the preparation of ultrahigh-purity metals for technology industries have demanded that new and highly selective separation and purification processes be developed. Given that chemical separations are involved in nearly every major industrial process, there will continue to be increasing regulatory and economic incentives to improve efficiencies, incorporate pollution prevention strategies, and minimize adverse environmental impacts. Separations science must advance to meet these needs, and a review of the ongoing research in this field indicates that it is indeed evolving to meet its economic, regulatory, and societal drivers.

Due to the multitude of techniques available to the separation scientist, a detailed discussion of the factors influencing metal ion separations and their applications is beyond the scope of this chapter. Furthermore, these topics have been adequately covered in general texts on separations science (*24,25*) and in more focused treatises on solvent extraction (*26-30*), ion exchange (*28,31-33*), membrane-based processes (*15-17*), and supercritical fluid extraction (*18-20*). Some understanding of the unifying themes common to most metal ion separations is useful, however, as it will aid in describing the organization of this book.

Giddings has defined the process of separation in the most general terms: "... a group of components, originally intermixed, are forced into different spatial locations by the process of separation" (*24*). This definition is depicted in Figure 1, in which a mixture of cationic and anionic components and unspecified matrix species are segregated by an arbitrary separation method. Numerous means of separating metal ions using such a broad definition exist, and each obviously has thermodynamic and kinetic aspects. Other features common to many separation processes include the role of the solvent matrix (most often aqueous) and its constituents; extractant design, synthesis, and incorporation into a diluent or a polymeric support; and application-directed testing. These quite general but common features have provided the platform upon which this book is organized. Chapters 2 and 3 are overviews of the status of metal ion separations in nuclear energy and defense-related systems and are followed by chapters discussing the role of the aqueous phase in separations (Chapter 4) and the use of aqueous two-phase systems for metal ion separations (Chapter 5). The focus then turns to the design and synthesis of ion-selective extractants in Chapters 6-8 and to development and testing of various liquid-liquid (Chapters 9-10) and solid-liquid (Chapters 11-18) separation modes. Finally, specific techniques including membrane filtration (Chapters 19-20), countercurrent and centrifugal solvent extraction (Chapters 21 and 22, respectively), and separations using supercritical fluids (Chapter 23) are presented. Such an organization proceeds from the influence of the often encountered aqueous phase to extractant design and preparation, and culminates with fundamental and application-directed studies. For the topical reader, this "modular" organization permits the independent study of major research areas. For example, solid-liquid-based separations of metal ions are grouped together and are categorized by cation and anion exchange, extraction chromatography, membrane disks, and biologically-derived solids. Our goal was to provide a timely account of advances in the field of metal ion separation and preconcentration and for this work to serve as a resource for both the novice and experienced practitioner.

Overview of Chapters

Chapter 2 by Choppin provides a historical perspective on the origin of the most significant radioactive wastes in the United States Department of Energy inventory. These wastes were generated primarily from uranium ore processing, isotopic enrichment by gaseous diffusion, and chemical separations of fissile plutonium from irradiated uranium targets. Factors motivating the remediation of such wastes include the hazards associated with long-term tank storage of radioactive liquids and the realization that the nation's facilities for the storage of spent fuel elements and the associated wastes (e.g., fuel ponds, cooling water, etc.) from energy uses are reaching capacity. This latter point is of particular relevance considering that the Department of Energy is slated to begin accepting spent fuel rods from civilian reactors in the very near future. The factors important in the design of actinide-specific ligands and various facets of actinide chemistry (e.g., oxidation/reduction and complexation chemistry) important to their separations chemistry are discussed.

An overview of candidate treatment options covers aqueous processing techniques including solvent extraction and ion-exchange strategies, while the discussion of nonaqueous processes is limited to volatility-based separations and

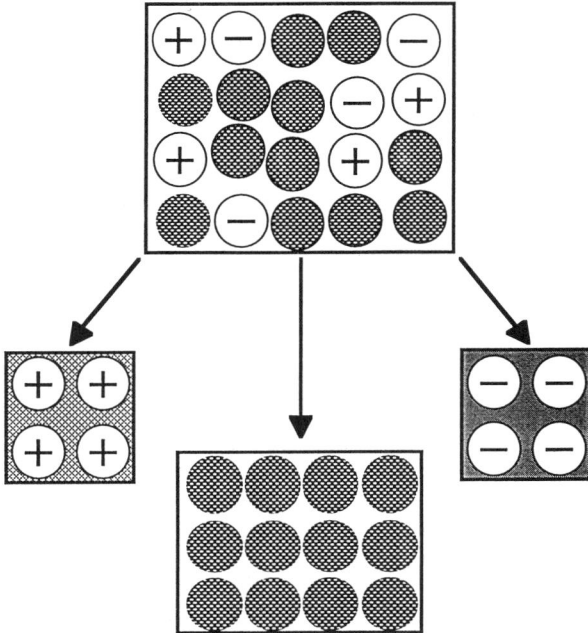

Figure 1. Schematic representation of a generic separation of monovalent ions proposed by Giddings' definition (24).

pyrochemical processes. The use of natural sequestering agents, which continues to attract attention, is also reviewed.

Horwitz and Schulz address the status of solvent extraction in the processing of radioactive acidic liquid wastes in Chapter 3. Flowsheets that have been tested for the separation of U, Np, Pu, Am, and Cm (the latter for civilian reactor wastes) and processes for the removal of the ^{90}Sr and ^{137}Cs fission products are presented.

Dialkylphosphoric acids, neutral bifunctional carbamoylmethylphosphine oxides, and neutral bifunctional malonamides have all been tested for actinide removal from acidic wastes or representative simulants. The advantages and disadvantages of each process are discussed in both fundamental chemical and engineering terms. In general, the limiting criterion for success of processes employing these extractants is the distribution ratio for Am^{3+}, and a comparison of D_{Am} vs. [HNO_3] is presented for the favored extractants in the Chinese, French, Russian, and American flowsheets.

Solvent extraction processes for the removal of ^{90}Sr and ^{137}Cs fission products are limited to cobalt dicarbollide and crown ether-based flowsheets. The advantages and limitations of each are addressed, as are their compatibility with subsequent actinide separation systems.

The influence of the aqueous phase on metal ion separations is discussed by Nash in Chapter 4. The thermodynamic and kinetic factors affecting the transfer of a metal ion from an aqueous to an organic medium are generalized for application to liquid-liquid, extraction chromatographic, and solid-liquid extraction systems. The overall focus is on separations of f-elements, and their hydrolysis and complexation chemistries are discussed. Ion solvation and short and long-range water structuring are presented in the context of metal ion extraction and, often, dehydration upon transfer to an organic phase. Use of soft donor ligands and/or aqueous phase complexation in the separation of lanthanides from trivalent actinides provides the framework for discussion of the fundamentals of aqueous phase complexation.

Chapter 5 presents work by Rogers et al. aimed at identifying the major factors influencing solute partitioning in polyethylene glycol-based liquid-liquid and solid-liquid aqueous biphasic extraction systems. Once these factors are identified and their relative contributions understood, models capable of predicting the partitioning of solutes in these systems can be developed. The authors discuss the role of polymer molecular weight, salt type, and ion hydration thermodynamics (ΔG_{hyd}) in partitioning and have extended these concepts into predictive instruments. Partitioning of the pertechnetate anion (TcO_4^-) has been successfully correlated with the ΔG_{hyd} of the salt-phase forming components and with specialized phase incompatibility parameters. Similarities in the partitioning mechanism between liquid-liquid aqueous biphasic separations and solid-liquid aqueous biphasic extraction chromatography (ABEC) are also discussed.

The section on extractant design and synthesis commences with Chapter 6 by Hay, in which the correlation of macrocyclic ligand strain energies with metal ion binding affinities is presented. A general algorithm accounting for energy changes due to conformational rearrangement upon complexation is discussed and tested for five series of ligands. The primary differences within each series of ligands are the location, regiochemistry, and/or type of alkyl substituent, while the chelate ring size and donors generally remain identical. Complex formation constants, extraction constants, and distribution ratios are correlated with ligand strain energies in a series of oxygen donor macrocycles with different alkyl groups and rigidities. In each case, correlations of the ligand reorganization energies with metal ion affinities and/or distribution are observed. The importance of ligand preorganization in initiating complexation and conformational change within the ligand during coordination to the metal ion are discussed for each series of ligands.

Chapter 7 by Sachleben and Moyer discusses the many variables involved in the design of 14-membered macrocyclic ligands selective for Li^+. From the development of novel synthetic strategies for 14-crown-4 derivatives to uptake and molecular modeling results, the influence of ligand substituents, solvation, and diluent effects are detailed. The monomethyl and tetracyclopentyl-substituted 14-crown-4 extractants show enhanced uptake and selectivity for Li^+ over other alkali metal matrix cations. The importance of diluent effects on solvation and distribution of metal ions and metal-extractant complexes is addressed, with a particular focus on their importance in ligand design. Correlation of metal-crown ether complex stability with ligand strain energies calculated by molecular mechanics is discussed and presented as a useful tool in extractant design studies.

Chapter 8 by Bradshaw et al. presents the synthesis and solution thermodynamic properties of a series of nitrogen-containing macrocycles, cryptands, and hemispherands. The synthetic methodology generally involves reaction of an azamacrocycle with formaldehyde and methanol followed by coupling with Lewis base donor-substituted aromatic compounds. This aminomethylation route simplifies the synthesis of N-aryl-functionalized ligands by using existing azamacrocycles and regioselective addition to aromatic donors. The synthesis of a series of aryl-substituted 15- and 18-membered macrocycles and bi- and tricycles containing various nitrogen and oxygen donor arrays is detailed. The complex formation constants for aza-15-crown-5, aza-18-crown-6, diaza-18-crown-6, and their various N-aryl-substituted analogs are presented. One group of compounds, differing only in the point of attachment of the aryl substituent to the macrocycle and the resulting location of donor atoms in the aryl arm, shows marked differences in formation constants and quite different selectivities over alkali and alkaline-earth matrix cations.

Development and testing of separation systems begins in Chapter 9 by Bartsch with a description of alkali metal ion extraction using proton ionizable lariat ethers. The importance of electroneutrality in solvent extraction and the difficulty often associated with co-ion transport has motivated the development of macrocyclic extractants possessing acidic functionalities: carboxylate-containing lariat ethers in this case. Combining an acidic functionality with the trademark size selectivity of crown ethers has resulted in enhanced extraction and selectivity in alkali cation partitioning into chloroform. Examples of the role of macrocycle cavity size, spacer length separating the acidic group from the macrocycle, site of attachment of lipophilic groups, macrocycle rigidity, and the nature of the acidic moiety in the extraction of alkali metal cations are addressed.

The concepts of redox-recyclable extraction and recovery processes are introduced by Strauss in Chapter 10. At the center of such systems are transition metal-containing extractants capable of undergoing one-electron oxidation or reduction and exhibiting stability at the pH extremes common to many process waste streams. The general mechanism of extraction involves oxidation or reduction of the extractant followed by ion exchange with the solute. Stripping is effected by a redox reaction converting the charged extractant back to a neutral moiety that subsequently releases the extracted solute to a new aqueous phase. Several separation systems using substituted ferrocenes and lithium-intercalated MoS_2 operating in liquid-liquid, physisorbed or extraction chromatographic, or solid inorganic ion-exchange modes are presented.

Structure/function relationships in inorganic ion-exchange materials are discussed by Clearfield et al. in Chapter 11, in which the coverage of solid-liquid separations begins. Several tunnel-type titanosilicates and the related pharmacosiderites have been structurally characterized in both the acid and alkali metal-substituted forms. Structural features including bond distances, ionic diameters, and tunnel cavity sizes have been used to explain the observed stoichiometries and metal ion uptake properties of these materials. The titanosilicates exhibit strong binding and good selectivity for Cs^+ and K^+ over Li^+ and Na^+, but the observed capacity for Cs^+ is only $\approx 25\%$ of the theoretical value. Structural analyses show that the low capacity is due to a combination of steric, bonding, and electrostatic repulsion interactions that impede diffusion of Cs^+ into the tunnels of the ion-exchange material. The development of such structure/function relationships has proven useful in understanding the uptake properties of these ion exchangers and is important in the design of new materials.

In Chapter 12, Bartsch and Hayashita describe metal ion separations using lariat ether cation-exchange resins. These materials are prepared using condensation polymerization techniques wherein dibenzo polyethers are copolymerized in the presence of formaldehyde and acid. Alkali metal selectivity and uptake by a series of acyclic and cyclic polyethers are compared, and the influence of *n*-alkyl substitution on the carbon atom *geminal* to the lariat functionality is discussed. Dramatic differences in selectivity and uptake are observed when the sterics of the alkyl group force the acidic functionality to reside over the macrocyclic cavity, thereby favoring complexation. Separations of alkali, alkaline earth, transition, and main group cations by lariat ether carboxylic, phosphonic monoester, and sulfonic acids are discussed.

Chapter 13 describes the development of novel ion-exchange resins. In this work, Alexandratos and Hussain summarize early studies of ion-exchange resins that have provided the foundation for more recent developments in this field. Separations with traditional organic ion-exchange resins are often plagued by a lack of selectivity and the inability to adequately retain the target solutes in the presence of high concentrations of matrix ions or at pH extremes. Advances in resin design and preparation have largely overcome these obstacles, primarily by the development of chelating resins and by the incorporation of phosphorus acids into resin beads. The phosphonic acids, whose pK_a values are intermediate between the strong sulfonic and the weak carboxylic acids, offer enhanced selectivity and good uptake at intermediate to low pH values. Chelating diphosphonic acid-based resins, which provide exceptional uptake and selectivity even at low pH, are discussed as are the design of new reactive and dual bifunctional polymers. The latter materials possess improved uptake mechanisms that allow more rapid penetration of ions into the resin matrix and offer oxidation/reduction, ion-exchange, or precipitation reactions to immobilize the sorbed species.

Chapter 14 by Horwitz and Chiarizia et al. describes the diversification of the Diphonix class of chelating ion-exchange resins to include inorganic substrates and new cation- and anion-exchange functionalities. The cornerstone *gem*-diphosphonic acid moieties present in the sulfonated polystyrene-divinylbenzene copolymer, termed Regular Diphonix, have been grafted to a silica-based substrate to yield a resin somewhat more resistant to radiation damage and better suited to radioactive waste treatment applications than its precursor. Phenol-formaldehyde copolymer chains have been added to Regular Diphonix to yield Diphonix-CS, which effectively binds Cs^+ and Sr^{2+} for potential alkaline radioactive waste treatment applications. Incorporation of both *N*-methylpyridinium and diphosphonic acid moieties into a polymeric backbone provides another unique material capable of both cation and anion exchange. The major process-scale application of the Diphonix resins is Fe^{3+} removal from copper electrowinning bleed streams, and this application and its process chemistry are presented.

Separation of TcO_4^- from alkaline radioactive wastes by columns of Reillex-HPQ anion-exchange resin is detailed in Chapter 15 by Ashley and Schroeder et al. The origin and initial uses of the *N*-methylpyridinium-containing resin are discussed and lead into the testing of this resin for $^{99}TcO_4^-$ separations from alkaline media. Factors important to the use of Reillex-HPQ resins specifically for alkaline radioactive waste treatment are also discussed. Batch uptake kinetics studies were used to determine column residence times and flowrates for subsequent column efficiency and cycling experiments. Technetium is reductively stripped from the column using a combination of Sn^{2+}, ethylenediamine, and NaOH. This approach is extremely effective, yielding 99.4% of the reduced technetium in 1.5 bed volumes with an overall recovery of > 98%.

Advances in the development and application of extraction chromatographic resins are discussed by Dietz et al. in Chapter 16. Background details on the nature of the extractant and diluent interactions with the support are accompanied by a discussion of the factors controlling the efficiency of extraction chromatographic systems.

Recent developments have derived largely from the adaptation of new ion-specific extractants, originally designed for use in solvent extraction, to the extraction chromatographic mode. Several specific examples including radiostrontium separations using crown ethers and actinide separations using organophosphorus extractants are detailed. The influence of the support in extraction chromatographic systems is discussed with respect to its "inertness" (i.e., lack of participation in the extraction reaction, as for polystyrene divinylbenzene copolymers) or "active" (e.g., an extractant impregnated in an ion-exchange resin) role in such separations. Current applications of extraction chromatography are limited to analytical-scale separations, as loss of extractant and/or diluent to the mobile phase results in performance irregularities that preclude its use on the process-scale.

Chapter 17 by Izatt et al. describes the development and application of commercial ion-selective disks and cartridges. A variety of disks and methods for Hg^{2+}, Pb^{2+}, Sr^{2+}, and Ra^{2+} analyses have been developed. High selectivities over commonly encountered matrix ions have been achieved, and interference levels for these systems are discussed. Separations using the ion-selective disks are amenable to a variety of quantitation methods, as the solutes may be stripped and analyzed or the disks themselves may be counted if the analytes are radioactive.

The use of metal ion sorbents derived from biological materials is discussed in Chapter 18 by Rayson et al. Variables important in the design and use of separations employing biosorbents as well as the reactions influencing metal ion uptake by living and nonliving materials are covered. One such biosorbent, *Datura innoxia*, has been studied in the solution and solid states, and as a solid distributed on various organic and inorganic supports. Details regarding the number and type of binding sites have been derived from pH variation studies, chemical deactivation of specific binding sites, Eu^{3+} luminescence measurements, ^{113}Cd and ^{27}Al NMR, and affinity chromatographic experiments. *Datura innoxia* was shown to exhibit a range of mono- through tridentate carboxylate coordination modes and possibly some participation from native sulfonate groups.

Coverage of membrane filtration methods begins with Chapter 19 by Taylor and Scamehorn et al., where the technique of ligand-modified micellar-enhanced ultrafiltration (LM-MEUF) is presented. Incorporation of the 4-hexadecyloxybenzyliminodiacetic acid ligand into MEUF systems containing the cationic surfactant *N*-hexadecylpyridinium chloride has led to enhanced retention and selectivity for Cu^{2+} over Ca^{2+}. The influence of several variables, including the feed compositions of Cu^{2+} and Ca^{2+} and the ligand and surfactant concentrations, were studied. From this data, the metal:ligand stoichiometry was determined and optimal conditions were defined for selective retention of a copper chelate complex over calcium ions in the matrix.

Smith and Jarvinen et al. present an overview and detailed applications of water-soluble metal-binding polymers coupled with ultrafiltration techniques in Chapter 20. The various modes of ultrafiltration, their operating parameters, and variables pertinent to process scale-up are discussed. The applications range from analytical and bench-scale separation and concentration of actinides to pilot-scale gross α concentration from radioactive liquid waste streams. Net α activities were reduced from $> 10^5$ pCi/L to < 30 pCi/L in this demonstration. Two industrial applications are under development, including electroplating waste processing for both cations and anions and treatment of acid mine run-off, including that from the Berkeley Pit in Butte, MT.

In Chapter 21, Spivakov et al. discuss the design and development of countercurrent chromatographic (CCC) systems for metal ion separations. In addition to describing the chemical properties of the solvent systems (extractant and diluent), the important physical parameters affecting stationary phase retention in CCC separations are introduced. These factors affect the separation efficiency, as they relate to the volume of stationary phase that can be employed (influencing loading capacity). Several applications are presented in which separation and/or preconcentration of analytes is carried out. Isolation of lanthanides and yttrium from geological samples by stationary phases containing organophosphorus extractants show reasonable selectivity and good recoveries. Trivalent lanthanide and actinide separations; concentration of Zr, Hf, Nb, and Ta; and Cs and Sr separations from terrestrial matrices by a variety of organophosphorus, crown ether, and other chelating extractants are described.

The fundamental kinetic aspects of centrifugal partition chromatography (CPC) are detailed by Muralidharan and Freiser in Chapter 22. Separations of platinum group metals by anion exchange or trialkylphosphine oxide solvating extractants, Ni(II) by chelating ligands possessing different interfacial surface activities, and lanthanide cations using acylpyrazolones are characterized in terms of their extraction kinetics, mechanisms, and chromatographic efficiencies in CPC. It is shown that the efficiency of the CPC separations of these metals is limited by back extraction kinetics occurring both in the bulk aqueous and interfacial regimes and that the reduced plate heights correlate with the half-lives of the rate-determining chemical reactions.

Chapter 23 by Wai et al. is devoted to the supercritical fluid extraction (SFE) of f-element chelates. Factors important in the extraction and stripping of solutes from liquid and solid phases and the variables important to SFE of metal ions are addressed. Extraction of UO_2^{2+} from citric acid media by tri-n-butyl phosphate in supercritical CO_2 is shown to proceed similarly to its extraction into the hydrocarbon diluents used in PUREX processing. Intralanthanide separations using SFE with CO_2 and alkylphosphoric acids are also described. Extraction by di-(2,4,4-trimethylpentyl)phosphoric acid (Cyanex 272) shows the expected increase in extraction efficiency that accompanies an increase in the lanthanide atomic number. The mono- and dithiophosphinic acids (Cyanex 302 and Cyanex 301, respectively) show appreciably greater extraction for Dy-Lu than for La and Ce-Tb. An interesting potential application of metal ion separations using SFE involves the leaching of UO_2^{2+} from solid mine tailing samples by thenoyltrifluoroacetone.

Future Prospects

Identification of prospective research areas in the field of metal ion separations is complicated by the fact that the future needs are intimately related to technological advances in other fields, many of which cannot be readily foreseen. It is evident, however, that while certain goals of the 1987 National Research Council report on separations have been achieved, much remains to be accomplished. For example, the opportunities identified in the NRC report must now be balanced with the need for environmentally sustainable technologies. New industrial processes must weigh the need for improvements in efficiency with requirements for pollution reduction. Clearly, an opportunity exists for metal ion separations to play a major role in future pollution prevention strategies.

More recently, the Technology Vision 2020 statement (*34*) has identified a number of separations-related issues facing the chemical industry, but its goals of environmental stewardship are applicable to any manufacturing industry. This report asserts that environmentally sustainable technologies are not only a desirable goal, but an achievable one. Separation science will continue to play a major role in environmental remediation, but perhaps the most significant future challenges come from utilizing separations science in pollution prevention.

More focused fundamental investigations and application-directed studies also await the attention of separation scientists. For example, many chemical separation processes comprise the steps of extraction, rinse/scrub, and stripping or regeneration. The extraction stage has received the greatest attention to date, as the fundamentals of extractant design, synthesis, and studies of metal ion uptake properties are involved. Stripping and regeneration have received considerably less attention, despite the fact that they are often the major source of energy consumption in large-scale separations. Thus, efforts targeted at easing the demands of back extraction or regeneration would yield separations with significantly enhanced energy efficiencies.

Along these same lines, an understanding of the fundamental behavior of a given separation medium is often obtained using idealized or convenient laboratory conditions. To permit an accurate performance assessment and achieve efficient large-scale separations, the evaluation must be extended to more realistic process-specific conditions. That this transition is often difficult is demonstrated by the observation that advances are still to be made in the design of selective separation processes effective at high acidities (e.g., on sulfuric acid leach solutions in hydrometallurgy), and at high ionic strengths (e.g., alkaline radioactive wastes and solutions from certain environmental remediation operations). Fundamental speciation studies and an understanding of the behavior of extraction systems at high metal loading (which most industrial-scale separations approach before stripping or regeneration) are also required for existing processes to meet continually increasing economic and energy efficiency standards.

Finally, increasingly stringent environmental regulations dictate that the number of separation systems generating minimal or nonhazardous secondary waste streams be expanded. Specifically, improvements are needed to reduce process solvent entrainment in aqueous raffinates, decrease resin or process solvent attrition rates, reduce energy and chemical consumption during all separation stages, and decrease waste byproduct formation.

Acknowledgments

This work was performed under the auspices of the Office of Basic Energy Sciences, Division of Chemical Sciences, U.S. Department of Energy, under contract number W-31-109-ENG-38.

Literature Cited

(1) Rydberg, J. In *Principles and Practices of Solvent Extraction*; Rydberg, J.; Musikas, C.; Choppin, G. R., Eds.; Marcel Dekker: New York, 1992; p 1.
(2) Cunningham, B. B.; Werner, L. B. *J. Am. Chem. Soc.* **1949**, *71*, 1521.
(3) Wallmann, J. C. *J. Chem. Ed.* **1959**, *36*, 340.
(4) Seaborg, G. T. In *Transuranium Elements: A Half Century*; Morss, L. R.; Fuger, J., Eds.; American Chemical Society: Washington, DC, 1992; p 10.
(5) Musikas, C.; Schulz, W. W. In *Principles and Practices of Solvent Extraction*; Rydberg, J.; Musikas, C.; Choppin, G. R., Eds.; Marcel Dekker: New York, 1992; p 413.
(6) *Science and Technology of Tributyl Phosphate. Volume I, Synthesis, Properties, Reactions, and Analysis*; Schulz, W. W.; Navratil, J. D.; Talbot, A. E., Eds.; CRC Press: Boca Raton, FL, 1984; Vol. I.
(7) McKibben, J. M. "Chemistry of the PUREX Process"; DPSPU-83-272-1; Savannah River Plant, 1983.
(8) Flett, D. S. In *Hydrometallurgy: Research, Development, and Plant Practice*; Osseo-Asare, K.; Miller, J. D., Eds.; American Institute of Mining, Metallurgical, and Petroleum Engineers: New York, 1982; p 39.
(9) Ritcey, G. M.; Ashbrook, A. W. *Solvent Extraction: Principles and Applications to Process Metallurgy*; Elsevier Science Publishers: Amsterdam, 1984.
(10) Cox, M. In *Principles and Practices of Solvent Extraction*; Rydberg, J.; Musikas, C.; Choppin, G. R., Eds.; Marcel Dekker: New York, 1992; p 357.
(11) *Solvent Extraction and Ion Exchange in the Nuclear Fuel Cycle*; Logsdail, D. H.; Mills, A. L., Eds.; Ellis Horwood, Ltd.: Chichester, UK, 1985.

(12) *Chemical Pretreatment of Nuclear Waste for Disposal*; Schulz, W. W.; Horwitz, E. P., Eds.; Plenum Press: New York, 1994.
(13) Harris, D. C. *Quantitative Chemical Analysis;* Second ed.; W. H. Freeman and Co.: New York, 1987.
(14) Choppin, G. R.; Nash, K. L. *Radiochim. Acta* **1995**, *70/71*, 225.
(15) Frankenfeld, J. W.; Li, N. N. In *Handbook of Separation Process Technology*; Rousseau, R. W., Eds.; John Wiley and Sons: New York, 1987; p 840.
(16) Noble, R. D.; Stern, S. A. *Membrane Separations Technology: Principles and Applications*; Elsevier: New York, 1995.
(17) *Chemical Separations with Liquid Membranes*; Bartsch, R. A.; Way, J. D., Eds.; American Chemical Society: Washington, DC, 1996.
(18) McHugh, M. A.; Krukonis, V. J. *Supercritical Fluid Extraction: Principles and Practice*; Butterworths: Boston, MA, 1986.
(19) Wenclawiak, B. *Analysis with Supercritical Fluids: Extraction and Chromatography*; Springer Verlag: New York, 1992.
(20) *Innovations in Supercritical Fluids: Science and Technology*; Hutchenson, K. W.; Foster, N. R., Eds.; American Chemical Society: Washington, DC, 1995.
(21) "Separation and Purification: Critical Needs and Opportunities"; National Research Council, 1987.
(22) King, C. J. In *Separation Technology*; Li, N. N.; Strathmann, H., Eds.; Engineering Foundation: New York, 1988; p 1.
(23) Izatt, R. M.; Bradshaw, J. S.; Bruening, R. L.; Tarbet, B. J.; Krakowiak, K. E. In *Emerging Separation Technologies for Metals and Fuels*; Lakshmanan, V. I.; Bautista, R. G.; Somasundaran, P., Eds.; The Minerals, Metals, and Materials Society: Warrendale, PA, 1993; p 67.
(24) Giddings, J. C. *Unified Separation Science*; John Wiley and Sons: New York, 1991.
(25) *Handbook of Separation Process Technology*; Rousseau, R. W., Ed.; John Wiley and Sons: New York, 1987.
(26) Sekine, T.; Hasegawa, Y. *Solvent Extraction Chemistry: Fundamentals and Applications*; Marcel Dekker: New York, 1977.
(27) *Principles and Practices of Solvent Extraction*; Rydberg, J.; Musikas, C.; Choppin, G. R., Eds.; Marcel Dekker: New York, 1992.
(28) *Ion Exchange and Solvent Extraction*; Marinsky, J. A.; Marcus, Y., Eds.; Marcel Dekker: New York, 1966-1997; Vols. 1-13.
(29) *Modern Countercurrent Chromatography*; Conway, W. D.; Petroski, R. J., Eds.; American Chemical Society: Washington, DC, 1995.
(30) *Centrifugal Partition Chromatography*; Foucault, A. P., Ed.; Marcel Dekker: New York, 1995.
(31) Helfferich, F. G. *Ion Exchange*; McGraw-Hill: New York, 1962.
(32) Diamond, R. M.; Whitney, D. C. In *Ion Exchange*; Marinsky, J. A., Ed.; Marcel Dekker: New York, 1966; Vol. 1; p 277.
(33) Massart, D. L. "Nuclear Science Series, Radiochemical Techniques: Cation-Exchange Techniques in Radiochemistry"; NAS-NS 3113; National Academy of Sciences, 1971.
(34) "Technology Vision 2020"; The U. S. Chemical Industry, American Chemical Society, American Institute of Chemical Engineers, The Chemical Manufacturer's Association, The Council for Chemical Research, The Synthetic Organic Chemical Manufacturer's Association, 1996.

Chapter 2

Future of Separation Science in U.S. Department of Energy Processing and Remediation

Gregory R. Choppin

Department of Chemistry, Florida State University, Tallahassee, FL32306–4390

The Department of Energy has stored large quantities of nuclear wastes from defense activities in many sites around the U.S., and is beginning to process them for final disposition as well as to clean the associated environmental contamination. The magnitude and diversity of the wastes and its radioactive nature present major challenges to separation technologies, and the waste processing and land remediation is expected to require decades and large costs. A number of actinide separation methods, both in use or in the R & D stage, are discussed with emphasis on metal specific separation agents, extractants, natural and biological agents, and some nonaqueous systems. The principles, advantages, and problems of these separation methods are the focus of the review.

Historical Review

Separations of radioactive elements began with the discovery of radioactivity. W. Crookes and H. Becquerel found that carbonate anions kept dissolved uranium in solution as a soluble uranyl carbonate complex, allowing purification from insoluble carbonates. The Curies separated components of pitchblende to isolate the radioactive elements polonium and radium. The methods used by these pioneers depended on separation by precipitation, which remained the dominant separation technique in radiochemistry until the Manhattan Project of World War II.

Radiochemical separation technology on the kg scale was developed as part of the wartime effort to separate plutonium from irradiated uranium and its fission and decay products. The first such separation process was based on the use of bismuth phosphate precipitation (*1*) as a carrier for Pu^{3+} and Pu^{4+}. This precipitation technique was replaced by solvent extraction methods in which both uranium and plutonium were isolated from fission products. However, the underlying principle of reduction and oxidation employed in the bismuth phosphate process was retained in the solvent extraction processes. Among the extraction systems developed, the PUREX process, using tributyl phosphate (TBP) as the organic phase, became the most used and remains today the international choice for nuclear fuel reprocessing (*2*).

©1999 American Chemical Society

Several nonaqueous processes have been developed for separation purposes in the nuclear fuel cycle. For example, the uranium hexafluoride production processes, based on the volatility of UF_6, were used on a large scale to produce the feed material for the enrichment of ^{235}U by gaseous diffusion. Plutonium also forms a volatile hexafluoride, PuF_6, which allows separation of uranium and plutonium from bulk impurities or fission products by volatilization. Extractions with molten salts and molten metal purification by electrolysis are examples of other nonaqueous processes that have been used for some separation systems. These methods have the advantage compared to aqueous systems of higher radiation resistance and much smaller secondary waste streams.

Present Technology Drivers

Research and development in separation technology has been stimulated in recent years by the need to initiate treatment leading to the ultimate disposal of nuclear wastes in a safe and cost-effective manner. While the greater volume of these radioactive wastes in the U. S. has been generated in the nuclear weapons program (defense wastes), the spent fuel from the operation of nuclear power plants which has accumulated has almost filled the temporary storage facilities and must be treated in the near future for permanent disposal.

The defense wastes were generated mainly in the reprocessing of irradiated uranium to produce plutonium. The radioactive components are fission products, transuranic elements and their decay products, and nuclides resulting from the neutron activation of structural materials such as cladding. The U. S. has large volumes of these stored defense radioactive wastes as well as contaminated equipment, soils, etc. at a number of sites. The largest volume at a single site is at the Hanford area in the state of Washington where about 2.6 x 10^8 liters of waste, as a mixture of liquid, sludge, and solid, are stored in 177 buried tanks. In the storage process sodium hydroxide was added to neutralize the nitric acid (from the original PUREX processing) to minimize corrosion of the mild steel tanks, resulting in precipitation of salts and sludge and considerably increasing the volume and chemical complexity of the wastes.

A number of methods have been proposed for separating the intensely radioactive (e.g., Cs, Sr) and longest lived (e.g., the actinides) components from the bulk of the wastes. Many of these technologies are being studied in laboratories while others are ready to be or are being tested in pilot plant operations. In this paper, some of the methods, with emphasis on those for actinide separations, are reviewed and their present state of development assessed.

Important Characteristics of Actinides

To understand the science of actinide separation chemistry, it is necessary to be familiar with certain characteristic properties of actinide elements. The more important of these are discussed subsequently while more complete discussions can be found in reviews and books (e.g., Chapter 16 of ref. 1).

The actinides of Z < 96 can exist in oxidation states of III to VI in solution, some in 2 or 3 such states in equilibrium. The change between III and IV and between V and VI is rapid, involving only electron exchange. However, between III, IV and V, VI, the rate is slower as chemical bonds must be broken or made in addition to electron exchange since the An(V) and An(VI) species exist as linear dioxo cations. In all oxidation states, the actinide cations are hard acids and show little interaction with soft base donors such as S, P, etc. in aqueous media. Also, as expected for hard-hard interactions, the bonding to donors is primarily ionic.

An(III) and An(IV) cations have coordination numbers (CN) of 6 to 12 while the dioxo actinyls have CN = 4 - 6. These ranges in CN reflect the ionic bonding which results in the coordination number and symmetry being dominated by steric and net charge effects. The complexation strength normally follows the pattern:

$$An^{4+} > AnO_2^{2+} > An^{3+} > AnO_2^+$$

which has been attributed to an effective charge on the An in AnO_2^{2+} of 3.3 ± 0.1 and in AnO_2^+ of 2.2 ± 0.1 (3).

In designing specific ligands, it is helpful to be aware of the following:
i. An-N bonds are longer-lived than An-O bonds.
ii. An-O bonds promote and may be necessary to the formation of An-N bonds.
iii. Five-membered chelate rings are the most stable.
iv. Prearranged structures more stable.
v. If too rigid, prearranged structures may be too slow to form.
vi. Bulky steric groups can slow dissociation kinetics.
vii. Redox, steric effects, high CN, and weak bond covalency (in extractant ligands) can be exploited to improve specificity in separations.

Aqueous Processes

Aqueous processes have played important roles in the separation of irradiated uranium and are likely, over the near future, to be the principal separation processes for the treatment of nuclear wastes. The separation of radioactive elements in these processes is based on the differences in such chemical properties of the dissolved species as redox potentials, complexation strength with various ligands, affinity to extractants and/or to ion exchange resins or inorganic ion exchangers, transport behavior through membranes or in an electric field, etc. Variations within these processes are necessary, however, to treat successfully wastes of different compositions, pH, etc.

Solvent Extraction. In the PUREX process, uranium and plutonium are coextracted (as tetravalent cations) into the organic phase from an aqueous nitric acid solution (2 to 3 M) by TBP while the fission and other (e.g., Am, Np) products remain in the aqueous phase. The plutonium is subsequently separated from the uranium by selectively reducing it to Pu^{3+}, which is stripped into the aqueous phase. Depending on the operating conditions, neptunium may remain in the uranium streams and can be separated from uranium by adjusting extraction conditions in subsequent steps. This process could be adapted to treat the defense wastes in order to concentrate the actinides into fractions of relatively small volumes for storage or for destruction by fissioning in reactors. However, depending on the composition of the wastes, additional separations (e.g., solvent extraction or ion exchange) might have to be added to the PUREX process to obtain sufficient separation of the actinides from other radioactive species in the wastes.

The TRUEX process uses CMPO (octyl(phenyl)-N, N-diisobutylcarbamoyl-methylphosphine oxide) to separate the transuranium elements from acidic high level waste (HLW) solutions (4). This process has been demonstrated in the laboratory. Diphenyl dibutylcarbamoyl methyl phosphine oxide in polar inorganic solvents is also a promising extractant (5). Both extractants can be sorbed on a solid porous support and used in a liquid chromatographic elution system.

The Talspeak process (6, 7) is based on separation of lanthanides from trivalent actinides by extraction of the former into di(2-ethylhexyl) phosphoric acid (HDEHP) solution from an aqueous phase of lactic + DTPA (diethylenetriaminepentaacetic acid) acids at pH 2.5 - 3.0. Subsequently, 6 M HNO_3 is used to strip the Ln(III). A modification - the "reverse" Talspeak process in which the An(III) + Ln(III) elements are extracted into an organic phase by HDEHP then stripped into an aqueous phase of lactate + DTPA has also been proposed (8). Since the Talspeak process operation may be adversely affected by radiation damage, testing at full radiation levels in pilot plant operations is needed for full evaluation of the process.

A number of bidentate extractants have been studied. Diphosphine dioxides (9) and diamides (10) show good extraction and radiation resistant properties but, like CMPO, show poor separation of An(III) and Ln(III). In general, they show extraction properties similar to that of CMPO and warrant further evaluation as alternatives to CMPO. The diamides have the advantage that they do not cause formation of phosphate which can be a problem in the processing of the waste for final disposal.

Good separation factors have been reported between trivalent lanthanides and trivalent actinides for solvent extraction systems based on complexants with soft donor groups (e.g., N, S). In general, the complexants are based on amide functional groups or on sulfur-based β-diketones (11). These systems show good promise for separations of the trivalent cations. However, redox-based separations can easily separate the actinides of $Z \leq 95$ more simply and, usually, more efficiently from the Ln(III) fission products. Consequently, there is little incentive to develop soft donor ligand systems for nontrivalent actinide separations.

Bis-dicarbollycobaltate ((π-(3)-(1,2-$C_2B_9H_{11})_2Co^-$) or dicarbollide) has been shown to have very high selectivity and efficiency for extraction of Cs(I) and Sr(II) from 2 - 3 M HNO_3 (12). Efficient extraction of Cs, Sr, Ba, and An(III) from deacidified PUREX wastes has been achieved using a solution of dicarbollide with p-nonylphenolnonaethyleneoxide with subsequent separation of Ln(III) and An(III) species by stripping into nitric acid. Dicarbollides have desirable radiation resistant properties.

Crown ethers, cryptands, podands, and similar macrocyclic ligands with cavities of specific size and shape can show high selectivities for atoms or ions which match the cavity size and shape. Studies on these and other stereospecific extractants could be of great value for isolation of radioactive elements such as cesium, strontium, technetium, and plutonium. A crown ether is the extractant used in the SREX process for strontium isolation (13). Specific chelating ligands for various actinides have been developed and tested in the laboratory (14), but have not been used in the treatment of nuclear waste streams. The efficiency of regeneration for recycling can be a problem if the specificity is due to the rigidity and, hence, the inertness of the encapsulating ligand structure. In treatment of nuclear wastes, the role of selective extractants would be in the later parts of the processing sequence where it is desirable to remove and/or isolate a particular element. The potential value of reversible, highly specific, and recyclable extractants offers great opportunities for use in efficient waste treatment and justifies a major research effort; however, selective specificity is not the only criterion to be used in assessing the value.

Ion Exchange Materials. A number of interesting new solid ion exchange materials are in various stages of development. One of the more promising is a substituted diphosphonic acid resin, "Diphonix", which is commercially available and relatively inexpensive. This resin is a very strong complexing agent, removes actinides from 10 M HNO_3 solutions, and is effective for removing a wide variety of toxic heavy metals (including Pb, Hg, Cd, Zn, Ni, Co, and Cr) from waste water. It has promise for use in removing certain radionuclides from nuclear wastes.

Some high temperature zeolites, such as titanium phosphate, may prove useful for isolation of specific cations such as Cs^+. Silicon titanates have very high capacity for removing Cs and Sr from radioactive solutions of high salt content such as that in the Hanford tanks. The clay sodium fluorophogopite mica is reported (15) to be superior to the zeolite clinoptilolite for removing Sr from nuclear waste and also can remove Cs from solutions of high sodium content. The high capacity and the chemical radiation stability of these inorganic exchangers also make them interesting candidates for study as materials within which to incorporate the concentrated radionuclides for final geologic storage.

Nonaqueous Processes

Nonaqueous processes, based on the difference in such properties as volatility of various compounds or redox thermodynamics of elements in molten-salt media, have been used over many years in uranium isotope separation, electrorefining of plutonium metal, and production of metallic fuel for advanced nuclear reactors. There is interest in conducting research of nonaqueous processes as separation technologies for treatment of nuclear wastes. These processes have the advantage of relatively high insensitivity to radiation effects - in contrast to aqueous processes for which radiolysis can be a serious problem, causing degradation of the organic extractants and changing the aqueous-phase chemistry through the radiolysis of water.

Volatility Processes. Uranium hexafluoride has been used for 50 years in the gaseous diffusion process for uranium isotopic enrichment (1). Volatility techniques with fluorides have also been used to purify plutonium in isotope separation plants (16) and were studied for use in fuel processing in the molten-salt reactor project at Oak Ridge National Laboratory. The separation of uranium and plutonium from fission products is limited in these processes by the fact that volatile fluorides are formed by several fission products, and, in particular, by iodine and tellurium. However, iodine and tellurium can be separated from uranium and plutonium by distillation after oxidation. The decontamination of technetium remains a difficult task in the fluoride volatility process because its fluoride diffuses with the UF_6 and PuF_6 streams.

Other volatility separation processes may have promise for separating particular elements from certain types of wastes, but they are not as well developed as the fluoride volatility process. For example, the volatility of $ZrCl_4$ could be used to remove the zirconium cladding on spent fuel elements.

The β-diketone complexes of trivalent actinides are volatile but their possible use in separations needs more research (17). In a proposed scheme for transmutation of technetium to the nonradioactive ruthenium, the product Ru is converted to RuO_4 by ozonolysis and separated from the remaining Tc_2O_7, using the higher volatility of RuO_4 (18). These volatility processes have not been studied sufficiently to evaluate them for practical use in full-scale separation systems.

Pyrochemical Processes. Molten salt (19) and molten metal (20) systems are among the pyroprocesses investigated as possible technologies to treat spent fuel. These types of processes could be considered for treatment of nuclear wastes but since the latter are usually in wet, oxic conditions, the pretreatment to prepare them for treatment in the anhydrous, anoxic pyrochemical systems would be a major disadvantage to use of such methods. An advantage of these nonaqueous systems is the much reduced (relative to aqueous processes) volume of secondary, low level wastes resulting from the treatment of the high level wastes.

Natural Agents

The use of specific microbial siderophores is receiving attention since these reagents could result in important separation and concentration applications in waste treatment and/or land remediation. It may be possible to use siderophores (microbially produced chelating agents) to sequester species such as Pu(IV) from the environment (21).

In contact with jimson weed radioactive sludges have plutonium removed via binding to cell walls (22). Moreover, the jimson weed cells sorbed the plutonium when dead and when alive. It has been known for some time that such cells accumulate uranium from waste streams by hydrolytic sorption on the walls. An interesting possibility for use of biological material is the use of a derivative of chitin, in the form of porous beads, to remove heavy metals from ground water. The beads would be collected with a magnetic field and the sorbed heavy metal stripped. These processes are only at the level of laboratory study presently. A disadvantage may be the sensitivity of biological materials to destruction in high radiation fields.

Summary

Many novel and promising separation agents and methods have been proposed. The new initiative by the USDOE of the Environmental Management Science Program can be expected to continue the progress in this area in developing more efficient, more specific, and more economical separation techniques. It is necessary that the development of new specific ligands and extractants in research laboratories be followed by the R & D in radioactive facilities to demonstrate their applicability in high radiation fields, complex waste mixtures, etc.

Literature Cited

(1) Choppin, G. R.; Rydberg, J.; Liljenzin, J. O. *Radiochemistry and Nuclear Chemistry;* Butterworth-Heinemann Ltd: Oxford, 1995.
(2) Culler, F. L. In *Progress in Nuclear Energy, Series III, Process Chemistry;* Bruce, F. R.; Fletcher, I. M.; Hyman, H. H.; Katz, J. J., Eds.; McGraw-Hill: New York, 1956; Vol. 1, Ch. 5.2.
(3) Choppin, G. R.; Rao, L. F. *Radiochim. Acta* **1984**, *37*, 143.
(4) Schulz, W. W.; Horwitz, E. P. *Sep. Sci. Technol.* **1988**, *23*, 1191.
(5) Dzekun, E. G., et al. In *Proc. Symp. On Waste Management* Post, R. G.; Wacks, M. E., Eds.; Arizona Board of Regents: Tucson, AZ, 1992.
(6) Weaver, B.; Kappelmann, F. A. *Talspeak: A New Method of Separating Americium and Curium from Lanthanides by Extraction from an Aqueous Solution of Aminopolyacetic Acid Complex with a Monoacidic Phosphate or Phosphonate*; Report ORNL-3559, Oak Ridge National Laboratory: Oak Ridge, TN, 1964.
(7) Kolarik, E.; Koch, G.; Kuesel, H. H.; Fritsch, J. *Separation of Am and Cm from Highly Radioactive Waste Solution*, KFK-1533, Karlsruhe Nucl. Res. Center: Germany, 1972.
(8) Persson, G. E.; Svantesson, S.; Wingefors, S.; Liljenzin, J. O. *Solvent Extr. Ion Exch.* **1984**, *2*, 89.

(9) Rosen, A. M.; Nikolotova, Z. I. *Radiokhimiya* **1991**, *33*, 1.
(10) Musikas, C.; Hubert, H. *Solvent Extr. Ion Exch.* **1987**, *5,* 877.
(11) Ensor, D. D.; Jarvinen, G. D.; Smith, B. F. *Solvent Extr. Ion Exch.* **1988**, *6*, 439.
(12) Esimantoviskii, D.; Romanovskii, V.; Shishkin, N., Dzekun, E. G., Proc. Symp. On Waste Management, Tucson, 1992, Arizona Board of Regents.
(13) Horwitz, E. P.; Dietz, M. L.; Fisher, D. E. *Solvent Extr. Ion Exch.* **1980**, *8*, 557.
(14) Kappel, M. J.; Nitsche, H.; Raymond, K. N. *Inorg. Chem.* **1985**, *24*, 605.
(15) Paulus, W. J.; Komarmeni, S.; Roy, R. *Nature* **1992**, *357*, 571.
(16) Hyman, H. H.; Vogel, R.; Katz, J. J. In *Progress in Nuclear Energy, Series III, Process Chemistry;* Bruce, F. R.; Fletcher, I. M.; Hyman, H. H.; Katz, J. J., Eds.; McGraw-Hill: New York, 1956; Vol. 1, Ch. 6.1.
(17) Steinberg, M.; Powell, J. R.; Takahashi, H. *Nucl. Tech.* **1982**, *58*, 437.
(18) Dewey, H. J.; Jarvinen, G. D.; Marsh, S. F.; Marsh, N. C.; Schroeder, N. C.; Smith, B. F.; Villareal, R.; Walker, R. B.; Yarbro, S. L.; Yates, M. A. *Status of Development of Actinide Blanket Processing Flowsheets for Accelerator Transmutation of Nuclear Waste*, Report LA-UR-93-2944, Los Alamos National Laboratory: Los Alamos, NM, 1993.
(19) Steunenberg, R. K.; Pierce, R. D.; Johnson, I. *Symp. on Reproc. Nucl. Fuels*; CONF-690801; USAEC: Washington, D.C.; Vol. 15.
(20) Coops, M. S.; Knighton, J. B.; Mullins, L. J. In *Plutonium Chemistry;* Carnall, W. T.; Choppin, G. R., Eds.; ACS Symp. Ser. 216, Amer. Chem. Society: Washington, D. C., 1983.
(21) Wildung, R. E.; Garland, T. R.; Rogers, J. E. In *Environmental Research on Actinide Elements*; Pinder, J. E., et al., Eds.; OSTI; U. S. Department of Energy, 1987.
(22) *Nuclear Wastes: Technologies for Separations and Transmutation*; National Research Council; National Academy Press: Washington, D. C., 1995; p 177.

(HDEHP) (*9-13*), diisodecyl phosphoric acid (DIDPA) (*14-17*) and bis(hexoxyethyl) phosphoric acid (HDHoEP) (*18*).

Swedish CTH Process. A Swedish team developed and hot tested a process utilizing HDEHP to treat commercial HLW (*9-12*). The process (called CTH for Chalmers Tekniska Högskola) begins with a HLW solution prepared by adjusting PUREX aqueous raffinate to 6 M HNO_3 and then treating with NO_x. A 1 M HDEHP solution in kerosene is used in the first SX cycle to remove U, Np and Pu together with most of the Fe, Zr, Nb and Mo. The loaded organic phase is first washed with an HNO_3-HF solution to remove Zr and Nb and then stripped with ammonium carbonate. A mixture of ammonia and manitol are added to the organic phase before stripping to avoid precipitation of iron. The strip solution is evaporated to recover the ammonium carbonate for recycle and then acidified to recover the U, Np and Pu. Test runs demonstrated that 99.99% of U, 98.9% of the Np and >99.99% of the Pu can be recovered in the first extraction cycle.

The second extraction cycle in the CTH process involves the use of 50% TBP in kerosene to reduce the acidity of the aqueous phase to prepare it for a second low-acid HDEHP cycle. In addition to extracting HNO_3, TBP also extracts Pd, Ru and Tc leaving Am and Cm plus the light lanthanides in a 0.1 M HNO_3 aqueous raffinate. Nitric acid is recovered by scrubbing with water and then evaporated and distilled to produce ~9 M HNO_3. The HNO_3 is then reused to prepare the feed for the first extraction cycle. The Tc, Ru and Pd can be recovered from the strip solution by anion exchange.

The third and final extraction cycle in the CTH process consists of the extraction of Am and Cm, together with the light lanthanides, from the raffinate of the second extraction cycle using a solution of 1 M HDEHP in kerosene. At this stage the trivalent actinides can be selectively stripped from the light lanthanides using a diethylenetriamine pentaacetic acid (DTPA)-lactic acid mixture, or the actinides together with the lanthanides can be stripped using 6 M HNO_3. If the trivalent actinides are partitioned from the lanthanides, a final HDEHP extraction cycle is required to remove the trivalent actinides from the DTPA-lactic acid solution. Another feature of the flowsheet is that Sr and Cs in the raffinate from the third extraction cycle can be sorbed on inorganic ion exchangers such as titanates and zeolites, respectively, for disposal.

The entire CTH process has been successfully tested by processing 20 liters of synthetic HLW feed using a number of small-scale (150 mL) mixer-settlers. Analytical results showed the process behaved as predicted. The major advantage of the CTH process is that it utilizes two commercially available, low cost and well characterized extractants. The major disadvantages of the process are that multi-extraction cycles (at least three) using two different extractants are required and that all strip solutions require further processing to recover the TRUs. In addition, a significant acidity adjustment is required before trivalent actinides can be extracted. A process this complex would be very costly and difficult to operate on a plant-scale.

JRC-ISPRA/HDEHP Process. A team of chemists working at the Joint Research Centre-Ispra Establishment, Italy, have also developed and tested a process to treat commercial HLW (*13*). The JRC-ISPRA process is somewhat similar to the CTH process. The process begins by performing either an exhaustive TBP extraction or HDEHP extraction of a PUREX raffinate that has been concentrated 10-fold. (Residual U, Np and Pu are removed by the TBP or HDEHP.) The raffinate from the above front-end extraction is denitrated to pH 2 using formic acid. After a clarification step, Am and Cm, together with the light lanthanides, are extracted with 0.3 M HDEHP - 0.2 M TBP in dodecane. Partitioning of Am and Cm from light lanthanides is performed by a Reverse TALSPEAK process, or the actinides and lanthanides can be stripped together using 4 M HNO_3. Batch extraction tests with actual HLW demonstrated >99% removal of Am and Cm after a single extraction from a feed solution with a pH of 2.76 (*13*). The Pu left in the precipitate formed during the denitration step after formic and nitric acid washings varied from 2 to 0.6% (*13*).

The JRC-ISPRA process suffers from the same problems as the CTH process. A multi-cycle process of this type would be most difficult and very costly to operate on a plant-scale. Formation of precipitates of varying alpha contamination is also a drawback.

Japanese/DIDPA Process. The third SX system using an acidic extractant has been developed by a team of chemists from the Japan Atomic Energy Research Institute (*14-17*). These chemists have utilized a less common acidic extractant called diisodecyl phosphoric acid (DIDPA). The major difference between DIDPA and HDEHP is that the former can extract light trivalent actinides from higher acidities, for example, 0.5 M rather than ≤0.1 M HNO_3 required by HDEHP. As with the CTH process, the major motivation is to reduce the quantity of HLW for disposal. The HLW is generated by reprocessing fuel from commercial nuclear power plants.

The process consists of four steps. The first step is an SX cycle using TBP to remove the small amount of U and Pu left in the raffinate from PUREX. The second step involves a denitration of the raffinate from the first extraction with formic acid. The second step has two objectives; first, to reduce the acidity to pH 0.5 to allow the efficient extraction of Am and Cm with 0.5 M DIDPA - 0.1 M TBP-dodecane; and second, to precipitate zirconium and molybdenum (as $Zr(MoO_4)_2$) that interfere with the subsequent extraction of Am and Cm. The third step involves the extraction of Am and Cm with 0.5 M DIDPA - 0.1 M TBP-dodecane. If the feed for the Am/Cm SX step is made 0.5 to 1 M in H_2O_2, at least 99% of the Np can also be removed. The fourth and final step of the process involves the selective stripping of Am and Cm with 0.05 M DTPA - 1 M lactic acid. As with the CTH process, the raffinate from step 3 can be treated with inorganic ion exchange materials to remove Sr and Cs. The flowsheet has been tested in a 16-stage bank of mixer settlers using actual commercial HLW. Greater than 99.99% of the Am and Cm were extracted with DIDPA. Overall results showed that the process behaved as predicted.

The same advantages and disadvantages listed for the CTH and JRC-ISPRA processes apply equally to the Japanese process. In addition, an eight-fold dilution of the HLW may be required to prevent formation of a gelatinous precipitate in the DIDPA process solvent (*15*) and the percentage of TRU precipitated during the formic acid denitration step can be significant if the pH increases to 1.0 (*14*). Both of the last two points are major drawbacks to implementing such a process on a plant-scale.

ANL/HDHoEP Process. The Argonne National Laboratory (ANL) process utilizes the strongest acidic organophosphorus extractant of the four systems (*18*). Bis-(hexoxyethyl) phosphoric acid ($(C_6H_{13}OC_2H_4O)_2PO(OH)$), abbreviated HDHoEP, requires no adjustment of acidity up to a nitric acid concentration of 2.4 M. The flowsheet developed and cold-tested by the Argonne Group was based on a PUREX aqueous raffinate generated by processing a light water reactor fuel with a burnup of 3300 megawatt-days per ton (*19*). A HLW simulant was prepared containing thirty components. The process solvent consisted of 0.5 M HDHoEP in diethylbenzene. The organic to aqueous phase ratio (O/A) in the extraction stages was one. After a 1 M HNO_3 scrub (O/A = 4), Am and Cm were stripped with 6 M HNO_3 (O/A = 2) and Np and Pu together with Zr, Nb, Mo and Fe were stripped with a mixture of 0.35 M oxalic acid and 0.35 M tetramethylammonium hydrogen oxalate (O/A = 2). The process solvent was washed with 8 M H_3PO_4 to remove any U and degradation products. The entire process was operated at 50°C. A three extraction stage batch countercurrent test run removed 93% of the Am and Cm, 97% of the Np and 99.7% of the Pu. Using eight extraction stages, the calculated amount of Am, Np and Pu extracted from the feed would be 99.8, 99.8 and 99.99%, respectively.

The ANL process flowsheet is much simpler than the flowsheets based on HDEHP and DIDPA because it has only one extraction cycle and requires no acidity adjustment of the feed, provided that the HNO_3 concentration in the feed is not above 2.4 M. Nevertheless, the process has several serious drawbacks. For example, the process solvent uses diethylbenzene as a diluent, which is not looked upon favorably because of its low flashpoint. The Am and Cm are recovered in 6 M HNO_3, which would require extensive evaporation and recovery of concentrated nitric acid. An even bigger drawback of the ANL process is that the Np and Pu fraction would require further processing to remove macroquantities of Fe, Zr and Mo and the 8 M H_3PO_4 solvent wash will be contaminated with U and Y fission products. These last two features make the process much more complicated than it first appears.

Conclusion on the Use of Acidic Extractants. All of the acidic organophosphorus extractants discussed above have major drawbacks. Most important are that the extractants have insufficient selectivity for TRUs (especially trivalent TRUs) over a number of fission products and inert constituents that are present in HLW and that the acidic extractants require low acidities to efficiently extract trivalent actinides. Because the trivalent actinides are the major source of alpha radiation in HLW, this latter drawback is perhaps the most serious of all. Both of these shortcomings lead to complex multicycle flowsheets and frequently to unfavorable stripping conditions.

Processes Based on Neutral Extractants. Neutral extractants, particularly bifunctional neutral extractants, were developed to overcome the major drawbacks of acidic extractants. Neutral extractants remove metal ions from aqueous nitrate media by the formation of a bond between the electron donor group of the extractant and the nitrate-complexed metal ion. The resultant complexes are usually lipophilic because the donor groups of the extractant enter the inner coordination sphere of the metal ion displacing waters of hydration (20,21). Lipophilic complex formation may also occur by the formation of an ion pair between the extractant and the metal ion nitrato complex. One of the most noteworthy and useful properties of the actinides is their propensity for the formation of nitrato complexes. (The extraction of the nitrato complexes of tetra- and hexavalent actinides is the basis of the PUREX process.) The extraction of trivalent actinide-nitrato complexes using TBP is much more difficult than the corresponding extraction of tetra-and hexavalent actinides. High concentrations of salting out agents, such as aluminum or lithium nitrate, and low acidity are required for efficient extraction of Am(III) by TBP (13).

In 1963-64 Sidall reported the synthesis and characterization of several new classes of compounds that showed vastly improved efficiencies for the extraction of trivalent actinides and lanthanides from high nitric acid concentrations (22,23). Although Sidall's compounds were impure, and in most cases the improvement in extraction efficiencies was insufficient for practical applications, his studies nevertheless opened up a new area of actinide separation chemistry that eventually led to some of the most successful waste processing flowsheets.

Early studies utilizing neutral bifunctional extractants focused largely on dihexyl-N,N-diethylcarbamoylmethylphosphonate (DHDECMP) (24-28). (The structure of DHDECMP is shown in Figure 1.) These studies utilized a 30% solution of DHDECMP in diisopropylbenzene (DIPB). Feed solutions were defense HLW at least 3 M in HNO_3. Although these early studies were plagued by the use of an impure extractant (~85% DHDECMP), which led to difficulties with stripping, and by a superficial knowledge of the chemical behavior of this new class of extractants (24,26), the foundation was laid for the development of a more advanced system.

TRUEX Process. The TRUEX process is a generic TRU element extraction/recovery process based on the use of a carbamoylmethylphosphoryl type of extractant, called octyl (phenyl)-N,N-diisobutylcarbamoylmethylphosphine oxide

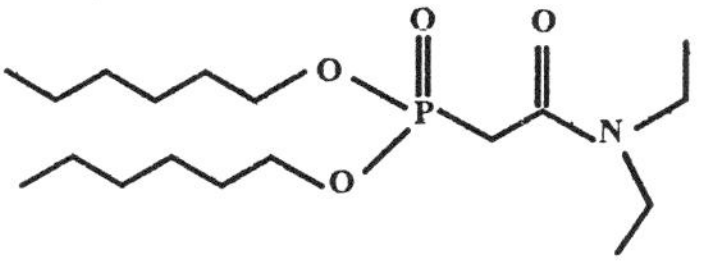

Dihexyl-*N,N*-diethylcarbamoylmethylphosphonate (DHDECMP)

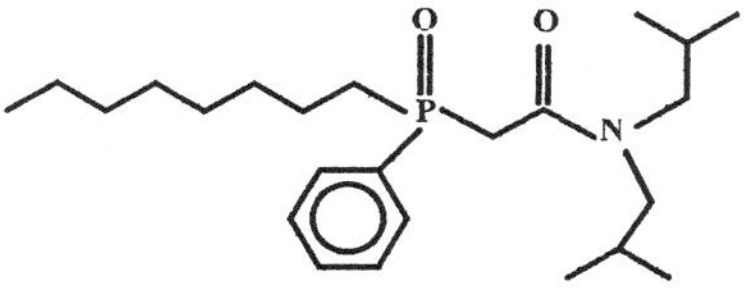

Octyl(phenyl)-*N,N*-diisobutylcarbamoylmethylphosphine oxide (O(Φ)DiBCMPO)

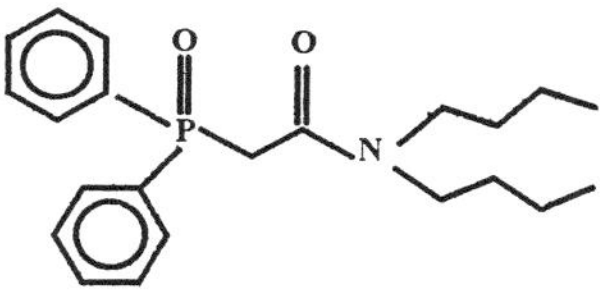

Diphenyl-*N,N*-di-*n*-butylcarbamoylmethylphosphine oxide (DΦDBCMPO)

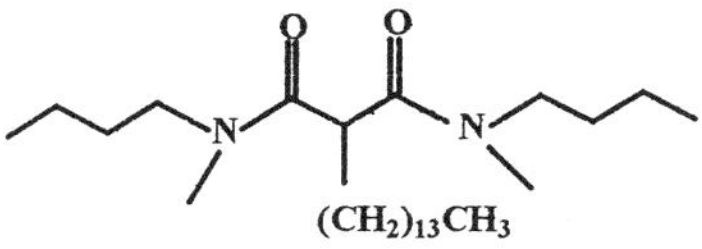

Dimethyldibutyltetradecylmalonamide (DMDBTDMA)

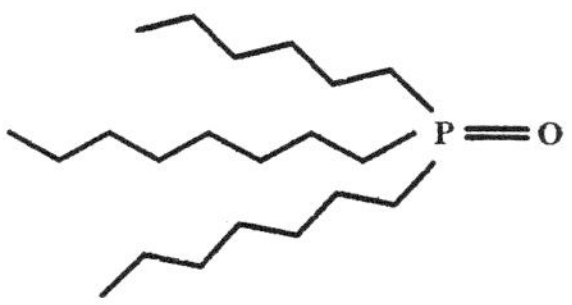

Trialkylphosphine oxide (TRPO)

Figure 1. Structures of neutral bifunctional and monofunctional extractants.

(abbreviated O(ø)DiBCMPO or CMPO for short), dissolved in PUREX process solvent (29-33). The CMPO molecule contains a unique combination of substituents that enable it to extract trivalent actinides, as well as tetra- and hexavalent actinides, over a wide range of nitric acid concentrations. (The structure of O(ø)DiBCMPO is shown in Figure 1.) The chemistry of CMPO with particular relevance to the TRUEX process has been discussed (33, and references therein). TRUEX was developed primarily to treat a wide variety of defense high-level and TRU wastes generated at the DOE defense establishments (29-32,34). In order for a TRU extraction/recovery process to be applicable for the treatment of defense waste, it must meet a number of criteria. First, it must be capable of efficiently extracting all TRUs from a wide range of nitric acid concentrations, for example, 1 to 6 M HNO_3, in a single extraction cycle. Second, the process must be selective for actinides in the tri-, tetra- and hexavalent oxidation states over most of the fission products (lanthanide fission products excluded) and inert constituents. Third, stripping of the TRUs should not require high concentrations of acid or complexants. Last but not least, the process must be adaptable to existing processing facilities. The TRUEX process meets all of the above criteria.

The key feature in the TRUEX process is the process solvent formulation. TRUEX process solvent is very similar to PUREX process solvent, with the only major difference being the presence of CMPO in the TRUEX solvent formulation. TRUEX process solvent contains 0.20 M CMPO - 1.2 to 1.4 M TBP in a paraffinic hydrocarbon with a carbon chain length of twelve to thirteen. The quantity of TBP used is a function of the chain length and branching of the paraffinic hydrocarbon. If the average chain length of the hydrocarbon is thirteen and contains no branching, that is, an n-paraffin, then the higher TBP concentration is utilized (29). It is not surprising, therefore, that the physical properties of PUREX and TRUEX process solvents are similar.

One of the major reasons for selecting the octyl (phenyl)-N,N-diisobutyl CMPO derivative over the corresponding diphenyl derivative was the solubility of the former in PUREX process solvent. This property of the octyl (phenyl) derivative was considered more important than the fact that the diphenyl derivative has a somewhat more favorable nitric acid dependency for the extraction of Am(III) as shown in Figure 2, which compares the D_{Am} versus nitric acid concentration curves for DHDECMP and the N,N-diisobutyl derivatives of DøCMPO and O(ø)CMPO. Because of solubility constraints, the three extractants were compared in 0.75 M TBP-CCl_4 media. Another reason for selecting the octyl (phenyl) derivative was the rather remarkable beneficial effect that TBP has on the behavior of O(ø)DiBCMPO (33, and references therein). The presence of TBP in CMPO-diluent mixtures increases D_{Am} at high acidities (extracting conditions), decreases D_{Am} at low acidities (stripping conditions), decreases the variation of D_{Am} over the range of 0.75 to 6 M HNO_3, increases the solubility of CMPO in paraffinic hydrocarbons, decreases third phase formation at high loading and increases radiolytic stability. The nitric acid dependency curve for the extraction of Am using a typical TRUEX process solvent formulation is shown in Figure 3. Conoco (C_{12}-C_{14}) was used as the diluent. (Conoco C_{12}-C_{14} is a mixture of normal paraffinic hydrocarbons with an average carbon-chain length of thirteen.) It is noteworthy that D_{Am} values of ten or greater can be obtained from 0.5 to 7.0 M HNO_3 and that D_{Am} values from 1.0 M to 7.0 M are very insensitive to nitric acid concentration.

Comparisons of DHDECMP and O(ø)DiBCMPO in TBP-paraffinic hydrocarbon mixtures were also performed in some of the early studies on TRUEX development (34). Although DHDECMP has a number of favorable properties, its much weaker extractant strength prohibits exploiting the beneficial effects of high TBP to extractant concentration ratios. To be comparable to CMPO, the concentration of DHDECMP must be increased to at least 0.5 M to compensate for low D_{Am} values. A 0.5 M solution of DHDECMP would require a TBP concentration of 2.5 to 3 M to

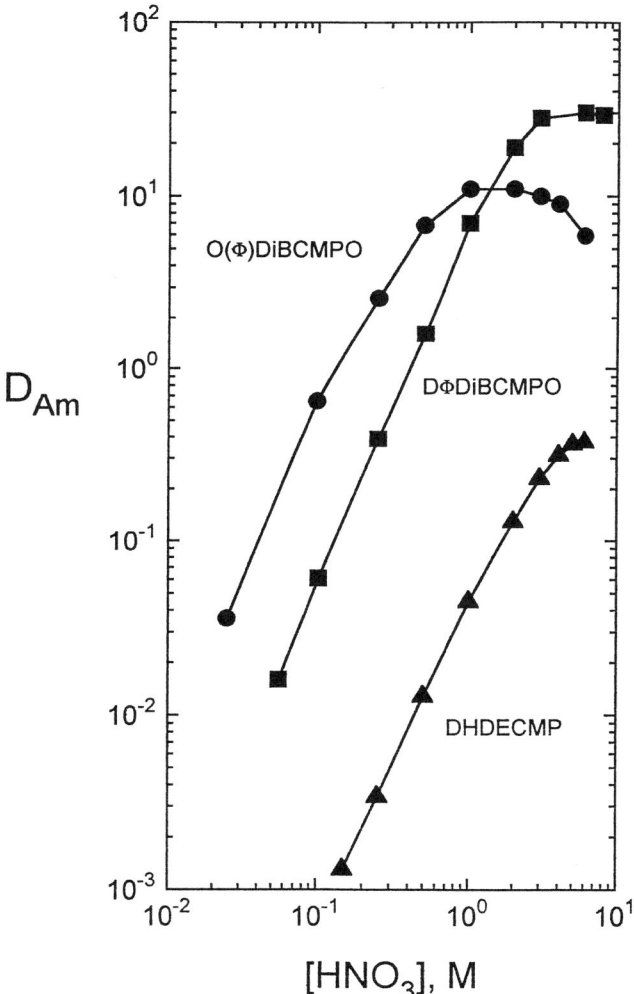

Figure 2. Comparison of octyl (phenyl)- and diphenyl-DiBuCMPO and DHDECMP in the presence of TBP at 25°C as extractants for Am(III) in nitric acid (Adapted from ref. *33*.)

0.25 M CMP or CMPO - 0.75 M TBP - CCl_4.

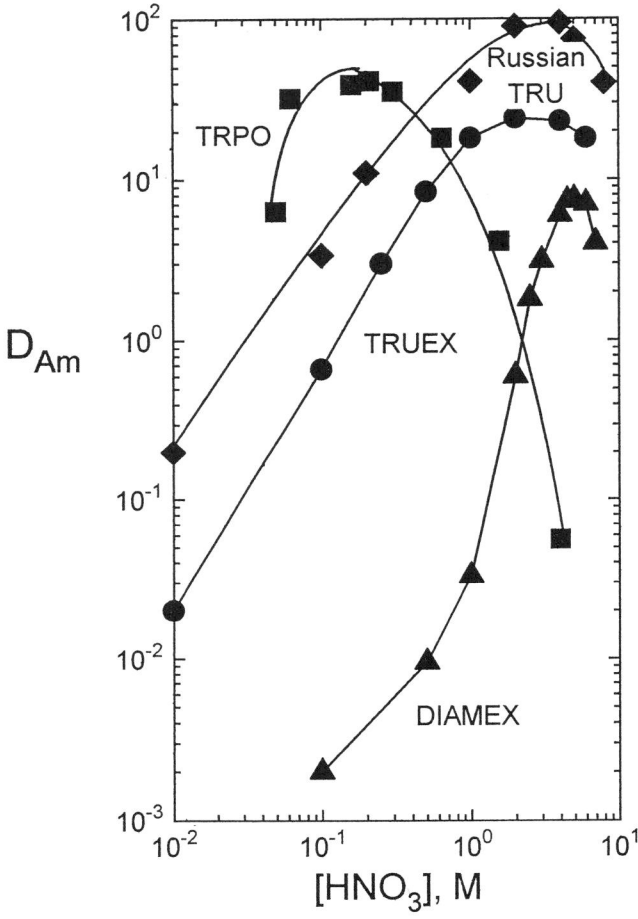

Figure 3. Comparison of TRUEX, Russian TRU, DIAMEX and TRPO process solvents as extractants for Am(III) in nitric acid
(Adapted from refs. 33, 60, 63, 69.)

TRUEX process solvent (0.02 M O(ø)DiBCMPO - 1.4 M TBP-Conoco (C_{12}-C_{14})), T = 30°C
Russian TRU process solvent (0.05 M DøDBCMPO-Fluoropol-732), T = 23°C
DIAMEX process solvent (0.5 M DMDBTDMA-TPH), T = 25°C
TRPO process solvent (30 vol % TRPO-Kerosene), T = 25°C

eliminate third phase formation. Such a mixture has no room for a diluent (*34*). Therefore, DHDECMP is most commonly utilized as a 30% solution in DIPB (*26,27*).

TRUEX and PUREX process flowsheets are very similar, that is, extraction is carried out at high acidities and stripping is carried out at low acidities. However, unlike PUREX, TRUEX can treat feed solutions that vary widely in both acidity and composition. Depending on the composition of the streams, oxalic acid is sometimes added to suppress the extraction of Fe, Zr, Mo and Pd. Reducing agents, such as ferrous sulfamate, may be added to ensure that neptunium is in the tetravalent oxidation state. (It makes no difference if Pu is in the trivalent or tetravalent oxidation state because both are extractable.) Solvent scrubs are usually 0.25 to 1 M HNO_3 or 1.5 M HNO_3 - 0.03 M $H_2C_2O_4$, but if an unusually high concentration of Zr is present in the feed, different scrubbing regimes may be required. Trivalent actinides, together with light lanthanides, are stripped with dilute HNO_3, whereas tetravalent actinides are stripped with a dilute complexing agent such as 0.05 M HNO_3 - 0.05 M HF or ammonium oxalate to solubilize macroquantities of Pu. Uranyl ion and pertechnetate are removed from the process solvent during solvent washing with dilute carbonate solution. Thus, in one extraction cycle TRUEX can achieve what required multiple steps to achieve using acidic extractants. In addition, trivalent actinides are separated from tetra- and hexavalent actinides and tetravalent actinides may be partitioned from hexavalent actinides using relatively dilute reagents. In most cases Tc also may be recovered, but its recovery, although adequate, is not as high as for the actinides.

To aid in designing site- and feed-specific TRUEX flowsheets and in estimating the space and cost requirements for installing a TRUEX process, chemists and engineers at ANL have developed a Generic TRUEX Model (GTM) (*35*). The GTM is composed of four major sections: (1) SASSE (Spreadsheet Algorithm for Stagewise Solvent Extraction), (2) SASPE (Spreadsheet Algorithm for Speciation and Partitioning Equilibria), (3) SPACE (Size of Plant and Cost Estimates) and (4) INPUT/OUTPUT. All four sections interact together and are executed by Microsoft Excel software. SASSE calculates multistage, countercurrent flowsheets based on distribution ratios calculated in the SASPE section. SPACE enables one to estimate the space and cost requirements for installing a specific TRUEX process using a variety of different equipment and processing facilities. INPUT/OUTPUT is a menu-driven interface that allows the user to choose which option to run, prompts the user for all the information needed by the GTM before calculations begin and generates reports of results. The GTM has been utilized in designing a number of TRUEX flowsheets described in the following sections.

Experience with TRUEX. A number of laboratories, both in the USA and abroad, have carried out hot test runs with actual waste solutions using the TRUEX process. Although this chapter specifically focuses on HLW, some of the results of test runs on TRU wastes have been included below because they have demonstrated the capability of TRUEX and the utility of the GTM. Table I lists (alphabetically) the laboratories in the USA that have tested TRUEX and the foreign countries that are currently performing studies on TRUEX. The types of waste treated are also shown.

ANL Experience. Except for the initial development studies on TRUEX, which utilized a dissolved sludge waste that simulated the insoluble HLW sludge from a Hanford storage tank (*29*), all additional studies with real waste solutions carried out at ANL involved testing the TRUEX process by treating the large volumes of TRU analytical waste generated at ANL and the New Brunswick Laboratory. A total of 118 L of TRU nitric acid waste solutions were treated in four process runs (*36*). A 20-stage bank of 4 cm centrifugal contactors housed in a plutonium glovebox was used to process the wastes. The nitric acid concentration in the feed varied from 1.7 to 4.5 M. Various amounts of other acids, such as HCl, H_2SO_4 and H_3PO_4, were also present in some of the solutions. For example, the phosphoric acid concentration was as high as 1.3 M in one of the waste solutions.

TABLE I. Experience with TRUEX

USA

Laboratory	Type of Waste Processed or Studied
Argonne National Laboratory (ANL)	TRU-Analytical Waste
Lockheed Martin Idaho Technologies Co. (LMITCO)	Sodium Bearing Waste, Calcine Waste
Oak Ridge National Laboratory (ORNL)	Irradiated ^{242}Pu Targets
Pacific Northwest National Laboratory (PNNL) and Westinghouse Hanford Co. (WHC)	Neutralized Cladding Removal Waste (NCRW) and Plutonium Finishing Plant (PFP) Waste

INTERNATIONAL

Country	Laboratory	Type of Waste
India	Bhabha Atomic Research Centre	Commercial and Defense HLW
Japan	Power Reactor and Nuclear Fuel Development Corp.	Commercial HLW

The flowsheets utilized in each of the four processing runs were designed using the GTM. The TRUEX process solvent formulation consisted of 0.2 M CMPO - 1.4 M TBP-n-dodecane. Alpha decontamination factors, D.F.s, of the waste solutions ranged from 4.0×10^3 to 6.5×10^4, which allowed disposal of the process raffinates as low-level wastes. Approximately 18 g of U, 84 g of Pu and 200 mg of Am were recovered. In addition to the high alpha D.F.s achieved, several other features of these processing runs are noteworthy. In one processing run, the Am product contained 99.1% of the Am and only 0.2% of the Pu whereas the Pu product contained 99.8% of the Pu and only 0.9% of the Am. An innovative scrub/Am strip section was designed and tested to reduce the number of stages required for stripping and at the same time to concentrate Am by a factor of ~17. A 0.28 M $(NH_4)_2C_2O_4$ solution was used to strip Pu and prevent its precipitation. Details of all of the processing runs and flowsheets have been reported (*36*).

LMITCO Experience. The TRUEX process is being evaluated at the Lockheed Martin Idaho Technologies Co. (LMITCO) for the separation of the actinides from acidic HLW (*8,37-40*). The treated wastes resulting from TRUEX and subsequent fission product separation processes (discussed below) are anticipated to be grouted and disposed of as non-TRU low-level wastes. The TRUs and lanthanide fission products will be vitrified and disposed of as HLW. Several TRUEX demonstration runs have been performed on Sodium Bearing Waste (SBW) (*38*, and references therein). The SBW is a secondary acidic HLW (1.5 to 1.7 M HNO_3) containing twelve inert constituents, including Pb and Hg, and the following radioactive constituents: ^{90}Sr, ^{99}Tc, ^{137}Cs, $^{235,238}U$, $^{238,239}Pu$ and ^{241}Am. Based on preliminary studies of TRUEX, an optimized TRUEX flowsheet was tested in shielded hot cells at the ICPP Remote Analytical Laboratory using a 20-stage bank of 2 cm diameter centrifugal contactors. The flowsheet, which was designed using the GTM, consisted of six stages of extraction (O/A=0.33), four 0.01 M HNO_3 scrub stages (O/A=1.5), six 0.01 M 1-hydroxyethane-1,1-diphosphonic acid (HEDPA) strip stages (O/A=1.0), two 0.25 M Na_2CO_3 solvent wash stages (O/A=1.0) and two 0.1 M HNO_3 rinse stages (O/A=6.0). HEDPA is a powerful complexing agent that will effectively remove actinides (III, IV, VI) from TRUEX process solvent (*41*). The process solvent formulation consisted of 0.2 M CMPO - 1.4 M TBP-in Isopar-L. (Isopar-L is a mixture of isoparaffinic hydrocarbons with an average of 12 carbons per molecule.)

In the optimized demonstration run with SBW, 99.79% of the actinides were removed from the waste, resulting in a reduction of alpha activity from 540 nCi/g in the feed to 0.90 nCi/g in the aqueous raffinate (*38*). This reduction is well below the NRC Class A low-level waste requirement of 10 nCi/g for non-TRU waste. Removal efficiencies of 99.84% for ^{241}Am, 99.97% for $^{238,239}Pu$ and 99.80% for $^{235,238}U$ were obtained (*38*). Although some iron was extracted by TRUEX solvent, it was effectively scrubbed from the solvent, resulting in only 0.7% of the Fe existing in the TRU strip product. Seventy-four percent of the Hg was extracted by the TRUEX solvent but was effectively stripped by the 0.25 M Na_2CO_3 wash. The 0.01 M HEDPA solution back-extracted 99.4% of the actinides, but some additional adjustments in the strip section need to be made to eliminate a slight precipitate from forming. Details of the entire demonstration run have been reported (*38*).

TRUEX demonstration runs have also been carried out on dissolved zirconium calcine simulant using 2 cm centrifugal contactors (*42*). Zirconium type calcine was produced from reprocessing zirconium clad fuel. Zirc-Calcine comprises the majority of calcine stored at ICPP. Two liters of spiked simulant were treated in six extraction stages using the same composition process solvent used for SBW. The four-stage scrub section utilized a 0.2 M NH_4F - 1.0 M HNO_3 solution (O/A = 3) to remove extracted Zr. A 0.004 M HEDPA solution was used to concomitantly strip U, Np, Pu, Am and lanthanides. Detailed results of the preliminary demonstration run have been published (*42*). The major problem encountered in the test run was precipitate formation in the strip section. The precipitate was determined to be zirconium phosphate, which was formed most likely from the excessive amount of Zr carried into the strip section and from the small amount of H_3PO_4 usually present in commercial grade HEDPA. In spite of the problems encountered in the stripping section, >99.2% of the Am was removed from the feed in the extraction section. Although this level of Am removal is not sufficient to ensure the <10 nCi/g TRU content in the low-level waste raffinate, modifications in the flowsheet are being made to suppress the extraction of Zr into the TRUEX process solvent. Based on the highly successful demonstration runs using TRUEX to treat wastes at the Idaho plant, consideration is being given to implementing TRUEX after year 2000. However, further experimental studies need to be carried out to address solvent cleanup and recycle issues and to refine flowsheets.

ORNL Experience. The Radiochemical Engineering Development Center (REDC) at Oak Ridge National Laboratory has tested TRUEX for separation and recovery of macroquantities of Am and Cm from irradiated Mark 42 PuO_2 targets (*43*). These targets were highly irradiated (>87% fission), therefore, the test runs with TRUEX were carried out in a shielded hot cell. Three banks of 16-stage mixer-settler contactors were used to conduct three test runs using TRUEX. The objectives of the test runs were twofold: (1) to test the performance of the TRUEX process when used to treat a highly radioactive feed solution containing macroquantities of fission products and TRUs and (2) to verify the GTM using gram quantities of Am and Cm and dekagram quantities of Pu. The TRUEX process solvent composition was 0.2 M CMPO - 1.2 M TBP-Norpar-12. (Norpar-12 is a mixture of normal paraffinic hydrocarbons with an average chain length of twelve.)

In general, the design criteria for the process flowsheet were met. The Am-Cm product contained <0.1% of the Pu, and the Pu product contained <1% of the Am-Cm. The losses of TRUs to the aqueous raffinate were low as predicted by the GTM. The major problem area revolved around the stripping section for Pu. In all three test runs, a significant Pu concentration remained in the stripped process solvent. It is believed that the unstripped Pu is most likely retained by acidic degradation products formed in the process solvent.

PNNL/WHC Experience. The first TRUEX demonstration runs performed outside of ANL and the first demonstration runs performed with real waste solutions were carried out at the Plutonium Reclamation Facility (PRF) located at the Hanford Site (*30-32*). As mentioned above, TRUEX was developed primarily to address the problem of treating the vast quantities and varieties of defense waste generated and stored at the Hanford Site. The first type of waste stream selected for a demonstration run was a TRU waste referred to as Plutonium Finishing Plant (PFP) waste, that was generated at the PRF. The PFP waste consisted primarily of aqueous raffinate from a SX process using TBP in CCl_4 to recover Pu from scrap materials. In addition, PFP waste may also contain carbonate scrub solution from solvent cleanup operations, distillates from evaporators, HF scrubber waste and miscellaneous laboratory waste. (The PFP waste used for the demonstration did not contain any solvent scrub waste.) The major constituents in PFP waste were Al (0.5 M), Na (0.05 M), Mg (0.3 M), Ca (0.2 M), Fe (0.05 M) and F (0.4 M). The nitric acid concentration ranged from 1.5 to 4.0 M. Typical Pu and Am concentrations in PFP were 0.1 g/L (7,500 µCi/L) and 0.05 g/L (170,000 µCi/L), respectively. Because no flammable solvents were allowed in the PRF plant, the TRUEX process solvent formulation consisted of 0.25 M CMPO - 0.75 M TBP-tetrachloroethylene. This formulation is equivalent, with respect to TRU distribution ratios, to the corresponding formulations that utilize paraffinic hydrocarbon diluents.

Four centimeter centrifugal contactors were used to carry out the solvent extractions. The entire setup including solid-liquid separation equipment, associated tankage and transfer lines was housed in a large glovebox train (*31*). Five extraction stages, three scrub stages, five dilute nitric acid strip stages and three dilute HNO_3-HF strip stages were employed in the demonstration runs. The feed/scrub/solvent flows were 250/50/100, respectively. A total of 40 L of clarified PFP waste was treated in four separate runs. In each of the runs, 10 liters of clarified feed were processed through the centrifugal contactor equipment in about 40 minutes. The TRU content of all aqueous raffinates was reduced to well below the 100 nCi/g goal. The Am product contained 96% of the Am and only 7% of the Pu, whereas the Pu product contained 88% of the Pu and only 1% of the Am. (Note that even better Am/Pu partitioning was achieved in the ANL and ORNL demonstration runs that were performed at a later date (*36,43*).)

Several references (*31,44-46*) describe other Hanford HLW that are amenable to treatment by TRUEX. They include Complexant Concentrate (CC), Neutralized Cladding Removal Waste (NCRW) and Single-Shell Tank (SST) wastes. Chemists at PNNL have recently studied the feasibility of treating NCRW by TRUEX processing (*46*). The NCRW consists of the solids (principally $ZrO_2 \cdot xH_2O$) that precipitated when the spent NH_4F-NH_4NO_3 waste produced during chemical decladding of Zircalloy-Clad N-Reactor Fuel was made alkaline. There are currently about 2,300 m^3 of NCRW stored at the Hanford site (*7,31*). A sample of actual NCRW sludge was used to demonstrate TRUEX. The sludge was first washed with 0.1 M NaOH and then dissolved using 12 M HNO_3 and 10 M HF. The dissolved sludge (1.9 M in HNO_3, 1.0 M in HF and 0.16 M in Zr plus TRUs and fission products) was subjected to a series of batch SX contacts, first using TBP to separate U and then using TRUEX process solvent (0.2 M CMPO - 1.4 M TBP-dodecane) to separate TRUs. Approximately 98% of the U was separated from the TRUs in three stages using 30% TBP allowing the U to be handled as low-level waste. Greater than 99% of the TRUs, together with La, Ce and Nd were extracted in three stages. Three stages of scrub were carried out to remove excess HNO_3 from the solvent. The feed to scrub numbers 1, 2 and 3 to solvent flows ratios were 100/16, 16, 32/48. Stripping was achieved using two stages of 0.01 M $H_2C_2O_4$ (O/A = 3). No attempt was made to partition Am from Pu. A final 0.25 M Na_2CO_3 wash (O/A = 9) was used before recycling solvent. Using the proposed PUREX/TRUEX flowsheet, the number of HLW glass canisters would be reduced from 2,400, if only sludge washing was performed, to approximately 500.

Although test runs using TRUEX to treat a variety of wastes at the Hanford Site have been very successful, no plans have been made to implement the process. The overall problem of waste management at the Hanford Site is under intensive study. TRUEX processing is a possible option but not a favored one at this time.

Japanese Experience/PNC. During the last ten years, researchers from the Power Reactor and Nuclear Fuel Development Corp. (PNC) of Japan have been studying actinide partitioning as part of an advanced reprocessing system. Ozawa et al. (*47,48*) have been evaluating the option of separating TRUs from PUREX raffinate using the TRUEX process. These studies, which have been performed at the Chemical Tokai-works Processing Facility, have utilized actual HLW generated from an advanced PUREX process. (The PUREX process was carried out on fast breeder reactor spent fuel burned up to 54,000 megawatt-days per ton and cooled for 2 to 4 years.) The HLW was used without adjusting acidity, but small amounts of Pu and oxalic acid were added. Oxalic acid was added to suppress the extraction of Zr and Mo (*29*). Countercurrent experiments were conducted in mixer settlers with holdup volumes of 23 mL per stage. Nineteen stages for extraction and scrubbing and 16 to 19 stages for stripping and solvent regeneration were used to test the process flowsheets. The process solvent formulation consisted of 0.2 M CMPO - 1.0 M TBP-dodecane. (Note that a lower concentration of TBP was used than with previously employed process solvent formulations.)

Three countercurrent mixer-settler runs were performed in the early 1990s. The acidity of the feed solutions varied from 4.0 M to 7 M HNO_3. Scrub solutions varied from 0.3 M HNO_3 - 0.1 M $H_2C_2O_4$ to 7.7 M HNO_3 - 0.03 M $H_2C_2O_4$ to 0.3 M HNO_3. Strip solutions varied from a single 0.01 M HNO_3 strip to multiple strips using first 0.01 M HNO_3 followed by 0.3 M HNO_3 - 0.1 M hydroxylammonium nitrate (HAN) and then 0.5 M HNO_3 - 0.1 M $H_2C_2O_4$. A final solvent wash utilized 0.1 M Na_2CO_3. In all test runs, quantitative extraction of actinides resulted in removal of alpha-emitting nuclides with D.F.s >10^3. In the stripping section, Am and Cm were effectively stripped with dilute HNO_3; however, significant retention of Pu and Ru by the solvent was observed until the solvent cleanup section, where both are reported to strip very well. Dual scrubbing improved Ru removal from the solvent, but 6% of the initial value was still retained.

The GTM was used to predict concentration profiles for the three flowsheet demonstration runs (49). The agreement between experimental results and those calculated with GTM were quite good, especially for Am, Cm and lanthanide fission products, which once again proved the value of GTM as a computational tool for designing and simulating TRUEX flowsheets.

Recent studies by researchers at PNC have focused on performing a selective partition of trivalent actinides from light lanthanides using TRUEX process solvent (50). The process is called SETFICS, which is an acronym derived from Solvent Extraction for Trivalent f-elements Intra-group separation in CMPO-complexant System (50). A demonstration of SETFICS was carried out in a countercurrent mode using real TRUEX process solvent that was obtained in previous hot experiments. Americium and curium were partitioned from light lanthanides using 0.05 M DTPA - 4 M $NaNO_3$. The $^{144}Ce/^{241}Am$ decontamination factor was 72. Although 80% of the lanthanides were rejected from the Am and Cm products, very little of the Sm and Eu were removed from the actinide(III) product. The yield of Am and Cm was rather low because of the lack of achievement of steady state.

India Experience. Scientists at the Bhabha Atomic Research Centre have tested the TRUEX process using HLW generated from PUREX processing of thermal reactor fuels (51-53). Two process solvent formulations were used: 0.2 M CMPO - 1.2 M TBP-dodecane and 0.2 M CMPO - 1.4 M TBP-dodecane. Feed solutions were approximately 2 M in HNO_3. Two demonstration test runs were carried out. In the first, TRUEX processing was carried out directly on the HLW, whereas, in the second test run PUREX processing preceded TRUEX processing. In both sets of experiments TRUEX effectively removed the alpha activity from the feed. For example, in the second test run experiments, the alpha activity left in the raffinate after four contacts (O/A = 0.5) was less than 0.06% of the total in the feed. Stripping in succession with 0.04 M HNO_3, 0.05 M HNO_3 - 0.05 M HF and 0.25 M Na_2CO_3 resulted in the usual actinide partitioning. The stripped and washed process solvent did not contain any detectable activity above background.

Summary of TRUEX. The TRUEX process has been demonstrated in five laboratories in the USA and in two laboratories in foreign countries. It has been shown to be applicable to both defense and commercial TRU and HLW waste. Particularly noteworthy is its adaptability to treat a variety of defense wastes with wide ranging compositions. TRUEX can achieve in a single extraction cycle what requires multiple extraction cycles and cumbersome pH adjustments with acidic extractants. Another major feature of TRUEX is that stripping does not require high concentrations of acids or undesirable complexants. In addition, trivalent actinides can be concentrated many fold through innovative design of stripping sections.

One of the disadvantages of TRUEX is the cost of the CMPO, although manufacturing the extractant on a large-scale would reduce the cost to an acceptable level. Another disadvantage of the process is that extensive hydrolytic and radiolytic degradation of the process solvent reduces stripping efficiency. Solvent degradation of CMPO and TRUEX process solvent has been investigated at ANL and several methods have been developed and tested to restore the solvent to near pristine conditions (33, and references therein). Another drawback to TRUEX is the poor selectivity for trivalent actinides over Zr and Mo and, to some extent, Fe. However, the addition of oxalic or hydrofluoric acid or the use of special scrubbing reagents improves the selectivity to acceptable levels. It is interesting to note, as will be discussed below, that both acidic and neutral extractants utilized in TRU extraction processes have the same deficiency with regard to the selectivity of actinide(III) over Fe, Zr and Mo.

Russian TRU Extraction Process. Russian chemists have been studying neutral bifunctional extractants for a number of years (*54-58*). These studies have paralleled somewhat similar investigations carried out in the USA. To extract TRUs from HLW, Russian chemists have adopted a different carbamoylmethylphosphine oxide (CMPO) derivative than used in TRUEX, namely, diphenyl-*N,N*-di-*n*-butyl CMPO which is abbreviated DøDBCMPO. The structure of DøDBCMPO is shown in Figure 1. As discussed in the TRUEX section, the diphenyl CMPO derivative is insufficiently soluble in paraffinic hydrocarbon diluents, even in the presence of excess TBP, to be of practical use. Furthermore, diphenyl CMPO derivatives have a strong propensity towards third phase formation. However, Russian chemists found that by using a fluoroether, called Fluoropol-732, as a diluent for the diphenyl CMPO, the unfavorable solubility and third phase formation properties of this derivative could be overcome (*59*). The D_{Am} versus aqueous HNO_3 concentration curve using a 0.05 M DøDBCMPO solution in Fluoropol-732 is shown in Figure 3 (*60*). The data in Figure 3 show that the values of D_{Am} obtained with the DøDBCMPO-Fluoropol system are significantly higher, using only one-fourth the concentration of CMPO, than those obtained with TRUEX process solvent over the entire nitric acid concentration range.

The Russian TRU extraction process uses a 0.1 M solution of DøDBCMPO in Fluoropol-732 as the process solvent (*59*). An 18-stage bank of centrifugal contactors was used to test the TRU extraction flowsheet. Feed solution consisted of a HLW simulant, 5 M in HNO_3, containing more than 13 g/L of lanthanides and actinides. An interesting feature of the flowsheet is the use of acetohydroxamic acid (AHA) to strip Fe(III), Zr(IV) and Mo(VI), which also extract along with the transplutonium elements (TPEs). A solution of 2 M HNO_3 - 10 g/L AHA was employed for this purpose. The AHA strip solution is apparently contacted with fresh process solvent to remove any traces of TPEs and possibly Pu (*59*). TPEs and lanthanides were stripped from the process solvent using 0.01 M HNO_3. Greater than 99.5% of the actinides and lanthanides were recovered and concentrated by a factor of four to six. The reduction of Fe, Zr and Mo from the TPE fraction was >50. Efforts are currently underway to apply the process to a plant-scale operation (*60*).

A number of the favorable features outlined for TRUEX also apply to the Russian TRU/SX process, namely, efficient extraction of Am(III) over a wide range of HNO_3 concentrations and stripping with low concentrations of acid. The Russian TRU process has the added advantage of using a lower concentration of a less expensive extractant. Radiolytic and hydrolytic degradation is probably less with the DøDBCMPO-Fluoropol system than with TRUEX process solvent because of the absence of TBP in the former.

There are, however, some major concerns that arise regarding the Russian process. These concerns center around the use of the Fluoropol diluent. Introducing a diluent with physical properties that are radically different from the physical properties of paraffinic hydrocarbons could cause major problems in implementing this process in existing processing facilities that were designed for PUREX or are currently being used to carry out PUREX processing. Other concerns regarding Fluoropol are its radiolytic stability, its corrosive properties in high radiation fields, the adequacy of practical process solvent cleanup techniques, and the environmental issues created when the spent solvent must be discarded.

There are other features of the DøDBCMPO-Fluoropol system that need comment. For example, because of the very high distribution ratios of Am over a wide range of HNO_3 concentrations, back-extraction of Am requires unusually low acid concentrations as shown in Figure 3. Low acid concentrations may not be easily attainable under plant operating conditions. Note that the Russian process data in Figure 3 apply to a 0.05 M solution of extractant whereas the process solvent utilizes a 0.1 M solution (59). Assuming at least a second power extractant dependency, the D_{Am} values for the process solvent will be four times higher than the D_{Am} values shown in Figure 3. Most likely, additional low acid scrubs will be required to help reduce the HNO_3 concentration in the organic phase prior to stripping. Even the advantages of the high D_{Am} values attainable at high HNO_3 concentrations are partially negated by the limitations in stage efficiency. In addition, using low extractant concentrations such as 0.05 M to 0.1 M results in a significantly reduced loading capacity of the process solvent. It is important to keep in mind that in evaluating different extractant systems, low distribution ratios in the extraction stages can be compensated for by the addition of one or two extra stages and/or a higher organic to aqueous flow ratio.

French DIAMEX Process. French chemists at Fontenay-aux-Roses have investigated a number of new amidic extractants as part of a broad strategy for improving the management of radioactive waste (61-65). Two classes of amidic extractants have been investigated. The first class are monoamides of the general structure, RCONR'R", where R, R' and R" are alkyl groups. Monoamides are monofunctional extractants that only effectively extract tetra- and hexavalent actinides. The second class of extractants are bifunctional amides of the general structure, $(RR'NCO)_2CHR"$, where R and R' are alkyl groups and R" is an alkyl or oxyalkyl group. The bifunctional compounds are able to extract all actinides, including trivalent actinides, from acidic nitrate media. Only the bifunctional amidic extractants will be discussed in this chapter since they are the basis of the French DIAMEX process (61-64).

French chemists have synthesized and characterized a number of the 1,3-propanediamide (or malonamide) derivatives (62,63). Characterization studies have involved, in addition to the typical SX parameters, speciation and third phase formation (62,63,65). An interesting finding of the speciation studies is that the malonamide compounds form neutral complexes with U(VI) and Pu(IV) nitrates, whereas with Nd(III), a standin for Am(III), both neutral and acidic ion pair complexes are formed, with the latter predominating as the acidity increases (63,65). This behavior with Nd(III) is different from the behavior of the CMPO class of extractants with Nd(III), where only neutral complexes form between the actinide(III) nitrate and the extractant (33, and references therein). As a result of these basic studies, the N,N'-dimethyl-N,N'-dibutyl tetradecylmalonamide (DMDBTDMA) was selected as the extractant of choice for the DIAMEX process. The structure of DMDBTDMA is shown in Figure 1.

The process solvent in DIAMEX consists of a 0.5 M solution of DMDBTDMA in the aliphatic diluent TPH. The distribution ratio of Am(III) as a function of aqueous nitric acid concentration for DIAMEX process solvent is shown in Figure 3 (63). These data show that DMDBTDMA requires very high nitric acid concentrations (≥3 M) to extract Am(III). If 0.05 to 0.2 M $H_2C_2O_4$ is present in the feed to complex Fe, Zr and Mo, the HNO_3 concentration should be 5 M (63).

The DIAMEX process flowsheet is similar to, but somewhat simpler than, the TRUEX and the Russian TRU process flowsheets (61-64). Feed solutions are 5 M in HNO_3 with oxalic acid usually present to complex Fe, Zr and Mo. The O/A in the extraction stages is one. Scrub solution consists of 2 M HNO_3 and stripping is achieved using 0.1 M HNO_3 - 0.2 M hydroxylammonium nitrate, O/A = 1, for Am and Pu, 0.5 M $H_2C_2O_4$ for Np(V), and 0.01 M HNO_3 for U.

The most noteworthy features of the DIAMEX process are that the process solvent is incinerable and that back-extraction is extremely simple and efficient. Most impressive is the fact that radiolytic and hydrolytic degradation of the process solvent do not interfere with stripping (*63*). Therefore, no involved periodic solvent cleanup step would be required, as with the CMPO extractants. Also, ruthenium extraction was not a problem as in the TRUEX process.

The major disadvantage of the DIAMEX process is that very high HNO_3 concentrations are required to obtain efficient extraction of Am. Furthermore, the acid dependency of D_{Am}, as depicted in Figure 3, is extremely sensitive to the HNO_3 concentration below 4 M. This feature could cause operational problems when carried out on a plant-scale. However, because the French utilize high acidities in their PUREX processing flowsheets or would utilize equally high acidities in flowsheets where TBP is replaced by a monofunctional amidic extractant, the high HNO_3 concentrations required for feed solution in the DIAMEX process should not be a problem. Another disadvantage of the DIAMEX process is that the extraction of Tc along with the actinides does not appear to occur, as was the case with TRUEX. This feature is most likely due to the high HNO_3 concentration of the feed solutions. The cost of DMDBTDMA relative to the cost of the CMPO class of extractants is difficult to assess because none of these compounds has been prepared on a large scale. Although significant improvements have been made in the preparation of DMDBTDMA, the final step involves purification of the extractant on an adsorbent to eliminate traces of a phase transfer catalyst (*62*). On the other hand, both of the CMPO compounds are readily purified by crystallizations.

TRPO Process. In 1989 a cooperative program was established between Tsinghua University, Beijing, China and the European Institute for Transuranium Elements, Karlsruhe, Germany to develop a new TRU solvent extraction process to treat HLW (*66*). The key ingredient in the new process is a mixture of trialkylphosphine oxide extractants (abbreviated TRPO) (*66-69*). The TRPO mixture consists of seven different, primarily straight chain, phosphine oxides with six to eight carbon atoms per chain. The most abundant phosphine oxides are $(C_7H_{15})_3PO$, 27%, $(C_7H_{15})_2(C_8H_{17})PO$, 24%, and $(C_7H_{15})(C_8H_{17})_2PO$, 16%. A general structure for the TRPO is shown in Figure 1.

The TRPO-dodecane system has been extensively characterized with respect to the uptake of actinides (III, IV, V, VI) and with respect to selected fission products and Fe. In addition, speciation, loading capacity, third phase formation and radiolytic stability have been studied. Details of all these studies have been reported (*69*, and references therein). The most important data from these studies are the distribution ratio of Am(III) as a function of aqueous nitric concentration, which is shown in Figure 3 for 30 vol % TRPO in kerosene. The data show that Am is only effectively extracted between 0.05 M to 1 M HNO_3. The extraction of Am(III) may best be represented by the formation of a neutral complex containing three TRPO molecules coordinated to $Am(NO_3)_3$ (*69*). The behavior of TRPO with respect to speciation more closely resembles that of the CMPOs than that of the DMDBTDMA. This is not surprising since the TRPOs and CMPOs are both phosphine oxides.

Hot demonstration runs of the TRPO process were carried out on real commercial HLW originating from WAK, Karlsruhe. Composition of the HLW can be found in (*70*). The HLW solutions were centrifuged to remove solids, treated with Fe(II) sulfamate along with HAN to reduce Np(V) to Np(IV) and diluted ten-fold with H_2O. The nitric acid concentration was adjusted to 0.7 M for one test run and 1.4 M for another test run. Process solvent consisted of 30 vol % TRPO in kerosene.

The demonstration runs were carried out in two separate parts. The first consisted of the extraction and stripping of Am. After the first part reached steady state, 200 mL of loaded organic phase (with Am removed) was fed into clean centrifugal contactors to carry out the second part of the process, which consisted of stripping Np, Pu and U. In all, 22 stages were used for the entire process.

The phase ratios in the extraction and 1 M HNO_3 scrub stages were 0.44 and 3.0, respectively. Americium was stripped with 5.5 M HNO_3, O/A=1. Np and Pu were stripped with 0.6 M $H_2C_2O_4$, O/A=1, after the organic phase was given a preliminary scrub with 0.1 M HNO_3 to reduce the concentration of HNO_3 in the process solvent. Finally, U was stripped with 5% Na_2CO_3, O/A=1.

More than 99.97% of the Am was extracted in seven stages. All TRUs were effectively stripped from the loaded organic phase. Cross contamination of the TRU elements among the three different product solutions was very low. Technetium-99 was also effectively extracted from the feed with D.F.s of >1400. However, Zr, Mo and, of course, lanthanides were also extracted. The lanthanides were stripped along with Am, Zr and Mo were stripped with Np and Pu and Tc most likely was stripped with U (70).

Additional hot demonstration runs were carried out with a highly salted HLW that was obtained from reprocessing facilities in China (71,72). Nonradioactive nuclides such as Na, Al, Cr, Fe and Ni contributed 88% of the total mass, whereas the TRUs and ^{99}Tc contributed only 0.31% of the total mass. Greater than 99% of the total alpha activity was from Am and Pu and <1% of the alpha activity was from Np. Test runs were carried out in a hot cell using a 50-stage bank of 1-cm miniature centrifugal contactors. To avoid third phase formation due to the enhanced extraction of Fe from salting-out effects, the feed solution was diluted 2.7 times. The final HNO_3 concentration of the feed was 1.2 M. The flowsheet used to extract and recover TRUs from highly salted HLW was essentially the same as the flowsheet used for the extraction and recovery of TRUs from commercial HLW, except for the number of stages employed (71). The decontamination factors achieved were 650-700 for Am and Pu, 15 for Np, 10^4 for U and 125 for Tc. Additional processing of the raffinate from the TRPO process for Sr recovery was also carried out. This subject will be discussed below under Sr and Cs Extraction/Recovery.

Noteworthy features of the TRPO process are: (1) it utilizes a relatively inexpensive extractant, (2) the process solvent has very high radiolytic and hydrolytic stability, (3) the partitioning of tri-, tetra- and hexavalent actinides from each other is relatively simple and easy to control and (4) very high removal of Tc is achievable. Disadvantages of the process are: (1) Am is not efficiently extracted from even moderately high HNO_3 concentrations, (2) Am back-extraction requires very high HNO_3 concentrations and (3) Am is not significantly concentrated in the stripping section relative to its concentration in the feed. Because of the first disadvantage, most HLW streams would require adjustments in acidity, primarily dilutions. These adjustments would most likely cause precipitations to occur, especially in attempting to treat dissolved HLW sludges. The second and third disadvantages would create major problems because the Am product would have to be significantly concentrated and the nitric acid recovered before further treatment, such as Am/rare earth separation or vitrification, could be carried out.

Conclusion on the Use of Neutral Extractants. All of the neutral extractants, whether monofunctional or bifunctional, have the same major advantages over acidic extractants. Foremost among these is that neutral extractants (with the exception of TRPO) are able to extract TRUs from much higher acidities and release them at low acidities. This behavior is due to the fact that neutral extractants coordinate to TRU nitrates, therefore, extraction is directly dependent on nitrate (or nitric acid) concentration. Acidic extractants, on the other hand, are inversely dependent on hydrogen ion concentration and coordinate to the bare (dehydrated) metal ion. An outgrowth of this difference in complexing behavior of the two classes of extractants is responsible for another major advantage of neutral extractants, namely, their significantly superior selectivity for TRUs over fission products (lanthanides excluded) and over the inert constituents usually present in HLW.

Although the four different types of neutral extractants, namely, O(ø)DiBCMPO, DøDBCMPO, DMDBTDMA and TRPO, show major differences in D_{Am} vs. nitric acid dependencies, they all can be effectively utilized in TRU extraction processes. Frequently, the specific types of HLW treatment problems that are being addressed, for example, commercial vs. defense wastes or highly acidic raffinates vs. dissolved sludges, dictate the choice of the extractant.

Strontium and Cesium Extraction

Strontium-90 and cesium-137 are the two major generators of heat and ß/γ radiation in HLW. Their presence, therefore, greatly complicates waste handling and disposal. The state of separations technology for the removal of ^{90}Sr and ^{137}Cs prior to the mid 1980s has been summarized (73). The conclusion stated in (73) was that improved extraction/recovery processes for Sr and Cs from acidic waste streams are sorely needed. It is not the objective of this chapter to review all of the prior systems that have the potential to extract Sr and Cs from acidic media. Our objective is to focus on those systems that have actually been demonstrated in process flowsheets to remove Sr and Cs from moderate to highly acidic HLW.

Strontium(II) and cesium(I) are large cations with low charge densities. Consequently, there are relatively few systems capable of efficiently extracting these cations selectively from highly acidic aqueous solution into a nonaqueous medium. Two systems that can achieve this extraction and serve as the basis of process flowsheets are: the bis-dicarbollylcobaltate anion (referred to hereinafter as cobalt(III) dicarbollide or dicarbollide) and the macrocyclic polyethers (referred to hereinafter as crown ethers). The structures of cobalt(III) dicarbollide and a crown ether used for Sr extraction are shown in Figure 4.

Cobalt(III) dicarbollide. Fundamental studies on the extraction of Cs and Sr by cobalt(III) dicarbollide have been carried out primarily in the Czech Republic and are summarized elsewhere (74, and references therein). Application of the dicarbollide technology to HLW processing has been performed in Russia (75,76). Figure 5 shows the nitric acid dependencies for the extraction of Sr and Cs using a 0.01 M solution of cobalt(III) dicarbollide and 0.01 M polyethylene glycol (PEG-400) in nitrobenzene (77).

Plant-scale demonstration runs utilizing a process solvent consisting of 0.15 M chlorinated cobalt dicarbollide (ChCoDiC), 3% PEG-400 dissolved in *m*-nitrobenzotrifluoride were carried out in August 1996, at Mayak, UE-35 Facility, Chelyabinsk, Russia (78,79). The feed was a HLW solution with 2 M HNO$_3$. In all, 320 m^3 of waste solution was treated.

Demonstration runs have also been carried out in the USA at INEEL, by a team of chemists and engineers from the Khlopin Radium Institute, St. Petersburg, Russia and LMITCO (78,79). Initial investigations were performed using a sodium bearing waste (SBW) simulant as the feed and a 0.15 M ChCoDiC - 3% PEG-400 in Fluoropol (78). Two flowsheets were tested on a continuous basis with a 24-stage bank of 2 cm centrifugal contactors. Eleven extraction stages followed by two stages of solvent scrub with 1.0 M HNO$_3$ were used. The stripping section was comprised of six stages of a proprietary reagent followed by five stages of solvent wash. Cesium and Sr removal efficiencies were 98.6% to 98.9% and 99.89%, respectively (78).

Additional demonstration runs were carried out with actual SBW using 0.15 M ChCoDiC in Fluoropol. Polyethylene glycol, which is added to the dicarbollide solution to enhance Sr extraction, was not added to the process solvent. This was done to evaluate the process strictly as a Cs removal technology (79). The flowsheet consisted of twelve extraction stages, two scrub stages, six strip stages, three concentrated HNO$_3$ wash stages, and one dilute HNO$_3$ wash stage. One molar HNO$_3$ was used for the scrub and a proprietary reagent was used for the strip. The activity of ^{137}Cs was reduced from 236 Ci/m^3 in the feed to <0.003 Ci/m^3 in the aqueous

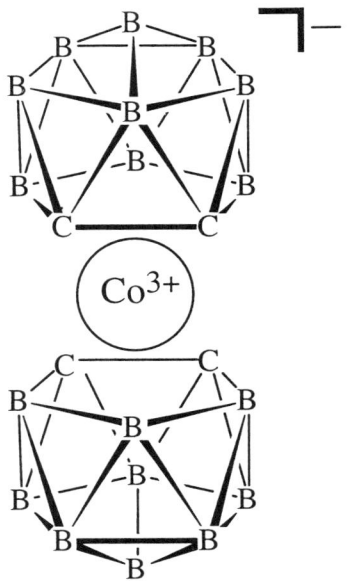

Cobalt(III) Dicarbollide Anion

Bis-4,4'(5')-(*t*-butylcyclohexano)-18-crown-6

Figure 4. Structures of Cs and Sr extractants. (Adapted from refs. 77,92.)

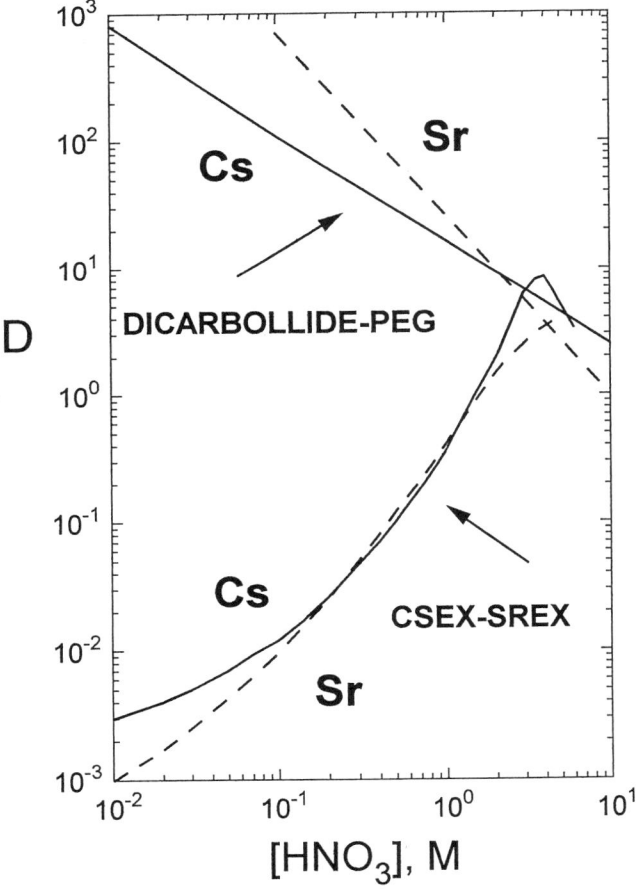

Figure 5. Nitric acid dependencies of Cs and Sr extraction using dicarbollide and crown ether extractants. (Adapted from refs. *77,92,93*.)

Dicarbollide (0.01 M cobalt(III) dicarbollide - 0.01 M PEG-400-nitrobenzene)
CSEX-SREX (0.1 M Cs extractant - 0.05 M Sr extractant - 1.2 M TBP-Isopar-L - 5 vol % lauronitrile)

raffinate. Other components in the waste that also extracted were K (50% extracted) and Hg (34% extracted), but both were effectively stripped along with the Cs.

The cobalt(III) dicarbollide and its chlorinated derivative are obviously very effective Cs extractants. With the addition of PEG-400, the ChCoDiC is also an effective Sr extractant. The dicarbollides have high selectivity over Na, which is the major metal ion constituent in SBW, and have high chemical and radiation stability (74,77, and references therein). The cost of the extractant appears to be quite low (77). The major drawbacks of the dicarbollides are the poor solubility in paraffinic hydrocarbons and the need to use stripping reagents that would be looked upon very unfavorably in most countries. Although the use of a Fluoropol diluent is an improvement over m-nitrobenzotrifluoride, the concerns raised about Fluoropols in the Russian TRU process also apply with the dicarbollide system. The fact that dicarbollides function as acidic extractants, and thus have inverse hydrogen ion dependencies, causes D_{Sr} and D_{Cs} to increase as acidity decreases. Stripping, therefore, has to be carried out at high nitric acid concentrations or by utilizing mass action effects of other chemicals. Both of these approaches give solutions of Cs and Sr that will require further treatment.

Crown Ethers/SREX. A new Sr extraction/recovery process was developed by chemists at ANL in the early 1990s for the removal of Sr from HLW (80). The new process is based on the use of an 18-crown-6 derivative dissolved in 1-octanol. The first version of the process utilized a 0.2 M solution of dicyclohexano-18-crown-6, DCH18C6, (primarily the *cis-syn-cis* isomer) dissolved in 1-octanol as the process solvent (80). Efficient extraction occurs from 3 M to 6 M HNO_3 and efficient stripping from 0.01 to 0.1 M HNO_3. The crown ether-octanol process solvent formulation was tested in a countercurrent mode by Chinese chemists at Tsinghua University in Beijing, China using a HLW simulant (71,72). The feed solution contained only 1 M HNO_3 but was very high in nitrate salt. The composition of the high salt waste solution is reported in (71). The process solvent consisted of a 0.1 M solution of DCH18C6 in 1-octanol. Approximately 99% of the Sr was removed from the feed in ten stages (O/A = 0.67).

A second generation Sr extraction/recovery process called SREX was developed by the ANL team to overcome the problem of aqueous solubility of DCH18C6. The SREX process solvent contains a mixture of isomers of 4,4'(5')-di-*t*-butylcyclohexano-18-crown-6 (DtBuCH18C6) dissolved in 1-octanol (81). The structure of DtBuCH18C6 is shown in Figure 4. Once again the most important isomer of the crown is the *cis-syn-cis* derivative (81). Chemists at the Idaho Chemical Processing Plant (ICPP) batch-tested SREX on a SBW simulant. A 0.15 M solution of DtBuCH18C6 in 1-octanol was used as the process solvent (82). Although the results were very favorable, 1-octanol was not looked upon favorably by engineers at ICPP. Meanwhile, further improvements were being made with the SREX process solvent (83). The new process solvent formulation is a modified PUREX process solvent of the following composition: 0.2 M DtBuCH18C6 - 1.2 M TBP-Isopar-L. The nitric acid dependency for the extraction of Sr using 0.05 M DtBuCH18C6 - 1.2 M TBP-Isopar-L is shown in Figure 5. Although the data shown in Figure 5 were obtained using a mixture of the Sr extractant and a Cs extractant, to be discussed below, D_{Sr} is not significantly affected by the presence of the Cs extractant.

During the last few years, chemists at ICPP have performed a number of investigations on the SREX process using the modified PUREX solvent (84-90). These studies focused on the behavior of matrix components including Na, K, Ca, Fe, Zr, Hg and Pb (85-87) and on preliminary flowsheet demonstration runs with SBW simulant (87). The process solvent consisted of 0.15 M DtBuCH18C6 - 1.2 M TBP-Isopar-L. The preliminary studies demonstrated proof-of-principle and identified potential stripping problems with Pb and Hg, which are extracted along with Sr (87). Improvements in the SREX flowsheet were then made and tested in a countercurrent mode to gain valuable operational data for more elaborate flowsheet testing to be carried out in hot cells (*88,89*).

Radiolytic degradation of SREX process solvent exposed to gamma radiation has also been studied at ICPP. Exposures varied from low doses up to 1000 KGy (*90*). It was found that radiation had very little effect on the extraction and stripping performance of the process solvent. Exposure of the process solvent to gamma radiation in the presence of solutions of nitric acid and simulated waste solutions also resulted in no change in D_{Sr}. Calculations show that SREX process solvent in continuous contact with SBW would receive a total dose of 8.8 KGy/year (*90*).

Based on recent studies (*84-90*), more elaborate countercurrent flowsheet testing of SREX process was performed at ICPP using 24 stages of 2 cm diameter centrifugal contactors that were installed in the Remote Analytical Laboratory hot cell (*91*). The process solvent consisted of 0.15 M DtBuCH18C6 - 1.5 M TBP-Isopar-L. Demonstration runs were carried out with both SBW simulant and actual SBW. The flowsheet consisted of ten extraction stages (O/A=1), two 2 M HNO_3 scrub stages (O/A=4), four 0.05 M HNO_3 strip stages (O/A=0.5), four 0.1 M ammonium citrate strip stages (O/A=1) and four 3.0 M HNO_3 solvent wash stages (*90*). The extraction/stripping behavior of actinides, fission products and numerous non-radioactive elements was evaluated. Removal efficiencies of 99.995% for Sr and >94% for Pb were obtained with actual SBW (*90*), which resulted in reducing the activity of ^{90}Sr in the raffinate to 0.0089 Ci/m^3. This level of ^{90}Sr in the raffinate is well below the NRC Class A low-level waste limit of 0.04 Ci/m^3. Further ^{90}Sr decontamination would have been achieved if the centrifugal contactors used for testing did not contain residual contamination from previous actinide flowsheet test runs (*91*). The strip sections of the flowsheet also performed extremely well, with 99.99% of the Sr and <6% of the Pb exiting with the 0.05 M HNO_3 strip and 0.007% of the ^{90}Sr and 93% of the Pb exiting with the 0.1 M ammonium citrate strip (*90,91*).

The behavior of actinides using the modified PUREX solvent is also very interesting. Only 1.9% of the ^{241}Am extracted with the SREX process solvent whereas 99.94% of the Pu and 99.6% of the U were extracted. The extraction of the Pu and U was due to the presence of TBP. In all, 94% of the total alpha activity was removed from the actual SBW (*90,91*). Other constituents that extracted to some degree are Na (0.5%), K (37%), Ba (64%), Zr (>82%) and Hg (>99%) (*90,91*). Sodium and K, and most likely Ba, were effectively stripped along with Sr, whereas only 9.5% of the Hg was stripped from the solvent by 0.05 M HNO_3 and the 0.1 M ammonium citrate. The remaining 90.5% of the Hg remained in the solvent.

References (*90,91*) summarize all the work performed at ICPP on the SREX process. The SREX flowsheet has been recommended to serve as the basis for further feasibility studies and/or facility design studies (*91*). Further improvements in the flowsheet need to be developed with regard to back-extracting Hg from the process solvent and possibly minimizing K levels in the Sr strip (*91*).

Combined CSEX-SREX. Efforts have been underway during the last few years to combine SREX with a Cs extraction/recovery process (*92,93*). The key ingredient in the Cs extraction process, referred to hereinafter as CSEX, is a dibenzo-18-crown-6 derivative of proprietary composition. The distribution ratios of Sr and Cs as a function of aqueous nitric acid concentration using the Combined CSEX-SREX process solvent are shown in Figure 5 (*92,93*). The Combined CSEX-SREX process solvent consists of 0.1 M Cs extractant - 0.05 M Sr extractant - 1.2 M TBP in Isopar-L. Five volume percent of lauronitrile is added to the process solvent to solubilize precipitates that sometimes form from the interaction of macroconcentrations of Sr and Ba with DtBuCH18C6. The data in Figure 5 show that extraction occurs at high acidity and stripping at low acidity. The acid dependency curves for both Cs and Sr are typical of situations in which nitrato complexes of metal ions are extracted by neutral extractants.

The Combined CSEX-SREX process offers several features of potential importance in chemical processing schemes for high-level waste treatment. First, if the Combined CSEX-SREX process is applied as the first step in chemical pretreatment, the radiation level for all subsequent processing steps (e.g., TRUEX) will be significantly reduced. Thus, TRUEX could most likely be carried out in a glovebox. Second, the recovered Cs-Sr fraction could be partitioned from the TRUs and therefore will decay to low-level waste after a few hundred years. Finally, combining Cs and Sr extraction into a single process will reduce the amount of equipment and space required to pretreat the waste. This latter point should reduce the cost of pretreating HLW.

The Combined CSEX-SREX process was tested at ANL in a batch countercurrent mode in 1995 using a zirconia calcine waste simulant obtained from chemical engineers at INEEL (*92*). The results were sufficiently favorable to carry out a hot test run in a continuous countercurrent mode using centrifugal contactors (*93*). A 24-stage bank of 2 cm contactors housed in a plutonium glovebox was used for the test run. The Zirc-Calcine waste simulant was 3.8 M in nitric acid. Millicurie quantities per liter of ^{85}Sr, ^{99m}Tc, ^{137}Cs and ^{241}Am were added to the waste simulant. The composition of the waste and the exact concentrations of radioisotopes added to the waste are reported (*93*). The flowsheet consists of two stages of solvent preconditioning with 4.0 M HNO_3 (O/A=1), nine extraction stages (O/A=2.4), four 4 M HNO_3 scrub stages (O/A=4.3), eight 0.1 M HNO_3 strip stages (O/A=1.3) and one carbonate wash stage (O/A=3.2).

Overall, the hot test continuous countercurrent run was highly successful. The removal efficiencies of Cs and Sr from the raffinate were 4.5×10^5 and $>2.6 \times 10^5$, respectively. Greater than 99.99% of the Cs and Sr were stripped from the organic phase. As expected, ^{241}Am remained entirely in the aqueous raffinate. The concentration of ^{241}Am in the Cs and Sr strip solution was $<1 \times 10^{-9}$ M. The distribution of Tc throughout the flowsheet is difficult to describe because Tc never appeared to reach steady state. The results indicate, however, that the combined process will not effectively remove Tc from the feed solution.

In summary, SREX and CSEX-SREX are highly effective systems for the extraction and recovery of Sr and Cs and Sr, respectively, from acidic HLW. The similarities between the CSEX-SREX, SREX, TRUEX and PUREX processes are noteworthy. All involve extraction from high nitric acid concentrations and back-extraction from low nitric acid concentrations. All process solvents contain TBP in a similar concentration range, utilize a paraffinic hydrocarbon diluent and have similar physical properties. Radiolytic degradation is somewhat similar for all four process solvents because TBP is a major constituent in each. Although TBP is not noted for its radiolytic and hydrolytic stability, the technology for its cleanup from degradation products is well established.

The major drawback of SREX and CSEX-SREX is the cost of the crown ethers. Both crown ethers are commercially available and are quite costly. However, the price of the crown ethers would decrease substantially if production were increased. It is unlikely that the cost of the extractant would be a significant item in the overall cost of pretreating waste. Another disadvantage of SREX and CSEX-SREX processes is that the process solvents are multicomponent systems. PUREX process solvent is a simple two component mixture but the CSEX-SREX process solvent has five components, namely, two extractants, two phase modifiers, and a diluent. Maintaining control of the composition of such a complex mixture on a plant-scale could present a major problem.

Combined TRU-Fission Product Processes

The significant success achieved during the last ten years in developing efficient TRU, Sr, Tc and Cs extraction/recovery processes has led quite naturally to attempts to develop systems that achieve the simultaneous extraction of TRUs, lanthanides, Sr and Cs (94). For example, chemical engineers at ICPP in collaboration with Russian chemists have tested, with some success, a process solvent containing 0.08 M Co(III) dicarbollide - 0.05 M DøDBCMPO in xylene (79). Studies at ANL have focused on developing a TOtal Radionuclide EXtraction/recovery process, called TOREX, based on a modified PUREX solvent (95). The entire topic of combining SX processes for actinide and selected fission product separations has been discussed (95). At this time it is debatable whether such processes simplify the overall pretreatment scheme. Two separate processes carried out in tandem may be simpler to operate on a plant-scale, particularly if the combined process requires multipartitioning steps to separate, for example, TRUs from U, Pu from Am and Sr and Cs from TRUs.

Acknowledgment

The authors wish to thank Dr. Boris Myaseodov of the V. I. Vernadsky Institute of Geochemistry and Analytical Chemistry, Moscow, Russia and Dr. Yongjun Zhu of the Institute of Nuclear Energy Technology, Tsinghua University, Beijing, China for furnishing up-to-date information on their respective TRU extraction processes. The authors also wish to thank Mr. Terry Todd of the Lockheed Martin Idaho Technologies Company, Idaho Falls, USA for supplying up-to-date reports on flowsheet testing at ICPP and for helpful discussions.

This manuscript was prepared under the auspices of the Office of Basic Energy Sciences, Division of Chemical Sciences, U. S. Department of Energy, under contract number W-31-109-ENG-38.

Literature Cited

(1) *New Separation Chemistry Techniques for Radioactive Waste and Other Specific Applications*; Cecille, L.; Casarci, M.; Pietrelli, L., Eds.; Elsevier Applied Science: New York, 1991.
(2) *Chemical Pretreatment of Nuclear Waste for Disposal*; Schulz, W. W.; Horwitz, E. P., Eds.; Plenum Press: New York, 1994.
(3) *Separation Techniques in Nuclear Waste Management*; Carleson, T. E.; Chipman, N. A.; Wai, C. M., Eds.; CRC Press: New York, 1995.
(4) *Global 93: Future Nuclear Systems: Emerging Fuel Cycles and Waste Disposal Options*; American Nuclear Society: La Grange Park, IL, 1993.
(5) *Proceedings of the 4th International Conference on Nuclear Fuel Reprocessing and Waste Management*; London, 1994.
(6) *International Conference on Evaluation of Emerging Nuclear Fuel Cycle Systems: Global 1995*; Versailles, France, 1995.
(7) Gephart, R. E.; Lundgren, R. E. "Hanford Tank Clean Up: A Guide to Understanding the Technical Issues"; PNL-10773; Pacific Northwest Laboratory, 1996.
(8) Todd, T. A.; Olson, A. L.; Palmer, W. B. In *Abstracts of Papers from the 214th American Chemical Society National Meeting*; American Chemical Society: Washington, DC, 1997; I&EC 141.
(9) Svantesson, I.; Hagstrom, I.; Persson, G.; Liljenzin, J. *J. Inorg. Nucl. Chem.* **1979**, *41*, 383.
(10) Svantesson, I.; Hagstrom, I.; Persson, G.; Liljenzin, J. *Radiochem. Radioanal. Lett.* **1979**, *37*, 215.

(11) Svantesson, I.; Hagstrom, I.; Persson, G.; Liljenzin, J. *J. Inorg. Nucl. Chem.* **1980**, *42*, 1037.
(12) Liljenzin, J. O.; Rydberg, J.; Skarnemark, G. *Sep. Sci. Technol.* **1980**, *15*, 799.
(13) Cecille, L.; Dworschak, H.; Girardi, F.; Hunt, B. A.; Mannove, F.; Mousty, F. In *Actinide Separations, ACS Symposium Series 117*; Navratil, J. D.; Schulz, W. W., Eds.; American Chemical Society: Washington, DC, 1980; p 427.
(14) Kubota, M.; Nakamura, H.; Tachimori, S.; Abe, T.; Amano, H. "Removal of Transplutonium Elements from High Level Waste"; IAEA-SM-246/24; International Atomic Energy Agency, 1981.
(15) Morita, Y.; Kubota, M. *J. Nucl. Sci. Technol.* **1985**, *22*, 658.
(16) Kubota, M.; Yamaguchi, I.; Morita, Y.; Kondow, Y.; Shirahasi, K.; Yamagishi, I.; Fujiwara, T. In *Global '93: Future Nuclear Systems: Emerging Fuel Cycles and Waste Disposal Options*; American Nuclear Society: La Grange Park, IL, 1993; p 588.
(17) Kubota, M. *Radiochim. Acta* **1993**, *63*, 91.
(18) Horwitz, E. P.; Delphin, W. H.; Mason, G. W.; Steindler, M. In *Actinide Partitioning and Transmutation Program Progress Report*; Oak Ridge National Laboratory: Oak Ridge, TN, 1978; p 101.
(19) Bond, W. D.; Leuze, R. E. "Feasibility Studies of the Partitioning of Commercial High-Level Waste Generated in Spent Nuclear Fuel Reprocessing: Annual Report for FY-1974"; ORNL-5012; Oak Ridge National Laboratory, 1975.
(20) Horwitz, E. P.; Kalina, D. G.; Muscatello, A. C. *Sep. Sci. Technol.* **1981**, *16*, 403.
(21) Horwitz, E. P.; Muscatello, A. C.; Kalina, D. G.; Kaplan, L. *Sep. Sci. Technol.* **1981**, *16*, 417.
(22) Sidall III, T. H. *J. Inorg. Nucl. Chem.* **1963**, *25*, 883.
(23) Sidall III, T. H. *J. Inorg. Nucl. Chem.* **1964**, *26*, 1991.
(24) Schulz, W. W.; McIsaac, L. D. In *Proceedings of the International Solvent Extraction Conference*; Canadian Institute of Mining and Metallurgy: Montreal, Quebec, 1979; p 619.
(25) Schulz, W. W.; Navratil, J. D. In *Recent Developments in Separation Science*; Li, N. N., Eds.; CRC Press: Boca Raton, FL, 1981; Vol. VII; p 31.
(26) McIsaac, L. D.; Baker, J. D.; Krupa, J. F.; Meikrantz, D. H.; Schroeder, N. C. In *Actinide Separations, ACS Symposium Series 117*; Navratil, J. D.; Schulz, W. W., Eds.; American Chemical Society: Washington, DC, 1980; p 395.
(27) Bond, W. D.; Leuze, R. E. In *Actinide Separations, ACS Symposium Series 117*; Navratil, J. D.; Schulz, W. W., Eds.; American Chemical Society: Washington, DC, 1980; p 441.
(28) Navratil, J. D.; Martella, L. L. In *Actinide Recovery from Waste and Low-Grade Sources*; Navratil, J. D.; Schulz, W. W., Eds.; Harwood Academic Publishers: New York, 1982; p 27.
(29) Horwitz, E. P.; Kalina, D. G.; Diamond, H.; Vandegrift, G. F.; Schulz, W. W. *Solvent Extr. Ion Exch.* **1985**, *3*, 75.
(30) Horwitz, E. P.; Schulz, W. W. In *Solvent Extraction and Ion Exchange in the Nuclear Fuel Cycle*; Logsdail, D. H.; Mills, A. L., Eds.; Ellis Horwood, Ltd.: Chichester, UK, 1985; p 137.
(31) Schulz, W. W.; Horwitz, E. P. *Sep. Sci. Technol.* **1988**, *23*, 1191.
(32) Naser, A. J.; Barney, G. S.; Escobar, G. A. "TRUEX Processing of Hanford Plutonium Finishing Plant Liquid Waste"; WHC-SA-0313-FP; Westinghouse Hanford Co., 1988.

(33) Horwitz, E. P.; Chiarizia, R. In *Separation Techniques in Nuclear Waste Management*; Carleson, T. E.; Chipman, N. A.; Wai, C. M., Eds.; CRC Press: New York, 1995; p 3.
(34) Vandegrift, G. F.; Leonard, R. A.; Steindler, M. W.; Horwitz, E. P.; Basile, L. J.; Diamond, H.; Kalina, D. G.; Kaplan, L. "Transuranic Decontamination of Nitric Acid Solutions by the TRUEX Solvent Extraction Process - Preliminary Development Studies"; ANL-84-85; Argonne National Laboratory, 1984.
(35) Vandegrift, G. F.; Chamberlain, D. B.; Conner, C.; Copple, J. M.; Dow, J. A.; Everson, L.; Hutter, J. C.; Leonard, R. A.; Nunez, L.; Regalbuto, M. C.; Sedlet, J.; Srinivasan, B.; Weber, S.; Wygmans, D. G. In *Proceedings of the Symposium on Waste Management*; Tucson, AZ, 1993; p 1045.
(36) Chamberlain, D. B.; Conner, C.; Hutter, J. C.; Leonard, R. A.; Wygmans, D. G.; Vandegrift, G. F. *Sep. Sci. Technol.* **1997**, *32*, 303.
(37) Brewer, K. N.; Herbst, R. S.; Tranter, T. J.; Todd, T. A. *Solvent Extr. Ion Exch.* **1995**, *13*, 447.
(38) Law, J. D.; Brewer, K. N.; Herbst, R. S.; Todd, T. A.; Olsen, L. G. "Demonstration of Optimized TRUEX Flowsheet for Partitioning of Actinides from Actual ICPP Sodium-Bearing Waste Using Centrifugal Contactors in a Shielded Cell Facility"; INEL/EXT-98-00004; Idaho National Engineering Laboratory, 1998.
(39) Brewer, K. N.; Herbst, R. S.; Todd, T. A. In *Abstracts of Papers from the 211th American Chemical Society National Meeting*; American Chemical Society: Washington, DC, 1996; I&EC-209.
(40) Law, J. D.; Brewer, K. N.; Todd, T. A. In *Abstracts of Papers from the 214th American Chemical Society National Meeting*; American Chemical Society: Washington, DC, 1997; I&EC-136.
(41) Horwitz, E. P.; Diamond, H.; Gatrone, R. C.; Nash, K. L.; Rickert, P. G. In *Proceedings of the International Solvent Extraction Conference, ISEC-90*; Sekine, T., Ed.; Elsevier Science Publishers: Amsterdam, 1992; p 357.
(42) Law, J. D.; Brewer, K. N.; Herbst, R. S.; Todd, T. A. "TRUEX Flowsheet Testing for the Removal of Actinides from Dissolved ICPP Zirconium Calcine Using Centrifugal Contactors"; INEEL/EXT--97-00837; Idaho National Environmental Engineering Laboratory, 1998.
(43) Felker, L. K.; Benker, D. E. "Application of the TRUEX Process to Highly Irradiated Targets"; ORNL/TM-12784; Oak Ridge National Laboratory, 1995.
(44) Lumetta, G. J.; Swanson, J. L. *Sep. Sci. Technol.* **1993**, *28*, 43.
(45) Lumetta, G. J.; Wagner, M. J.; Barrington, R. J.; Rapko, B. M.; Carlson, C. D. "Sludge Treatment and Extraction Technology Development: Results of FY1993 Studies"; PNL-9387; Pacific Northwest Laboratory, 1994.
(46) Lumetta, G. J. "Pretreatment of Neutralized Cladding Removal Waste Sludge: Results of the Second Design Basis Experiment"; PNL-9747/UC-721; Pacific Northwest Laboratory, 1994.
(47) Ozawa, M.; Nemoto, S.; Togashi, A.; Kawata, T.; Onishi, K. *Solvent Extr. Ion Exch.* **1992**, *10*, 829.
(48) Nomura, K.; Koma, Y.; Nemoto, S.; Ozawa, M.; Kawata, T. In *Global 93: Future Nuclear Systems: Emerging Fuel Cycles and Waste Disposal Options*; American Nuclear Society: La Grange Park, IL, 1993; p 595.

(49) Vandegrift, G. F.; Regalbuto, M. C. In *Proceedings of the Fifth International Conference on Radioactive Waste Management and Environmental Remediation, ICEM'95*; Slate, S.; Feizollahi, F.; Creer, J., Eds.; The American Society of Mechanical Engineers: New York, 1995; Vol. 1; p 457.
(50) Koma, Y.; Watanabe, M.; Nemoto, S.; Tanaka, Y. *Solvent Extr. Ion Exch.* **1998**, *16*, in press.
(51) Mathur, J. N.; Murali, M. S.; Natarajan, P. R.; Badheka, L. P.; Banerji, A. "Extraction of Actinides from High Level Waste Streams of Purex Process Using Mixtures of CMPO and TBP in Dodecane"; BARC/1992/E/009; Bhabha Atomic Research Centre, 1992.
(52) Mathur, J. N.; Murali, M. S.; Natarajan, P. R.; Badheka, L. P.; Banerji, A. *Talanta* **1992**, *39*, 493.
(53) Mathur, J. N.; Murali, M. S.; Iyer, R. H.; Badheka, L. P.; Banerji, A. In *Global 93: Future Nuclear Systems: Emerging Fuel Cycles and Waste Disposal Options*; American Nuclear Society: La Grange Park, IL, 1993; p 601.
(54) Chmutova, M. K.; Kochetkova, N. E.; Myasoedov, B. F. *J. Inorg. Nucl. Chem.* **1980**, *42*, 897.
(55) Chmutova, M. K.; Kochetkova, N. E.; Koiro, O. E.; Myasoedov, B. F.; Medved', T. Y.; Nesterova, N. P.; Kabachnik, M. I. *J. Radioanal. Nucl. Chem.* **1983**, *80*, 63.
(56) Myasoedov, B. F.; Chmutova, M. K.; Kochetkova, N. E.; Koiro, O. E.; Pribylova, G. A.; Nesterova, N. P.; Medved', T. Y.; Kabachnik, M. I. *Solvent Extr. Ion Exch.* **1986**, *4*, 61.
(57) Chmutova, M. K.; Litvina, M. N.; Nesterova, N. P.; Myasoedov, B. F.; Kabachnik, M. I. *Radiokhimiya* **1989**, *31*, 73.
(58) Pribylova, G. A.; Chmutova, M. K.; Nesterova, N. P.; Myasoedov, B. F.; Kabachnik, M. I. *Radiokhimiya* **1991**, *33*, 70.
(59) Myasoedov, B. F.; Chmutova, M. K.; Smirnov, I. V.; Shadrin, A. U. In *Global '93: Future Nuclear Systems: Emerging Fuel Cycles and Waste Disposal Options*; American Nuclear Society: La Grange Park, IL, 1993; p 581.
(60) Myasoedov, B. F., Vernadsky Institute of Geochemistry and Analytical Chemistry, Russian Academy of Sciences, personal communication, 1997.
(61) Musikas, C.; Cuillerdier, C.; Condamines, N. In *New Separation Techniques for Radioactive Waste and Other Specific Applications*; Cecille, L.; Casarci, M.; Pietrelli, L., Eds.; Elsevier Applied Science: New York, 1991; p 49.
(62) Cuillerdier, C.; Musikas, C.; Nigond, L. *Sep. Sci. Technol.* **1993**, *28*, 155.
(63) Nigond, L.; Condamines, N.; Cordier, P. Y.; Livet, J.; Madic, C.; Cuillerdier, C.; Musikas, C. *Sep. Sci. Technol.* **1995**, *30*, 2075.
(64) Jubin, R.; Baron, P.; Madic, C.; Nichol, C.; Hudson, M. In *Abstracts of Papers from the 211th American Chemical Society National Meeting*; American Chemical Society: Washington, DC, 1996; I&EC-210.
(65) Chan, G. Y. S.; Drew, M. G. B.; Hudson, M. J.; Iveson, P. B.; Liljenzin, J.-O.; Skaalberg, M.; Spjuth, L.; Madic, C. *J. Chem. Soc., Dalton Trans.* **1997**, 649.
(66) Apostolidis, C.; DeMeister, R.; Koch, L.; Molinet, R.; Liang, J.; Zhu, Y. In *New Separation Chemistry Techniques for Radioactive Waste and Other Specific Applications*; Cecille, L.; Casarci, M.; Pietrelli, L., Eds.; Elsevier Applied Science: New York, 1991; p 80.

(67) Zhu, Y.; Jiao, R. In *Global 93: Future Nuclear Systems: Emerging Fuel Cycles and Waste Disposal Options*; American Nuclear Society: La Grange Park, IL, 1993; p 44.
(68) Song, C.; Glatz, J.-P.; He, X.; Bokelund, H.; Koch, L. In *Proceedings of the 4th International Conference on Nuclear Fuel Reprocessing and Waste Management*; London, 1994.
(69) Zhu, Y.; Jiao, R. *Nucl. Technol.* **1994**, *108*, 361.
(70) Glatz, J.-P.; Song, C.; Koch, L.; Bokelund, H.; He, X. In *International Conference on Evaluation of Emerging Nuclear Fuel Cycle Systems: Global 1995*; Versailles, France, 1995; p 548.
(71) Chongli, S.; Jianchen, W.; Junfu, L. In *Global 1997,* Yokohama, Japan, 1997; in press.
(72) Zhu, Y., Institute of Nuclear Energy Technology, Tsinghua University, personal communication, 1997.
(73) Schulz, W. W.; Bray, L. A. *Sep. Sci. Technol.* **1987**, *22*, 191.
(74) Kyrs, M. *J. Radioanal. Nucl. Chem., Lett.* **1994**, *187*, 185.
(75) Egorov, N. N.; Kudryavtsev, E. G.; Lazarev, L. N.; Romanovskii, V. N. In *Proceedings of the Symposium on Waste Management*; Tucson, AZ, 1991; p 671.
(76) Esimantovskii, V. M.; Galkin, B. Y.; Dzekun, E. G.; Lazarev, L. N.; Ljubtsev, R. I.; Romanovskii, V. N.; Shishkin, D. N. In *Proceedings of the Symposium on Waste Management*; Tucson, AZ, 1992; p 22.
(77) Reilly, S. D.; Mason, C. F. V.; Smith, P. H. "Cobalt(III) Dicarbollide: A Potential ^{137}Cs and ^{90}Sr Waste Extraction Agent"; LA-11695, UC-701; Los Alamos National Laboratory, 1990.
(78) Law, J. D.; Herbst, R. S.; Todd, T. A.; Brewer, K. N.; Romanovsky, V. N.; Esimantovskiy, V. M.; Smirnov, I. V.; Babain, V. A.; Zaitsev, B. N.; Dzekun, E. G. In *Proceedings of the International Topical Meeting Nuclear Hazardous Waste Management, SPECTRUM '96*; American Nuclear Society: La Grange Park, IL, 1996; p 2308.
(79) Todd, T. A., Lockheed Martin Idaho Technologies Co., personal communication, 1997.
(80) Horwitz, E. P.; Dietz, M. L.; Fisher, D. E. *Solvent Extr. Ion Exch.* **1990**, *8*, 557.
(81) Horwitz, E. P.; Dietz, M. L.; Fisher, D. E. *Solvent Extr. Ion Exch.* **1991**, *9*, 1.
(82) Wood, D. J.; Tranter, T. J.; Todd, T. A. *Solvent Extr. Ion Exch.* **1995**, *13*, 829.
(83) Dietz, M. L.; Horwitz, E. P.; Rogers, R. D. *Solvent Extr. Ion Exch.* **1995**, *13*, 1.
(84) Wood, D. J.; Todd, T. A.; Atkinson, D. A.; Mincher, B. A. In *Abstracts of Papers from the 213th American Chemical Society National Meeting*; American Chemical Society: Washington, DC, 1996; ANYL-073.
(85) Wood, D. J. In *Abstracts of Papers from the 211th American Chemical Society National Meeting*; American Chemical Society: Washington, DC, 1996; I&EC-170.
(86) Wood, D. J.; Law, J. D. *Sep. Sci. Technol.* **1997**, *32*, 241.
(87) Law, J. D.; Wood, D. J.; Herbst, R. S. *Sep. Sci. Technol.* **1997**, *32*, 223.
(88) Law, J. D.; Wood, D. J. "Development and Testing of a SREX Flowsheet for the Partitioning of Strontium and Lead from Simulated ICPP Sodium-Bearing Waste"; INEL-96/0437; Idaho National Engineering Laboratory, 1996.
(89) Wood, D. J.; Law, J. D. *Solvent Extr. Ion Exch.* **1997**, *15*, 65.

(90) Wood, D. J.; Law, J. D.; Garn, T. G.; Tillotson, R. D.; Tullock, P. A.; Todd, T. A. "Development of the SREX Process for the Treatment of ICPP Liquid Wastes"; INEEL/EXT-97-00831; Idaho National Engineering Laboratory, 1997.

(91) Law, J. D.; Wood, D. J.; Olson, L. G.; Todd, T. A. "Demonstration of a SREX Flowsheet for the Partitioning of Strontium and Lead from Actual ICPP Sodium-Bearing Waste"; INEEL/EXT-97-00832; Idaho National Engineering Laboratory, 1997.

(92) Horwitz, E. P.; Dietz, M. L.; Jensen, M. P. In *Proceedings of the International Solvent Extraction Conference (ISEC'96)*; University of Melbourne: Melbourne, Australia, 1996; Vol. 2; p 1285.

(93) Horwitz, E. P.; Dietz, M. L.; Leonard, R. A. "Efficient Separations and Processing Crosscutting Program, 1997 Technical Exchange Meeting"; PNNL-SA-28461; Pacific Northwest National Laboratory, 1997.

(94) Smirnov, I. V. In *Proceedings of the International Topical Meeting Nuclear Hazardous Waste Management, SPECTRUM '96*; American Nuclear Society: La Grange Park, IL, 1996; p 2115.

(95) Dietz, M. L.; Horwitz, E. P. In *Science and Technology for Disposal of Radioactive Wastes*; Lombardo, N. J.; Schulz, W. W., Eds.; Plenum Press: New York, 1998; in press.

Aqueous Systems

Chapter 4

Aqueous Complexes in f-Element Separation Science

Kenneth L. Nash

Chemistry Division, Argonne National Laboratory, 9700 South Cass Avenue, Argonne, IL 60439-4831

The focus of this chapter is on the role of the aqueous phase and reaction that occur in aqueous media in defining separation efficiency and metal ion selectivity. As our programmatic emphasis is on actinide solution chemistry, the separations chemistry of the f-elements will be used to illustrate the principal role of aqueous chemistry in metal ion separations. Most of the arguments developed apply to metal ion separations chemistry and processes in general. The discussion will consider the role of aqueous complexes that remain in the aqueous phase, aqueous complexants that are extracted, and the effect of properties of the aqueous medium on separation efficiency and selectivity. Historically important separations processes will be discussed along with the results of recent efforts in our laboratories to design and characterize new water soluble complexants for improved f-element separations.

The process of metal ion separations in both ion exchange and solvent extraction consists, in its most elementary form, of the transfer of a charged metal ion from a polar aqueous phase to an immiscible phase (with different solvating properties) with concomitant charge neutralization. Metal complexation reactions (in both phases), solvation reactions of complexes, metal ions, and complexants, and the general nature of both media all impact the equilibrium condition of the system. Each of these factors can be manipulated to affect the desired phase transfer and recovery, for both transport of the metal ion into the counter-phase and its recovery from that phase are important considerations in the design of most separation processes. The effectiveness of any separations process is a function of the ability of these reactions to accomplish phase transfer and of the relative affinity of the counter-phase for the species to be separated. Such flexibility derives from the small energy differences required to affect separation. Free energy differences of only 2.7 kilocalories/mole (11.4 kJ/mole), equivalent to the energy associated with one hydrogen bond, are adequate to produce a change in distribution ratio from 0.1 to 10. This example represents a separation factor of 100, implying 99% mutual separation of species in a single contact.

The thermodynamics of both solvent extraction and ion exchange have been discussed extensively and elaborated with elegant arguments previously (*1, 2*). The details of those discussions will not be repeated here. However, an outline of the basic concepts is useful to provide a framework for the discussion that follows. The equilibria involved in the extraction of metal ions in a solvent extraction process are shown schematically in Figure 1. The energetics of each of these processes impact the separation equilibrium. Marcus (*1*) considered metal ion separation by solvent extraction using thermodynamic cycles to demonstrate (not surprisingly) that the principal barrier to extraction of a polyvalent cation is the energy of hydration of the cation.

In the aqueous medium, cations are solvated by some number of water molecules. For d-transition metal ions, the number of water molecules in the primary coordination sphere (A-zone) is determined by the strength of orbital overlap between the metal ion and H_2O, crystal field stabilization effects, and cationic charge. Other species (e.g., alkaline earths, rare earths) interact with solvent via ion-dipole forces, and their solvation numbers are determined predominantly by steric factors. In addition to this inner solvation shell, all cations in the aqueous medium organize solvent water in a second coordination sphere (B-zone), the volume of which is strongly a function of the charge/radius ratio of the cation. Rizkalla and Choppin (*3*) suggests that there is a third, disordered (C) zone of water molecules surrounding the ion wherein water structure is intermediate between the ion-dipole ordered structure and the tetrahedral arrangement that represents the bulk solvent. The necessary anions accompanying the cation are not typically strongly hydrated, though O, N, and F engage in hydrogen bonding interactions to modify water structure. Different classes of ions have different effects on the net structure of water (discussed below). Because of the high dielectric constant of H_2O, close association of cations with anions is not required. As a result, free hydrated cations and anions diffuse freely and independently through the solution, though ion pairing becomes more important at high concentrations of salts. Another consequence of the high dielectric constant of water is that, although charge is conserved in the solution, stable metal complexes may carry a formal charge.

The hydrated cations in the aqueous medium are therefore free to interact with any species present in the aqueous medium. Among the most important species in separations chemistry are water soluble chelating agents. Such species, designated as H_nX in Figure 1, typically exhibit little tendency to distribute into the less polar medium (organic solvent or ion exchange resin). They respond to the acidity and ionic strength of the solution and to the presence of polyvalent metal ions. The species H_nX are ionized forming various anionic species and complexes with the metal ions in the solution. They can also interact with the solvent (H_2O). Besides the chelating agents H_nX, the metal ion may form complexes with the background electrolyte anions (Y^-) and with that fraction of the lipophilic extractant (HL) dissolved in the aqueous medium.

For a target separations problem, there will almost always be multiple metal cations present in the aqueous medium. Water soluble chelating agents are typically introduced because they interact more strongly with one metallic component than another. They can therefore be used to control selectivity. This chemistry applies with little modification equally to both ion exchange and solvent extraction separations. It is in fact the basis of the

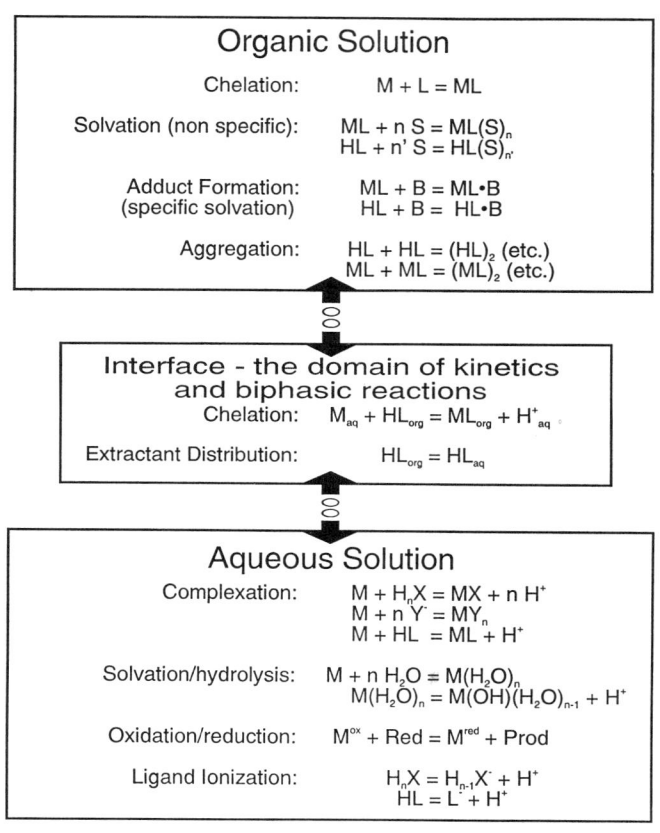

Figure 1. Schematic representation of pertinent equilibria in solvent extraction processes.

experimental approach to determination of metal complex stability constants in aqueous media by separations methods (4).

The reactions accounting for the transfer of the target metal ion to the counter phase require that the metal ion be "transformed" into a less polar species that is uncomfortable in water. In solvent extraction, the phase transfer reaction is analogous to precipitation (in the sense that a charge neutral compound is formed which leaves the polar aqueous medium). Extractant molecules exhibit some degree of interfacial activity in solvent extraction. The rates of the phase transfer reactions are typically controlled by reactions occurring at the solution interface. Interfacial reactions are not relevant to the overall thermodynamics of extraction, but they are of overriding importance in understanding the rates of phase transfer reactions. Aqueous complexants employed in cation separation are typically not interfacially active.

Basic Equations and Equilibria

The observable parameter most commonly used to describe metal ion separations is the distribution ratio, simply the ratio of the metal ion concentrations in the organic (or resin) phase to that in the aqueous phase. For process design purposes, mass transfer must take into account phase ratios, but since the present focus is on the thermodynamics of extraction (rather than process design per se) the following development of equilibria will always consider the phase ratio to be 1:1 (either by volume for solvent extraction or mass:volume for ion exchange). For all systems, therefore, the distribution ratio is:

$$D = [M]_{org}/[M]_{aq} \qquad (1)$$

The specific species represented by $[M]_{org}$ and $[M]_{aq}$ will be governed by overlapping equilibria in the aqueous and organic phases. In the following paragraphs, the appropriate equilibria will be described for different classes of separation reactions.

Lipophilic acidic chelating agents which readily exchange H^+ for the target metal ion are one class of ligands. Such reagents may exhibit some aqueous solubility and thus function to a degree as the water soluble chelating agents, but the most effective extractants do not distribute substantially to the aqueous phase. The tendency of such reagents to distribute to the aqueous phase is also a function of the solvating power of the organic medium. The principal equilibria necessary to describe metal ion extraction in such systems are (taking M^{3+} as an example):

$$HL_{org} = HL_{aq} \qquad (2)$$

$$HL_{aq} = H^+_{aq} + L^-_{aq} \qquad (3)$$

$$M^{3+}_{aq} + 3\,HL_{org} \overset{K_{ex}}{=} (ML_3)_{org} + 3\,H^+_{aq} \qquad (4)$$

The first two equilibria describe the distribution of the extractant between aqueous and organic phases and the acid-base equilibrium of the extractant (which is taken to occur in the aqueous phase). Equation 4 is the principal equilibrium expression governing metal ion extraction. The requirement of charge neutrality of the extracted metal complex is met for all extracted complexes, irrespective of the mode of phase transfer. The distribution ratio for the metal ion is described by the equilibrium expression (assuming that HL does not distribute appreciably to the aqueous phase):

$$D = [ML_3]_{org}/[M^{3+}]_{aq} = K_{ex} [HL]^3_{org}/[H^+]^3_{aq} \tag{5}$$

For certain types of extractants (e.g., carboxylic, organophosphorus, and sulfonic acids), dimerization and higher order aggregation of the extractant in the organic phase must also be included in a complete description of the thermodynamics of the system. At high metal loading of the organic phase, the extracted metal complexes tend to aggregate as well.

A second class of extractant molecules are those that accomplish cation phase transfer by solvation of a neutral metal complex. Charge neutrality of the metal complex may be achieved through formation of complexes with simple water soluble anions like nitrate (present either by purposeful addition or as the counter ion for preparation of the initial metal ion solution):

$$M^{3+}_{aq} + 3\,Y^-_{aq} + 2\,S_{org} \overset{K_{ex}}{=} (MY_3S_2)_{org} \tag{6}$$

the distribution ratio for the metal ion is defined as:

$$D = [MY_3S_2]_{org}/[M^{3+}]_{aq} = K_{ex}[Y^-]^3_{aq}[S]^2_{org} \tag{7}$$

unless Y^- also participates in complex formation in the aqueous phase in which case $[M]_{aq} = [M^{3+}] + [MY^{2+}] + [MY_2^+] + ...$ and the distribution ratio becomes:

$$D = [MY_3S_2]_{org}/([M^{3+}] + [MY^{2+}] + [MY_2^+] + ...) = K_{ex}[Y^-]^3[S]^2_{org} \tag{8}$$

Equation 8 can be rewritten in terms of the equilibrium constants for the formation of aqueous complexes, which will be discussed below. Solvating extractants (S) are also known to extract mineral acids from the aqueous phase, and occasionally these equilibria must be taken into consideration when describing the thermodynamics of extraction. Charge neutralization can also be achieved through the agency of an acidic extractant molecule as was discussed in the previous paragraph:

$$M^{3+}_{aq} + 3\,HL_{org} + 2\,S_{org} = (ML_3S_2)_{org} + 3\,H^+_{aq} \tag{9}$$

Successful combination of a solvating extractant with an acidic chelating extractant results in enhanced extraction efficiency. Such systems are called synergistic, as extraction by the sum of the two is greater than that of either extractant functioning separately.

A related system is based on the application of ternary or quaternary amine extractants to extract metal ions as ion pairs. With this type of extraction system, the metal ion must associate with an excess of anions to form an anionic complex:

$$M^{3+}{}_{aq} + 3\,Y^-{}_{aq} + (A^+Y^-)_{org} \overset{K_{ex}}{=} (MY_4A)_{org} \qquad (10)$$

The distribution ratio for the metal ion is defined as:

$$D = [MY_4A]_{org}/[M^{3+}]_{aq} = K_{ex}[Y^-]^3[A^+Y^-] \qquad (11)$$

This equilibrium expression is also affected by the existence of aqueous complexes $MY_n{}^{3-n}$ and the ion pairing equilibrium for the extractant. For many d-transition metal ions, the species extracted by amines are the thermodynamically stable anionic complexes that exist in the aqueous phase in the absence of the lipophilic reagent. For other classes of metal ions (e.g., rare earths), the extracted anionic complexes often are not important aqueous species. The presence of the amine is necessary to encourage the net phase transfer reaction through formation of the anionic complex. This class of extractants is the solvent extraction equivalent of anion exchange resin, and the same equilibria describe both.

A fourth class of extractant molecules, comparable to cation exchange resins, are surfactants. These extractants are strongly acidic and highly aggregated in the organic phase forming reverse micelles that sequester the extracted metal ions in the hydrophilic inner region of the reverse micelle. Their metal ion extraction equilibria are comparable to those of acidic extractants except that the stoichiometry with respect to the analytical concentration of the extractant is typically reduced to a nominal 1:1 with the micelle performing the role of the extractant. Charge neutralization is attained by the expulsion of H^+ (or other monovalent cations, e.g., Na^+) from the micelle. Such extractants exhibit good selectivity for metal ions of significantly different properties (e.g., oxidation state/formal charge) but almost none for cations with similar properties.

Of these four general classes of phase transfer reactions, only those systems involving the application of lipophilic chelating agents exhibit any inherent tendency toward cation selectivity for cations like the trivalent lanthanides. Solvating extractants (e.g., tributyl phosphate, TBP), ion pair forming extractants (e.g., quaternary amines), and micelle forming extractants typically extract a given class of metal ions indiscriminately (though they are sensitive to the oxidation state of the metal). Such reagents (and their ion exchange resin equivalents) must rely on changes in aqueous chemistry for selectivity. In effect, such systems serve as "platforms" for selective separations based on differences in the aqueous chemistry of the system. The variation in aqueous chemistry may involve changes in the oxidation state of the metal ion, complexation, or more subtle alteration of the aqueous medium.

As a production-scale separation process is typically operated in a stage-wise manner, the extraction equilibria are readjusted to transfer the metal ion back to a "clean" aqueous solution thus separating the metal ion from the contaminants present initially and the extractant solution/resin. The recovered metal ion can then be transformed, for example, by precipitation or electrochemically (called electrowinning) to produce the final purified material. This process creates additional opportunities to increase the purity of the recovered metal ion. In most production-scale processes, it is desirable to recycle the extractant solution/resin. Clean up of the these materials typically relies on contact with an appropriate aqueous solution. Analytical separations do not ordinarily employ such processes because the small scale reduces the requirement for recycle of expensive reagents. The focus of the following discussion will be on the various ways in which changes in the composition of the aqueous medium affect extraction efficiency and selectivity in metal ion separations.

Early Separations of f-Elements

To elaborate the role of the aqueous solution on the separation of metal ions, the f-elements provide a particularly relevant framework for discussion. Their chemistry is diverse providing an opportunity to examine a variety of different aspects of metal ion separations while narrowly focusing on a particular class of metal ions. At the same time, the similarity of the chemical behavior of trivalent actinides and lanthanides represents a severe test of the viability of any separation scheme. It is also particularly appropriate to pursue this discussion based on the chemistry of lanthanides and actinides, since no class of metal ions owes more to modern separation science. The availability of macroscopic amounts of pure lanthanides for technological applications and our knowledge of the properties of the actinides relies heavily on ion exchange and solvent extraction separations, as the following discussion will illustrate.

Fractional crystallization is the most straightforward solid-liquid separation method and was the technique of choice in the early days of the investigation of the chemistry of the lanthanides. The separation of individual lanthanide ions by this method relied on extremely small differences in solubility which therefore demanded hundreds or even thousands of repetitions to achieve useful separations of the elements (5). The introduction of first ion exchange and later solvent extraction procedures greatly increased capacities for the production of individual lanthanides and ultimately made their practical application possible.

Our present state of knowledge regarding the basic science of the actinides is even more intricately bound to their separations chemistry. The work of Becquerel and the Curies that led to the discovery of radioactivity and new elements relied heavily on precipitation techniques. The discovery of fission and the production of the first transuranic elements likewise required physical separation methods to isolate small amounts of radioactive metals from complex matrices. Most separations were based on precipitation, though ether extraction of selected metal complexes also figured prominently.

During World War II and for more than 40 years after, plutonium production was accomplished at the Hanford site on the Columbia River near Richland, Washington (6). The isolation of plutonium from uranium and fission products was initially accomplished by precipitation with $BiPO_4$. The process, pioneered by S. G. Thompson, involves coprecipitation of Pu(IV) by $BiPO_4$ followed by oxidation to Pu(VI), which is readily resolubilized from the $BiPO_4$ matrix. This process was soon replaced by solvent extraction using methyl isobutyl ketone (REDOX Process) and later tributyl phosphate (PUREX Process). PUREX remains the principal method for processing spent reactor fuel today.

Following the discovery of neptunium and plutonium in 1940, a major research effort was launched to synthesize and determine the properties of the transplutonium elements. Seaborg proposed that these elements represented the 5f series, analogous to the lanthanides, and headed a team that synthesized and characterized the remaining nine members of the series during the period of 1944 - 1961. The irradiation methods used to produce the transuranium elements are always accompanied by some fission. Most important among fission products from a separations perspective are lanthanides, whose solution chemistry closely resembles that of transplutonium actinides. Identification of the new transplutonium elements therefore required efficient separations methods not only for actinides from actinides but also for actinides from lanthanides. Several of these techniques will be discussed below.

f-Element Solution Chemistry

To facilitate the following discussion, a brief overview of the solution chemistry of f-elements will be useful. The predominant stable oxidation state for lanthanide ions is the trivalent. Only Eu^{2+} and Ce^{4+}, potent reducing and oxidizing agents, respectively, are encountered in "normal" aqueous solutions. Because the spatial extension of the 4f (and 5f) valence orbitals is slight, addition of valence electrons cannot effectively shield the increasing nuclear charge as the atomic number increases across the series. As a result, the ionic radii of trivalent lanthanide ions decrease more-or-less regularly (by about 20%) from La to Lu. The radii of trivalent actinides behave similarly with $r_{Am} \approx r_{Nd}$. This is an important characteristic in the separations chemistry of f-elements.

For the elements with Z > 94 (except for nobelium, Z=102), the trivalent oxidation state is the most stable, although Am(V) and Bk(IV) have been utilized in separations in basic systems. Therefore, the solution chemistry of the transplutonium elements strongly resembles that of the trivalent lanthanides. For thorium, only the tetravalent oxidation state is important. For U, Np, and Pu, the redox chemistry is varied and different oxidation states are of use in separation schemes. The lower oxidation states (III and IV) exist as hydrated cations in aqueous solutions while the upper oxidation states (V and VI) are linear dioxocations having formal +1 and +2 charges. In general, acidic solutions favor lower oxidation states while basic media promote the stability of oxidized species. The most important species in actinide processing are U^{4+}, UO_2^{2+}, Np^{4+}, NpO_2^+, NpO_2^{2+}, Pu^{3+}, Pu^{4+}, PuO_2^{2+}, Am^{3+}, and Cm^{3+}. The middle oxidation states (IV and V) of U, Np, Pu, and Am are prone to disproportionation at moderate concentrations in acidic solutions. The multiplicity of readily available oxidation states for these elements is of major significance in their process chemistry.

Lanthanides and actinides (in all oxidation states) form weak complexes with halides (except F^-) and moderate to strong complexes with oxygen donor ligands like aminopolycarboxylates and polycarboxylic acids. The relative order of complex stability is typically $An^{4+} > AnO_2^{2+} > An^{3+} > AnO_2^+$, though chelating agents with unfavorable coordination geometries can reduce the relative stability of AnO_2^+ and AnO_2^{2+} complexes. The coordination/hydration numbers for these ions in solution are variable, reflecting the strongly ionic nature of the bonding and the general absence of directed valence effects: 9-12 for An^{4+}, 7-9 for An^{3+}, and 4-6 for $AnO_2^{+/2+}$ (considering equatorial coordination only). They are readily hydrolyzed (hydroxides precipitate at pH 1-2 for An^{4+}, pH 5 for AnO_2^{2+}, pH 7 for An^{3+}, pH 9 for AnO_2^+) and generally insoluble in basic media in the absence of complexing agents. The actinides exhibit a slightly greater tendency to interact with soft donor atoms (sulfur, chloride, nitrogen) than analogous lanthanides. The redox chemistry, solvation effects, and strength of soft donor interactions all are important in the separation chemistry of these elements.

To affect a mutual separation of individual actinides lighter than Am, or of these metal ions from the lanthanides, separation systems sensitive to the oxidation state of the metal ion is needed. The above mentioned REDOX and PUREX processes achieve good selectivity for Pu and U through the simple expediency of oxidation state control. For example, in the PUREX process, Pu(IV) and U(VI) are extracted from nitric acid. Selective stripping of Pu is accomplished by reduction with Fe^{2+} or U^{4+}.

For the transplutonium elements, more subtle differences between metal ions must be employed for a successful separation as the dominant oxidation state is the trivalent. Decreasing cation radii with increasing atomic number and the slightly greater affinity exhibited by actinide cations for soft donor ligands form the basis for these separations. The latter characteristic is the central determinant in the separation of trivalent actinides from lanthanides. The shrinking cation radius and the concordant increase in the interaction strength (largely electrostatic in nature) of these metal ions with ligand donor atoms is used to isolate individual members of the respective series. The multiple interactions involved in the phase transfer process limit the predictability of this effect, as will be discussed below.

In the following section, the impact of chemical processes occurring in the aqueous phase on the separations chemistry of lanthanide and actinide ions will be presented. It is intended that the generic role of aqueous phase reactions in metal ion separations will emerge from this discussion. The discussion will include both historically important processes and the results of research in our laboratories on the search for greater understanding of these phenomena.

Medium Effects and Solvation

Choppin and co-workers (7) have investigated the effect of various solutes on the hydrogen bonded structure of water. Their studies considered the impact of electrolyte composition and concentration, and that of non-aqueous solvents on the hydrogen bonded structure of water. Infrared spectroscopic data were interpreted in terms of the number of hydrogen bonds to individual water molecules. Liquid water can be considered to consist of molecules having 0, 1, or 2 water molecules hydrogen bonded to the protons. A higher percentage of more highly hydrogen bonded species implies greater three dimensional structure in the solvent. The energetics of the fit of the metal ion into this structure constitutes an entropy contribution to the overall extraction reaction.

The solutes that make up aqueous solutions can either promote or disrupt the 3-D structure of water. Hydrogen ions (H^+) and hydroxide ions (OH^-) are the ultimate structure makers in aqueous solutions, as they fit perfectly into the water structure. Therefore, pH must have an effect on water structure, at least at the extremes. Water miscible solvents (e.g., methanol, ethanol, acetone, DMSO) generally reduce hydration energies of solutes by interfering with hydrogen bonding. Small, hard sphere metal cations tend to promote order in the solution, while large cations tend to disrupt the structure. Among typical anions, fluoride is a strong structure maker, sulfate, phosphate, and nitrate less efficient structure makers, while the heavy halides and thiocyanate disrupt the water structure. Among anions typically encountered in separations, one of the most effective at disrupting water structure is perchlorate. Though such effects are difficult to quantify, the observation of a perchlorate effect is very likely the result of a more favorable net entropy contribution growing out of the structure breaking nature of the anion. Though this effect increases overall extraction efficiency, its effect on selectivity is more difficult to predict.

Lincoln (8) has discussed both the rates and the solvation numbers of the lanthanide cations, observing that there is a great variety in both solvent exchange rates and coordination numbers in non-aqueous media. The kinetic parameters for water exchange (9) are consistent with a concerted associative mechanism, (dependent on [H_2O]) characterized by rapid exchange. Partial substitution of non-aqueous solvents for water results in a net decrease in the hydration of the cation, thereby reducing the energetic requirements for desolvation and promoting phase transfer. Several important early investigations of the chemistry of actinides relied heavily on this effect for separation of lanthanides from the transplutonium actinides.

Beginning with the work of Street and Seaborg (10) in 1950 and followed by Diamond, Street and Seaborg (11) in 1954, methods for the separation of trivalent actinides from lanthanides by ion exchange techniques were developed. These authors found that below 6 M HCl, trivalent actinides and lanthanides eluted from a cation exchange column together. At concentrations of HCl greater than 6 M, actinides were eluted from the column while the lanthanides were retained. Diamond et al. offered a reasonable argument that behavior of Pm^{3+} in this system was "normal", that is, the distribution ratio should level off or increase slightly due to changes in the internal structure of the resin phase as acidity increases beyond 6 M. The observation of steadily decreasing distribution ratios for Am^{3+} were taken to indicate the existence of more stable Am-Cl complexes based on a covalent interaction between Am^{3+} and Cl^- (specifically involving the participation of the 5f valence orbitals of Am). Recent examination of actinide/lanthanide complexation has failed to establish a thermodynamic basis for a covalent component in actinide bonding to soft donor atoms. The lack of such proof should not be too surprising as the covalent contribution is likely to be less than 10% of the total bond strength (under the best of circumstances), the energy differences are slight, and cation hydration (for which there is minimal likelihood of a covalent contribution) dominates the thermodynamics of these systems.

An alternative approach to reducing the energy required to accomplish partial dehydration of the cation (and hence allow the actinide ions to interact more strongly with weak ligands like Cl^-) is to substitute a weaker solvent for a portion of the water comprising the solution. Street and Seaborg (10) reported that group separation of the lanthanides and trivalent actinides could be achieved by cation exchange eluting with 20% ethanol saturated with HCl. The presence of the alcohol enhances the difference in chloride complex stability between the lanthanides and actinides as a result of partial dehydration of the cation in the 20% ethanol solution. Development of processes based on partial substitution of ethanol, acetone, or acetonitrile for water in both cation and anion exchange continues today.

Anion exchange employing either lithium chloride or ammonium thiocyanate were reported as efficient methods for the separation of lanthanides from trivalent actinides. SCN^- coordinates with the actinide through the harder N atom (rather than S), and much lower concentrations are needed than are required for LiCl. The actinides are preferentially sorbed by the anion exchange resin due to stronger complexes with Cl^-, leaving the lanthanide ions in the aqueous solution. This separation has technological importance. Anion exchange from LiCl for separation of trivalent actinides from lanthanides is a critical step in the production of transamericium elements for research purposes at the Radiological Engineering Development Center (REDC) at ORNL.

Remembering the A-B-C zone model for cation hydration discussed above, a simple calculation reveals how high salt concentrations contribute to this separation. In 10 M LiCl (density = 1.1812 g/cc, 8.47 molal solution, CRC Handbook of Chemistry and Physics, 67th edition) the nominal water concentration is reduced to 36.1 molal. However, if we consider only the primary hydration sphere of the tetrahydrated lithium cation ($Li(H_2O)_4^+$), the "free" water concentration is reduced to about 2 molal. The chloride ion, also 8.47 molal, tends to promote the disordered C zone to further disrupt the normal hydrogen bonded structure of water. The polyvalent lanthanide and actinide cations, which are present at very low concentrations, thus are surrounded by fewer and less organized free water molecules. As a result of the decreased hydration, Cl^- can compete more favorably for the available cation coordination sites and the stronger soft-donor interaction of the actinide is manifested.

An additional phenomenon related to the effect of the anion on water structure is a so-called "perchlorate effect" (*12*). It has been often observed that extraction of metal ions from perchlorate media is greater than that from equivalent nitrate or chloride solutions (independent of the class of extractant). In the case of separations based on solvating extractants (e.g., TBP), it is at first glance surprising that lanthanide/actinide extraction should be stronger from perchlorate than from nitrate media, as ClO_4^- is a notoriously weak ligand while NO_3^- forms comparatively strong complexes with polyvalent cations. An explanation for the phenomenon, and for the relative ease of extraction of metal ions from salt solutions (relative to that from equivalent acids or from mixed aqueous media), may lie in the effect of the solutes on water structure.

The results reported by Sekine (*13*) show the combined effect of soft-donor ligands and of perchlorate on Am/Eu separation using a solvating extractant. The extraction/separation of americium and europium with 5% TBP/hexane from 5.0 M $NaClO_4/NaSCN$, and from NaSCN solutions without supporting electrolyte (pH 4-5) exhibit strong (but different) dependences on the concentration of SCN^- (Figure 2). In the absence of perchlorate, the separation factors are higher at low thiocyanate concentrations, and decline as [SCN^-] increases. At constant ionic strength (5.0 M), separation factors (S_{Eu}^{Am}) change very little at higher concentrations of SCN^-. Distribution ratios are dramatically higher in the presence of perchlorate. The free energy for extraction of both Am and Eu is 12-17 kJ/mol more favorable in the presence of perchlorate at [SCN^-] < 1 M. The ions are poorly separated at [SCN^-] > 5 M.

One of the most interesting examples of a medium effect in liquid-liquid separations

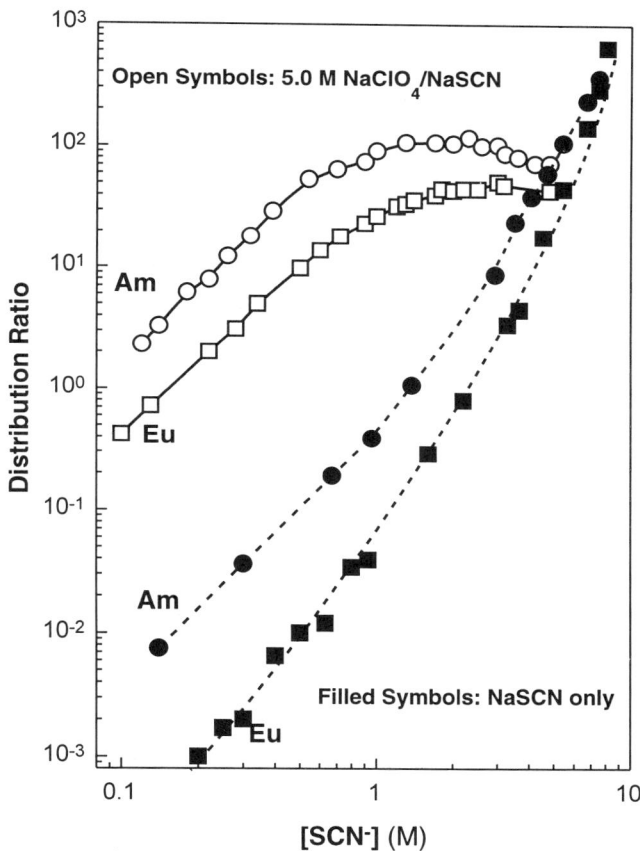

Figure 2. Solvent extraction separation of americium and europium using 5 % tributyl phosphate in hexane from NaSCN and 5.0 M NaSCN/NaClO$_4$ mixtures at pH 4-5 (Reference 13).

of f-elements is the so-called "aqueous biphasic" systems. Polyethylene glycols have been demonstrated to form separate phases from concentrated aqueous salt solutions. The effect relies heavily on the ability of certain salts to order the structure of water. Sulfate, carbonate, and hydroxide are particularly effective in this role. A key characteristic of these systems is that the immiscible phases are still largely aqueous in nature. Myasoedov and co-workers have reported conditions for the separation of transplutonium elements from uranium, thorium, and lanthanides (*14*). The aqueous biphase itself does not extract actinides strongly, but Arsenazo III is a highly effective carrier for these metal ions. There are similarities between the aqueous biphasic hydration environment and that existing in the internal structure of ion exchange resins.

Aqueous Complexation in Lanthanide/Actinide Separations

Medium effects of the type described above typically have their greatest impact on enhancement of extraction efficiency by reducing the dehydration energy barrier to phase transfer. The enhanced selectivity observed in lanthanide/actinide group separations from SCN^- solutions rely on the combined effects of reduced hydration energy and complex formation. The role of aqueous complexation in f-element separations will now be considered in more detail.

For the simplest case in which the aqueous metal complexes are not extracted into the counter phase, the observed distribution ratio represents a balance between the two phase extraction reaction and the homogeneous complexation equilibria in the aqueous phase. If we return to the earlier example of an acidic chelating extractant, the distribution ratio for trivalent metal ion extraction in the absence of aqueous complexation reactions is governed by the equilibrium shown in equation 4. The thermodynamic equilibrium constant for this reaction must include activity coefficients. However, activity coefficients are seldom known with sufficient accuracy to be included in a description of the thermodynamics of separations. It is expedient for this discussion to consider equilibrium quotients in which activity coefficients are considered to be constant and included in the constant. Providing that a standard state is defined and maintained and the concentration ranges do not vary over an extensive range, such a simplification is ordinarily justified.

Introduction of a water soluble complexing agent causes a redistribution of the metal ion among the extracted complex (ML_3), the free metal ion (M^{3+}) and the various water soluble complexes (MX_n). Assuming that the aqueous complexant does not participate in the phase transfer reaction, the distribution ratio is:

$$D = [ML_3]_{org}/([M^{3+}] + [MX^{2+}] + [MX_2^+] + ...) \qquad (12)$$

which can be rewritten in terms of the homogeneous (aqueous) phase equilibrium reaction for complex formation as:

$$D = [ML_3]_{org}/([M^{3+}](1 + \beta_1[X^-] + \beta_2[X^-]^2 + ...)) \qquad (13)$$

In equation 5, the ratio $[ML_3]_{org}/[M^{3+}]_{aq}$ was established as equivalent to $K_{ex}[H^+]^3/[HL]^3$ for the case of an acidic extractant and a trivalent cation, so:

$$D = K_{ex}[HL]^3/[H^+]^3/(1 + \Sigma \beta_i[X^-]^i) \tag{14}$$

The object of most metal ion separations processes is to isolate one metal ion from several related species. The separation factor for mutual isolation of two metal ions is a measure of the effectiveness of the process. It is defined as the ratio of the respective distribution coefficients (D's) or:

$$S^m_{m'} = \frac{D^m}{D^{m'}} = \frac{K_{ex}^m/(1 + \Sigma(\beta_i^m[X]^i))}{K_{ex}^{m'}/(1 + \Sigma(\beta_i^{m'}[X]^i))} \tag{15}$$

The $[HL]^3/[H^+]^3$ terms cancel providing the target metal ions are extracted with the same stoichiometry. If the extraction stoichiometries differ, the separation factor also will be a function of acidity and extractant concentration. The separation factor for the two metal ions is directly proportional to the relative extraction efficiency of the metals ($K_{ex}^m/K_{ex}^{m'}$), but inversely proportional to the relative stability of the aqueous complexes, though potentially in a complicated fashion. It is further possible to include a second aqueous complexant to enhance the difference between the two metal ions. An example of such a system will be presented below.

According to equation 15, the most effective separations will be achieved in those systems in which the target metal ion interacts more strongly with the extractant HL but is complexed less strongly by the aqueous ligand X. The need for complementarity of extractant and complexant has been illustrated in a previous study (15) of the effect of the water-soluble phosphonate complexant phosphonoacetic acid (PAA) on Am/Eu separation from three different extraction platforms: an acidic extractant (bis(2-ethylhexyl) phosphoric acid - HDEHP), a micellar extractant (dinonylnapthalene sulfonic acid - HDNNS), and a neutral bifunctional extractant (octyl(phenyl)-N,N-diisobutyl-carbamoylmethylphosphine oxide - CMPO) from nitrate and thiocyanate media (Table I). In the case of HDEHP, Eu is extracted about 30 times more efficiently than Am (log K_{ex}^{Eu} = -1.06, log K_{ex}^{Am} = -2.61, S_{Eu}^{Am} = 0.028 after correction for nitrate complexes). Since the Eu-PAA complexes are more stable than those of Am, the extractant and complexant ligands in this system work in the opposite sense resulting in lower separation factors when the complexant is present than when it is absent and the PAA thus reduces the separation efficiency. In HDNNS/toluene as the extractant, the separation factor was S_{Eu}^{Am} = 0.83 in the absence of PAA and S_{Eu}^{Am} = 1.81 for extraction from 0.34 M PAA at $[H^+]$ = 0.02 M and 0.5 M NaNO$_3$. Similarly, neutral bifunctional extractants like CMPO exhibit little lanthanide/actinide selectivity when extracting these metal ions from nitrate media (S_{Eu}^{Am} = 1.03). When 0.2 M PAA is introduced into the aqueous phase, the separation factor increases to S_{Eu}^{Am} = 2.07, which is consistent with calculations based on the respective aqueous stability constants of Am^{3+} and Eu^{3+} (15).

With thiocyanate as the counter ion in the CMPO system, americium is preferentially extracted and the separation factor S_{Eu}^{Am} is about 7 (16). As the Eu complexes are more stable than those for Am, both the K_{ex}^{Am}/K_{ex}^{Eu} and $(1+\Sigma\beta_{mhl}[H^+]^h[X^{3-}]^l)_{Eu}/(1+\Sigma\beta_{mhl}[H^+]^h[X^{3-}]^l)_{Am}$ terms of equation 15 are favorable. For extraction from 0.1 or 0.2

Table I. Americium/Europium Separation Factors for Extraction from Nitrate/Thiocyanate Solutions in the Presence of Phosphonoacetic Acid (pH 2)

Extractant	[PAA]	[SCN⁻]	[NO₃⁻]	S_{Eu}^{Am}
HDEHP	0.0	0.0	0.5	0.032
HDEHP	0.2	0.0	0.5	0.051
HDNNS	0.0	0.0	0.5	0.83
HDNNS	0.24	0.0	0.5	1.67
CMPO (0.1 M)	0.0	0.0	0.5	0.77
CMPO (0.1M)	0.2	0.0	0.5	1.43
CMPO (0.5 M)	0.0	0.0	0.5	1.03
CMPO (0.5M)	0.2	0.0	0.5	2.07
CMPO (0.05 M)	0.0	0.5	0.0	5.9
CMPO (0.05 M)	0.2	0.5	0.0	15

M PAA/0.5 M KSCN at pH 2, the separation factors increase to 12 and 15, respectively. This system has been applied to determine that the separation factor for Am over the entire lanthanide series ranges between 9 to 25, as shown in Figure 3. This combination of aqueous reagents and extractant has been validated for both solvent extraction and extraction chromatographic separations.

There are important examples from the literature that demonstrate the effect of aqueous complexants on f-element separations. In the nucleosynthesis of transplutonium elements, actinides were separated from trivalent lanthanide fission products using chloride or thiocyanate ion exchange as described above. Another system useful for group separations is TALSPEAK (Trivalent Actinide Lanthanide Separation by Phosphorus reagent Extraction from Aqueous Komplexes) (17). This extraction system is based on the use of acidic organophosphorus extractants (like HDEHP), which are excellent reagents for separation of individual trivalent lanthanides/actinides from a mixture (Figure 4a). Weaver and Kappelmann (17) reported that replacement of the mineral acid solution used in the intragroup separation with various carboxylic acids, and mixtures of carboxylic and aminopolycarboxylic acids results in a very satisfactory group separation. Extraction from 1 M carboxylic acid (pH 1.8) solutions depresses the extraction of Am relative to the lanthanides. At higher pH's ($\approx$ 3) the separation factors are increased, with the most consistent enhancement observed for lactic acid. Addition of only 0.05 M diethylenetriamine-N,N,N',N",N"-pentaacetic acid (DTPA) to the solutions of carboxylic acids at pH 3 resulted in dramatically improved separation factors. For extraction from 1 M

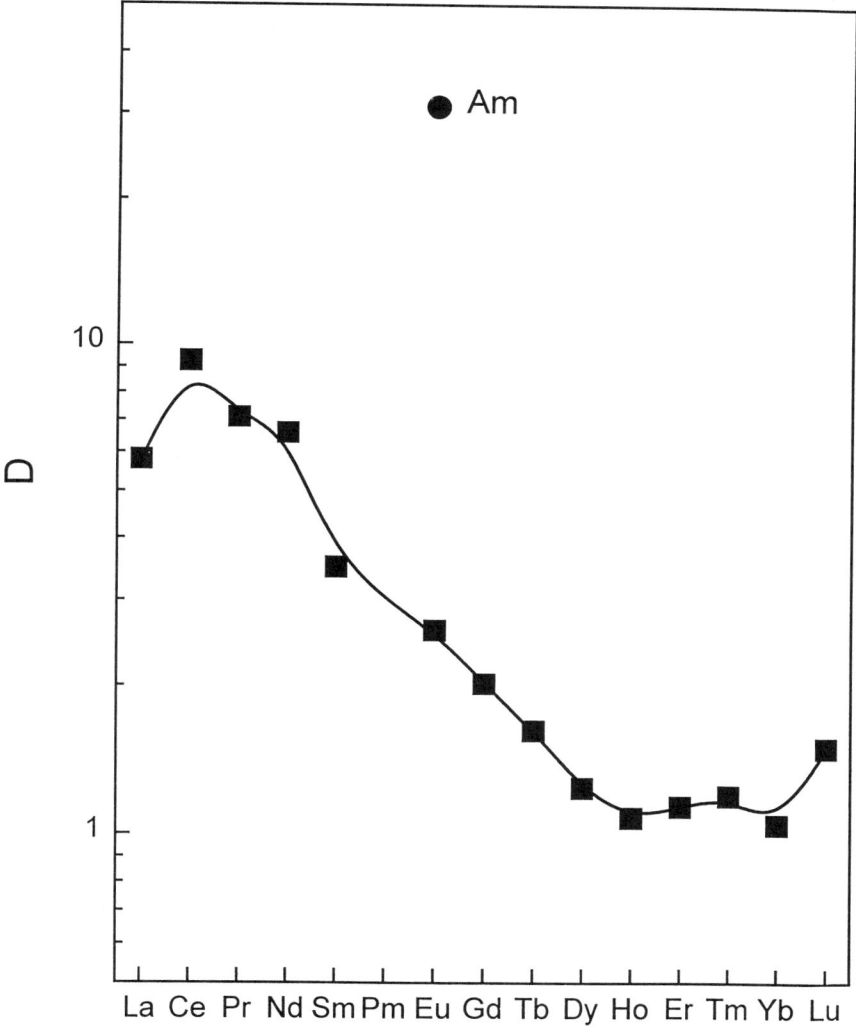

Figure 3. Separation of Am^{3+} (●) from the lanthanides (■) by a mixture of CMPO and TBP in Isopar L from a solution containing 0.5 M NaSCN and 0.2 M phosphonoacetic acid at pH 2.

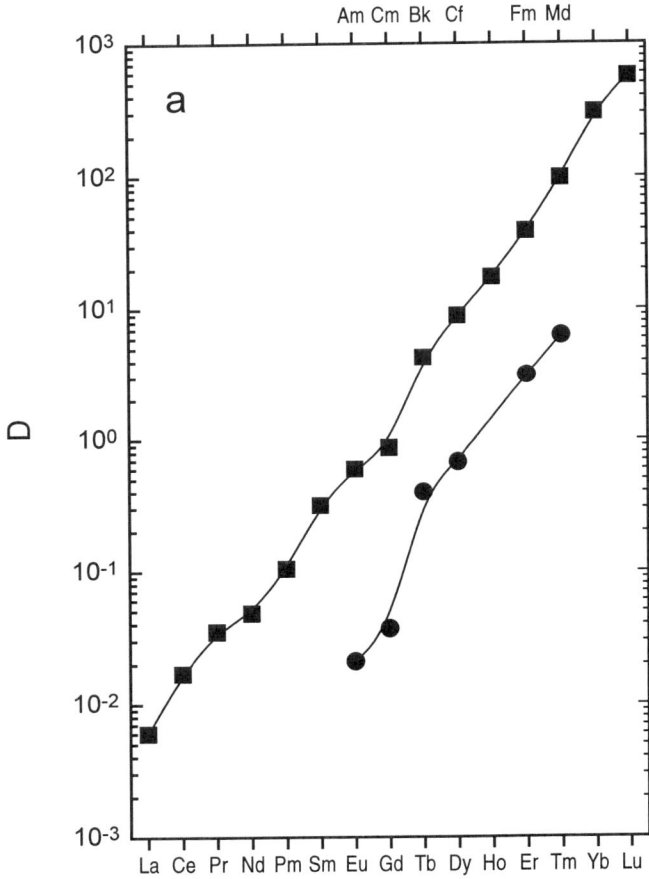

Figure 4. a.) Distribution ratios for trivalent lanthanide (■) and actinide (●) extraction by HDEHP/toluene from aqueous nitric acid solutions, b.). Distribution ratios for trivalent lanthanide and actinide cations by 0.3 M HDEHP/diisopropylbenzene from 1.0 M lactic acid/0.05 M DTPA at pH 3 (TALSPEAK).

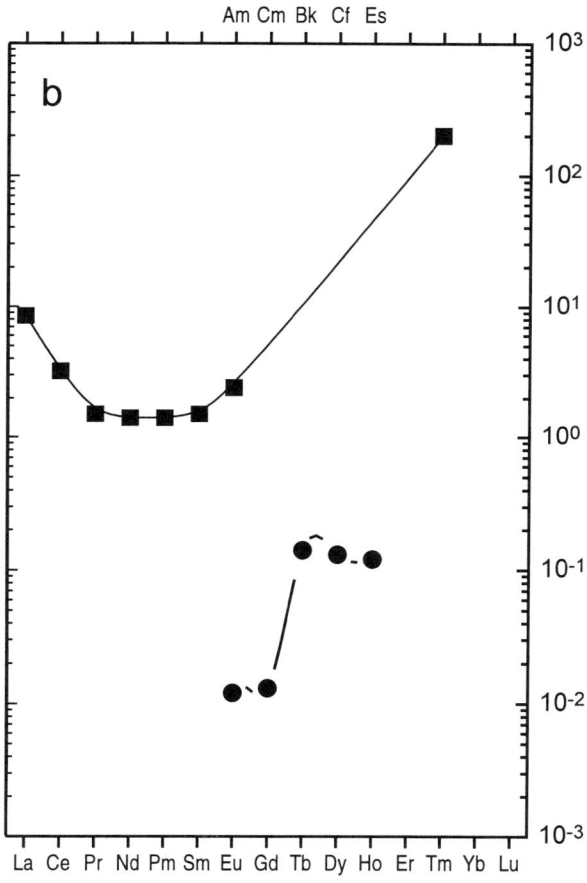

Figure 4. *Continued.*

lactic acid/0.05 M DTPA/pH 3 with 0.3 M HDEHP/diisopropylbenzene (DIPB), the worst actinide/lanthanide separation factor is for $S_{Nd}^{Bk} \approx 10$ (Figure 4b).

The combination of water soluble chelating agents and cation exchange for phase separation led to the development of the first efficient separations methods for isolation of transplutonium elements (18). Am, Cm, Bk, Cf, Es, and Fm were eluted in reverse order from Dowex 50 cation exchange resin when the eluting solution was 0.25 M NH$_4$(citrate) at pH 3.0-3.5 or 0.4 M NH$_4$-lactate at pH 4.0-4.5. Aminopolycarboxylic acid ligands like EDTA also demonstrated good separation factors but suffered from slower equilibration rates. Because transuranic actinides were often produced a few atoms at a time and had unknown properties, fast separations techniques with predictable behavior were required. More selective separations with better kinetics were observed with the introduction of α-hydroxyisobutyric acid (19), which provides average separation factors for adjacent lanthanides or trivalent actinides of about 1.3-1.5. In these systems, separation is achieved almost solely as a result of the difference in stability of the aqueous complexes of the metal ions.

One might guess that since the lanthanide cationic radii change consistently across the series that there should be a variety of chelating agents as effective as α-hydroxyisobutyric acid for accomplishing the isolation of individual lanthanide ions. Examination of the extensive database of critically evaluated stability constants for lanthanide complexes reveals that the aqueous chemistry of the system is much more complicated (and there are in fact very few systems that exhibit as consistent a trend). This fact becomes abundantly clear in a comparison of the relative free energy of complexation of lanthanide complexes with a series of potentially useful ligands.

In Figure 5a is shown the relative stability of a variety of lanthanide complexes (normalized to La^{3+}) with several carboxylic acids for which stability constants are known. The free energies for formation of complexes of lanthanides with acetate, oxydiacetate, and citrate show regular changes in lanthanide stability between La and Nd but are independent of radii or even reverse for complexes of the heavier lanthanide ions with some ligands. The relative free energies for α-hydroxyisobutyric acid are remarkably consistent across the series, which explains its unique behavior in trivalent f-element separations. The structurally restricted analog for oxydiacetic acid (ODA) (tetrahydrofuran-2,3,4,5-tetracarboxylic acid, THFTCA) exhibits greater selectivity from La to Dy, but is not effective beyond Dy. Thermodynamic data suggest that this selectivity is less pronounced in the corresponding 1:2 complexes (20). The relative free energies of lanthanide aminopolycarboxylate ligands (Figure 5b) change more consistently with cation radius than the polycarboxylates. Calculated separation factors for the structurally restricted ligands DCPA, 2,6-dicarboxypiperidine acetic acid (as compared with NTA, nitrilotriacetic acid) and DCTA, trans-1,2-diaminocyclohexane-N,N,N',N'-tetraacetic acid (compared with EDTA, ethylenediamine-N,N,N',N'-tetraacetic acid) offer steeper slopes than their non-constrained analogs. However, the planar arrangement of donor atoms in dipicolinic acid (dipic - 2,6-dicarboxypyridine) is not favorable for lanthanide separations across the series. Clearly, constraint of ligand donor atoms is a useful but not universally successful technique for designing new separation reagents. For successful ligand design, the ligand structure must accommodate the steric demands of the metal ion.

Because there are substantial amounts of actinides present in alkaline media in underground storage tanks at former actinide production facilities, there has been some

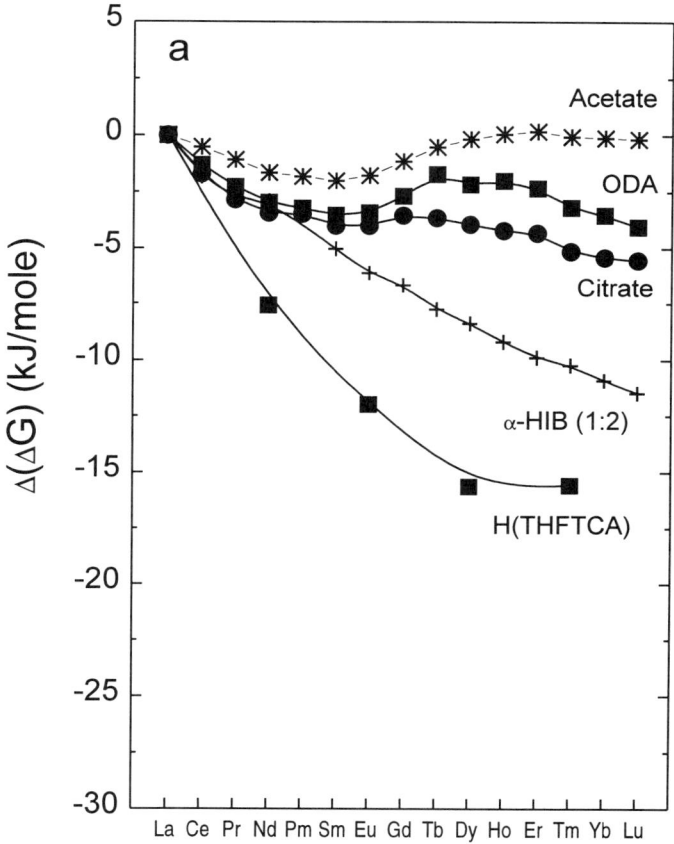

Figure 5. Relative free energies of lanthanide complexes (normalized to La^{3+}) with a) polycarboxylic acids (ODA - oxydiacetic acid, -HIB - -hydroxyisobutyric acid, H(THFTCA) - monoprotonated form of tetrahydrofuran-2,3,4,5-tetracarboxylic acid) and b) aminopolycarboxylic acids (NTA - nitrilotriacetic acid, DCPA - 2,6-dicarboxypiperidine acetic acid, EDTA - ethylenediamine-N,N,N',N'-tetraacetic acid, DCTA - trans-1,2-diaminocyclohexane-N,N,N',N'-tetraacetic acid, dipic - 2,6-dicarboxypyridine).

Continued on next page.

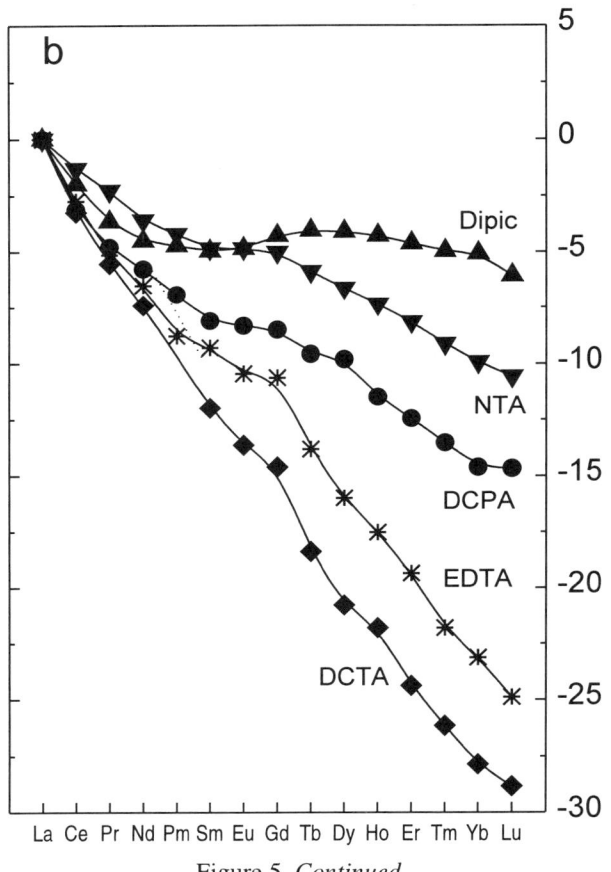

Figure 5. *Continued.*

focus on the development of separations techniques for actinides in alkaline solution. Under these conditions, water soluble chelating agents are required to maintain solubility of the readily hydrolyzed actinide cations. Such separations have been reviewed by Karalova et al. (*21*). Both solvating and chelating extractants have been used in these studies. Primary and quaternary amines, alkylpyrocatechols, β-diketones, pyrazolones, and N-alkyl derivatives of aminoalcohols are the extractants indicated as useful for alkaline extraction processes. The aqueous complexing agents employed include EDTA, DTPA, and DTPMPA (diethylenetriamine-N,N,N',N'',N'''-pentamethylenephosphonic acid). The separation factors are based mainly on the difference in the rates of the metal complexation equilibria for Eu and Am. Such kinetic-based separations represent a dramatically new approach to metal ion separations.

Recent Results on New Reagent/Process Development

Our research on the coordination complexes of f-elements with derivatives of methanediphosphonic acid (*15, 22-26*) has established that these complexing agents are very efficient general stripping agents for actinide processing. Stability constants for Eu(III) (*22*), Cm(III) (*23*), Am(III) (*15*), Th(IV) (*24*) and U(VI) (*25*) complexes with a wide variety of substituted diphosphonic acids have been measured using solvent extraction distribution methods. We have also investigated certain aspects of the kinetics of their complexation reactions for potential application in separation. Thermodynamic results and laser-induced fluorescence decay studies of europium complexes strongly suggest intramolecular hydrogen bonding contributes to the unusual stability of the f-element complexes in acid solutions (*26*). Our research effort has also established that diphosphonate complexing agents can be designed to be readily decomposed by mild treatment to facilitate waste disposal (*27*). We have also suggested an approach to oxidation state-specific separation of actinides within the general framework of TRUEX solvent extraction through application of diphosphonates in the stripping stage (*27*). The CMPO/SCN⁻/PAA separation of Am from lanthanides described above is another example of a separation enhanced by the application of phosphonate complexants.

The TRUEX process for extraction of actinides and lanthanides is now a well-known and accepted technology for total actinide recovery (*28*). More recent research has led to the development of the SREX process for extraction of Sr^{2+} (*29*). This solvent extraction process employs the crown ether extractant di(*t*-butyl cyclohexano)-18-crown-6 (di-tBuCH18C6) as the Sr selective extractant with 1-octanol as the organic diluent. These two solvent extraction systems function in a complementary fashion when the diluent system is a normal paraffin hydrocarbon with a simple phosphonate or phosphate ester present as a phase modifier. This combined process solvent will efficiently extract lanthanides, actinides and strontium from nitric acid solutions.

One ubiquitous component of typical radioactive waste streams which does not require burial in a geologic repository is uranium. By virtue of the low specific activity of its principal isotopes ($SA(^{238}U)$ = 0.746 dpm/μg, $SA(^{235}U)$ = 4.80 dpm/μg), uranium could be safely disposed of in near surface burial using a concrete waste form. However, the stable oxidation states of uranium (U(IV) and U(VI)) are both strongly extracted by CMPO from nitric acid solutions. Selective separation of uranium from trivalent and tetravalent actinide cations is highly desirable. However, UO_2^{2+} extraction is typically intermediate between that of the trivalent and tetravalent actinides, so selective separation must rely on the use of a reagent which can distinguish the unique geometry of the linear dioxouranyl cation.

Such a separation system has been identified (*30*). The extractant solution combining CMPO for actinide extraction, diamyl(amyl) phosphonate (DA(A)P) for enhanced phase compatibility, and di-tBuCH18C6 for Sr^{2+} extraction in Isopar L is designed for removal of Sr, An, Ln, and Tc from nuclear wastes. The aqueous complexant THFTCA (as the disodium salt), provides the desired separation of UO_2^{2+} from the other actinides. Effectively, THFTCA complexes An(III) and An(IV) holding them in the aqueous phase while UO_2^{2+} is extracted by CMPO. Extraction of trivalent, tetravalent and hexavalent actinides by CMPO from HNO_3 and from THFTCA are shown in Figure 6a and 6b respectively. Studies of the thermodynamics of complexation of uranyl and lanthanide complexes with THFTCA suggest that anomalously weak complexes are formed between UO_2^{2+} and THFTCA to account for the unique selectivity of this system.

Conclusions

In the above discussion, numerous examples have been given along with a theoretical framework to illustrate the impact of the composition and characteristics of aqueous media on metal ion separations. Though the discussion has emphasized separations of f-elements, most of the concepts presented apply with minor modifications to metal ion separations in general. These observations, for the most part, apply equally to separations based on liquid-liquid extraction, ion exchange, extraction chromatography, and membrane-based processes. They serve to illustrate that the low energetic requirements of a selective separation offer many opportunities to achieve a desired separation, and that great flexibility can be gained through modifications of the aqueous medium.

Areas for emphasis in future research include the following: design of aqueous complexants capable of distinguishing the size and shape of cations, improved understanding of solvation phenomena in both aqueous and mixed aqueous/organic media, development of aqueous complexants that do not create environmental hazards or waste disposal complications when their utility is concluded, and studies of the kinetics of metal ion complexation processes both in homogeneous solutions and at interfaces. Objectives specific to actinide/lanthanide separations include the design of new soft-donor or mixed hard/soft donor ligands (both water soluble and lipophilic species) to enhance lanthanide /actinide separations. Many of these suggested research areas require a serious commitment to fundamental research in coordination chemistry as well as separation science.

Acknowledgments

Work performed under the auspices of the U.S. Department of Energy, Office of Basic Energy Sciences, Division of Chemical Sciences under contract number W-31-109-ENG-38

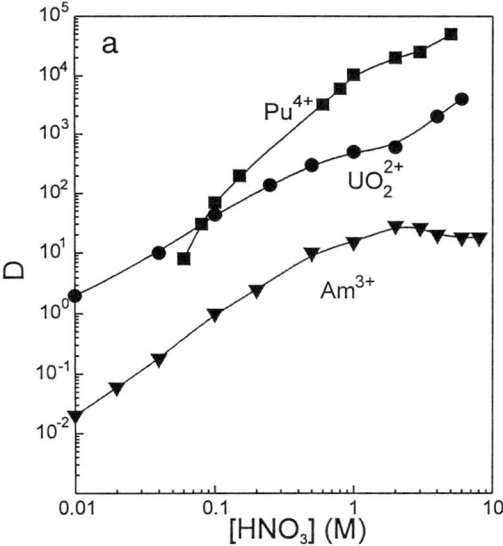

Figure 6. a) Extraction of trivalent, tetravalent and hexavalent actinides by TRUEX Process Solvent (0.2 M CMPO/1.2 M TBP/dodecane b) Extraction of trivalent, tetravalent and hexavalent actinides by the Combined Process Solvent (0.2 M CMPO/1.2 M DA(A)P/0.05 M di-tBuCH18C6/Isopar L from 0.05 M $Na_2THFTCA/HNO_3$.

Continued on next page.

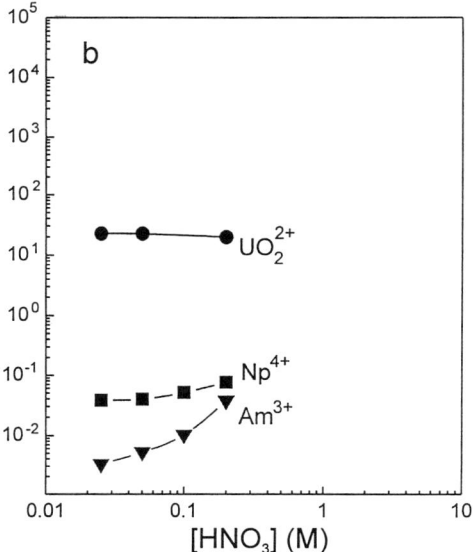

Figure 6. *Continued.*

Literature Cited

1. Marcus, Y.; Kertes, A. S. *Ion Exchange and Solvent Extraction of Metal Complexes*; Wiley Interscience: London, 1969.
2. F. Helferrich *Ion Exchange*; McGraw-Hill: New York, 1962.
3. Rizkalla, E. N.; Choppin, G. R., in *Handbook on the Physics and Chemistry of Rare Earths, Volume 18*; Gschneidner, K. A., Jr.; Eyring, L.; Choppin, G. R.; Lander, G., Eds.; North Holland: Amsterdam, 1994; pp. 529-558.
4. Schubert, J. *J. Phys. Coll. Chem.* **1948**, *52*, 340.
5. Moeller, T. *The Chemistry of the Lanthanides*; Reinhold: New York, 1963.
6. Gerber M. S. *Legend and Legacy: Fifty Years of Defense Production at the Hanford Site*, Westinghouse Hanford Report WHC-MR-0293, **1992**.
7. Choppin, G. R. *J. Molecular Structure* **1978**, *45*, 39.
8. Lincoln, S. F. In *Advances in Inorganic and Bioinorganic Mechanisms, Vol. 4*; Sykes, A. G. Ed.; Academic Press: London, 1986; pp 217-287.
9. Cossy, C.; Helm, L.; Merbach, A. E. *Inorg. Chem.* **1989**, *28*, 2699.
10. Street, K., Jr.; Seaborg, G. T. *J. Am. Chem. Soc.* **1950**, *72*, 2790.
11. Diamond, R. M.; Street, K., Jr.; Seaborg, G. T. *J. Am. Chem. Soc.* **1954**, *76*, 1461.
12. Gmelin Handbook of Inorganic Chemistry, 8th Edition *Sc, Y, La-Lu Rare Earth Elements, Part D 6, Ion Exchange and Solvent Extraction Reactions, Organometallic Compounds*; Springer-Verlag: Berlin, 1983, pp 1-136.
13. Sekine, T. *Bull. Chem. Soc. Jpn.* **1965**, *38*, 1972.
14. Myasoedov, B. F.; Chmutova, M. K. In *Separations of f-Elements*, Nash, K. L.; Choppin, G. R., Eds.; Plenum Press: New York, 1995, pp 11-29.
15. Ensor, D. D.; Nash, K. L. In *f-element Separations*; Nash, K. L.; Choppin, G. R., Eds.; Plenum Press: New York, 1995, pp 143-152.
16. Muscatello, A. C.; Horwitz, E. P.; Kalina, D. G.; Kaplan, L. *Sep. Sci. Technol.* **1982**, *17*, 859; Horwitz, E. P.; Muscatello, A. C., Argonne National Laboratory,Unpublished work, **1981**.
17. Weaver, B.; Kappelmann, F. A. *J. Inorg. Nucl. Chem.* **1968**, *30*, 263.
18. Thompson, S. G.; Harvey, B. G.; Choppin G. R.; Seaborg, G. T. *J. Am. Chem. Soc.* **1954**, *76*, 6229.
19. Choppin, G. R.; Silva, R. J. *J. Inorg. Nucl. Chem.* **1956**, *3*, 153.
20. Feil Jenkins J. F.; Nash, K. L.; Rogers, R. D. *Inorg. Chim. Acta* **1995**, *236*, 67.
21. Karalova, Z. K.; Myasoedov, B. F.; Bukhina, T. I.; Lavrinovich, E. A. *Solvent Extr. Ion Exch.* **1988**, *6*, 1109.
22. Nash, K. L.; Horwitz, E. P. *Inorg. Chim. Acta* **1990**, *169*, 245.
23. Jensen, M. P.; Rickert, P. G.; Schmidt, M. A.; Nash, K. L. *J. Alloys and Cmpnds.* **1997**, *249/250*, 86.
24. Nash, K. L. *Radiochim. Acta* **1991**, *54*, 171.
25. Nash, K. L. *Radiochim. Acta* **1993**, *61*, 14.
26. Nash, K. L.; Rao, L. F.; Choppin, G. R. *Inorg. Chem.* **1995**, *34*, 2753.

27. Nash, K. L.; Rickert, P. G. *Sep. Sci. Technol.* **1993**, *28*, 25.
28. Schulz, W. W.; Horwitz, E. P. *Sep. Sci. Technol.* **1988**, *23*, 1191.
29. Horwitz, E. P.; Dietz, M. L.; Fisher, D. E. *Solvent Extr. Ion Exch.* **1991**, *9*, 1.
30. Nash, K. L.; Horwitz, E. P.; Diamond, H.; Rickert, P. G.; Muntean, J. V.; Mendoza, M. D.; di Giuseppe, G. *Solvent Extr. Ion Exch.* **1996**, *14*, 13.

Chapter 5

Metal Ion Separations in Aqueous Biphasic Systems and Using Aqueous Biphasic Extraction Chromatography

Jonathan G. Huddleston, Scott T. Griffin, Jinhua Zhang, Heather D. Willauer, and Robin D. Rogers[1]

Department of Chemistry, The University of Alabama, Tuscaloosa, AL 35487

> Polyethylene glycol-based aqueous biphasic systems (ABS) and the complementary aqueous biphasic extraction chromatographic (ABEC) resins are capable of selectively removing metal ions from complex solutions, such as the radioactive Hanford tank waste supernates. These aqueous separations methods have the potential to eliminate the use of volatile organic solvents in many separations, yet their utility in more conventional solvent extraction processes has until recently received scant attention. This paper reviews the nature of ABS and ABEC separations, categorizes the types of possible metal ion separations, and discusses where these techniques may find practical application. The relationship between the liquid/liquid ABS separations and the chromatographic ABEC separations is discussed in detail.

"Greening the Chemical Industry." Current thinking in industry, academe, and government acknowledges that the sustained growth, profitability, and technological development of the chemical industry may only be achieved by stressing the importance of the health and safety of employees, consumers, and the general public and by fully embracing the idea of a continuing responsibility for environmental stewardship (1). To achieve this, established products and processes may have to be reengineered and new products and processes, to be adopted at all, will have to be environmentally benign at the outset. There will be pressure to eliminate the generation of toxic waste and secondary waste products during a chemical process. In this context, separations processes will have an increasingly important role in a situation in which the costs of raw discharge and effluent treatment must inevitably rise.

[1] Corresponding author.

Uniquely amongst currently available waste treatment options, separations processes offer the potential for closed processing and product recycling. However, liquid/liquid separations processes seem problematic in this context since the vast majority currently involve the use of toxic and flammable volatile organic compounds (VOCs) (2). Nevertheless, the widespread adoption, in general where distillation is not an option, of liquid/liquid extraction using VOCs testifies to its unique advantages as a unit operation in separations processing. Liquid/liquid extraction can be adapted to the selective separation of a wide variety of solutes through compatibility with a range of diluents and extractants. Extraction kinetics are usually rapid, thus enabling high throughput and large scales of operation. Additionally, with suitably designed multi-stage contactors, extractions may be optimized for high selectivity and efficiency. Frequently, the back extraction or stripping of the organic phase may be accomplished easily, either by oxidation/reduction or by manipulation of the charge state of the solute through a pH change. However, the need to utilize VOCs as diluents in these processes brings with it a number of significant disadvantages. Costs of diluents and extractants may be high and there is significant capital cost associated with the safe engineering of unit operations involving volatile and flammable solvent systems. Disposal of spent diluent and extractants will also incur significant cost and be increasingly impacted by environmental regulations.

Aqueous Alternatives to Solvent Extraction. Despite their forty-year history (3), little attention has been paid by the chemical engineering community to the existence of a class of liquid/liquid extraction systems whose nature is entirely aqueous. These so-called aqueous two-phase systems or aqueous biphasic systems (ABS) are formed when certain water-soluble polymers are combined with one another or with certain inorganic salts at specific concentrations in aqueous solution (3). ABS have been shown to be effective for the liquid/liquid extraction of a wide variety of solutes including biological macromolecules and particles (3-5) and, more recently, for the extraction of metal ions and small organic molecules (6-23). The metal ion extractions may be of three types (6-9). The metal ion may be extracted into the PEG-rich phase without the addition of any extractant (10-15). The extraction may proceed through the formation of negatively charged inorganic anionic complexes (16) or by the use of a water-soluble organic complexant (17-21). The latter extraction technology heightens interest in the partitioning of small organic molecules in ABS. However, such extractions may find important applications in their own right (22,23). ABS retain all the practical advantages of traditional liquid/liquid extraction schemes, and also possess a number of unique advantages, due, in large part, to their wholly aqueous nature. ABS based on polyethylene glycol (PEG) are virtually non-toxic and the components are inexpensive bulk commodities. The physical properties of such systems are sufficiently close to those of traditional liquid/liquid extraction systems, that for the most part, common plant may be used in the engineering design of the extraction process (24).

PEG/salt-based ABS are formed when aqueous solutions of high molecular weight PEG are salted-out by specific salt solutions producing two immiscible, but wholly aqueous phases. The equilibrium phase diagram of a typical aqueous two-phase system based on PEG monomethylether-5000 (M-PEG-5000) and $(NH_4)_2SO_4$ is shown in Figure 1. Mixture compositions to the left of the binodal curve are monophasic whilst systems to the right form biphasic systems. Mixtures having the overall compositions B on the tie lines ABC, form phases having the compositions indicated by the nodes A and C. It can be seen that the light phase (A) is composed primarily of PEG and the lower, heavy phase (C), primarily of $(NH_4)_2SO_4$. The figure also shows that systems lying on longer tie lines form phases of increasingly divergent composition. It is from this difference in the compositions of the two phases that the selectivity of the system arises.

A variety of salts may be used to form ABS with PEG (25), a number of which are shown in Table I. The ability of salts to form ABS with PEG has been related (6) to their Gibbs free energy of hydration (Table II; ΔG_{hyd} equivalent to $\Delta_{hyd}G^*_{calc}$ in reference 26). Both the cation and anion contribute to this effect but the anion dominates. The more negative the ΔG_{hyd} of an ion, the greater its salting-out effect for PEG. ΔG_{hyd} of the salt seems to be the most important factor in determining the choice of salt used to form the biphase. The more negative the ΔG_{hyd} of the salt, the lower will be the concentration of PEG and salt required to form a biphasic system. For particular combinations of anions and cations, the important factor appears to be their combined free energy of hydration which seems to be simply additive in its effect on water structure and salting-out of PEG.

In cases where metal salts are present in complex matrices (10,11), such as concentrated solutions of NaOH as in the Hanford waste tanks (11), the effectiveness of the separation may be qualitatively estimated from the ΔG_{hyd} of the matrix ions and that of the ions to be extracted. It has been demonstrated, for example, that chaotropic ions with small negative ΔG_{hyd}, such as TcO_4^-, partition quantitatively to the PEG-rich phase of a suitable ABS (11). In an ABS prepared by mixing equal volumes of 3.5 M $(NH_4)_2SO_4$ and 40 % w/w PEG-2000, sodium salts having anions with ΔG_{hyd} equal to about -310 kJ/mol, have a distribution value close to 1 (10). The presence of sodium halide salts having anions with $-\Delta G_{hyd} < 310$ kJ/mol (Cl^-, Br^-, I^-) in this ABS, prefer the PEG-rich phase and have the effect of depressing the distribution ratio of the pertechnetate anion. Sodium salts with anions having $-\Delta G_{hyd}$ greater than this (e.g., F^-) add to the salting-out effect and thus increase the pertechnetate distribution ratios. Sodium salts of HCO_3^- (ΔG_{hyd} -310 kJ/mol) have a negligible effect on the distribution of TcO_4^- in this ABS (10).

Overcoming the Perceived Disadvantages of ABS. Traditional solvent extraction processes often involve the back extraction of the extracted species into a fresh aqueous phase following the initial forward extraction into the organic phase (27). This is often a relatively straightforward step involving, for example, reduction or change in pH to create a charged species with enhanced solubility in the aqueous phase. Following this, spent solvent is recycled, usually after regeneration by distillation or secondary extraction (27). Currently, this appears much more difficult to achieve with ABS. Back extraction steps may be more difficult to design as molar concentrations of certain salts are required to maintain a biphasic system and recycling is hampered by difficulties associated with preparing the PEG-rich phase for recycling.

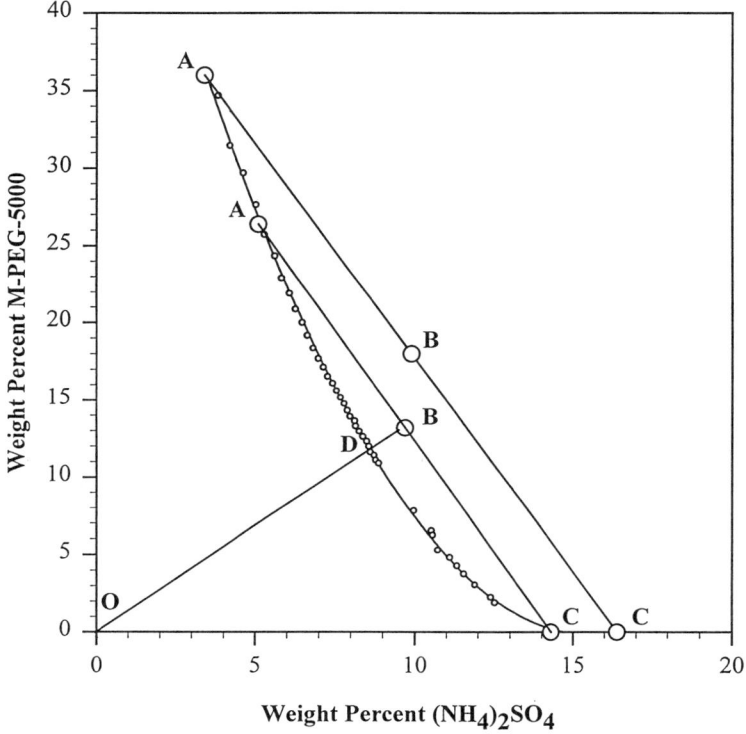

Figure 1. Equilibrium composition of an ABS composed of M-PEG-5000 and $(NH_4)_2SO_4$ showing the binodal curve bounding the points AADCC, the tie lines (ABC) connecting the nodes (A,C), and also a line (ODB) from which the System Stability (ST) may be derived (see text).

Table I. A Selection of Salts Forming ABS with Polyethylene Glycol (25)

Univalent Anions	Divalent Anions		Trivalent Anions	Tetravalent Anions
NaOH	Na_2CO_3	$ZnSO_4$	Na_3PO_4	Na_4SiO_4
KOH	K_2CO_3	Alum	K_3PO_4	$Na_4(HEDPA)^a$
RbOH	$(NH_4)_2CO_3$	Na_2SeO_4	Na_3VO_4	
CsOH	Rb_2CO_3	Na_2CrO_4	Na_3(citrate)	
NaF	Li_2SO_4	Na_2MoO_4	$(NH_4)_3$(citrate)	
Na(formate)	Na_2SO_4	Na_2WO_4		
	$(NH_4)_2SO_4$	K_2HPO_4		
	Rb_2SO_4	Na_2SO_3		
	Cs_2SO_4	Na_2SiO_3		
	$MgSO_4$	Na_2S		
	$Al_2(SO_4)_3$	Na_2(succinate)		
	$FeSO_4$	Na_2(tartrate)		
	$CuSO_4$			

^{a}1-hydroxyethane-1,1,-diphosphonic acid.

Table II. The Gibbs Free Energies^a of a Selection of Anions and Cations

Anion	ΔG_{hyd} (kJ/mol)	Cation	ΔG_{hyd} (kJ/mol)
ReO_4^-	-170	Cs^+	-245
I^-	-220	Rb^+	-285
Br^-	-250	NH_4^+	-285
Cl^-	-270	K^+	-305
HCO_3^-	-310	Na^+	-385
F^-	-345	Li^+	-510
OH^-	-345	Fe^{2+}	-1825
SeO_4^{2-}	-1110	Zn^{2+}	-1880
CrO_4^{2-}	-1120	Cu^{2+}	-1920
SO_4^{2-}	-1145	Mg^{2+}	-1940
SO_3^{2-}	-1230	Al^{3+}	-5450
S^{2-}	-1280		
CO_3^{2-}	-1300		
PO_4^{3-}	-2835		

^{a}Adapted from reference 26.

Covalent attachment of PEG onto a solid support has overcome many of these limitations since forward extraction conditions may be achieved by addition of salt and back extraction conditions simply by adding water (9-12,28). Thus, a simple process may be envisaged in which the metal ion is extracted into the covalently bonded PEG-rich phase, under the appropriate conditions of salt, pH, etc. Recovery is achieved simply by elution with water, thus overcoming most of the limitations of ABS operation associated with the recycling of the polymer-rich phase. It will be shown in the present chapter that applications developed in ABS can be directly transferred to ABEC. The distribution of solutes between the liquid phases of ABS and the liquid and solid phase of ABEC is directly comparable under identical salt solution conditions.

Experimental

Complete details of the experimental methods involving partitioning of pertechnetate may be found in reference *28*. $(NH_4)_2SO_4$, K_2CO_3, K_3PO_4, NaOH, and PEG were of reagent grade and obtained from Aldrich (Milwaukee, WI, USA). $NH_4^{99}TcO_4$ was obtained from Isotope Products Laboratories (Burbank, California, USA) and was diluted with deionized water to an activity of 0.06 - 0.08 µCi/µL for use in the experiments. Ultima Gold Scintillation Cocktail (Packard Instrument Co., Downers Grove IL, USA) and a Packard Tri-Carb 1900 TR Liquid Scintillation Analyzer were used in the standard liquid scintillation assays.

Polymer and salt stock solutions were prepared on a weight percent or molar basis and the compositions quoted refer to preequilibrium stock solution concentrations. Distribution ratios were determined by mixing 1 mL of a 40% w/w PEG solution with salt stock solution of known concentration. Mixtures were then vortex mixed for 2 min and centrifuged (2000 x g) for 2 min. Mixtures were then spiked with the metal ion tracer and centrifuged (2 min, 2000 x g) and then vortexed for 2 min. The coexisting phases were then disengaged by further centrifugation (2 min, 2000 x g). Aliquots were removed from each phase for liquid scintillation analysis. Since equal aliquots of the phases were analyzed, the distribution of the metal ion tracer in the phases could be defined as:

$$D = \frac{\text{Activity in counts per minute PEG - rich phase}}{\text{Activity in counts per minute salt - rich phase}}$$

For the determination of the distribution of pertechnetate to solid-phase conjugated PEG, EIChroM Iodine resin (ABEC-5000, 100-200 mesh) was used (EIChroM Industries, Darien, IL). Full details of the preparation of this resin may be found in reference *28*. Salt solutions were prepared in the same way as for the ABS experiments. The distribution ratio for the metal ions onto the ABEC resin was determined as follows. 15 to 20 mg of resin was added to each of two vials and the weight recorded. 1 mL of a TcO_4^--spiked salt solution was added to each adsorbent containing tube. The tubes were briefly centrifuged and mixed by magnetic stirring for 30 min. The adsorbent was separated from the supernatant by centrifugation (2000 x g, 2 min). 100 µL of adsorbent free supernatant was transferred to a vial containing liquid scintillation cocktail and counted. The distribution of pertechnetate was calculated using:

$$D_W = \frac{A_i - A_f}{A_f} \cdot \frac{\text{contact volume (mL)}}{\text{wt. of resin (g)} * dwcf}$$

where A_i is the activity in counts per minute in the solution prior to contact with the adsorbent and A_f is the activity in counts per minute after contact with the adsorbent. The contact volume is the total volume of the adsorbate containing solution used and *dwcf* is the dry weight conversion factor of the resin. The latter is determined from the weight of the resin dried to constant weight at 110 °C in an oven.

Results and Discussion

Extraction of Metal Ions in ABS. The distribution of metal ion species within ABS is controlled by their preference for hydration in one phase or the other. The degree to which the ion behaves as a chaotrope or a chosmotrope and the difference in hydrogen bond orientation and water structure between the two phases, one rich in salt and the other rich in PEG, seem to be the dominant factors in the resulting distribution. Pertechnetate, in common with a number of similar metal anions, partitions preferentially to the PEG-rich phase because, in classical terms, it is a soft anion having large size and low charge density. It is these properties which are placed on a quantitative footing through the calculation of the free energy of hydration.

The most important application to date of ABS for metal ion extraction involves the quantitative recovery of the pertechnetate anion. This is of importance since the short lived ^{99m}Tc ($t_{1/2}$ = 6 h) is used in the majority of medical procedures involving radioisotopes *(29,30)*. Additionally, high levels of ^{99}TcO$_4^-$ are produced as by products of nuclear fission where its long half-life and environmental mobility give rise to some concern *(31)*. However, TcO$_4^-$ is also useful as a system probe to determine the factors of importance in controlling the distribution of metal anions in ABS, since its distribution may easily be followed by liquid scintillation counting. Thus, the effects on the relative composition of the phases, and thus on metal ion partitioning, of various ABS parameters, such as concentration and molecular mass of PEG, salt type and concentration, or the presence of matrix ions, may be determined by study of pertechnetate distribution ratios.

Figure 2 shows the distribution of TcO$_4^-$ at 25 °C in six different ABS formed with PEG-2000 and M-PEG-5000 and with the salts potassium carbonate, ammonium sulfate, and sodium hydroxide. As the salt concentration is increased at a fixed concentration of PEG (40% w/w), the phase incompatibility increases as the compositions of the phases diverges from a theoretically identical composition at the critical point. Consideration of Figure 1 shows that the concentration of PEG rises rapidly in the top phase with a concomitant decrease in the concentration of the phase-forming salt. For the lower, salt-rich, phase, the reverse takes place. As a consequence, the solute (TcO$_4^-$) becomes unequally distributed between phases of increasingly divergent composition, as shown in Figure 2. In general, higher molecular weights of PEG are salted-out by a given salt at lower concentrations of PEG and salt. Also, with only a few exceptions *(25)*, PEG of defined molecular weight is salted-out at lower concentrations of PEG and salt by salts having more negative ΔG_{hyd}. Thus, D_{Tc} is lower over the range of salt concentrations used for PEG-2000/(NH$_4$)$_2$SO$_4$ than for PEG-2000/K$_2$CO$_3$ and D_{Tc} is also lower for M-PEG-5000/(NH$_4$)$_2$SO$_4$ than M-PEG-5000/K$_2$CO$_3$. Additionally the M-PEG-5000 systems produce higher D_{Tc}s than the PEG-2000 systems. This exactly reflects the idea that

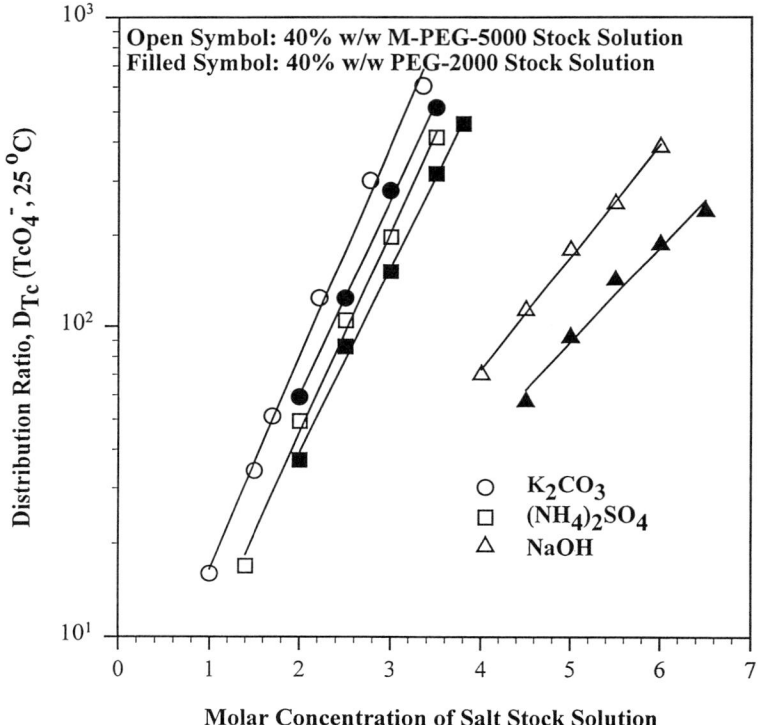

Figure 2. The distribution of TcO_4^- in ABS composed of M-PEG-5000 or PEG-2000 and the salts K_2CO_3, $(NH_4)_2SO_4$, and NaOH.

the divergence in composition of the phases of the system is responsible for the unequal distribution of the solutes and that this is greater for higher molecular weights of PEG and more strongly salting-out agents at defined salt and PEG concentration.

The data in the lower portion of Figure 3 shows the distribution of TcO_4^- in ABS prepared from 40% w/w M-PEG-5000 to which equal volumes of M_2SO_4 salt solutions (where M is either Na^+, NH_4^+, Rb^+, or Cs^+) having the salt concentrations shown in the figure were added. It is apparent that the distribution ratio of TcO_4^- depends upon ΔG_{hyd} of the cation so that the more negative the ΔG_{hyd} of the cation, the higher the distribution ratio of the TcO_4^-. This is clarified in Figure 4 where the distribution ratio of TcO_4^- is shown in relation to the ΔG_{hyd} of the cation. Since the systems are all composed of 1.4 M M_2SO_4, the ΔG_{hyd} of the anion is constant. Thus, the distribution ratio of TcO_4^- is dependent on ΔG_{hyd} of the phase-forming cation, increasing with cations having increasingly negative ΔG_{hyd}.

It must be conceded that in Figure 4, it is the experimental determination of ΔG_{hyd} (as given in reference 26) which is used and not, as in the other figures the theoretically calculated value (26). This requires some justification. Two aspects of the solvation of ions are specifically accounted for in the calculation of ΔG_{hyd}, the free energy of cavity formation and the free energy associated with the electrostriction of water in the hydration shell of the ion due to its electrical field. Ion dependent features included in this model of ionic hydration energy are the charge on the ion and the ionic radius, which for Rb^+ and NH_4^+ are practically identical (r = 0.149 and 0.148 nm, respectively) and thus, the calculated ΔG_{hyd} for these ions is the same (-285 kJ/mol). Thus, as mentioned, the experimental values (Rb^+ = -275 kJ/mol, NH_4^+ = -285 kJ/mol) have been preferred in Figure 4.

From this argument, it follows that partition in ABS should prove to be an excellent method for the estimation of ΔG_{hyd} of ionic solutes which would include in the estimate all factors involved in the aqueous solvation of the solute under the conditions used, such as speciation, complex formation, and specific hydrogen bonding effects. It would thus be of some interest to compare estimates so obtained to theoretical estimates and to estimates obtained by other experiments. For instance, it is interesting to note that the free energy of hydration of H^+ is given as -1015 kJ/mol (26) (-1050 kJ/mol, experimental). Considering the value of -310 kJ/mol for an anion having even distribution in the PEG-2000/$(NH_4)_2SO_4$ system mentioned above, this implies a preference of the H^+ ion for the lower phase. In turn, this implies that the pH of the lower phase of an ABS will be less than that of the upper phase which has been found, experimentally, to be the case (32). We are currently investigating these aspects of metal ion partitioning in ABS.

The effect of changing the anion may be examined analogously by the partition of TcO_4^- in systems composed of different salt anions but in which the cation is held constant. This is shown in the lower part of Figure 5 where the distribution ratio of TcO_4^- increases with increasingly negative free energies of hydration of the anion in the order $SeO_4^{2-} < SO_4^{2-} < CO_3^{2-}$.

These effects of ΔG_{hyd} of cation and anion may be combined to predict the distribution ratio of TcO_4^- in systems composed of salts differing in both anion and cation type where the system is also composed of a defined molecular weight of

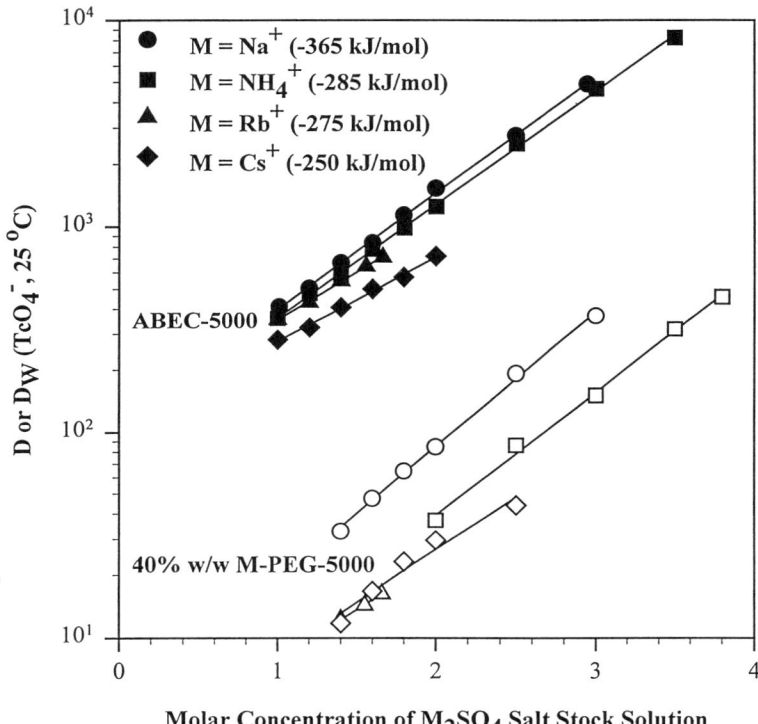

Figure 3. The distribution (D or D_W) of TcO_4^- in ABS composed of M-PEG-5000 or to ABEC-5000 with increasing concentrations (M) of salts of the form M_2SO_4; where M is Na^+, NH_4^+, Rb^+, or Cs^+. The Gibbs free energies of hydration indicated for the cations are experimental values from reference *26* (see text).

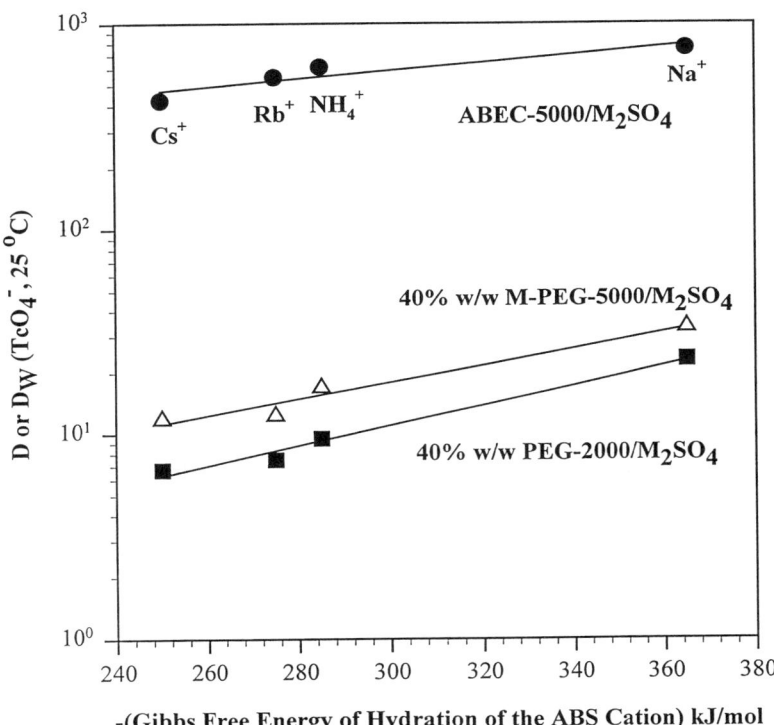

Figure 4. The distribution of pertechnetate (D or D_W) of TcO_4^- in ABS composed of M-PEG-5000 or to ABEC-5000 in the presence of different sulfate salts (1.4 M) in relation to the Gibbs free energy of hydration (experimental values from reference 26) of the ABS cation.

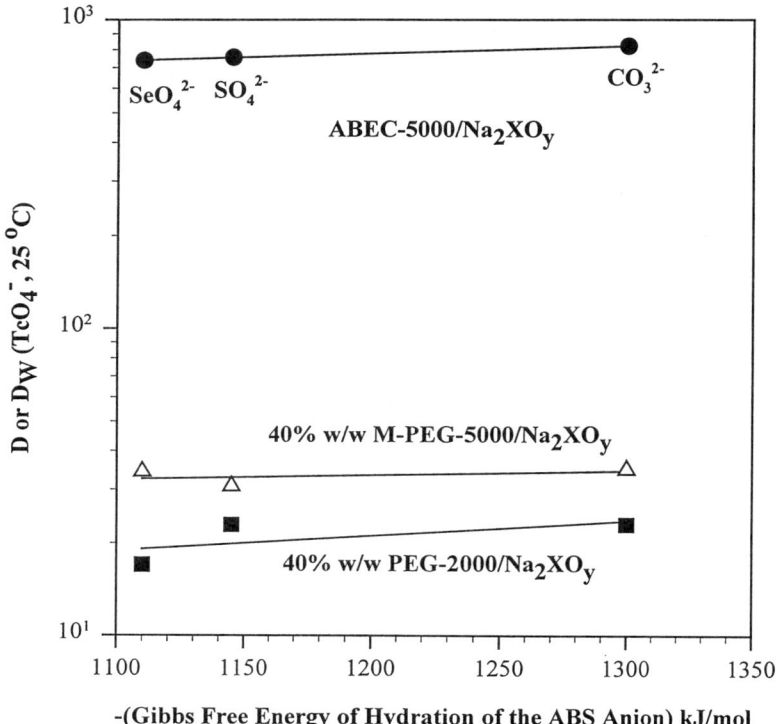

Figure 5. The distribution of pertechnetate (D or D_W) of TcO_4^- in ABS composed of M-PEG-5000 or to ABEC-5000 in the presence of different sulfate salts (1.4 M) in relation to the Gibbs free energy of hydration (from reference 26) of the ABS anion.

PEG. This is illustrated in Figure 6, which shows the distribution ratio of TcO$_4^-$ in several different ABS. The molecular weight and concentration of PEG used to form these systems is held constant at 40% w/w M-PEG-5000, but the salt type and concentration is varied as indicated in the figure. D_{Tc} is plotted against the total ΔG_{hyd} of the salt solution added to form the ABS. This is simply calculated as the sum of ΔG_{hyd} of the ions comprising the salt multiplied by their concentration to give the total ΔG_{hyd} of the salt stock solution in kJ/L. A good correlation is observed between the total ΔG_{hyd} of the stock salt solution and D_{Tc}.

Prediction of Metal Ion Partitioning from ABS Characteristics. It seems likely that it will be possible to predict the distribution of metal ions in PEG-salt ABS simply from knowledge of ΔG_{hyd} of the solute and the total ΔG_{hyd} of the salt solution used to form the ABS. This seems to rely on the fact that ΔG_{hyd} of the salt, determines the cloud point of PEG solutions and therefore the increasing divergence of the phase compositions of ABS. Since the cloud point of PEG is also dependent on the molecular weight of PEG (*33*), it should be possible to take this factor into account to improve the scope of the prediction to encompass systems composed of different molecular weights of PEG. It should be noted, however, that certain salts do not appear to follow the expected lyotropic sequence and thus these relationships cannot be universally applied. Examples of such salts (*25*) include CaCl$_2$ and AlCl$_3$ which do not form ABS, certain sulfates having multivalent cations, such as magnesium and zinc, and lithium salts (when compared to similar salts of other alkali metal cations) which do not form ABS at the expected concentrations of PEG and salt.

In biological application of ABS, it is usual to relate solute partitioning to the tie line length (TLL - see Figure 1) (*34*). TLL is simply the orthogonal sum of the difference in PEG (or salt) concentrations between the two coexisting phases, denoted by the nodes A and C, in Figure 1 and has the units w/w %.

$$TLL = \sqrt{\Delta(PEG)^2 + (\Delta SALT)^2}$$

The logarithm of the distribution coefficient is usually found to be a linear function of the TLL, at least close to the critical point (*35*), for a defined molecular weight of PEG and a particular salt. This requires considerable expenditure of effort to determine the appropriate phase diagrams and tie line relationships such as that shown in Figure 1.

Similarly, the distribution of partitioned species between the phases of an ABS may be described in terms of the difference in salt (or PEG) concentrations between the phases (Δsalt or ΔPEG). Figure 7 shows the distribution of TcO$_4^-$ in several PEG-2000/NaOH ABS in relation to Δ[NaOH] (the difference in NaOH concentration between the salt-rich and PEG-rich phases). The data shown in Figure 7 fit the relationship:

$$\log D_{Tc} = -0.174 + 0.678(\Delta[NaOH])$$

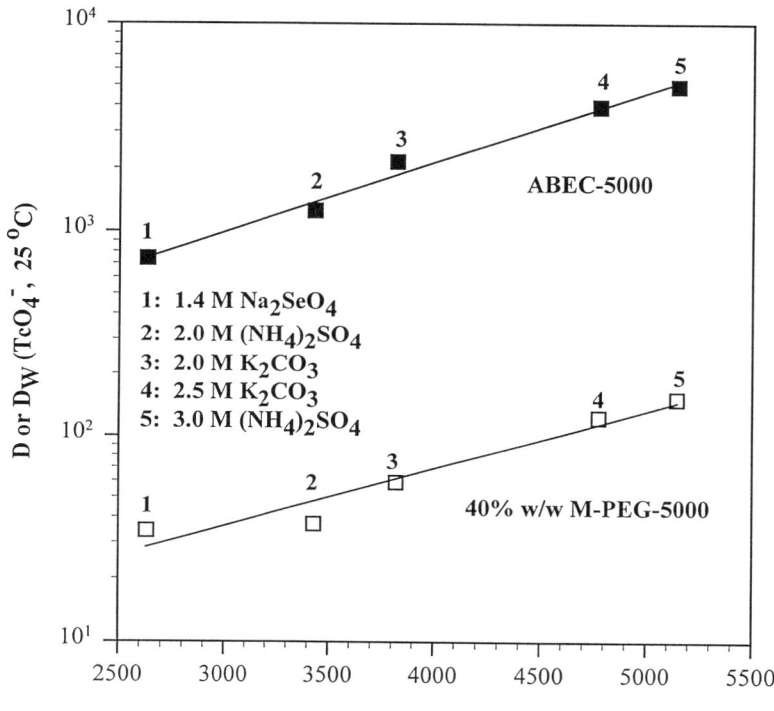

Figure 6. The distribution of pertechnetate (D or D_W) of TcO_4^- in ABS composed of M-PEG-5000 or to ABEC-5000 in the presence of various concentrations of different salts (as shown) in relation to the total Gibbs free energy of hydration of the salt (calculated as $2[M^+] \cdot \Delta G_{cat} + [R^{2-}] \cdot \Delta G_{an}$ where $[M^+]$ and $[R^{2-}]$ represent the concentrations of the cation and anion, and ΔG represents the Gibbs free energy of hydration of the ion (as calculated in reference 26).

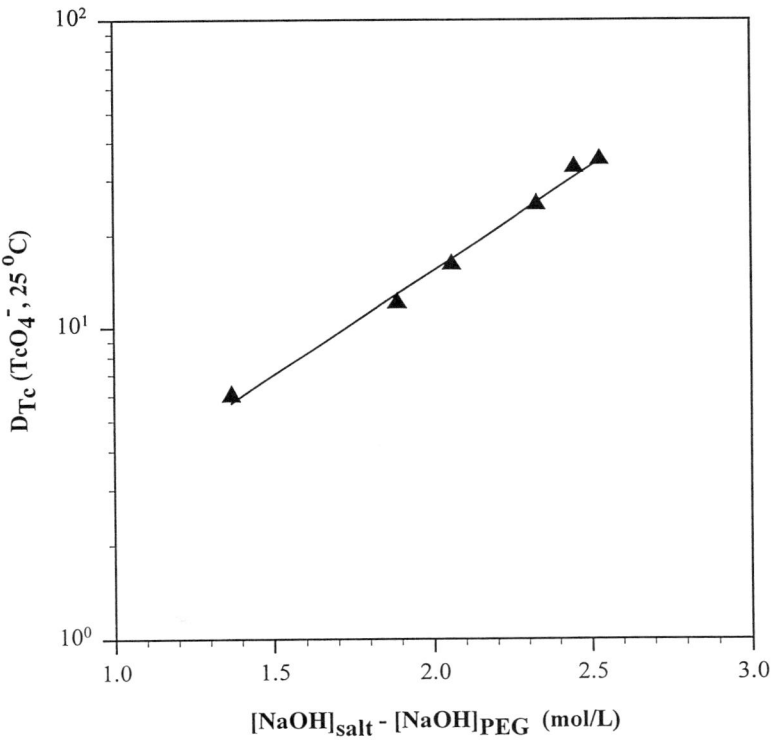

Figure 7. The distribution of TcO$_4^-$ in relation to the salt concentration difference between the top and bottom phases of a PEG-2000/NaOH ABS.

More generally, the following relationship may describe the distribution of TcO_4^- in PEG/salt ABS:

$$\log D_{Tc} = a + b(\Delta salt)$$

where a and b are constants for a given salt.

This suggests that the driving force for the distribution of a solute in an ABS results from the concentration difference of the phase-forming salt between the phases, or more generally, from the divergence of the phase compositions. Although Δ[salt] may be used to predict the distribution of TcO_4^- in ABS, the coefficients a and b must be determined for each salt. Additionally, a phase diagram must be constructed for each salt system employed.

Other methods for predicting solute distribution in ABS have been proposed including the "system stability" (ST) (24), which measures the distance of a particular system from the origin of the phase diagram relative to the distance of the binodal curve from the origin. This method of determining the system stability is shown in Figure 1 where constructing the line OB from the origin to the overall system composition results in a point D where the line intersects the binodal curve. The ratio BD/BO is defined as the system stability. Under most practical circumstances, this measure should be strongly correlated with TLL. However, at extremes of the phase diagram, that is, at high or low volume ratio, it is likely to deviate somewhat from this relationship. The distribution of TcO_4^- in relation to the system stability is shown in Figure 8 for biphasic systems formed with PEG-2000 and three different salts (NaOH, $(NH_4)_2SO_4$, and K_3PO_4). The system stability is closely correlated to the observed distribution coefficient of TcO_4^-, despite the great differences in the relative position of the binodal curves for these PEG/salt systems. The data shown in Figure 8 fit the relationship:

$$\log D_{Tc} = 0.725 + 3.82 ST$$

where ST is the system stability ratio defined above.

As mentioned earlier, changing the molecular weight of PEG used to form an ABS, or changing the temperature, or the salt used, changes the relative position of the binodal curve on the phase diagram and thus the phase incompatibility and the solute distribution ratio. System stability and D_{Tc} have therefore been examined in systems composed of different molecular weights of PEG and formed at different equilibrium temperatures. These systems, shown in Figure 9, were PEG-2000/$(NH_4)_2SO_4$ and PEG-3400/$(NH_4)_2SO_4$ at temperatures of 25 °C and 50 °C. An almost identical relationship is observed for these systems as was observed for the systems shown in Figure 8. It appears that ST may be used to describe the distribution of TcO_4^- in ABS composed of different salts and molecular weights of PEG and at different equilibrium temperatures.

As a measure of the relative divergence of the phase compositions of different biphasic systems and as a predictive tool for determining the partitioning behavior of solutes, system stability is most useful where detailed knowledge of the phase

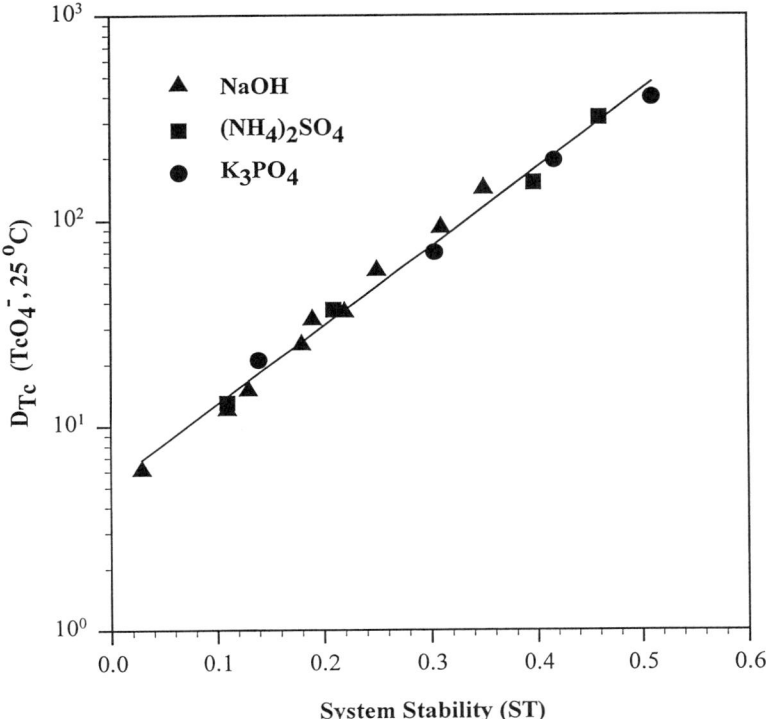

Figure 8. The distribution of TcO_4^- in relation to the System Stability (see text) for ABS composed of PEG-2000 with several different types and concentrations of salts.

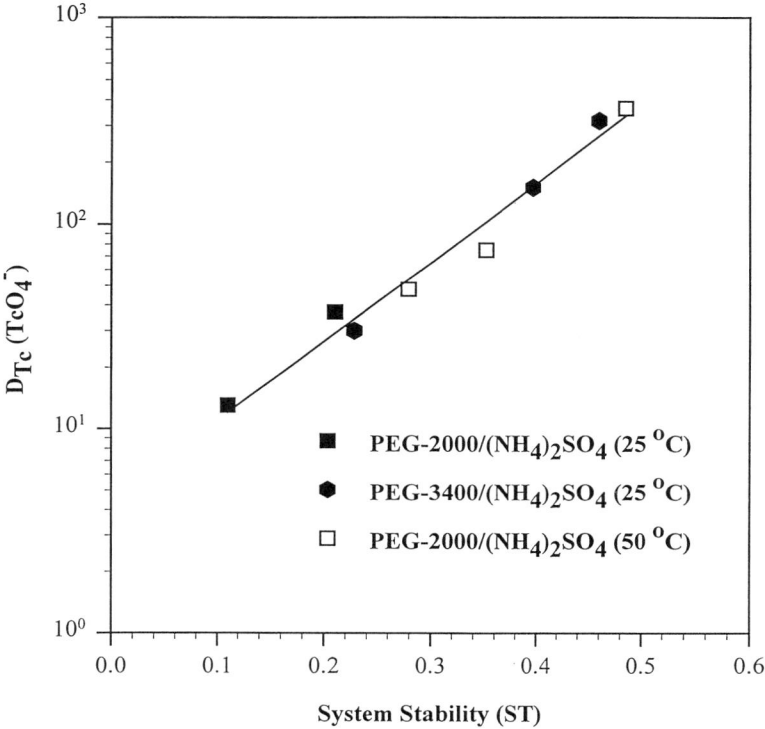

Figure 9. The distribution of TcO_4^- in relation to the System Stability (see text) for ABS composed of different molecular mass PEG fractions at different temperatures in the presence of $(NH_4)_2SO_4$.

diagram beyond a simple coexistence curve is not available. However, it is necessary to know the overall composition of the equilibrium system.

The present methodology, in which ABS are prepared from stock solutions of PEG and salt, provided that two phases are formed, requires knowledge of the phase diagram but not the tie line relationships, or the overall equilibrium composition of the partitioning system. In this case, an approximate correlation of the partitioning behavior with an increase in added salt concentration may be derived from a reformulation of the binodal curve in terms of the slope of the logarithm of the PEG concentration (% w/w) versus the salt concentration (M). The intercept of this line with the molar concentration of the salt gives a parameter M_0 which may be used to plot distribution data for different salt systems as M/M_0 (37). It appears that the latter two parameters (ST and M/M_0) may be independent of PEG molecular weight for metal ions and small organics, that is, where excluded volume effects (38) are relatively unimportant. This correlation is currently being investigated further. Nevertheless, understanding and predicting the performance of liquid/liquid ABS in metal ion partitioning has enabled the facile application of ABEC resins to these same problems as illustrated below.

Comparison of ABS and ABEC Processes for Metal Ion Extraction. Figures 3 to 6 show the distribution of TcO_4^- between the aqueous phases of an ABS composed of M-PEG-5000 and a variety of salts and between a solid phase bearing covalently attached M-PEG-5000 in contact with similar salt containing aqueous phases. Since the water content of these materials is very high (dwcf < 0.2) the distribution values for the ABEC resin may be expected to be almost an order of magnitude higher than for the corresponding ABS. This is because the distribution value for the chromatographic system is expressed in terms of the resin dry weight unlike the ABS distribution coefficient. In addition, Figure 2 shows that D_{Tc} is higher at constant salt concentration for ABS formed with M-PEG-5000 than with PEG-2000 by some smaller additional factor. Since the ABEC resin used here is derivatized with M-PEG-5000, a similar increment in distribution ratio may be expected over PEG-2000/salt systems. Nevertheless, Figures 3-6 show that the ABEC solid phase has a consistently higher uptake (D_W) for TcO_4^- than would be predicted from the distribution (D_{Tc}) in the corresponding liquid/liquid ABS. The reasons for this are not clear, but may, perhaps, be related to the density of the PEG ligands on the polymeric adsorbent surface. This matter is currently under investigation.

Apart from this apparent increase in the distribution ratio for the ABEC system over the corresponding ABS, the other factors of importance in determining the distribution of TcO_4^- in ABS seem to be exactly reproduced in relation to the distribution of TcO_4^- to the ABEC resin. Figure 3 shows the effect of varying the cation in ABS systems composed of M-PEG-5000/M_2SO_4 on D_{Tc}. Also shown in the figure is the D_W of TcO_4^- in contact with the ABEC resin under identical conditions. Allowing for the previously discussed difference in magnitude of the distribution ratios, the behavior of the ABEC system closely follows the behavior of the M-PEG-5000 system in its uptake of this anion.

Figure 4 shows the distribution ratio of TcO_4^- in relation to the ΔG_{hyd} of the cation for both ABEC and ABS systems for different cations present as 1.2 M

M_2SO_4. The distribution ratio of TcO_4^- is dependent on ΔG_{hyd} of the phase-forming cation, increasing with cations having increasingly negative ΔG_{hyd} in both the ABEC and ABS modes of operation. This similarity in partitioning behavior between the two modes of operation, ABS and ABEC, is again evident in Figure 5. Here the cation is held constant whilst the anion is changed. Under these circumstances the distribution ratio of TcO_4^- increases for both ABEC and ABS with increasingly negative free energies of hydration of the anion in exactly the same order $SeO_4^{2-} < SO_4^{2-} < CO_3^{2-}$. Finally, Figure 6 shows that the total ΔG_{hyd} of the stock salt solution used to form the biphase or promote the ABEC interaction may be used to predict the distribution of the TcO_4^- anion with both the ABEC resin and in the ABS. Here the distribution ratio for TcO_4^- is shown for several different ABS and for the ABEC resin. The molecular weight and concentration of PEG used to form the ABS and the amount of added ABEC is held constant whilst the type and concentration of added salt is varied as shown. There is a strong relationship between both D_{Tc} and D_W and the total free energy of hydration, calculated as the sum of the ΔG_{hyd} of the ions comprising the salt multiplied by their concentration. Not only is their a good correlation between the total ΔG_{hyd} of the salt stock solution and D_{Tc} and D_W, but the partitioning response is very similar in both situations. This is a useful simplification and implies the facile translation of ABS partitioning experiences to the design and operation of partitioning operations utilizing the ABEC resin.

Conclusions

Aqueous biphasic systems and aqueous biphasic extraction chromatography are applicable to many metal ion separations provided the free energy of hydration of the target species is appropriate. This has been demonstrated comprehensively for the pertechnetate anion. Recently, the partition of small organic molecules and industrial dye molecules has also been demonstrated (22,23). This allows the selection and design of a range of specific extractants suitable for use in ABS and ABEC paralleling those found in conventional solvent extraction processes, thus, further extending the range of solutes which may be extracted through the application of this technology. The process engineer will then be faced with a real choice in process design in the field of liquid/liquid extraction, between the traditional organic solvent-based approach and a wholly aqueous alternative. The advantages of ABEC systems over liquid/liquid ABS lie principally in the facile retention and reuse of the extracting phase. The essential similarity, if not complete identity, of the two processes has been demonstrated. However, it must be expected that there will be differences between the two operations in practice, due to differences in kinetics and mass transfer arising from the essential physical differences between a liquid phase and a porous solid support. These processes represent a wholly aqueous liquid/liquid extraction scheme for the recovery of a wide range of inorganic and organic species. As such, they have great potential for application to a wide range of separations problems, in particular, they may find application where the necessity to replace VOCs with alternative aqueous processes is a major concern.

Acknowledgments

This work is supported by the National Science Foundation (Grant CTS-9522159) and the Division of Chemical Sciences, Office of Basic Energy Sciences, Office of Energy Research, U.S. Department of Energy (Grant No. DE-FG02-96ER14673).

Literature Cited

(1) *Technology Vision 2020, The U.S. Chemical Industry;* American Chemical Society, American Institute of Chemical Engineers, The Chemical Manufacturers Association, The Council for Chemical Research, and The Synthetic Organic Chemical Manufacturers Association: Washington, DC, 1996.
(2) Sekine, T.; Hasegawa, Y. *Solvent Extraction Chemistry, Fundamentals and Applications*; Marcel Dekker: New York, 1977.
(3) Albertsson, P.-Å., *Nature* **1958**, *182*, 709.
(4) *Partitioning in Aqueous Two-Phase Systems*; Walter, H.; Brooks, D. E.; Fisher, D., Eds.; Academic Press: Orlando, FL, 1985.
(5) *Aqueous Two-Phase Systems;* Walter, H.; Johansson, G., Eds.; In Methods in Enzymology; Abelson, J. N.; Simon, M. I., Eds.; Academic Press: San Diego, CA 1994; Vol. 228.
(6) Rogers, R. D.; Bond, A. H.; Bauer, C. B.; Zhang, J.; Griffin, S. T. *J. Chromatogr., B: Biomed. Appl.* **1996**, *680*, 221.
(7) Rogers, R. D.; Bond, A. H.; Bauer, C. B. *Sep. Sci. Technol.* **1993**, *28*, 101.
(8) Rogers, R. D.; Bond, A. H.; Bauer, C. B.; Zhang, J.; Jezl, M. L.; Roden D. M.; Rein, S. D.; Chomko, R. R. In *Aqueous Biphasic Separations: Biomolecules to Metal Ions*; Rogers, R. D.; Eiteman M. A., Eds.; Plenum: New York, 1995; p. 1.
(9) Rogers, R. D.; Zhang, J. In *Ion Exchange and Solvent Extraction*; Marinsky, J. A.; Marcus, Y., Eds.; Marcel Dekker: New York, 1997, Vol. 13; p. 141.
(10) Rogers, R. D.; Zhang, J.; Griffin, S. T. *Sep. Sci. Technol.* **1997**, *32*, 699.
(11) Rogers, R. D.; Bond, A. H.; Bauer, C. B.; Zhang, J.; Rein, S. D.; Chomko, R. R.; Roden, D. M. *Solvent Extr. Ion. Exch.* **1995**, *13*, 689.
(12) Rogers, R. D.; Zhang, J.; Bond, A. H.; Bauer, C. B.; Jezl, M. L.; Roden, D. M. *Solvent Extr. Ion Exch.* **1995**, *13*, 665.
(13) Rogers, R. D.; Bond, A. H.; Zhang, J.; Bauer, C. B. *Appl. Radiat. Isot.* **1996**, *47*, 497.
(14) Rogers, R. D.; Zhang, J. *J. Chromatogr., B: Biomed. Appl.* **1996**, *680*, 231.
(15) Rogers, R. D.; Bond, A. H.; Zhang, J.; Horwitz, E. P. *Sep. Sci. Technol.* **1997**, *32*, 867.
(16) Rogers, R. D.; Bond, A. H.; Bauer, C. B. In *Solvent Extraction in the Process Industries, Proceedings of ISEC '93*; Logsdail D. H.; Slater, M. J., Eds.; Elsevier: London, 1993, Vol. 3; p. 1641.
(17) Rogers, R. D.; Bond, A. H.; Bauer, C. B. *Sep. Sci. Technol.* **1993**, *28*, 139.
(18) Rogers, R. D.; Bauer, C. B.; Bond, A. H. *J. Alloys Compd.* **1994**, *213/214*, 305.
(19) Rogers, R. D.; Bond, A. H.; Bauer, C. B. *Pure Appl. Chem.* **1993**, *65*, 567.

(20) Rogers, R. D.; Bauer, C. B.; Bond, A. H. *Sep. Sci. Technol.* **1995**, *30*, 1203.
(21) Rogers, R. D.; Bauer, C. B. *J. Chromatogr., B: Biomed. Appl.* **1996**, *680*, 237.
(22) Rogers, R, D.; Willauer, H. D.; Griffin, S. D.; Huddleston, J. G. *J. Chromatogr. B* **1997**, (in press).
(23) Huddleston, J. G.; Willauer, H. D.; Boaz, K.; Rogers, R. D. *J. Chromatogr. B* **1997**, (in press).
(24) Hustedt, H.; Kroner, K. H.; Kula, M.-R. In *Partitioning in Aqueous Two-Phase Systems*; Walter, H.; Brooks, D. E.; Fisher, D., Eds.; Academic Press: Orlando, FL, 1985; p. 529.
(25) Ananthapadmanabhan, K. P.; Goddard, E. D. *Langmuir* **1987**, *3*, 25.
(26) Marcus, Y. *J. Chem. Soc., Faraday Trans.* **1991**, *87*, 2995.
(27) Robbins, L. A. In *Handbook of Separation Techniques for Chemical Engineers;* Schweitzer, P. A., Ed.; McGraw-Hill: New York, 1996, 3rd Edition; p. 1-419.
(28) Huddleston, J. G.; Griffin, S. T.; Zhang, J.; Willauer, H. D.; Rogers, R. D. In *Aqueous Two-Phase Systems;* Kaul, R., Ed.; Methods in Biotechnology; Walker, J. M., Ed.; Humana Press: Totowa, NJ, 1997; (in press).
(29) Steigman, J.; Eckelman, W. C. *The Chemistry of Technetium in Medicine;* National Academy Press: Washington, DC, 1992.
(30) Boyd, R. E. *Radiochimica Acta* **1982**, *30*, 123.
(31) Jones, C. J. In *Comprehensive Coordination Chemistry;* Wilkinson, G.; Gillard, R. D.; McCleverty, J. A., Eds.; Springer-Verlag: Berlin, 1987 Vol. 6; p. 881.
(32) Eiteman, M. A.; Gainer, J. L. *Chem. Eng. Comm.* **1991**, *105*, 171.
(33) Bamberger, S. B.; Brooks, D. E.; Sharp, K. A.; Van Alstine, J. M.; Webber, T. J. In *Partitioning in Aqueous Two-Phase Systems. Theory, Methods, Uses, and Applications to Biotechnology;* Walter, H.; Brooks, D. E.; Fisher, D., Eds.; Academic Press: Orlando, FL, 1985, p. 85.
(34) Huddleston, J. G.; Lyddiatt, A. *Appl. Biochem. Biotechnol.* **1990**, *26*, 249.
(35) de Belval, S.; le Breton, B.; Huddleston, J.; Lyddiatt, A. *J. Chromatogr.B* **1997**, (in press).
(36) Asenjo, J. A.; Turner, R. E.; Mistry, S. L.; Kaul, A. *J. Chromatogr., A.* **1994**, *668*, 129.
(37) Huddleston, J. G.; Rogers, R. D. **1997**, (unpublished results).
(38) Abbott, N. L.; Blankschtein, D.; Hatton, T. A. *Bioseparation* **1990**, *1*, 191.

EXTRACTANT DESIGN AND SYNTHESIS

Chapter 6

A Molecular Mechanics Method for Predicting the Influence of Ligand Structure on Metal Ion Binding Affinity

Benjamin P. Hay

Environmental Molecular Sciences Laboratory, Pacific Northwest National Laboratory, Richland, WA 99352

This paper presents a molecular mechanics method for the quantitative analysis of ligand binding site organization for metal ion complexation. For series of polydentate ligands bearing a constant number and type of donor atom, the method can be used to identify the connectivity that will yield the highest affinity for a given metal ion. The utility of the method is demonstrated through correlations of alkali and alkaline earth cation binding affinities with the ligand strain energies for five series of multidentate ether ligands.

A new methodology for the application and interpretation of molecular mechanics calculations in the structural design of ligands has been described recently (1). In this paper, this molecular mechanics methodology is used to analyze the differences in metal ion binding affinity for five series of polydentate ligands. These ligand series were selected for two reasons. First, each series of ligands consists of members that contain the same number of ligating donor groups, but vary in the way that the donors are connected together. Second, relative metal ion binding affinities have been determined for each series of ligands under a constant set of experimental conditions providing a consistent set of data for the comparison. Under such constraints, it is reasonable to expect that differences in metal complexation would largely result from structural factors. Correlations between metal ion binding affinity and ligand strain will demonstrate this to be so.

A Description of the Methodology

Complexation of a metal ion to a ligand often causes changes to the ligand's structure that introduce steric strain within the ligand. A design criterion for ligands that will complex metal ions more strongly is to choose a ligand structure that minimizes this strain, that is, ligands that are sterically efficient (2). In the approach described here, molecular mechanics calculations are used to define the degree of ligand binding site organization in terms of two ligand strain energies. This involves a consideration of two types of structural change: conformational reorganization and distortion of the binding conformer.

The conformational mobility of the ligand can affect both its metal ion selectivity and binding affinity. As a result, the imposition of conformational constraints is an often-cited criterion for metal-selective ligand design (*2-6*). Many ligands exhibit more than one conformer (different conformers are distinguished by gross changes to torsion angles, e.g., from *gauche* to *anti*). Different conformers of the same ligand can possess quite different cavity sizes and shapes. In such cases, the metal ion selectivity of the ligand will be compromised if different conformers prefer different metal ions. In addition, if the ligand prefers one conformer when uncomplexed and a different conformer when in a metal complex, then the ligand must undergo a conformational reorganization to form the metal complex. This process costs energy, which can significantly lower the observed binding affinity. The strain energy associated with this conformer change can be quantified with molecular mechanics calculations. Herein, the quantity ΔU_{conf} is a strain energy defined as the change in ligand steric energy as it goes from the preferred uncomplexed conformer to the binding conformer.

Further destabilizing ligand strain may occur on metal ion complexation to the binding conformer. Structural distortion may result from movements of the ligand donor atoms in an attempt to satisfy metal ion - donor atom distance preferences (cavity size mismatch (*2,3,6*)), stereoelectronic preferences of the metal ion (topology mismatch (*4,6,7*)), or ligand donor atom geometry requirements (lone pair and/or dipole misalignment (*2,8*)). The degree of structural distortion to the binding conformation upon metal ion complexation provides a measure of the ligand's complementarity for the metal ion (*1*).

The principle of complementarity as originally stated is, "to complex, hosts must have binding sites which cooperatively contact and attract binding sites of guests without generating strong nonbonded repulsions" (*9*). A ligand with perfectly complementary binding sites will not experience unfavorable steric interactions (neither strong nonbonded repulsions nor distortions from preferred bond lengths, bond angles, and torsion angles) as a result of complexation. Metal ion complexation to noncomplementary binding sites will cause distortions to the ligand structure. This generates steric strain within the ligand. Herein, the quantity ΔU_{comp} is a strain energy defined as the change in ligand steric energy as it goes from the uncomplexed binding conformer to the metal complex.

The sum of the two structural reorganization components is defined as the ligand's reorganization energy, $\Delta U_{reorg} = \Delta U_{conf} + \Delta U_{comp}$. This quantity is a molecular mechanics strain energy that provides a quantitative measurement of the degree of a ligand's binding site organization for metal ion complexation. Ligands that are poorly organized for binding will exhibit large ΔU_{reorg} values; ligands highly organized for binding will exhibit small ΔU_{reorg} values. A ligand that is perfectly organized for binding, that is, constrained to a binding conformation with complementary sites, would exhibit a ΔU_{reorg} of zero.

Computational Methods

Molecular mechanics calculations were conducted with the MM3(96) program (*10*) using an extended parameter set developed and extensively validated for ethers and their complexes with the alkali and alkaline earth cations (*11-13*). Conformations for the free ligands and their metal complexes were taken from crystal structure data where available. In the absence of crystal structure data, minimum steric energy conformations were determined by conformational searches using methods described elsewhere (*14-16*).

Values of ΔU_{reorg} reported are the sum of ΔU_{conf} and ΔU_{comp} as defined above. In cases where no conformational reorganization is observed, ΔU_{conf} is zero. Otherwise, ΔU_{conf} is calculated as the difference in steric energy between the uncomplexed ligand conformer and the binding conformer of the ligand. ΔU_{comp} is calculated as the difference in steric energy between the binding conformer and the bound ligand. The latter quantity is computed by (i) optimizing the metal ion complex, (ii) removing the metal ion, and (iii) obtaining the initial energy of the resulting structure.

Results and Discussion

Potassium Formation Constants with Hexadentate Crown Ethers. The first series of ligands (**1 - 11**, Figure 1) were examined in a prior molecular mechanics study (*17*). Formation constants for potassium complexes (equation 1) have been reported for each of the ligands in the same solvent and at the same temperature (methanol solvent, 25 °C), providing a consistent set of data for comparison (*18*). These formation constants (log K values, Figure 1) span four orders of magnitude.

$$K^+ + L \rightleftharpoons KL^+ \quad (1)$$

In the prior study (*17*), a linear correlation was obtained by plotting log K values against the difference in steric energy between the uncomplexed ligand and the potassium complex. Following the current methodology, ΔU_{reorg} values (see Figure 1) have been computed for potassium complexation by **1 - 11** using conformations from

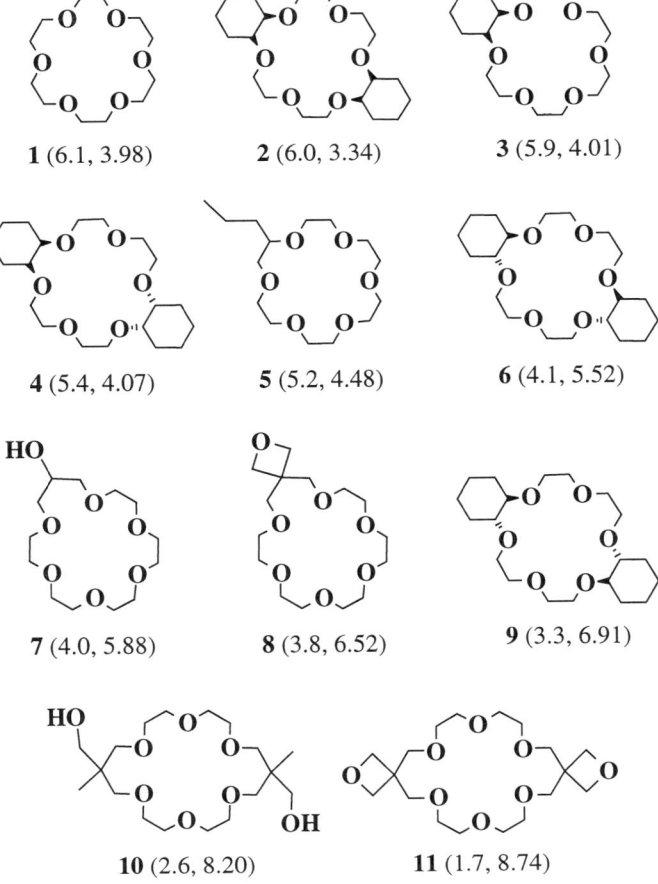

Figure 1. Series of eleven hexadentate ether ligands. Parentheses contain the log K value and ΔU_{reorg} (kcal/mol) for potassium complexation, respectively.

the prior study. A similar correlation is obtained when the log K values are plotted against ΔU_{reorg} values (Figure 2).

In the ΔU_{reorg} values for **1 - 11**, the majority of the ligand strain occurs in the ΔU_{comp} term. Structural distortions of the binding conformer on metal complexation result primarily from attempts to achieve preferred trigonal planar ether oxygen geometry rather than attempts to achieve preferred K-O distances (11). For example, while **1** has a cavity that gives near optimal K-O distances, the C-O-C groups do not point to the center of the cavity. In the D_{3d} conformation, the oxygens are positioned alternately above and below the plane of the ring. When potassium occupies the cavity, the oxygens all move in such a way as to be more in the plane of the macrocycle, that is, to attempt to achieve trigonal planar geometries with respect to the metal ion. This structural change generates 3.98 kcal/mol of destabilizing ligand strain. Thus, the D_{3d} conformation of **1** does not provide an optimum arrangement of ether donor groups for potassium complexation. Extrapolation of the linear correlation in Figure 2 to a ΔU_{reorg} value of zero suggests that a ligand with an optimum arrangement of six ether oxygen donor groups would exhibit a log K value of 8.8 in methanol, that is, a binding constant nearly three orders of magnitude higher than that of **1**.

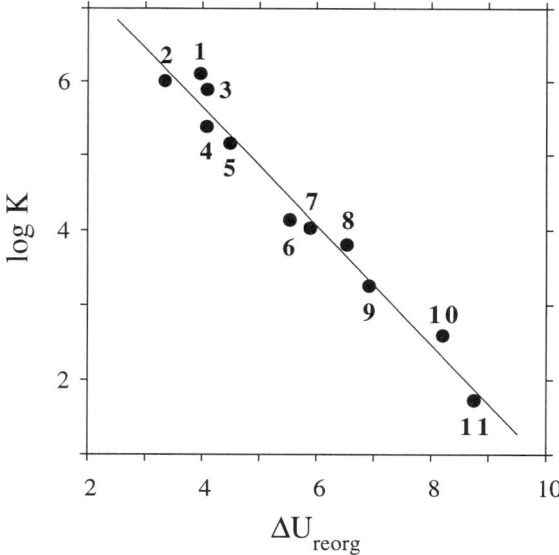

Figure 2. Plot of log K versus ΔU_{reorg} for potassium complexation by hexadentate ether ligands shown in Figure 1 (r = 0.986).

Lithium Extraction Constants with 14-Crown-4 Derivatives. The second series of ligands (**12 - 16**, Figure 3) consists of a set of five 14-crown-4 derivatives that differ in the alkylation of the ethylene and propylene bridges. Thermodynamic equilibrium constants (equation 2) have been determined for the transfer of $Li^+ClO_4^-$ from aqueous solution to nitrobenzene (Moyer, B. A.; Sachleben, R. A., Oak Ridge National Laboratory, unpublished results). These extraction constants (log K_{ex} values, Figure 3) span two orders of magnitude.

$$Li^+_{(aq)} + ClO_4^-{}_{(aq)} + L_{(org)} \rightleftharpoons LiL^+_{(org)} + ClO_4^-{}_{(org)} \qquad (2)$$

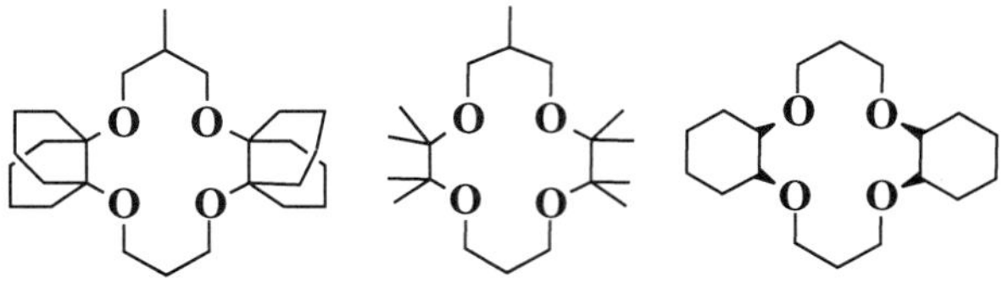

12 (0.81, 5.27)   **13** (0.10, 7.94)   **14** (-0.31, 8.93)

15 (-1.1, 10.99)   **16** (-1.3, 10.65)

Figure 3. Series of five 14-crown-4 ligands. Parentheses contain the log K_{ex} value and ΔU_{reorg} (kcal/mol) for lithium complexation, respectively.

Values of ΔU_{reorg} were calculated for this series (see Figure 3). In this case, conformations for the uncomplexed ligands **12** - **16** and for lithium complexes with **12** - **15** were taken from crystal structures (*19-21*). A conformational search was performed to locate the lowest steric energy lithium complex for **16**. A plot of log K_{ex} versus ΔU_{reorg} (Figure 4) provides a second example of how ligand strain correlates with metal ion binding affinity.

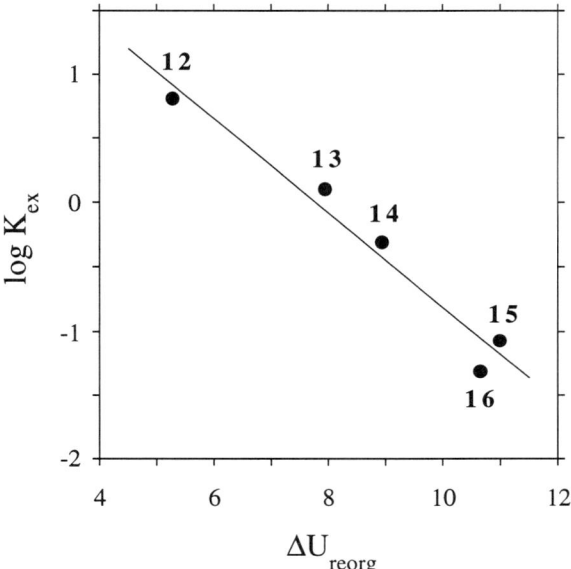

Figure 4. Plot of log K_{ex} versus ΔU_{reorg} for lithium complexation by the 14-crown-4 ligands shown in Figure 3 (r = 0.978).

In **12 - 16** both ΔU_{conf} and ΔU_{comp} contribute to the ΔU_{reorg} values. In all cases the ligand undergoes a conformational change prior to lithium complexation. The highest log K_{ex} occurs with **12**, which exhibits the lowest ΔU_{conf} (0.97 kcal/mol), while the lowest log K_{ex} values occur with **15** and **16**, which exhibit the highest values of ΔU_{conf} (3.88 and 6.43 kcal/mol, respectively). In addition to the ligand strain associated with conformational reorganization, lithium complexation to the binding conformer adds from 4.30 to 6.77 kcal/mol additional ligand strain in the ΔU_{comp} term. In the 14-crown-4 binding conformation, the lithium perches atop the four oxygens, and no change in ligand structure is required to attain preferred Li-O distances. Thus, as with the previous example, the ΔU_{comp} contribution primarily results from ligand distortion as the four ether donors attempt to achieve their desired orientation with respect to lithium. Extrapolation of the linear correlation in Figure 4 to a ΔU_{reorg} value of zero suggests that a ligand with an optimum arrangement of four ether donor groups would exhibit a log K_{ex} value of 2.8 under these conditions, that is, an extraction constant two orders of magnitude higher than that of the current best member of the series, **12**.

Lithium Formation Constants with 14-crown-4 Derivatives. The third series of ligands (**17 - 21**, Figure 5) is a set of five 14-crown-4 derivatives that differ in the alkyl substitution on one ethylene bridge. Solvent extraction studies showed that the extraction efficiency for lithium decreased in the order **17 > 18 > 19 > 20 > 21** (22). Formation constants for lithium complexation by **18 - 21** (equation 3) were determined in cyclohexanone at 80 °C (22). The ordering of the formation constants (log K, Figure 5), which span nearly two orders of magnitude, parallels that of the extraction efficiencies.

$$Li^+ + L \rightleftharpoons LiL^+ \qquad (3)$$

Values of ΔU_{reorg} were calculated for this series (see Figure 5). In the absence of crystal structure data, conformer searches were performed to identify the low energy forms of the uncomplexed ligands and their lithium complexes. A plot of log K versus ΔU_{reorg} (Figure 6) provides a third example of how ligand strain correlates with metal ion binding affinity. Although a log K value was not reported for the ligand with the lowest ΔU_{reorg}, **17**, calculation results are consistent with the observation that **17** gives the highest extraction efficiency for this series. The correlation shown in Figure 6 yields an estimate of log K = 4.5 for lithium complexation by **17** in cyclohexanone at 80 °C.

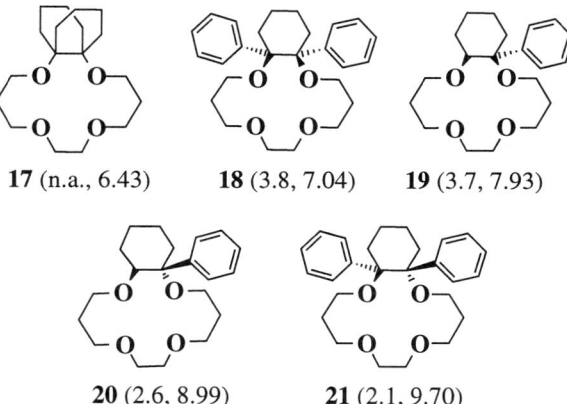

Figure 5. Series of five 14-crown-4 ligands. Parentheses contain the log K value and ΔU_{reorg} (kcal/mol) for lithium complexation, respectively.

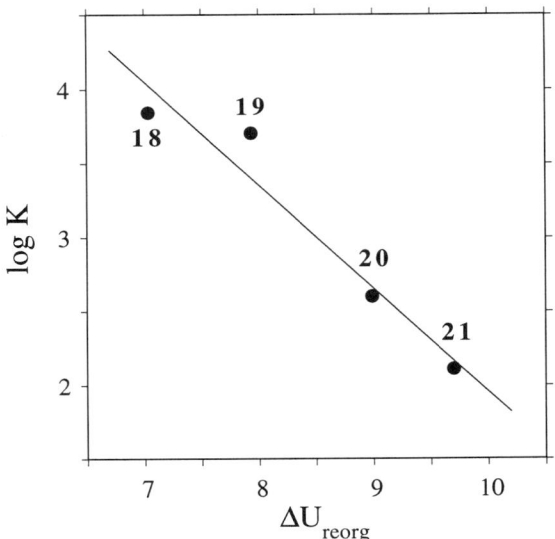

Figure 6. Plot of log K versus ΔU_{reorg} for lithium complexation by the ligands shown in Figure 5 (r = 0.968).

Strontium Distribution Coefficients with Dicyclohexano-18-Crown-6 Derivatives. The fourth series of ligands (**22** - **26**, Figure 7) are a set of five dicyclohexano-18-crown-6 derivatives that differ in the stereochemistry of the cyclohexyl rings and the placement of t-butyl substituents. Distribution coefficients (equation 4) have been determined for strontium extraction from 1 M nitric acid to 1-octanol (Dietz, M. L., Argonne National Laboratory, unpublished results). The distribution coefficients (log D_{Sr}, Figure 7), which span two orders of magnitude, were determined under identical conditions (solvent composition, reagent concentrations, and temperature) and were corrected for the formation of aqueous phase complexes.

$$D_{Sr} = [Sr^{2+}](org)/[Sr^{2+}](aq) \qquad (4)$$

Values of ΔU_{reorg} were calculated for this series (see Figure 7). Conformer searches were performed to identify the low energy forms of the uncomplexed ligands and their strontium complexes. In the case of **24** and **25**, calculated global minimum conformations for both the free ligands and their strontium complexes are fully consistent with available crystal structure data (*23-34*). A plot of log D_{Sr} versus ΔU_{reorg} (Figure 8) provides a fourth example of how ligand strain correlates with metal ion binding affinity.

In **22** - **26** both ΔU_{conf} and ΔU_{comp} contribute to the ΔU_{reorg} values. However, in all cases, it is the ΔU_{comp} term that dominates and, as with the potassium complex of **1**, structural distortion to the binding conformer arises primarily from attempts to meet the preferred planar orientation of the ether donor groups. Extrapolation of the linear correlation in Figure 8 to a ΔU_{reorg} value of zero suggests that a ligand with an optimum arrangement of six ether donor groups would exhibit a log D_{Sr} value of 6.8 under these conditions, that is, a D_{Sr} value five orders of magnitude higher than that of the current best member of the series, **22**.

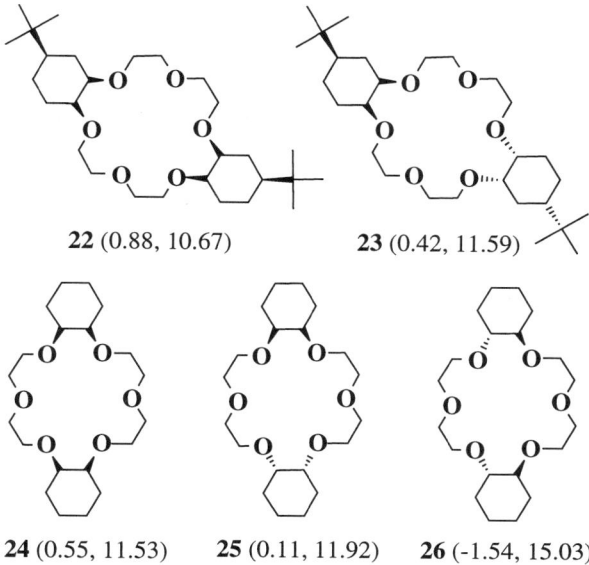

22 (0.88, 10.67)   **23** (0.42, 11.59)

24 (0.55, 11.53)   **25** (0.11, 11.92)   **26** (-1.54, 15.03)

Figure 7. Series of five dicyclohexano-18-crown-6 ligands. Parentheses contain the log D_{Sr} value and ΔU_{reorg} for strontium complexation, respectively.

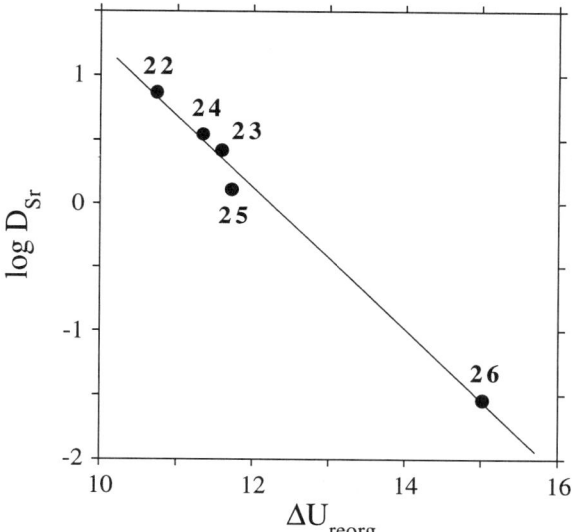

Figure 8. Plot of log D_{Sr} versus ΔU_{reorg} for strontium complexation by the ligands shown in Figure 7 (r = 0.995).

Sodium Formation Constants with Hexadentate Polyether Ligands. The fifth collection of ligands (**27** - **34**, Figure 9) is a set of eight hexadentate polyethers bearing combinations of aliphatic and methoxyaryl ether donors. Formation constants for sodium complexation (equation 5) were determined in water-saturated chloroform (*9,35-37*). These formation constants (log K values, Figure 9) span ten orders of magnitude.

$$Na^{+}\ picrate^{-} + L \rightleftharpoons (NaL)^{+}\ picrate^{-} \qquad (5)$$

Values of ΔU_{reorg} were calculated for this series (see Figure 9). With the exception of **27,** for which there are crystal structures of the free ligand and the sodium complex (*38*), conformer searches were performed to identify the low energy forms of the uncomplexed ligands and their sodium complexes. A plot of log K versus ΔU_{reorg} (Figure 10) provides a fifth example of how ligand strain correlates with metal ion binding affinity.

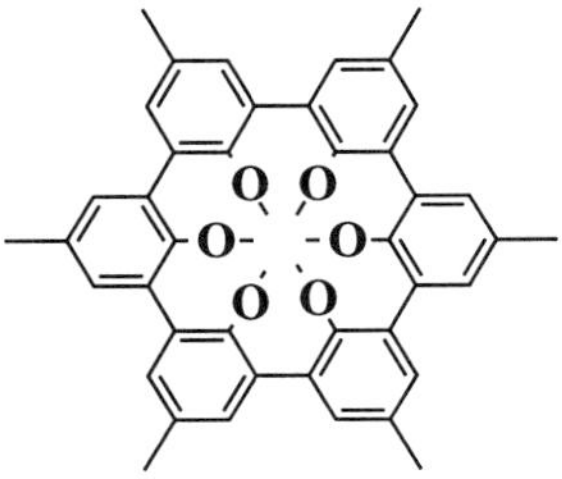

27 (14.1, 0.94)

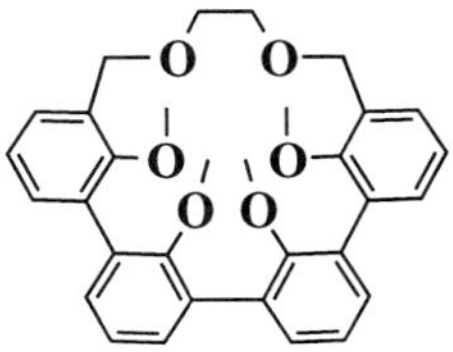

28 (9.9, 3.35)

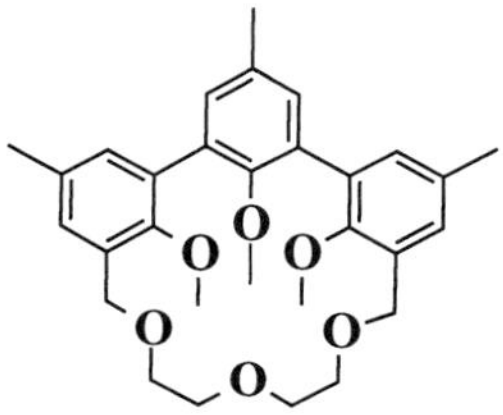

29 (9.2, 4.54)

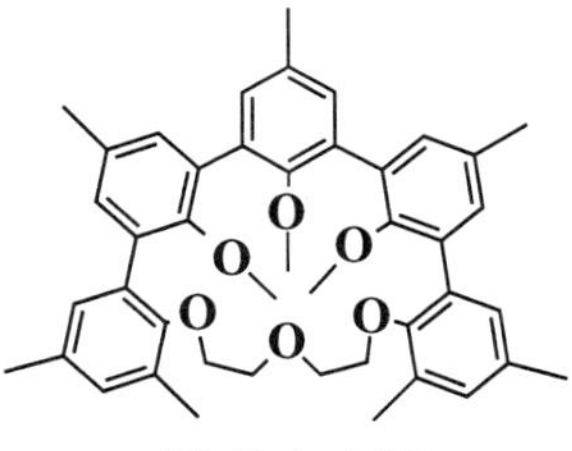

30 (8.6, 4.72)

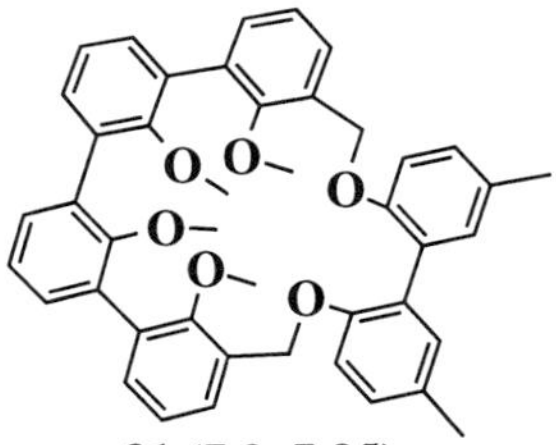

31 (7.2, 7.85)

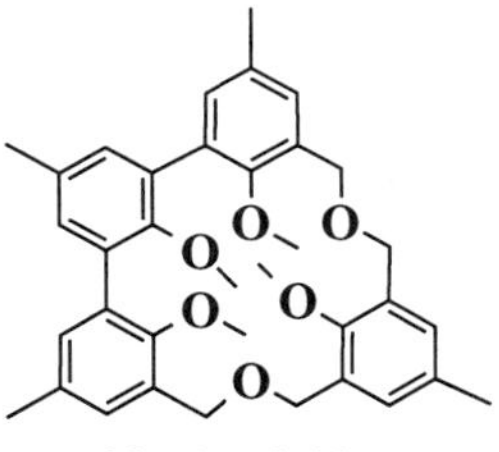

32 (5.7, 8.33)

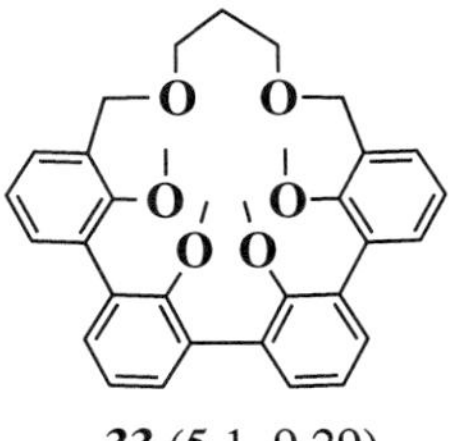

33 (5.1, 9.29)

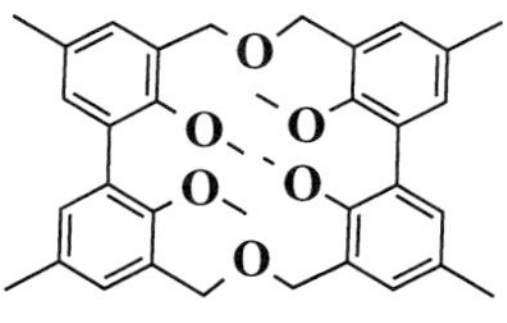

34 (3.4, 9.98)

Figure 9. Series of eight hexadentate ether ligands. Parentheses contain log K and ΔU_{reorg} (kcal/mol) for sodium complexation, respectively.

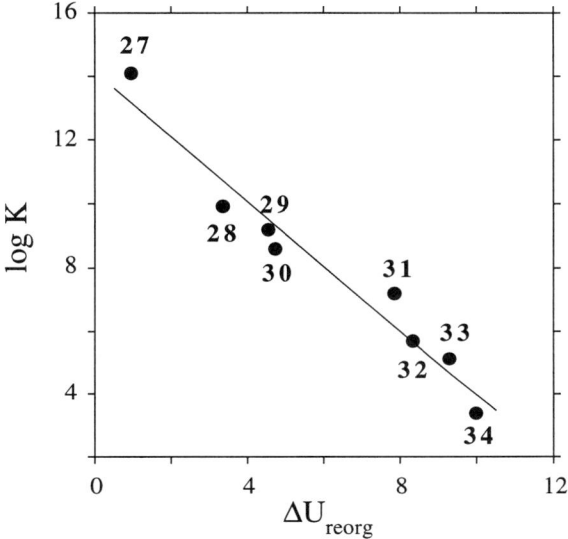

Figure 10. Plot of log K versus ΔU_{reorg} for sodium complexation by the ligands shown in Figure 9 (r = 0.975).

The correlation shown in Figure 10 strongly suggests that the different sodium binding affinities observed with this series of ligands are primarily due to steric rather than electronic effects. From this result it may be inferred that the two types of ether donor groups that occur in these ligands (i.e., aliphatic and conjugated), have a similar Lewis basicity. This inference is supported by the measured gas phase proton affinities for ethoxyethane (200.3 kcal/mol) and methoxybenzene (200.2 kcal/mol) (39).

While the modeling results indicate that the weakest ligands, **32 - 34**, undergo conformational reorganization prior to binding, **27 - 31** do not. In all cases, it is the ΔU_{comp} term that makes the largest contribution to ΔU_{reorg}. Unlike the preceding crown ether examples, ligand structural distortion on sodium complexation is caused by attempts to satisfy *both* Na-O distance *and* oxygen orientational preferences. This difference is caused by the fact that in **27 - 34** the metal ion resides in the center of a three-dimensional cavity that is not always sized correctly, whereas in **1 - 26**, the metal ions examined in this study either fit nicely within the two-dimensional cavity (K^+ and Sr^{2+}) or perch above it (Li^+).

Ligand **27**, often cited for its high degree of preorganization, exhibits the lowest value of ΔU_{reorg} observed of the 34 cases examined in this study, 0.94 kcal/mol. The correlation in Figure 10 suggests that it is not possible to achieve significant further enhancement of sodium binding affinity by altering the connectivity of six ether donor groups.

Conclusions

For the polyether ligands examined in this study, ligand strain may arise from both conformational reorganization and structural distortion of the binding conformer. The latter component is always present and is often the major contribution to the ligand strain. With the alkali and alkaline earth cations, structural distortion of the binding conformer occurs as the ligand attempts to satisfy M-O distance preferences and oxygen orientation preferences. As noted elsewhere (*11*), in crown ethers such as 1 - 26, it is the attempt to achieve trigonal planar oxygen geometry (i.e., the attempt to align the ether dipoles with the metal ion) that is the major cause of the structural distortion. Therefore, an important criterion in ligand design for enhanced complex stability is to choose a connecting structure that provides the preferred orientation of the donor atoms with respect to the metal ion.

When the binding conformation fails to provide preferred donor atom geometries, significant decreases in complex stability will result. The ethylene bridges that occur in the majority of polydentate ethers prevent the oxygen donor groups from achieving their preferred trigonal planar geometries (*1,8*). It can be concluded that the design of polydentate ether ligands to achieve the highest metal ion binding affinity must involve the use of other connecting structures, for example, the methoxyaryl groups in **27**.

This paper has presented a molecular mechanics method for the assessment of the degree of ligand binding site organization for metal ion complexation. The method involves the determination of ligand strain incurred on complexation of a metal ion. The five examples presented here demonstrate that the quantity ΔU_{reorg} is a very useful tool in predicting the relative metal ion binding affinity for series of polyether ligands bearing the same number of donor groups.

Acknowledgments

This research was supported in part by the Efficient Separations and Processing Crosscutting Program of the Office of Science and Technology, within the U. S. Department of Energy's Office of Environmental Management, and in part by the Environmental Management Science Program under direction of the U. S. Department of Energy's Office of Basic Energy Sciences (ER-14), Office of Energy Research and the Office of Science and Technology (EM-52), Office of Environmental Management. Pacific Northwest National Laboratory is operated for the U. S. Department of Energy by Battelle Memorial Institute under Contract DE-AC06-76RLO 1830.

Literature Cited

(1) Hay, B. P.; Zhang, D.; Rustad, J. R. *Inorg. Chem.* **1996**, *35*, 2650.
(2) Hancock, R. D.; Martell, A. E. *Chem. Rev.* **1989**, *89*, 1875.
(3) Cram, D. J. *Angew. Chem., Intl. Engl. Ed.* **1986**, *25*, 1039.
(4) Potvin, P. G.; Lehn, J.-M. *Prog. Macrocycl. Chem.* **1987**, *3*, 167.
(5) Garrett, T. M.; McMurray, T. J.; Hosseini, M. W.; Reyes, Z. E.; Hahn, F. E.; Raymond, K. N. *J. Am. Chem. Soc.* **1991**, *115*, 2965.
(6) Busch, D. H. In *Transition Metals in Supramolecular Chemistry*; Fabbrizzi, L.; Poggi, A. D., Eds.; Kluwer Academic Publishers: Dordrecht, The Netherlands, 1994, p 55.
(7) Pletnev, I. V. *Russ. J. Coord. Chem.* **1996**, *22*, 333.
(8) Hay, B. P.; Rustad, J. R. *Supramol. Chem.* **1996**, *6*, 383.
(9) Cram, D. J.; Lein, G. M. *J. Am. Chem. Soc.* **1985**, *107*, 3657.
(10) The program may be obtained from Tripos Associates, 1699 S. Hanley Road, St. Louis, MO 63144 for commercial users, and it may be obtained from the Quantum Chemistry Program Exchange, Mr. Richard Counts, QCPE, Indiana University, Bloomington, IN 47405, for non-commercial users.
(11) Hay, B. P.; Rustad, J. R. *J. Am. Chem. Soc.* **1994**, *116*, 6316.
(12) Hay, B. P.; Yang, L.; Lii, J.-H.; Allinger, N. L. *THEOCHEM, J. Mol. Struct.* **1997**, in press.
(13) Hay, B. P.; Yang, L.; Zhang, D.; Rustad, J. R.; Wasserman, E. *THEOCHEM, J. Mol. Struct.* **1997**, in press.

(14) Hay, B. P.; Rustad, J. R.; Zipperer, J. P.; Wester, D. W. *THEOCHEM, J. Mol. Struct.* **1995**, *337*, 39.
(15) Paulsen, M. D.; Hay, B. P. *THEOCHEM, J. Mol. Struct.* **1997**, in press.
(16) Paulsen, M. D.; Rustad, J. R.; Hay, B. P. *THEOCHEM, J. Mol. Struct.* **1997**, in press.
(17) Hay, B. P.; Rustad, J. R.; Hostetler, C. J. *J. Am. Chem. Soc.* **1993**, *115*, 11158.
(18) Izatt, R. M.; Bradshaw, J. S.; Nielsen, S. A.; Lamb, J. D.; Christensen, J. J. *Chem. Rev.* **1991**, *91*, 1721.
(19) Buchanan, G. W.; Kirby, R. A.; Charland, J. P. *J. Am. Chem. Soc.* **1988**, *110*, 2477.
(20) Burns, J. H.; Sachleben, R. A.; Davis, M. C. *Inorg. Chim. Acta* **1994**, *223*, 125.
(21) Sachleben, R. A.; Burns, J. H. *J. Chem. Soc., Perkin Trans. 2* **1992**, 1971.
(22) Kohiro, K.; Kaji, M.; Tsuzuke, S.; Tobe, Y.; Tuchiya, Y.; Naemura, K.; Suzuki, K. *Chem. Lett.* **1995**, 831.
(23) Burns, J. H.; Bryan, S. A. *Acta Cryst., Sect. C* **1988**, *44*, 1742.
(24) Damewood, J. R., Jr.; Urban, J. J.; Williamson, T. C.; Rheingold, A. L. *J. Org. Chem.* **1988**, *53*, 167.
(25) Dvorkin, A. A.; Fonar, M. S.; Malinovskii, S. T.; Ganin, E. V.; Simonov, Y. A.; Makarov, V. F.; Kotlyar, S. A.; Luk'yanenko, N. G. *Zh. Strukt. Khim.* **1989**, *30*, 96.
(26) Dvorkin, A. A.; Simonov, Y. A.; Suwinska, K.; Lipowski, J.; Malinovskii, T. I.; Ganin, E. V.; Kotlyar, S. A. *Kristallografiya* **1991**, *36*, 62.
(27) Dvorkin, A. A.; Fonar, M. S.; Ganin, E. V.; Simonov, Y. A.; Musienko, G. S. *Kristallografiya* **1991**, *36*, 70.
(28) Fonar, M. S.; Simonov, Y. A.; Dvorkin, A. A.; Malinovskii, T. I.; Ganin, E. V.; Kotlyar, S.; Makarov, V. F. *J. Inclusion Phenom. Molec. Recognit. Chem.* **1989**, *7*, 613.
(29) Fraser, M. E.; Fortier, S.; Markiewicz, M. K.; Rodrigue, A.; Bovenkamp, J. W. *Can. J. Chem.* **1987**, *65*, 2558.
(30) Krasnova, N. F.; Dvorkin, A. A.; Simonov, Y. A.; Abashkin, V. M.; Yakshin, V. V. *Kristallografia* **1985**, *30*, 86.
(31) Krasnova, N. F.; Simonov, Y. A.; Bel'skii, V. K.; Fedorova, A. T.; Yakshin, V. V. *Kristallografiya* **1986**, *31*, 1099.
(32) Navaza, A.; Villain, F.; Charpin, P. *Polyhedron* **1984**, *3*, 143.
(33) Rath, N. P.; Holt, E. M. *J. Chem. Soc., Chem. Commun.* **1986**, 311.
(34) Simonov, Y. A.; Malinovskii, T. I.; Ganin, E. V.; Kotlyar, S. A.; Bocelli, G.; Calestani, G.; Rizzoli, C. *J. Inclusion Phenom. Molec. Recognit. Chem.* **1990**, *8*, 349.
(35) Koenig, K. E.; Lein, G. M.; Stuckler, P.; Kaneda, T.; Cram, D. J. *J. Am. Chem. Soc.* **1979**, *101*, 3553.
(36) Katz, H. E.; Cram, D. J. *J. Am. Chem. Soc.* **1984**, *106*, 4977.
(37) Artz, S. P.; Cram, D. J. *J. Am. Chem. Soc.* **1984**, *106*, 2160.
(38) Trueblood, K. N.; Knobler, C. B.; Maverick, E. F.; Helgesen, R. C.; Brown, S. B.; Cram, D. J. *J. Am. Chem. Soc.* **1981**, *103*, 5594.
(39) Lias, S. G.; Liebman, J. F.; Levin, R. D. *J. Phys. Chem. Ref. Data* **1984**, *13*, 695.

Chapter 7

Ligand Design for Small Cations: The Li$^+$/14-Crown-4 System

Richard A. Sachleben and Bruce A. Moyer

Chemical and Analytical Sciences Division, Oak Ridge National Laboratory, P.O. Box 2008, Oak Ridge, TN 37831–6119

This review describes progress in the authors' research toward understanding factors related to the design of ligands for the solvent extraction of lithium. Recent work has demonstrated that 14-crown-4 ethers can be effective extractants for the separation of lithium from mixtures of alkali metal salts. It has also been shown that substituents on the macrocyclic ring of 14-crown-4 ethers can have a large effect on both extraction efficiency (i.e. strength) and selectivity (i.e. discrimination or recognition). The calculated strain induced in the ligand by the conformational reorganization of the ligand upon cation complexation correctly predicts the ordering of lithium extraction efficiencies in a series of substituted 14-crown-4 ethers. More detailed extraction experiments have revealed the component equilibrium processes underlying the overall extraction. A profound dependence of extraction efficiency on the nature of the diluent has been noted and related largely to the solvation of the coextracted anion. A thermochemical scheme has been proposed to rationalize the effect of ligand conformational strain on the net extraction process within the context of other terms sensitive to ligand structure and solvation.

The development of host-guest chemistry has had significant impact on separation science in recent years. The ultimate goal in this area is to achieve a level of understanding that allows the design of highly efficient and specific hosts for any particular ion or molecule of interest. Advances in host-guest chemistry (*1-5*) over the past 30 years have been significantly influenced by the development of crown ethers (*6*) and related macrocyclic compounds (*7-9*). Early work in crown ether coordination chemistry demonstrated the unique ability of these compounds to complex, and discriminate between, a wide variety of organic and inorganic cations (*6,10,11*). The observation that the number and type of coordinating groups and the size of the macrocyclic ring influenced both binding strength and specificity provided a basis for the development of ligand-design principles which are still under investigation today. In general, the larger macrocycles exhibited selectivities for larger cations, while smaller macrocycles preferred the smaller cations (*6,11*). However, exceptions to these general trends have been observed (*12-14*), and the "cavity-size/ion-size" concept must be used cautiously. Investigation into the factors that influence ion binding by crown ethers can provide information vital to the elucidation of design principles necessary for the development of more effective separation systems.

The complexation and separation of lithium ion is of interest both as an intellectual exercise and for several practical reasons (*15,16*). Lithium plays an important role in medicine, science, and technology, from its use in the treatment of manic-depressive illness to battery technologies and polymerization catalysis. Separation problems connected with these uses often entail dilute sources or matrices that contain relatively high concentrations of competing ions. The need therefore arises for high extraction *efficiency*, implying here strength of extraction, together with high *selectivity*, implying exclusivity or ability to reject other ions. The lithium ion is a small, charge-dense cation, which prefers hard oxygen donors (e.g. ethers, carbonyl) and low (4-6) coordination numbers (*17*). Small-ring crown ethers have shown promise for the detection and separation of lithium from mixtures of other cations (*18*) with 14-crown-4 appearing to be the best ring size for selective lithium complexation (*19-22*). Focusing then on the 14-crown-4 ring, the central ligand-design issue entails the relationship between ring substituents and properties such as lipophilicity, extraction efficiency, and selectivity. Insofar as extraction and transport require a high degree of lipophilicity, the 14-crown-4 ring must be substituted with appropriate hydrophobic groups, whose other functions are inductive and steric in nature. Dibenzo-14-crown-4 (**1a**, Figure 1) and derivatives of it have been widely studied for lithium extraction and transport (*23-25*). However, certain 14-crown-4 ethers bearing aliphatic substituents have been shown to extract lithium both more strongly and more selectively than the lipophilic dibenzo-14-crown-4 derivative **1b** (*26*). Interestingly, among the crown ethers bearing aliphatic substituents, wide variations in lithium extraction are observed, even between what seem to be structurally very similar compounds (*26*). For instance, didecalino-14-crown-4, **4a**, (*27*) and its methyl derivative **4b** have been shown to extract lithium salts with both good efficiency and selectivity. In a direct

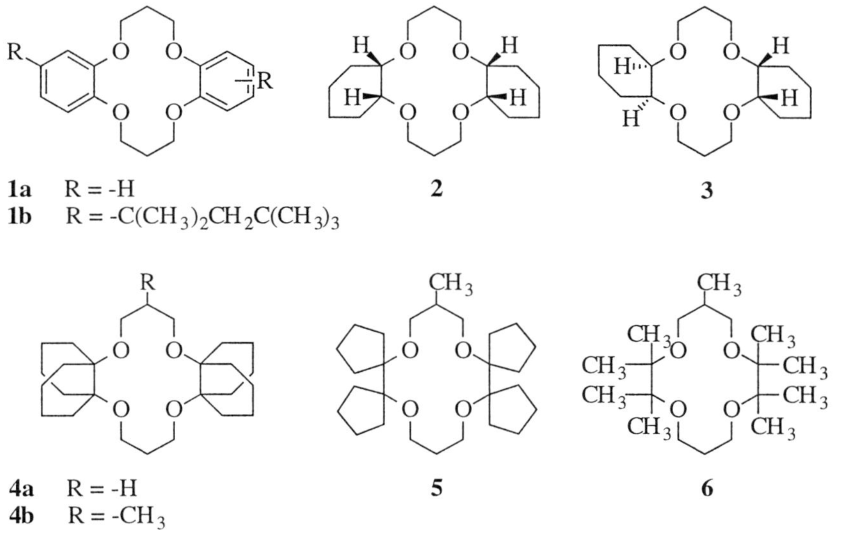

1a   R = -H
1b   R = -C(CH$_3$)$_2$CH$_2$C(CH$_3$)$_3$

2

3

4a   R = -H
4b   R = -CH$_3$

5

6

Figure 1. 14-Crown-4 ethers discussed in this paper as abbreviated by boldface number.

comparison, **4b** extracts lithium more than an order of magnitude more efficiently than **1b** (*26*). At first consideration, this might not seem surprising in view of the stronger basicity of aliphatic ethers compared to aromatic ethers (*28*) and the potentially encapsulating character of the large decalino substituents (*29,30*). However, the isomeric tetracyclopentyl-14-crown-4 (**5**) is a weaker extractant than either **1b** or **4b** (*26*). Clearly, some less-than-obvious factors are strongly influencing the extraction properties of these ligands.

Structural factors, involving both intra-ligand and metal-ligand interactions, as well as solvation and speciation can be expected to play a role in the complexation, extraction, and transport of cations by crown ethers. In the case of the 14-crown-4 ethers, numerous solid-state structural studies of crown ethers and their complexes with lithium salts have been reported (*17*), which can serve as a starting point for conformational analysis. Molecular modeling has been used to investigate the preferred, low-energy conformations of ligands and their complexes (*31*) and to quantitatively compare the effect of cation binding on the ligand conformational energies in a series of structurally-related crown ethers (*31,32*). The utility of such calculations to *predict* the binding properties of crown ethers, or any other ligand for that matter, will depend on the extent to which other factors, such as solvation and speciation, can be either included in the model, demonstrated to correlate with structural factors, or excluded as contributing insignificantly to the complexation process under consideration. In order to evaluate the role of speciation and solvation in the extraction process, detailed solution studies including equilibrium modeling are needed (*33*). These equilibrium studies require sufficient quantities of crown ethers to perform a large number of individual measurements; therefore appropriate synthetic methods are an important part of any such study.

We have been investigating the solvent extraction of alkali metal salts by crown ethers in order to understand the interrelated roles that speciation, solvation, and ligand structural factors play in cation and ion-pair recognition. The 14-crown-4 system has provided significant insight into how these various factors, separately and together, influence efficiency and selectivity in the extraction of lithium salts. In this paper, we review (a) the development of synthetic methodologies, (b) the structural properties of these ligands and their lithium complexes, and (c) previous surveys of their extraction behavior. We then introduce (d) the use of molecular modeling in an effort to correlate the structural and extraction results and present our most recent results on (e) the correlation of extraction strength with solvent parameters, (f) the role of solvent in determining speciation and the predominant extraction equilibria, (g) the correlation of extraction strength with ligand conformational strain upon complexation, and (h) a thermochemical analysis of the role of conformational effects in the net extraction process. Finally, we discuss how the results obtained so far pertain to ligand design and what remains to be done to obtain a useful predictive tool for development of an effective and efficient Li^+ separation system.

Synthesis

The practical utility of a crown ether as a separation reagent depends to some extent on its availability. The synthesis of crown ethers in general, and 14-crown-4 ethers in particular, typically involves the cyclization reaction of an α,ω-diol with an α,ω-dihalide or ditosylate. For example, dibenzo-14-crown-4, **1**, can be prepared as shown in Step 2 of Scheme 1 (*6*). This method is dependent on the availability of the requisite diols and dihalides (or ditosylates). For **1**, the diol can be readily obtained from catechol and 1,3-dibromopropane (*34*), and the synthesis proceeds in acceptable overall yield. The two dicyclohexano-14-crown-4 ethers, **2** and **3**, can be obtained by catalytic hydrogenation of **1** (Scheme 1) and separated by chromatography and crystallization (*6,35*). However, earlier literature methods for the preparation of highly-substituted alkyl-14-crown-4 ethers from tertiary alcohols (27,36) are not practical for the synthesis of sufficient quantities of crown ethers for detailed extraction, equilibrium, and structural studies, primarily because the tertiary diols that serve as starting materials for the syntheses are poor nucleophiles in the S_N2 reactions typically used (such as in Scheme 1). Therefore, we developed the synthesis shown in Scheme 2, which allowed us to prepare sufficient quantities (5-150 g) of crown ethers **4b-6** for the studies described below.

Scheme 1

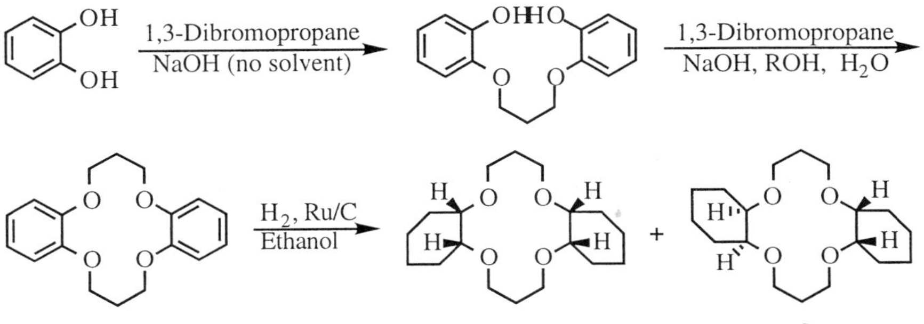

The two crucial points in Scheme 2 are the reductive ring-opening reaction in the synthesis of α,ω-diols, which avoids the substitution reaction altogether, and the use of the highly reactive, doubly allylic, methallyl dichloride in the cyclization step. Since this reagent lacks hydrogens on the central carbon of the three carbon unit, it cannot undergo the elimination side-reaction that often predominates over the S_N2 reaction with highly basic, hindered alkoxides. Reduction of the vinylic substituent provides the methyl-substituted crown ethers used in the studies discussed below. Alternatively, this site may be used for further functionalization of these compounds, including their incorporation into polymers and related materials.

Scheme 2

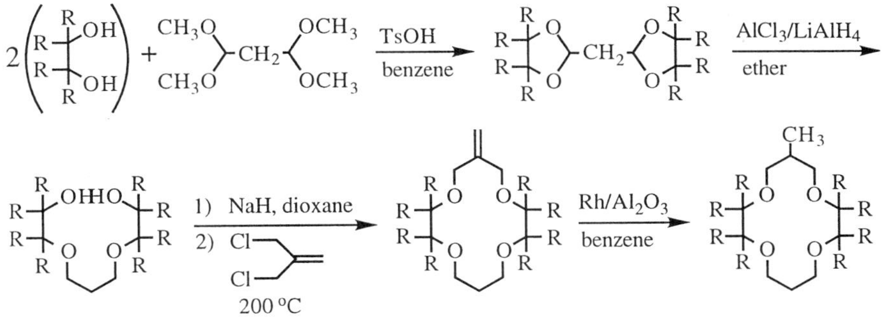

Structures

The solid-state structures of compounds **1a-3** and **4b-6** and the LiSCN complexes of **1a**, **2**, and **4b-6** have been reported (*37-40*). Whereas **1a** is highly preorganized for metal-ion complexation (*37,39*), the alkyl-substituted crown ethers **2**, **3**, **4b**, and **5** are more flexible and must undergo a conformational reorganization in order to complex Li^+ ion (*38,40*). The uncomplexed ligands **4b-6** adopt similar conformations in the solid state, while **2** and **3** adopt conformations different from this and from each other. However, **2** and **4b-6** adopt nearly the same conformation in their lithium complexes, which is very similar to that of **1a•LiSCN** (*37*), in that the spatial relationships between the lithium ion, the four crown ether oxygens, and the thiocyanate anion are nearly identical.

Extraction Survey

The extraction of alkali metal chloride and nitrate salts by the crown ethers **1b-3** and **4b-6** in 1-octanol at 25 °C is shown in Figure 2, along with the extraction of these metal salts by the solvent alone (*26,41*). In view of its good solvating ability for both cations and anions, 1-octanol was chosen as the water-immiscible organic diluent. Fourteen extraction experiments were performed in which the aqueous phase consisted of a mixture of the principal alkali metal cations, all as chloride (left frame in Figure 2) or all as nitrate salts (right frame), where $[Li^+] = [Na^+] = [K^+] = [Rb^+] = [Cs^+] =$ 0.4 mol/kg solution. These salt solutions were equilibrated at 25 °C at a 1:1 phase ratio with 1-octanol containing crown ether at an initial concentration of ~25 mmol/kg solution (~0.02 M); in blank runs, the crown ether was omitted. The results of each experiment are graphed as a grouping of five adjacent columns, each column indicating the extent of extraction of the corresponding alkali metal salt (see legend). To indicate the lower limit of lithium extraction expected in the presence of a crown ether, a horizontal reference line corresponding to the applicable lithium blank is given near the bottom of each frame in Figure 2. Assuming a 1:1 metal:ligand stoichiometry, the loading limit for each crown is given by the horizontal lines near the top of each frame.

The most striking feature of Figure 2 is the strong dependence of lithium extraction on ligand structure. The three ligands **2**, **4b**, and **6** all extract Li much more strongly than **1b** and to a significant degree of loading, while **3** and **5**, isomeric to **2** and **4b**, respectively, extract lithium salts only weakly above the level of the blanks. Apparently, the anion here (Cl^- vs. NO_3^-) has little influence on the ordering of the crown ethers, which may be given as **4b** > **6** > **2** > **1b** > **3** ~ **5**. The origin of these dramatic differences in extraction properties is not obvious by inspection.

From Figure 2, the main effect of substituting NO_3^- for Cl^- entails an increase in extractability of the alkali metals. This observation simply reflects the more extractable nature of the larger nitrate anion in general (*42,43*). As discussed in more detail below, the anion probably interacts only weakly with the cation in 1-octanol (*33*) and therefore controls *efficiency* in the extraction of alkali metal cations without significant effect on *selectivity*.

In the blank experiments shown in Figure 2, it was found that 1-octanol alone exhibits a measurable, though weak, ability to extract each of the alkali metals and is in fact somewhat lithium-selective. Defining the distribution ratio D_M of a metal ion as the ratio of the organic-phase metal concentration to that in the aqueous phase, the selectivity for, say, Li^+ vs. Na^+ ion may be conveniently discussed in terms of the selectivity ratio $S_{Li/Na} = D_{Li}/D_{Na}$. For 1-octanol used alone, $S_{Li/Na}$ has the value 3.6 for the chloride system and 3.7 for the nitrate system, essentially independent of the anion within experimental error (ca. ±5%). Although seemingly trivial, this result is important because it shows that unconstrained solvent cages of alcohol and water

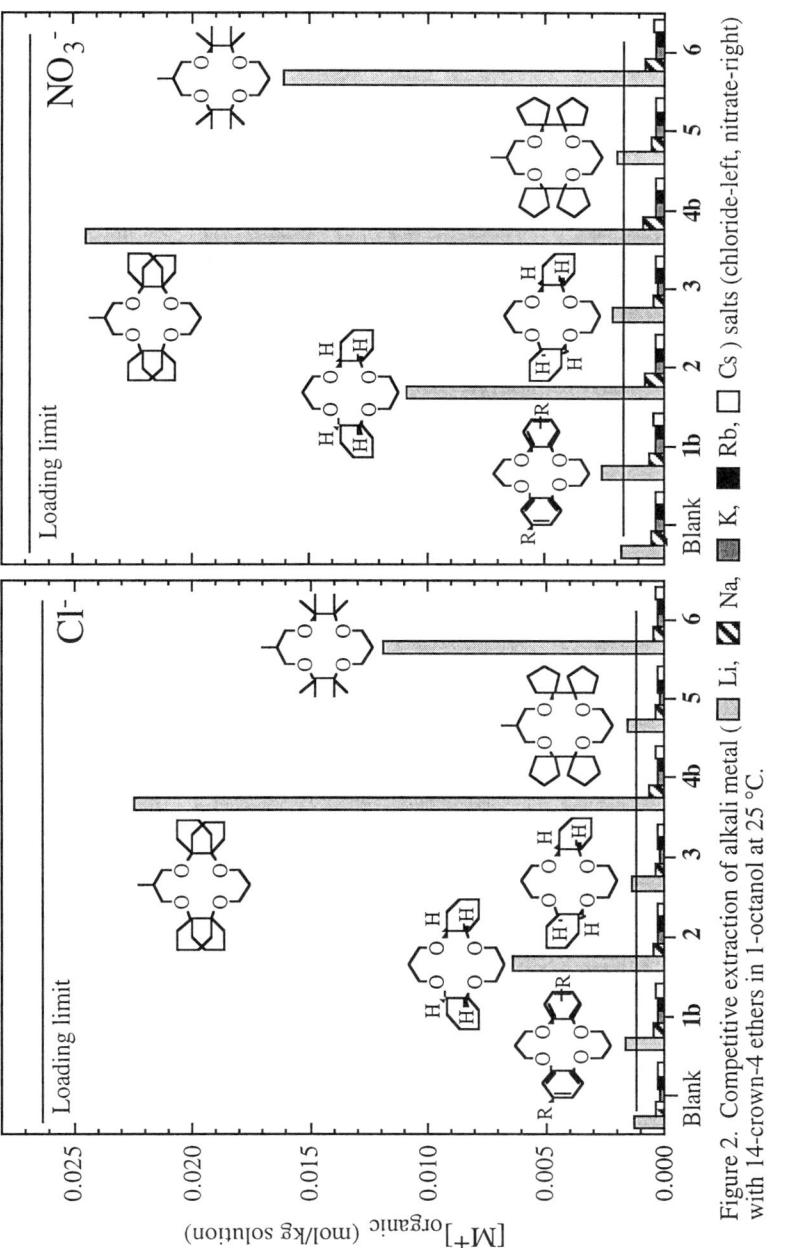

Figure 2. Competitive extraction of alkali metal ($\blacksquare$ Li; $\boxtimes$ Na, $\blacksquare$ K, $\blacksquare$ Rb, $\square$ Cs) salts (chloride-left, nitrate-right) with 14-crown-4 ethers in 1-octanol at 25 °C.

molecules as collective "hosts" for alkali metal ions already exhibit a preference for lithium. We are thus reminded that the total environment about a metal-ion guest includes not only the crown ether but additionally any bound solvent molecules. For 14-crown-4 ethers, solvent access to the Li^+ ion is limited, as shown by the structural evidence (*39,40*), but as the metal ion size increases, solvation of the exposed coordination sites obviously must become more important. Based upon principles of alkali metal extraction by solvation, the lithium selectivity of 1-octanol originates directly from the good electron-pair donor ability of the alcohol and water molecules present in the organic-phase (*44*). Ether atoms, also good electron-pair donors, therefore represent effective donor atoms for the construction of lithium-selective ligands.

As may be seen in Figure 2, all of the 14-crown-4 crown ethers tested exhibit selectivity for lithium. In fact, none of these crown ethers extracted Na^+, K^+, Rb^+, or Cs^+ ions much above the level obtained by the blank experiments. Overall selectivity ratios $S_{Li/Na}$ were essentially independent of anion and are given as follows: 4.1 (**1b**), 14.6 (**2**), 4.4 (**3**), 31.2 (**4b**), 4.0 (**5**), and 23.1 (**6**). These selectivity ratios include a contribution from the background extraction of salts by 1-octanol; this contribution dominates for the weaker crown ethers (**1b**, **3**, and **5**). In principle, a corrected selectivity ratio $S_{Li/Na}'$ corresponding to the selectivity conferred by the crown ether can be estimated from equation 1:

$$S_{Li/Na}' = ([Li]_{organic} - [Li]_{organic,free})/([Na]_{organic} - [Na]_{organic,free}) \quad (1)$$

For present purposes, we make the approximation $[M]_{organic,free} = [M]_{blank}$, where M = Li or Na. In reality, $[M]_{organic,free} < [M]_{blank}$ in 1-octanol, because a common-ion effect causes the concentration of free lithium in the organic phase to decrease somewhat as salt extraction by the crown ether increases (*33*). Owing to weak extraction, the values of $S_{Li/Na}'$ for the crown ethers **1b**, **3**, and **5** entailed high uncertainty and could not be reliably estimated from the data. In the cases of the more efficient crowns **2**, **4b**, and **6**, the respective values of $S_{Li/Na}'$ were found to be 39 ± 8, 60 ± 8, and 71 ± 16. These values reflect the good selectivity of related alkyl-substituted 14-crown-4 ethers (*18,27,29,30*).

For a more quantitative comparison of lithium extraction ability, the percent lithium loading $\%L_{Li}$ is given by:

$$\%L_{Li} = 100\% \times ([Li]_{organic} - [Li]_{organic,free})/[Crown]_{organic} \quad (2)$$

corresponding to the percent of the crown ether that is bound to lithium in the organic phase, assuming that lithium is bound only in 1:1 metal:ligand complexes and that all of the crown ether remains in the organic phase. Equation 2 represents the minimum loading, because again we have employed the approximation $[Li]_{organic,free} = [Li]_{blank}$, where in reality $[Li]_{organic,free} < [Li]_{blank}$ in 1-octanol (*33*). Values of $[Li]_{organic}$ and $\%L_{Li}$ are given in Table I. The same trends evident from Figure 2 may be observed. It may be seen that **3** and **5** have essentially the same extraction efficiencies within experimental error.

Structural Modeling

In an effort to determine to what extent structural factors contribute to the observed differences in extraction behavior of the alkyl-substituted 14-crown-4 ethers, molecular mechanics calculations were performed on crown ethers **2**, **3**, and **4b-6** (*45,46*). Ligand steric energies, U, were calculated using the MM3 force field modified for the treatment of metal complexes (*31*). Using the crystal-structure coordinates as a starting point for the calculations, energy-minimized steric-energy

Table I. Loading Behavior of Crown Ethers as Compared with Calculated Ligand Strain Energies[a]

Crown	Anion	[Li]organic	%L$_{Li}$	ΔU_{conf}	ΔU_{comp}	ΔU_{reorg}
4b	Cl$^-$	21.2	85 ± 4	0.97	4.30	5.27
	NO$_3^-$	22.7	91 ± 5			
6	Cl$^-$	10.7	44 ± 2	2.27	5.67	7.94
	NO$_3^-$	14.3	58 ± 3			
2	Cl$^-$	5.17	21 ± 1	2.54	6.39	8.93
	NO$_3^-$	9.12	37 ± 2			
3	Cl$^-$	0.13	0.6 ± 0.3	3.88	6.77	10.65
	NO$_3^-$	0.37	1.5 ± 0.4			
5	Cl$^-$	0.25	1.0 ± 0.3	6.43	4.56	10.99
	NO$_3^-$	0.26	0.9 ± 0.4			

[a]Extraction data were taken from Figure 2. %L$_{Li}$ is defined in equation 2. Ligand strain energies (ΔU) are defined in the text (equations 3-5) (45,46).

calculations were performed on each ligand in each of two conformations: the conformation found in the crystal of the uncomplexed ligand, to obtain U_{free}, and the conformation of the ligand starting with the conformation found in the crystal of the LiSCN•crown ether complex to provide U_{bind} (47). In addition, the steric energy of the ligand *in the complex*, U_{bound}, was calculated after minimization of the Li•crown ether complex (47). Operationally, the calculation procedure involves the following stepwise processes: (a) Obtain U_{free} by energy-minimizing the conformation of the free ligand starting from the conformation of the ligand in the crystal; (b) energy-minimize the conformation of the ligand in the complex starting from the conformation of the complex in the crystal; remove the metal from the resulting complex, but without further minimization, calculate U_{bound}; (c) obtain U_{bind} by energy-minimizing the conformation of the ligand (no metal present) in the previous step. Since the crystal structure of 3•LiSCN was not available, input data for 3 in the complex conformation were generated by deleting the appropriate carbon and hydrogen atoms from the structure of 4b and adding hydrogen atoms in the appropriate positions (33). The magnitude of U_{bind} and U_{bound} for 3 therefore cannot be considered as reliable as the steric energies of the other crown ethers.

The differences between the calculated steric energies for the free and bound ligands represent "strain" that is induced in the ligand by metal ion complexation. As described by Hay and Rustad (47), three strain energies may be considered:

$$\Delta U_{conf} = U_{bind} - U_{free} \qquad (3)$$

$$\Delta U_{comp} = U_{bound} - U_{bind} \qquad (4)$$

$$\Delta U_{reorg} = U_{bound} - U_{free} = \Delta U_{conf} + \Delta U_{comp} \qquad (5)$$

representing the steric energy changes that occur during the process of a host complexing a guest ion. For a preorganized (48) ligand, ΔU_{conf} will be small. For a ligand that is complementary (49) to its guest, ΔU_{comp} will be small. For a ligand that is *perfectly* preorganized for binding, $\Delta U_{reorg} = 0$ (47).

The ΔU_{reorg} can be related to $\Delta H°$ for complexation, and for a series of ligands with a fixed number and type of donor atoms interacting with a given guest under identical conditions, ΔU_{reorg} should be proportional to binding strength (47). The strain energies calculated for the series of crown ethers **2**, **3**, and **4b**-**6** are listed in Table I in order of decreasing lithium loading. It may be seen that the order of lithium salt extraction by these crown ethers in 1-octanol corresponds surprisingly well with the order of ΔU_{reorg}. The minimum reorganization energy is obtained for **4b**, which exhibits the strongest lithium extraction. This is followed in order by **6** and **2**. The crown ethers **5** and **3** have the greatest calculated reorganization energies and the weakest ability to extract lithium. For reasons given above, the magnitude of the ΔU_{reorg} for **3** must be considered speculative, though the predicted relatively weak extraction of this crown ether is indeed observed.

Solvent Effects

The choice of 1-octanol as the diluent in the survey experiments described above was initially an arbitrary one based largely on past experience indicating that hydroxylic solvents in particular facilitate the extraction of alkali metal salts of mineral acids (50). In fact, a wide range of solvents has been employed in ion-pair extraction by crown ethers (43,51), raising the question as to the sensitivity of extraction by the 14-crown-4 ethers to diluent properties. Regarding ligand design, with its reliance on calculations applicable to vacuum conditions (absence of solvation), the issue of the role of solvent environment remains an area of large uncertainty. In connection with lithium extraction by crown compounds, it has been shown that matrix effects strongly influence extraction behavior (41,52). The lack of correlation between the extraction of alkali metals in a picrate solvent extraction experiment and the results of ion-selective electrodes employing the same set of crown ethers is particularly puzzling (52). From such observations, one must ask whether the structural nature of the crown ether can be completely separated from matrix effects in ligand design.

To probe this question, a competitive extraction experiment similar to that shown in Figure 2 was performed using a single crown ether, nonamethyl-14-crown-4 (**6**) dissolved in a variety of diluents at a constant concentration of 0.020 M (Table II). Each solvent was contacted at a 1:1 phase ratio for at least 2 h with a mixture of alkali metal nitrate salts at 25 °C, where initially $[Li^+] = [Na^+] = [K^+] = [Rb^+] = [Cs^+] = 0.4$ mol/kg solution.

Blanks were run as before, where the crown ether was omitted from the solvent. Cyclohexanone was the only diluent other than 1-octanol for which the alkali metal nitrates were appreciably extracted in the absence of the crown ether. Both cyclohexanone and 1-octanol are classed as "wet" diluents (53), containing respectively 0.32 (approximate from densities at 20 °C assuming additivity of volumes) and 0.27 mole fraction water when saturated with pure water (unit water activity). All of the other diluents fall in the range of "dry" diluents, having less than 0.13 mole fraction water at saturation (53). Organic-phase metal-ion concentrations in units of mmol/kg solution for the cyclohexanone blank were found to be: Li, 2.03; Na, 0.93; K, 0.65; Rb, 0.56; and Cs, 0.49. For 1-octanol, the corresponding values are: 1.79, 0.48, 0.29, 0.29, 0.36. Again, the presence of good electron-pair donors in the solvent, either alcohol, ketone, or dissolved water molecules, confers a preference for lithium in the blank extractions (44).

As shown in Table II, the nature of the diluent profoundly influences the efficiency of extraction of lithium nitrate by the substituted 14-crown-4 ethers. It may

Table II. Effect of Diluent on Competitive Extraction of Alkali Metal Nitrates by Nonamethyl-14-crown-4 (6)[a]

Diluent	% Lithium loading[b] %L_{Li}	Li/Na selectivity ratio[c] $S_{Li/Na}'$	Diluent dielectric constant[d] ε
n-Octane	<0.4		1.9
Tetrachloromethane	<0.4		2.2
Tetrachloroethene	<0.4		2.3
Di-n-butyl ether	<0.4		3.1
o-Xylene	0.4		2.6
Benzene	0.6		2.3
1M 1-Octanol/n-octane	1.1		
1M 1-Octanol/o-xylene	7.4		
o-Dichlorobenzene	11.5		9.9
1,2-Dibromoethane	11.8		4.8
Trichloromethane	31.1	19	4.8
1-Octanol	58.2	67	10.3 (8.1)
Cyclohexanone	65.8	54	18.2
1,2-Dichloroethane	71.7	57	10.7
Nitrobenzene	73.3	53	34.8

[a]Experimental conditions are described in the text.
[b]%L_{Li} is defined in eq. 2 and corresponds to the percent (or minimum percent in the case of 1-octanol and cyclohexanone) of the crown ether that is bound to the metal ion in the organic phase. The effective limit of measurement of %L_{Li} here is estimated to be ca. 0.4.
[c]$S_{Li/Na}'$ is defined in equation 1 and is approximate for 1-octanol and cyclohexanone. Blanks in the table correspond to the condition %L_{Li} < 0.4 or %L_{Na} < 0.4.
[d]Dielectric constants for pure solvents at 25 °C taken from (54). For 1-octanol, it may be noted that the dielectric constant of the water-saturated diluent is 8.1 (55).

be seen that the percent loading (%L_{Li}) of the crown ether by lithium extends from below reliable measurement to appreciable saturation of the crown ether. Extraction of the other alkali metals was unimportant and below reliable measurement in most cases. In the few cases that sodium extraction could be measured, blank-corrected selectivity ratios $S_{Li/Na}'$ were found to lie in the range 19 - 67. Extraction selectivity therefore appears to depend at least somewhat on the diluent type.

As discussed in more detail below, the speciation of the extraction chemistry must be understood to fully interpret the diluent effect, but we may still profit from some qualitative reasoning. We may start by observing that the diluent effect may be divided into the solvation of the reactants vs. the solvation of the products. In this case, the solvation (i.e. hydration) of the reactant ions, lithium and nitrate, is constant for each system shown in Table II, since the initial aqueous phases are identical. Among the reactants, then, only the ligand, nonamethyl-14-crown-4 (6), undergoes changes in solvation as the diluent is varied. Since this solvation is governed to a large extent by the hydrocarbon bulk of the crown ether, it will be approximately canceled by the "neutral part" (distinguished from the electrostatic part) of the solvation of the crown ether complex with lithium. By reference to the crystal structures (39,40), the extraction complex consists of a large, hydrophobic cation in which the lithium ion is embedded within the crown ether. The organic-phase co-anion remains largely exposed to the external environment. The electrostatic solvation energies of the cationic complex and its co-anion, then, likely underlie the observed effect of the diluent on extraction efficiency.

Not surprisingly, lithium loading increases roughly with the dielectric constant, ε, of the diluent (Table II). Although the dielectric constant is notoriously unreliable as an absolute measure of solvation ability, the trend observed is indeed consistent with the solvation of ions or ion pairs in the solvent phase. Such solvation is generally understood on the basis of electrostatics, wherein the dielectric constant plays a primary role (44). In general, the charging energy of an ionic species is a positive quantity that decreases with increasing dielectric constant. It also decreases rapidly with increasing ion size. Thus, the solvation of the smaller ion, namely the nitrate co-anion vs. the large, hydrophobic complex cation, is expected to dominate the observed diluent effect. A more detailed analysis of diluent effects has been presented by Marcus, specifically pertaining to the case of potassium chloride extraction by dibenzo-18-crown-6 and dicyclohexano-18-crown-6 (50,56).

Diluent properties such as hydrogen bond acceptor (HBA) and donor (HBD) strength are thought to modify the simple dielectric effect (57-59). Although the interplay of diluent properties is not well understood, it arises from the simple reasoning described above that the solvation of the exposed co-anion should be strongly affected by the diluent properties, in particular the HBD strength. By contrast, the crown ether largely shields the lithium cation from the solvent, limiting coordination by the solvent (including water) to a single axial site and thereby attenuating the importance of electron-pair donor ability of the diluent. To test this reasoning, an extraction experiment was conducted employing 0.02 M nonamethyl-14-crown-4 (**6**) in six diluents to extract lithium chloride from an aqueous phase containing only 0.4 M LiCl. These conditions avoid complications due to high loading and coextraction of the other alkali metals. Figure 3 shows the results as plots of log D_{Li} vs. selected diluent parameters. Plot A gives the dependence of log D_{Li} on the inverse dielectric constant ($1/\varepsilon$), showing the expected trend, albeit roughly and with significant scatter ($r^2 = 0.53$). Correlation with the Dimroth-Reichardt E_T (60) parameter in plot B, however, is good ($r^2 = 0.92$). This parameter is thought to represent a measure of effective solvent polarity. Since the dye (No. 30) used to experimentally determine E_T contains a phenolate functionality, it is not surprising that E_T also correlates well with the ability of the solvent to donate hydrogen bonds (60) and to solvate anions (61). Hence, this correlation together with previous results (50,56,61) supports the hypothesis that the anion solvation in this case contributes to the diluent effect observed. Plot C shows that log D_{Li} correlates as well with the Shmidt-Marcus diluent parameter, DP*, which is empirically derived from the diluent effect observed in certain extraction reactions involving the formation of organic-phase ion pairs and other polar species (62,63). A good correlation with DP* follows from the correlation with E_T, since DP* correlates strongly with E_T (63).

Finally, plot D in Figure 3 raises the ubiquitous issue of the role of water in the organic phase. We note that log D_{Li} increases smoothly and steeply with the ability of the diluent to solubilize water, as given by the equilibrium water content of the pure diluent when saturated at unit water activity ([Water]$_{org}$). We also note that [Water]$_{org}$ correlates well with E_T ($r^2 = 0.95$), implying that the effective diluent polarity and HBD ability underlie the ability of the diluent to extract water. In searching for a cause-and-effect relationship, it therefore cannot be decided from the correlations alone whether organic-phase water plays a direct role in the extraction process or whether the organic-phase water content serves merely as an indicator of a more fundamental diluent property. We suggest, however, that the answer may be a combination of both explanations. Suppose the extraction process can be viewed hypothetically in two steps, first as transfer of completely dehydrated species to the organic phase followed in the second step by allowing water to associate with the dehydrated species in the organic phase. In the first transfer step, diluent polarity and HBD properties favor extraction in accord with the principles already outlined above. Borrowing from transfer thermodynamics (42,44), this expectation is valid even if the solvent phase contains no water at all. The second hypothetical step entails partitioning of water molecules from the aqueous phase and association with the dehydrated organic-phase species formed in the first step. Concerning solute hydration in organic solvents, the following facts may be noted: (a) metal:crown complexes in wet organic diluents are hydrated to a degree (43,51), and (b) small anions in wet organic diluents are also hydrated (42). In particular, chloride ion in water-saturated 1,2-dichloroethane, nitrobenzene, and chloroform associates with an average of 2.3, 3.3, and 1.0 water molecules (64), respectively. We may thus realistically expect that the cationic and anionic organic-phase species in our extraction of lithium salts by 14-crown-4 ethers are hydrated to a degree. As suggested previously (65) in rationalizing a correlation between cesium nitrate extraction by large crown ethers and diluent water content, it makes sense that diluents that provide a good solvation environment for water will also provide a good solvation environment in part for hydrated extraction species. Thus, the correlation of log D_{Li} with [Water]$_{org}$ in plot D of Figure 3 is understandable. Diluents that provide a good solvation environment for water are in fact those that are polar and have good HBD properties. Thus, both the first and second steps of our hypothetical process are favored by polar, HBD diluents, and both log D_{Li} and [Water]$_{org}$ accordingly correlate with E_T.

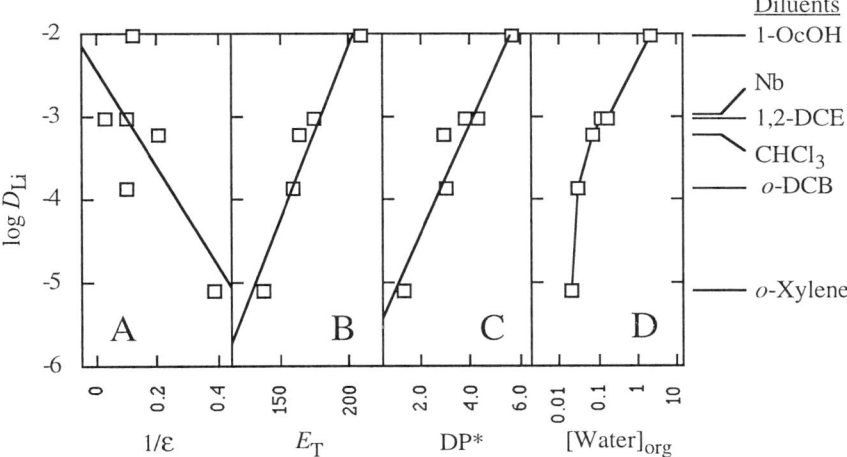

Figure 3. Extraction of lithium chloride vs. selected diluent properties, including the inverse dielectric constant 1/ε (plot A) (54), Dimroth-Reichardt E_T parameter (plot B) (60), Shmidt-Marcus diluent parameter, DP* (plot C) (62,63), and the organic-phase saturation water content, [Water]$_{org}$ in mol/L, (54) (plot D). Solvent abbreviations: 1–OcOH = 1-octanol, Nb = nitrobenzene, 1,2-DCE = 1,2-dichloroethane, CHCl$_3$ = chloroform, o-DCB = o-dichlorobenzene. In the case of 1-octanol, the values of ε (55), E_T (61), and DP* (63) correspond to the water-saturated diluent; [Water]$_{org}$ = 2.3 M (55). For all other diluents, the diluent parameters correspond to the pure diluent; this is acceptable, since the relatively low water contents classify these diluents as "dry."

Equilibrium Modeling and Speciation

To understand the relationship between lithium extraction efficiency and ligand reorganization energy and to further apply this understanding to ligand design requires knowledge of the speciation of the extracted lithium. Toward this end, we have been studying the extraction equilibria by measuring the distribution of both the lithium salt and crown ether between the aqueous phase and solvent for several solvents and several of the substituted 14-crown-4 ethers in Figure 1. Operationally, the distribution measurements are made over as wide a range as possible of the crown ether and lithium salt concentrations, and the results are subjected to a rigorous mass-action analysis employing the program SXLSQI of Baes (*33,44,66,67*). The analysis takes into account organic- and aqueous-phase activity effects and provides estimates of equilibrium constants referred to the state of infinite dilution. By measuring lithium distribution (via ion chromatography or ICP-AES), four equilibrium processes may be determined entailing reaction of the univalent aqueous metal ion M^+, univalent aqueous anion X^-, and organic-phase ligand (crown compound) L:

$$M^+ (aq) + X^- (aq) \rightleftharpoons MX (org) \qquad K_s \qquad (6)$$
$$M^+ (aq) + X^- (aq) \rightleftharpoons M^+ (org) + X^- (org) \qquad K_{s\pm} \qquad (7)$$
$$M^+ (aq) + X^- (aq) + L (org) \rightleftharpoons MLX (org) \qquad K_{ex} \qquad (8)$$
$$M^+ (aq) + X^- (aq) + L (org) \rightleftharpoons ML^+ (org) + X^- (org) \qquad K_{ex\pm} \qquad (9)$$

Equations 6 and 7 together with the salt partitioning constants K_s and $K_{s\pm}$ are examined by studying extraction in the absence of crown ether ("blank" extractions); such reactions have been recently reviewed (*44*).

All of the crown ethers shown in Figure 1 are lipophilic and predominantly distributed to the organic phase, even in the presence of aqueous lithium salts. However, the weak partitioning of the crown ethers to the aqueous phase as a function of aqueous lithium concentration can still be measured by GC, yielding a means to estimate the crown ether partition ratio K_p and the aqueous-phase complexation constant $K_{f,aq}$ (Sun, Y.; Moyer, B. A., Oak Ridge National Laboratory, unpublished data.). Considering that the crown ethers are lipophilic, the reactions in terms of the predominant reactant species may be written:

$$L (org) \rightleftharpoons L (aq) \qquad K_p^{-1} \qquad (10)$$
$$L (org) + M^+ (aq) \rightleftharpoons ML^+ (aq) \qquad K_{p,ML} \qquad (11)$$

Treating equations 6-11 as the experimentally observed distribution reactions, a number of fundamental reaction constants may be derived. Figure 4 shows the equilibrium scheme that comprehensively accounts for our observed extraction behavior. As discussed above, all of the species in the organic phase may be considered to be hydrated to a degree, though this aspect of the speciation has not yet been investigated. In all cases, only 1:1 lithium:crown complexes have been observed. For 1-octanol as the diluent, four organic-phase lithium species account for the observed extraction (*33*): free Li^+ ion, Li^+Cl^- ion pairs, the complex cation $Li(CE)^+$, and the complex ion pair $Li(CE)^+Cl^-$. By use of ^7Li NMR, the complexed and uncomplexed Li^+ species can be distinguished and correlated with the mass-action model, lending independent support for the modeling results (*33*).

Studies of other diluents and crown ethers are in progress and have so far corroborated the scheme in Figure 4. Although 1-octanol allowed the measurement of salt partitioning constants (K_s and $K_{s\pm}$), the other diluents, including 1,2-dichloroethane, 1,2-dichlorobenzene, and nitrobenzene, do not appreciably extract lithium salts in the absence of crown ether. Thus, the elimination of free salt extraction (equations 6 and 7) from the models greatly simplifies the modeling of the extraction data in these cases. The extent of ion-pair association in the solvent depends on the diluent dielectric constant. For intermediate dielectric constants such as that of 1-octanol (8.1 when water saturated), 1,2-dichloroethane (10.4), and 1,2-dichlorobenzene (9.9), systems exhibit behavior consistent with a state of partial ion-pairing. That is, free ions and their ion pairs coexist in appreciable concentrations throughout the experimental range. The resulting curvature in log-log plots of D_{Li} vs. compositional variables makes the normal slope analysis fruitless. However, fitting by use of SXLSQI enables one to determine the constants for both free complex ions ($K_{ex\pm}$) and their ion pairs (K_{ex}) in the organic phase (33,44,67). When the diluent dielectric constant is much higher or lower than 10, the complexity of the equilibria decreases considerably. For nitrobenzene ($\varepsilon = 34.8$), ion pairs are unimportant for the low organic-phase ionic strength that we have normally encountered ($I < 0.02$ M). For low-permittivity diluents such as o-xylene ($\varepsilon = 2.6$), ion-pairing is essentially complete.

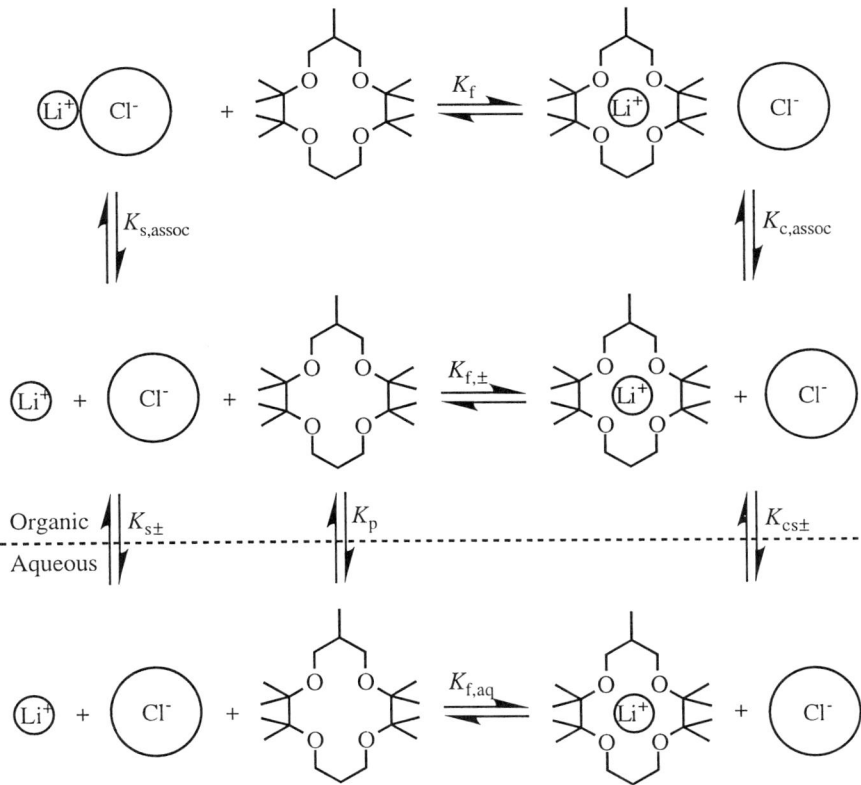

Figure 4. Equilibrium scheme for the extraction of lithium chloride by nonamethyl-14-crown-4 (**6**).

Knowledge of the complete speciation scheme over a range of solvents and 14-crown-4 ethers will ultimately allow the evaluation of the interplay between the structural and environmental factors in controlling the efficiency and selectivity of lithium extraction. Figure 5 provides a possible thermochemical analysis of the extraction process and how it relates to the gas-phase processes applicable to molecular mechanics. The net extraction equilibrium being depicted is the formation of the organic-phase complex MLX (eq. 8). Initial steps involve desolvation of the ligand ($-\Delta G_{solv,L}°$) and dehydration of the metal ion ($-\Delta G_{hyd,M}°$), whence the metal and ligand can react directly in the gas phase to give the gas-phase metal complex ML$^+$ ($-\Delta G_{cpx,L}°$). As was discussed in the Structural Modeling section above, the gas-phase complexation step may be further broken down to include individual steps corresponding to conformational reorganization of the ligand to its binding conformation prior to complexation ($\Delta G_{conf}°$), structural reorganization of the ligand to its bound structure but hypothetically without yet introducing the metal ion ($\Delta G_{comp}°$), and complexation of this prestrained ligand ($\Delta G_{cpx,Lb}°$). Final processes in the net formation of MLX include resolvation of the gas-phase complex cation to the organic phase ($\Delta G_{solv,ML}°$), partitioning of the aqueous anion X$^-$ to the organic phase ($\Delta G_{p,X}°$), and ion-pairing of the complex cation ML$^+$ with the organic-phase anion ($\Delta G_{ip}°$).

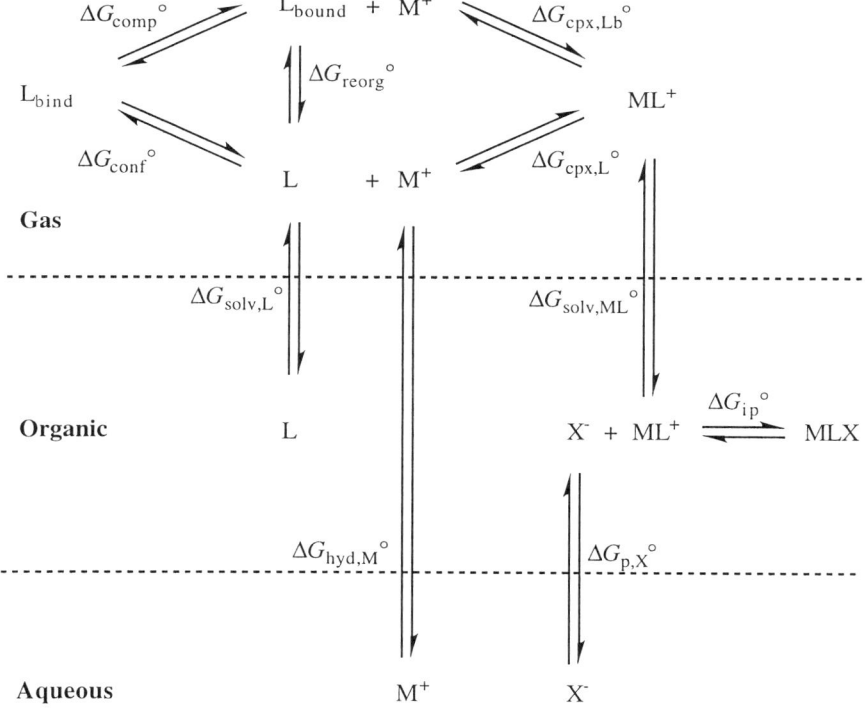

Figure 5. Hypothetical thermochemical cycle showing role of ligand strain.

With regard to ligand design, Figure 5 reveals the role of solvation effects as the structure of the ligand changes. By inspection, it may be seen that the diluent-dependent terms in Figure 5 include $-\Delta G_{solv,L}°$, $\Delta G_{solv,ML}°$, $\Delta G_{p,X}°$, and $\Delta G_{ip}°$. Among these terms, only $\Delta G_{p,X}°$ is independent of ligand structure. Thus, a quantitative approach to ligand design must include consideration of how changes in ligand structure influence the terms $-\Delta G_{solv,L}°$, $\Delta G_{solv,ML}°$, and $\Delta G_{ip}°$. To the extent that ligand-induced changes in these three terms are either small relative to, or functionally related to, ligand-reorganization energy, one may expect a correlation between the $\Delta G_{ex}°$ for the net extraction process and ligand reorganization energy. From the data presented in Table I, such a correlation has been found, suggesting that one of these conditions has been met. Qualitatively, we argued above that the ligand dependence of the terms $-\Delta G_{solv,L}°$, $\Delta G_{solv,ML}°$, and $\Delta G_{ip}°$ is likely small compared with changes in reorganization energy. That is because (a) the ion pairs in 1-octanol are likely solvent separated (33) and therefore insensitive to the ligand structure of the 14-crown-4 ethers that we tested, and (b) the charge neutral parts of the terms $-\Delta G_{solv,L}°$ and $\Delta G_{solv,ML}°$ largely cancel. To further examine this question without the problems of background extraction of lithium, high loading, and ion pairing, we have completed the determination of the equilibrium constants for the extraction of lithium chloride and perchlorate by our alkyl-substituted 14-crown-4 ethers in nitrobenzene (45). The results, which will be published in detail elsewhere, show that $\log K_{ex\pm}$ (or $\Delta G_{ex\pm}°$) indeed correlates linearly with Δu_{reorg}. [These results are also discussed in the chapter by Hay (46).] Certainly, if the MM approach has usefulness in general, such a correlation must survive changing the diluent, and experiments are in progress to test this expectation as well as to examine the diluent effect in terms of the component equilibria given in Figure 4.

Conclusions

The discovery (27,29,30,52) that certain alkyl substituents on the 14-crown-4 framework confer useful lithium selectivities stimulated questions concerning the origin of the effects of alkyl substitution on the extraction efficiency and selectivity of crown ethers and whether these effects could be predicted under the rubric of "ligand design." In pursuing these questions, we have developed synthetic routes to a set of new substituted 14-crown-4 ethers (Figure 1) (26) and have elucidated structural aspects of the free ligands and their lithium complexes (39,40). From the structural studies, it has become clear that the 14-crown-4 ligands are not preorganized for lithium complexation but must "untwist" to form the binding conformation in which the ether oxygen atoms are oriented in a square-planar configuration. The structural results enabled the calculation of the ligand reorganization energies (45,46), which explain the ordering of lithium extraction efficiencies from a mixture of alkali metals into 1-octanol solutions of the crown ethers. This observation suggested the usefulness of this approach for ligand design more generally, stimulating detailed extraction experiments aimed at understanding the component equilibrium processes underlying extraction of lithium salts by 14-crown-4 ethers together with the role of solvation effects on these component equilibria (33,41,45). It was observed that the nature of the diluent strongly influences extraction efficiency, and correlations with diluent properties suggest the predominance of anion solvation. Extraction behavior has been shown to be consistent with a set of component equilibria entailing 1:1 metal:crown complexation, ion-pairing equilibria, and background extraction by the diluent. A thermochemical scheme has been proposed to show how ligand strain energies calculated by molecular mechanics can be related to the net extraction process.

Acknowledgments

The authors would like to acknowledge the contributions of their co-workers whose efforts have contributed significantly in the development of new solvent extraction reagents for lithium separation and efforts to better understand the factors involved in lithium ion recognition: Charles F. Baes, Jr., James J. Bruce, John H. Burns, Kerri L. Cavanaugh, Zhihong Chen, Goutam Das, Matthew C. Davis, Jon L. Driver, Tamara H. Haverlock, Benjamin P. Hay (PNNL), Jerome M. Lavis, Erik S. Ripple, and Yunfu Sun. This research was sponsored by the Division of Chemical Sciences, Office of Basic Energy Sciences, U. S. Department of Energy, under contract number DE-AC05-96OR22464 with Oak Ridge National Laboratory, managed by Lockheed Martin Energy Research Corp.

Literature Cited

(1) Cram, D. J.; Cram, J. M. *Science* **1974**, *183*, 803.
(2) *Host-Guest Complex Chemistry I*; Vögtle, F., Ed.; Topics in Current Chemistry; Springer-Verlag: Berlin, 1981; Vol. 98.
(3) *Host-Guest Complex Chemistry II*; Vögtle, F., Ed.; Topics in Current Chemistry; Springer-Verlag: Berlin, 1982; Vol. 101.
(4) *Host-Guest Complex Chemistry III*; Vögtle, F.; Weber, E., Eds.; Topics in Current Chemistry; Springer-Verlag: Berlin, 1984; Vol.121.
(5) For current and comprehensive reviews see: *Molecular Recognition: Receptors for Cationic Guests*; Gokel, G. W., Ed.; Comprehensive Supramolecular Chemistry; Pergamon, Elsevier: New York, 1996; Vol. 1. *Molecular Recognition: Receptors for Molecular Guests*; Vögtle, F., Ed.; Comprehensive Supramolecular Chemistry; Pergamon, Elsevier: New York, 1996; Vol. 2.
(6) Pedersen, C. J. *J. Am. Chem. Soc.* **1967**, *89*, 7017.
(7) Dietrich, B.; Lehn, J.-M.; Sauvage, J.-P. *Tetrahedron Lett.* **1969**, 2885.
(8) Gokel, G. W.; Dishong, D. M.; Diamond, C. J. *J. Chem. Soc., Chem. Commun.* **1980**, 1053.
(9) Gokel, G. W. *Crown Ethers and Cryptands*; Royal Society of Chemistry: Cambridge, 1991; and references cited therein.
(10) Christensen, J. J.; Hill, J. O.; Izatt, R. M. *Science* **1971**, *174*, 459.
(11) Lamb, J. D.; Izatt, R. M.; Christensen, J. J.; Eatough, D. J. In *Coordination Chemistry of Macrocyclic Compounds*; Melson, G. A., Ed.; Plenum: New York, 1979; Chapter 3.
(12) Michaux, G.; Reese, J. *J. Am. Chem. Soc.* **1982**, *104*, 6895.
(13) Schults, R. A.; Dishong, D. M.; Gokel, G. W. *J. Am. Chem. Soc.* **1982**, *104*, 625.
(14) Gokel, G. W.; Goli, D. M.; Minganti, C.; Echegoyen, L. *J. Am. Chem. Soc.* **1983**, *105*, 6786.
(15) *Lithium - Current Application in Science, Medicine and Technology*, Bach, R. O., Ed.; Wiley: New York, 1985.
(16) *Lithium: Inorganic Pharmacology and Psychiatric Use*, Birch, N. J., Ed.; IRL Press: Oxford, 1988.
(17) Olsher, U.; Izatt, R. M.; Bradshaw, J. S.; Dalley, N. K. *Chem. Rev.* **1991**, *91*, 137.
(18) Bartsch, R. A.; Ramesh, V.; Bach, R. O.; Shono, T.; Kimura, K. In *Lithium Chemistry: An Experimental and Theoretical Overview*; Saspe, A.-M.; Von Ragué Schleyer, P., Eds.; John Wiley & Sons: New York, 1995; Ch. 10.
(19) Inoue, Y.; Hakushi, T.; Li, Y.; Tong, L-H. *J. Org. Chem.* **1993**, *58*, 5411.
(20) Czech, B. P.; Babb, D. A.; Son, B.; Bartsch, R. A. *J. Org. Chem.* **1984**, *49*, 4805.

(21) Bartsch, R. A.; Czech, B. P.; Kang, S. I.; Stuart, L. E.; Walkowiak, W.; Charewicz, W. A.; Heo, G. S.; Son, B. *J. Am. Chem. Soc.* **1985**, *107*, 4997.
(22) Wakita, R.; Yonetani, M.; Nakatsuji, Y.; Okahara, M. *J. Org. Chem.* **1990**, *55*, 2752.
(23) Olsher, U. *J. Am. Chem. Soc.* **1982**, *104*, 4006.
(24) Bartsch, R. A. *Solvent Extr. Ion Exch.* **1989**, *7*, 829.
(25) Sachleben, R. A.; Moyer, B. A.; Driver, J. L. *Sep. Sci. Technol.* **1995**, *30*, 1157.
(26) Sachleben, R. A.; Davis, M. C.; Bruce, J. J.; Ripple, E. S.; Driver, J. L.; Moyer, B. A. *Tetrahedron Lett.* **1993**, *34*, 5373.
(27) Kobiro, K.; Matsuoka, T.; Takada, S.; Kakiuchi, K.; Tobe, Y.; Odaira, Y. *Chem. Lett.* **1986**, 713.
(28) March, J. *Advanced Organic Chemistry*, 3rd ed.; John Wiley and Sons: New York, 1985; p 220.
(29) Suzuki, K.; Yamada, H.; Sato, K.; Watanabe, K.; Hisamoto, H.; Tobe, Y.; Kobiro, K. *Anal. Chem.* **1993**, *65*, 3404.
(30) Kobiro, K. *Coord. Chem. Rev.* **1996**, *148*, 135.
(31) Hay, B. P.; Rustad, J. R. *J. Am. Chem. Soc.* **1994**, *116*, 6316; and references cited therein.
(32) Hay, B. P.; Rustad, J. R.; Hostetler, C. J. *J. Am. Chem. Soc.* **1993**, *115*, 11158.
(33) Sun, Y; Chen, Z.; Cavenaugh, K. L.; Sachleben, R. A.; Moyer, B. A. *J. Phys. Chem.* **1996**, *100*, 9500.
(34) Shoham, G.; Christianson, D. W.; Bartsch, R. A.; Heo, G. S.; Olsher, U.; Lipscomb, W. N. *J. Am. Chem. Soc.* **1984**, *106*, 1280.
(35) Buchanan, G. W.; Kirby, R. A.; Charland, J. P. *J. Am. Chem. Soc.* **1988**, *110*, 2477.
(36) Kobiro, K.; Hiro, T.; Matsuoka, T.; Kakiuchi, K.; Tobe, Y.; Odaira, Y. *Bull. Chem. Soc. Jpn.* **1986**, *61*, 4164.
(37) Shoham, G.; Lipscomb, W. N.; Olsher, U. *J. Chem. Soc., Chem. Commun.* **1983**, 208.
(38) Buchanan, G. W.; Kirby, R. A.; Charland, J. P. *J. Am. Chem. Soc.* **1988**, *110*, 2477.
(39) Sachleben, R. A. ; Burns, J. H. *J. Chem. Soc., Perkin 2* **1992**, 1971.
(40) Burns, J. H.; Sachleben, R. A.; Davis, M. C. *Inorg. Chim. Acta* **1994**, *223*, 125.
(41) Moyer, B. A.; Sachleben, R. A.; Sun, Y.; Driver, J. L.; Chen, Z.; Cavanaugh, K. L.; Carter, R. W.; Baes, C. F., Jr., In *Value Adding Through Solvent Extraction: Proceedings of ISEC '96* (Proc. International Solvent Extraction Conference, Melbourne, Australia, Mar. 19-23, 1996) Shallcross, D. C.; Paimin, R.; Prvcic, L. M., Eds.; The University of Melbourne: Melbourne, Australia, 1996; p 359.
(42) Moyer, B. A.; Bonnesen, P. V. In *The Supramolecular Chemistry of Anions*; Bianchi, A.; Bowman-James, K.; Garcia-España, E., Eds.; VCH: Weinheim, 1997; Ch. 1, in press.
(43) Moyer, B. A. In *Molecular Recognition: Receptors for Cationic Guests*; Gokel, G. W., Ed.; *Comprehensive Supramolecular Chemistry*; Atwood, J. L.; Davies, J. E. D.; MacNicol, D. D.; Vögtle, F.; Lehn, J.-M., Eds.; Pergamon, Elsevier: Oxford, 1996; Vol. 1, Ch. 10.
(44) Moyer, B. A.; Sun, Y. In *Ion Exchange and Solvent Extraction*; Marcus, Y.; Marinsky, J. A., Eds.; Marcel Dekker: New York, 1997; Vol. 13, Ch. 6.
(45) Sachleben, R. A., Hay, B. P.; Sun, Y.; Moyer, B. A. **1997**, manuscript in preparation.

(46) Hay, B. P. In *Progress in Metal Ion Separation and Preconcentration*, Bond, A. H.; Dietz, M. L.; Rogers, R. D., Eds.; ACS Symposium Series; American Chemical Society: Washington, DC, 1997, in press.
(47) Hay, B. P.; Zhang, D.; Rustad, J. R. *Inorg. Chem.* **1996**, *35*, 2650.
(48) Cram, D. J. *Angew. Chem., Int. Ed. Engl.* **1986**, *25*, 1039.
(49) Cram, D. J.; Lein, G. M. *J. Am. Chem. Soc.* **1985**, *107*, 3657.
(50) Marcus, Y.; Asher, L. E. *J. Phys. Chem.* **1978**, *82*, 1246.
(51) Takeda, Y. *Host Guest Complex Chemistry III*; Vogtle, F.; Weber, E., Eds.; Topics in Current Chemistry; Springer-Verlag: Berlin, 1984; Vol. 121, Ch. 1.
(52) Bartsch, R. A.; Goo, M.; Christian, G. D.; Wen, X.; Czech, B. P.; Chapoteau, E.; Kumar, A. *Anal. Chim. Acta* **1993**, *272*, 285.
(53) Marcus, Y. *Solvent Extr. Ion Exch.* **1992**, *10*, 527.
(54) Riddick, J. A.; Bunger, W. B.; Sakano, T. K. *Organic Solvents: Physical Properties and Methods of Purification*; 4th ed.; Techniques of Chemistry; Weissberger, A., Ed.; Wiley-Interscience: New York, 1986; Vol. II.
(55) Westall, J. C.; Johnson, C. A.; Zhang, W. *Environ. Sci. Technol.* **1990**, *24*, 1803.
(56) Marcus, Y.; Pross, E.; Hormadaly, J. *J. Phys. Chem.* **1980**, *84*, 2708.
(57) Glikberg, S.; Marcus, Y. *J. Soln. Chem.* **1983**, *12*, 255.
(58) Marcus, Y.; Kamlet, M. J.; Taft, R. W. *J. Phys. Chem.* **1988**, *92*, 3613.
(59) Baes, C. F., Jr.; Moyer, B. A. *J. Phys. Chem.*, in press.
(60) Reichardt, C. *Solvents and Solvent Effects in Organic Chemistry*; 2nd ed.; VCH: Weinheim, 1990.
(61) Hormadaly, J.; Marcus, Y. *J. Phys. Chem.* **1979**, *83*, 2843.
(62) Shmidt, V. S. *Russ. Chem. Rev.* **1978**, *47*, 929-943; *Uspekhi Khimii* **1978**, *47*, 1730.
(63) Marcus, Y. *Solvent Extr. Ion Exch.* **1989**, *7*, 567.
(64) Kenjo, T.; Diamond, R. M. *J. Inorg. Nucl. Chem.* **1974**, 36, 183.
(65) Dietz, M. L.; Horwitz, E. P.; Rhoads, S.; Bartsch, R. A.; Krzykawski, J. *Solvent Extr. Ion Exch.* **1996**, *14*, 1.
(66) Baes, C. F., Jr.; Moyer, B. A.; Case, G. N.; Case, F. I. *Sep. Sci. Technol.* **1990**, *25*, 1675.
(67) Deng, Y.; Sachleben, R. A.; Moyer, B. A. *J. Chem. Soc., Faraday Trans.* **1995**, *91*, 4215.

Chapter 8

Synthesis of Novel Azamacrocyclic Metal Ion Receptors Using a Modified Mannich Aminomethylation Reaction

Jerald S. Bradshaw, Andrei V. Bordunov[1], Xian Xin Zhang, Victor N. Pastushok[2], and Reed M. Izatt

Department of Chemistry and Biochemistry, Brigham Young University, Provo, UT 84602

A modified Mannich aminomethylation reaction was used to prepare a series of N-substituted-phenol-containing azacrown ether ligands. Ligands synthesized include aza-15-crown-5, aza-18-crown-6, azapyridino-18-crown-6, diaza-18-crown-6, diaza-21-crown-7 and diaza-24-crown-8 containing various substituted phenols, salicylaldehyde, and 5-chloro-8-hydroxyquinoline (CHQ) groups as side arms. The modified Mannich reaction was also used to prepare bi- and tricyclic azamacroheterocycles containing phenol units, benzoazacrown ethers, benzocryptands, and cryptohemispherands. The phenol- and CHQ-substituted azacrown ligands interact more strongly with metal ions than do the parent unsubstituted azacrown ethers. Bis-CHQ-substituted diaza-18-crown-6, wherein the CHQ groups are attached through their 7 positions, are particularly selective for K^+ over other alkali metal ions and for Ba^{2+} over all other metal ions studied.

---

[1]Current address: Department of Chemistry, California Institute of Technology, Pasadena, CA 91125.
[2]Current address: A.V. Bogatsky Physico Chemical Institute, National Academy of Sciences of Ukraine, Odessa 270080, Ukraine.

A Mannich aminomethylation reaction, where an amine is reacted with an aldehyde and a C-H or N-H acid to form aminomethyl derivatives, has wide use in organic synthesis (*1*). We have explored this type of interaction for the synthesis of supramolecular metal ion complexing agents (*2*). This general approach to substituted azacrown ethers simplifies existing synthetic methodologies and, in contrast to conventional methods, allows the preparation of azamacrocyclic ligands in good to excellent yields in mostly one or two steps. The modified Mannich reaction expands considerably the functional and structural variety of synthesized azamacrocycles and facilitates their large-scale synthesis, particularly in the case of azacrown ethers and cryptands containing phenolic units. The modified Mannich reaction was also used to prepare bi- and tricyclic azamacroheterocycles, benzo- and bisazacrown ethers and cryptands. Lariat ether ligands synthesized include aza-15-crown-5, aza-18-crown-6, azapyridino-18-crown-6, diaza-18-crown-6, diaza-21-crown-7 and diaza-24-crown-8 lariat ether macrocycles functionalized with various substituted phenols, heterocycles, amides, imides, and sulfamides.

This synthetic method eliminates the need for benzyl halides or phenylacyl chlorides as starting building blocks for coupling with nitrogen-containing fragments. Aromatic azamacrocyclic frameworks can be constructed by Mannich-type condensation between *N*-methoxymethylamines and phenolic substances that are readily available or easily synthesized. Furthermore, Mannich condensation does not require protecting groups for the phenolic hydroxide under normal conditions. This reaction also allows benzylamine bond formation between phenols and secondary amines in the presence of other unprotected functional groups which can be used in further chemical transformations. Thus, phenolic hydroxides as well as carbonyl groups of aromatic fragments attached to the azamacrocyclic rings have been exploited to prepare ligands containing a greater number of rings (*3*) and Schiff base lariat azacrown ethers (*4*).

The basic idea of combining macrocyclic fragments with molecules of phenol-containing analytical reagents to improve the strength and, especially, selectivity of metal ion binding was realized via Mannich condensation of *N*-methoxymethylazacrown ethers with 5-chloro-8-hydroxyquinoline (CHQ) (*5*). A number of novel UV/fluorescent active metal ion receptors exhibit extremely high complexing ability and specificity upon binding cations. Highly specific reagents for K^+ and Ba^{2+} and chromogenic ligands for Mg^{2+} and Zn^{2+} have been found among CHQ-modified azacrown macrocycles (*5,6*). Several pyridinocrown ethers functionalized with phenol side arms show very high specificity towards Ag^+ over other monovalent cations (*7*). This chapter reviews the work done in our laboratory on the synthesis of macroheterocyclic ligands using the modified Mannich reaction. The interaction of some of these new ligands with various metal ions is also reviewed.

Phenol and 5-Chloro-8-hydroxyquinoline Armed Azacrown Ethers

A number of phenol-containing lariat azacrown ethers has been reported as selective chromogenic reagents for the extraction of alkali and alkaline earth metal ions from water to an organic phase (*8,9*). The usual method to prepare these lariat ethers is to react a hydroxy-substituted benzyl halide with the unsubstituted azacrown ether. The need to prepare suitable starting benzyl halides and the necessity to protect their phenolic hydroxy groups (*10*) often make this alkylation reaction difficult and time consuming. On the other hand, the aminomethylation reaction allows the preparation of *N*-phenol-substituted azacrown ethers in one step from inexpensive and accessible phenols containing electron donating as well as electron withdrawing groups on the substituent phenolic ring (*11*). Figure 1 shows the structures of some lariat ether metal ion receptors synthesized in our laboratory using the modified Mannich aminomethylation reaction. *N*-Methoxymethylazacrown ethers have been used as key intermediates in all chemical transformations leading to these materials.

Scheme 1 shows the typical synthesis of *N*-methoxymethyl derivatives of the azacrown ethers via treatment of the azamacrocycle with formaldehyde in methanol (*12,13*). *N*-Methoxymethylazacrown ethers were obtained in almost quantitative yields. The *N*-methoxymethyl-substituted azacrowns were then treated with phenolic compounds in non-polar solvents. Ligand **20**, for example, was prepared as follows (*13,14*). A solution of 1 g (3.8 mmol) of 1,10-diaza-18-crown-6 in 3 mL of pure methanol was added to a mixture of 0.23 g (7.6 mmol) of paraformaldehyde in 3 mL of methanol that was purified by refluxing in the presence of a trace of potassium hydroxide. The resulting methanol solution was left to stand for 12 h at 20 °C. The methanol was removed and the residue was dissolved in 5 mL of pure ether and the mixture was filtered. The filtrate was concentrated to 2 mL and cooled to -50 °C. The resulting crystals were rapidly filtered to give 0.83 g of *N,N*'-bis(methoxymethyl)diaza-18-crown-6; mp 36-37 °C. A further 0.26 g of product was obtained from the filtrate making a total of 1.09 g (82%) (*13*).

N,N'-Bis[(5-chloro-8-hydroxyquinolin-7-yl)methyl]-diaza-18-crown-6 (**20**) was prepared by treating 1 g (2.9 mmol) of the above *N,N*'-bis(methoxmethyl)diaza-18-crown-6 with 1 g (5.6 mmol) of 5-chloro-8-hydroxyquinoline in 30 mL of refluxing benzene for 10 h. The hot solution was filtered and the filtrate was evaporated under reduced pressure. The residue was mixed with 15 mL of a hot mixture of benzene and THF (1:1). Ligand **20** crystallized when the solution stood for 24 h to give 1.2 g (67%); mp 140-141 °C (*14*).

The aza- and diazacrown ethers are available but expensive. They can be prepared by a number of methods as we have outlined in a recent book (*15*). Nevertheless, the high yields for *N*-methoxymethylation and subsequent aminomethylation of the phenol make this method useful for the preparation of phenol-substituted azacrown ethers.

Unsubstituted phenols generally react in the *ortho*-position, possibly because of a six-membered transition state wherein the phenolic proton activates the aminomethylating reagent (*11*). When both *ortho*-positions are occupied, aminomethylation occurs on the unsubstituted *para*-position (*12,16*). Most of our synthetic transformations utilized *para*-substituted phenols or hydroxyquinolines as starting intermediates giving the products **1-6, 9-14, 16-20, 22**, and **23** shown in Figure 1.

CHQ derivatized azacrown ethers **3, 6, 7, 14, 15**, and **20-23** are the first members of a new family of metal ion receptors designed to have a combination of soft binding fragments of known analytical reagents and the hard donor atoms of crown ether moieties (*5,6,14*). The combination of these complexing subunits in one molecule provides a receptor with increased steric requirements caused by the geometry of the macrocyclic ring and the mode of attachment of the CHQ groups to the azacrown ether macrocycle.

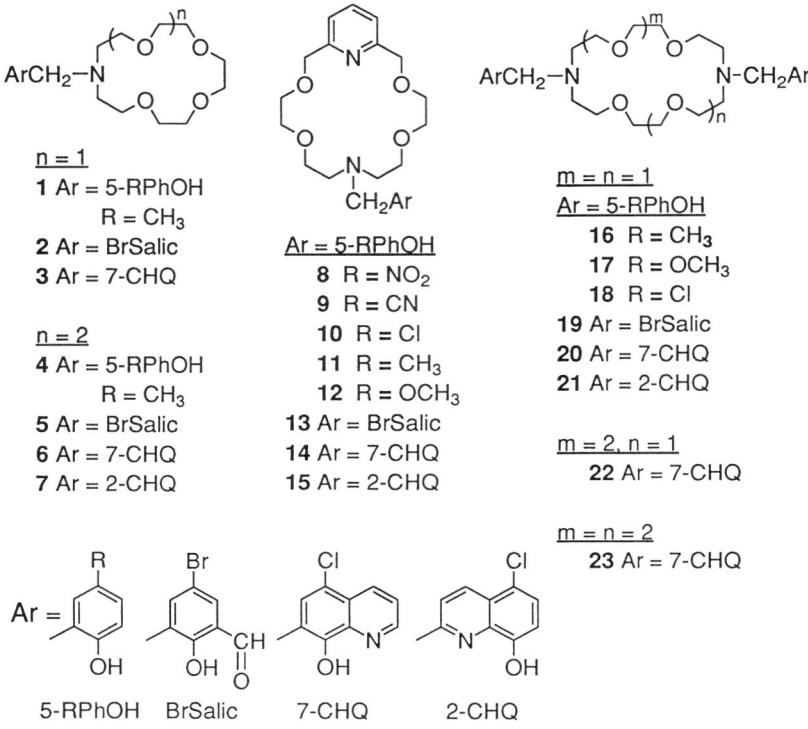

Figure 1. Azacrown ethers containing hydroxyaromatic substituents.

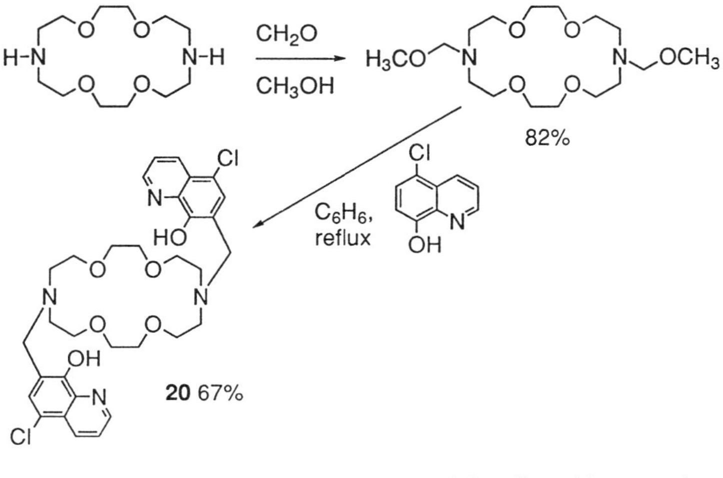

Scheme 1. Synthesis of bis-CHQ-containing diaza-18-crown-6.

Such cooperative coordination to metal ions increases both the binding affinity and selectivity of the ligand. Moreover, UV and/or fluorescent responses upon complexation due to the aromatic fragments of these molecules make them promising candidates to be metal ion sensing agents. The CHQ modified azacrown ethers **3, 6, 14, 20, 22,** and **23**, wherein the CHQ units are attached through their 7 positions were prepared as shown in Scheme 1 (*5*). The ligands with CHQ groups attached through their 2 positions (**7, 15,** and **21**) were prepared by alkylation of azamacrocycles with 2-bromomethyl-5-chloro-8-methoxyquinoline (*5,14*). The methyl protecting group was removed using lithium chloride in DMF at 130 °C. The attachment position of the azamacrocycle onto CHQ has a profound effect on the complexing ability of CHQ-armed azacrown ethers as discussed below.

Bi- and Tricyclic Azamacroheterocyclic Compounds Containing Phenol Units

The development of simple synthetic approaches for introducing phenolic fragments into a three-dimensional macrocyclic framework is important for at least four reasons. (1) Aromatic moieties provide the receptor with rigidity which reduces the entropy loss upon complexation with ions or guest molecules. (2) Intraannular hydroxy functions increase the ability of the ligand to interact with metal ions as well as with molecules that form hydrogen bonds with the host. Macrocycles containing phenol groups as a part of their framework exhibit higher affinities and selectivities for Ni^{2+}, Cu^{2+}, Zn^{2+}, Pb^{2+}, and Cd^{2+} compared to the aliphatic cryptands (*17*). (3) Electron rich phenolic fragments can create additional types of interactions such as π-π interactions, interactions with ammonium cations or highly polarizable metal ions (Cs^+). (4) Aromatic moieties incorporated into the macrocyclic framework are convenient sites for further chemical modification with chromogenic groups, polymerizable functions, etc.

Conventional approaches for the synthesis of phenol containing cryptands developed by Bartsch and coworkers (*17-19*) include several steps, protection and deprotection of phenolic OH groups, reduction with lithium aluminum hydride, and ring closure under high dilution conditions. The modified Mannich reaction allows the preparation of bicyclic cryptands with internal phenolic functional groups in one-step at relatively high concentrations of the cyclizing reagents without using any protecting groups (*20*) (see preparation of **26** from **24** and **25** as an example in Scheme 2). The modified Mannich reaction was also used to prepare tricyclic phenol-containing macroheterocyclic compounds (Scheme 3) (*20*). Precursor **16**, prepared by reacting **24** with *p*-cresol, was treated with **24** in refluxing xylene to form tricycle **27**. The second aminomethylation reaction requires a higher temperature. This temperature dependence allows the unsymmetric substitution of a phenol ring by introduction of different aminomethyl subunits onto the two *ortho*-positions of the phenol substrate (*21*).

The advantage of Mannich condensation can be seen by comparing the standard methods and aminomethylation cyclization for the synthesis of cryptohemispherands. Cryptohemispherand-type ligands, prepared by Cram and coworkers (*22*), show extremely high binding affinities and selectivities towards alkali metal ions, particularly Na^+ and Cs^+ (*23*). High selectivity allows these materials to be used in chemical sensing systems (*24*). The standard method of synthesizing cryptohemispherands requires several tedious steps to prepare either the bis(bromobenzyl)-containing aromatic fragments or

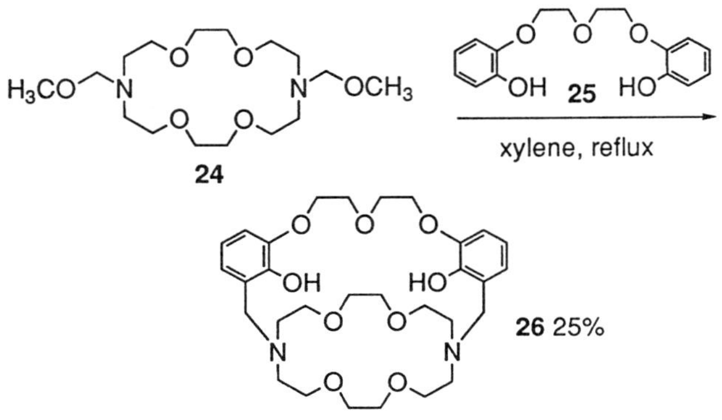

Scheme 2. Synthesis of a phenol-containing macrobicycle.

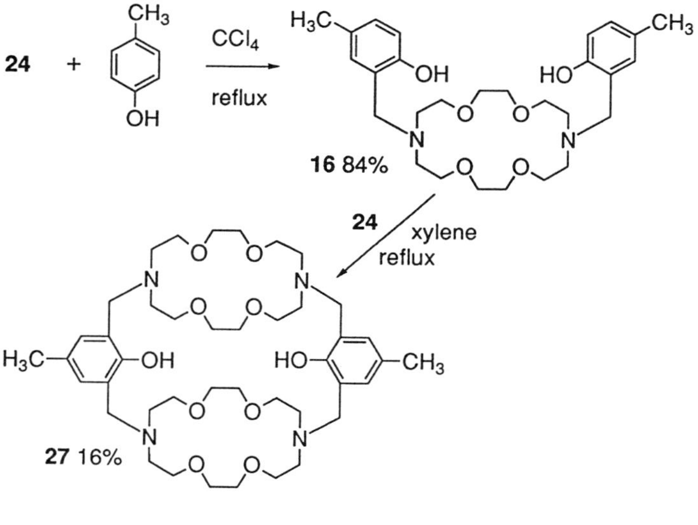

Scheme 3. Synthesis of phenol-containing macrotricycle **27**.

the diacid chloride derivatized trianisole precursors (*22,25*), as well as subsequent treatment of macrocyclic products (reduction of amide bonds and decomposition of borane complexes) after ring closure. Scheme 4 shows a one-step approach to the synthesis of a cryptohemispherand type molecule (**28**) from *N,N'*-bis(methoxymethyl)-diaza-18-crown-6 and the appropriate triphenolic substrate (*20*). The relatively high 45% yield of the final macrocycle was achieved in spite of using concentrated solutions of the reagents and the abscence of a template assembling matrix. The high efficiency of macrocyclization is probably due to self assembly of the reactant species through intramolecular hydrogen bonds (*20*).

Benzoazacrown Ethers and Cryptands

Introducing aromatic units into an azacrown ether framework has usually been done by reactions of functional groups already attached to the aromatic ring (*15*). These functional groups can be aminomethyl, halomethyl for alkylation of NHR end functions, or aromatic diacid chlorides for acylation of NHR end groups followed by reduction. Formation of cyclic bis-Schiff bases from aromatic dialdehydes followed by reduction is another method that has been used to form benzoazacrown ethers (*26, 27*). Most of these methods require additional chemical treatment after ring closure and often a several-step synthesis of starting difunctional aromatic compounds. The modified Mannich reaction using secondary diamines allows the construction of two benzylamine functionalities in one step. For example, treatment of two equivalents of *p*-chlorophenol with 1,7-dimethyl-4-oxa-1,7-diazaheptane, which had been pretreated with paraformaldehyde in water and dioxane, gave diazadiphenol **29** in an 88% yield (Scheme 5) (*3*). Synthon **29** was then reacted with 1,7-bis(methoxymethyl)-1,7-dimethyl-4-oxa-1,7-diazaheptane to give bisphenol-containing tetraaza-18-crown-6 **30** in 10% yield. Synthon **29** was also reacted with diethylene glycol ditosylate and potassium carbonate in refluxing acetonitrile to give dibenzodiazacrown **31** in a 43% yield. Bisphenol macrocycle **30** was likewise reacted with the same ditosylate to give dibenzo-macrotricycle **32** in a 71% yield. Scheme 5 shows just a few examples of the many types of benzoazamacrocycles that can be prepared using the modified Mannich aminomethylation reaction (*3*). A series of dibenzocryptands has also been prepared by treating bis(*p*-chlorophenol)-substituted diaza-18-crown-6 (**18**) with three ditosylate/-dihalide substrates (*28*).

Metal Ion Complexation By Phenol- and CHQ-substituted Azacrown Ethers

Some of the new phenol- and CHQ-substituted azacrown and pyridinoazacrown ethers shown in Figure 1 exhibit enhanced complexing ability and cation specificity. Table 1 lists log K values for the interaction of the azacrown ethers with some metal ions as determined by a calorimetric titration technique in methanol. Compared with the parent unsubstituted azacrown ethers, these new lariat macrocycles show increased binding constants for all metal ions studied. For example, the log K values for **3** interacting with Na^+ and K^+ are 3.00 and 3.17, respectively, while the parent A15C5 binds Na^+ and K^+ much more weakly (log K values of 1.70 and 1.60, respectively). Parent ligand PyA18C6 shows a very weak interaction with Na^+ (no reaction heat can be measured in MeOH).

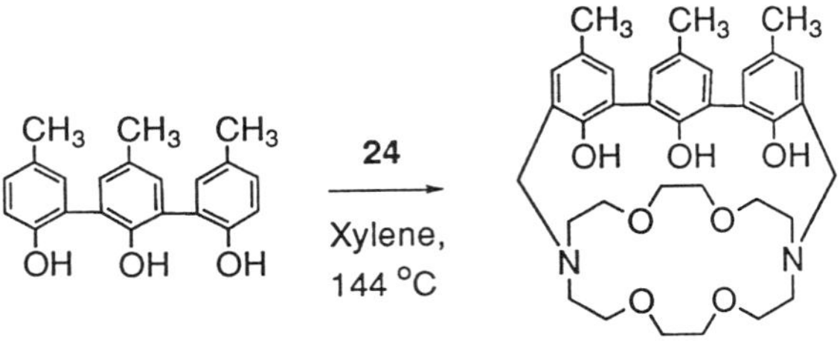

Scheme 4. Synthesis of cryptohemispherand **28**.

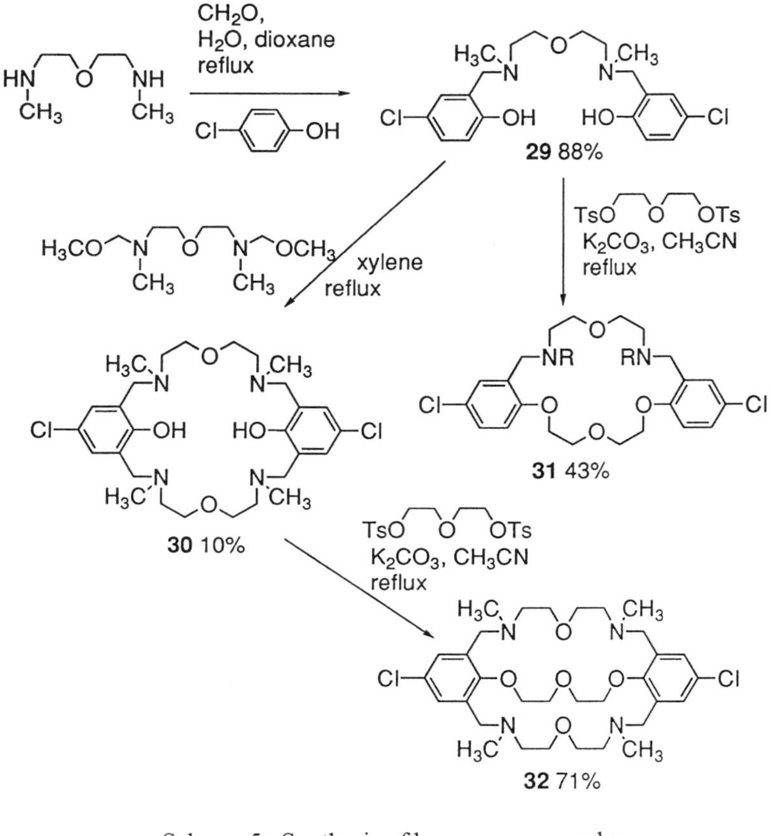

Scheme 5. Synthesis of benzoazamacrocyles.

Table 1. Log K Values[a] for the Interaction of Azacrown Ethers Containing Hydroxy-aromatic Sidearms With Various Metal Ions in Methanol at 25.0 °C.

Ligand	Na$^+$	K$^+$	Ba^{2+}	Cu^{2+}	Tl$^+$	Ag$^+$	Ref.
A15C5[b]	1.70	1.60					30
2	3.32	2.71					4
3	3.00	3.17	4.28	7.88			29
A18C6[b]	2.69						30
5	3.11	4.07					4
6	3.60	4.47	4.08	9.44			29
7	3.98	5.42	6.20	5.52			29
PyA18C6[b]	c						7
8	3.06	3.17			4.20	>8.5	7
9	3.10	3.27			4.22	>9	7
10	3.29	3.41			4.29	>9	7
11	3.49	3.53			4.34	>9	7
12	3.59	3.62			4.40	>9	7
13	3.21	3.40					4
14	3.85	4.01	4.12	8.12			29
15	4.20	5.16	5.49	3.72			29
DA18C6[b]	c	1.83	6.12	8.48	3.06		30
18	2.85	2.76	3.52	4.14			5
20	2.89	3.39	3.60	10.1			5
21	3.74	6.61	12.2	4.7			5

[a]Values determined by calorimetric titration.
[b]A15C5 = aza-15-crown-5, A18C6 = aza-18-crown-6, PyA18C6 = pyridinoaza-18-crown-6, DA18C6 = 1,10-diaza-18-crown-6.
[c]No measurable heat.

However, ligands **8-15** with different phenol, salicylaldehyde, and CHQ arms complex Na^+ with log K values of 3.06-4.20. CHQ modified azamcrocycles show especially interesting properties in terms of the correlation between binding ability and structures of ligands studied. Compounds **3, 6, 14,** and **20** form very stable complexes with transition metal ions (log K values for Cu^{2+} are shown as an example in Table 1). The enhanced copper binding specificity compared to the alkali and alkaline earth metal ions is due to participation of the CHQ moieties in complex formation. Ligand **18**, containing only chlorophenol sidearms, exhibits smaller log K values for interactions with metal ions. Ligand **18** forms a less stable complex with Cu^{2+} than does **20** by six orders of magnitude. In addition, **18** shows no special selectivity among the metal ions studied.

The position of attachment of the CHQ groups to the macroring has a significant effect on cation complexation. Ligands **3, 6, 14,** and **20** have CHQ attached through its 7 position (next to the OH group) and form more stable complexes with Cu^{2+} than with Na^+, K^+ and Ba^{2+} (5, 29). However, when CHQ is attached through its 2 position (next to the quinoline nitrogen), the resulting ligands **7, 15,** and **21** exhibit strong interactions with Na^+, K^+ and Ba^{2+} but decreased interactions with Cu^{2+}. Such reversed complexing ability is caused by different mutual positions of the quinoline OH groups and aliphatic nitrogen atoms of the azamacrocycles in 2-CHQ-substitututed (**7, 15,** and **21**) and 7-CHQ-substituted (**6, 14,** and **20**) ligands. Compounds **6, 14,** and **20** having hydroxy groups and azacrown nitrogens in a close proximity form less stable complexes with alkali metal ions and Ba^{2+} because of the deactivation of aliphatic nitrogen atoms through intramolecular hydrogen bonds with the quinoline OH groups (5). The hydroxy group also causes the soft nitrogen atom of the quinoline to be in a pseudo axial position and, thus, the nitrogen atom would not be available to metal ions complexed in the macroring.

Bis-CHQ-substituted **21** exhibits high selectivity for K^+ and Ba^{2+} over Na^+ and Cu^{2+}. Log K values for the formation of K^+ and Ba^{2+} complexes with **21** are larger than those for K^+ and Ba^{2+} complexes with all other lariat ethers (30). The log K value for the **21**-Ba^{2+} complex (12.2 in MeOH) is the same magnitude as that of the cryptand [2.2.2]-Ba^{2+} complex (12.9 in MeOH (29)). Selectivity factors for Ba^{2+} over other alkaline-earth cations and for K^+ over Na^+ are >10^7 and ~10^3, respectively (Table 1), which are the highest factors ever reported for lariat ethers. Moreover, the selectivity of **21** for Ba^{2+} is larger than that of any cryptand studied to date. The special complexing properties of **21** are related to its peculiar molecular structure. Through coordination with K^+ and Ba^{2+}, the two CHQ substitutents of **21** can overlap each other through π-π interaction so that a pseudo second macroring is formed. This effect results in a cryptate-like structure and, therefore, highly stable complexes (5,6).

High selectivity of CHQ-containing metal ion receptors is combined with their ability to perform UV/fluorescence response upon complexation. Ligand **20**, having high Mg^{2+} selectivity over other alkaline-earth, alkali metal, and zinc ions, also has a very specific absorption at 265 nm when it is bound to Mg^{2+} (5). Moreover, **20** fluoresces strongly in the presence of Zn^{2+}, but not with Na^+ or K^+. The chromogenic features of **20** allow the application of this receptor for the measurement of Mg^{2+} and Zn^{2+} concentrations in very dilute solutions and in mixtures with other metal ions.

Phenol-substituted pyridinoazacrown ethers **8-12** form stable complexes with Na^+, K^+, Tl^+, and Ag^+ in MeOH and show high selectivity for Ag^+ over the other cations studied. In each case, the log K values for complex formation increase in the order

Na$^+$<K$^+$<Tl$^+$<<Ag$^+$ (7). Selectivity factors of Ag$^+$ over Na$^+$, K$^+$, and Tl$^+$ by **8-12** are larger than four orders of magnitude. Furthermore, there is a correlation between binding ability of **8-12** and electron donating (withdrawing) character of the substituents in the *para*-position to phenolic hydroxide. The stability of metal ion complexes increases in the order **8**(NO$_2$)<**9**(CN)<**10**(Cl)<**11**(CH$_3$)<**12**(OCH$_3$). The linear dependence of binding and protonation-deprotonation constants on Hammett σ_p constants of the substituents have been observed for ligands **8-12** (7). This makes introducing phenolic groups into the macrocyclic structures a very promising approach for designing ligands with predicted complexing abilities.

Acknowledgement

U. S. Department of Energy, Chemical Sciences Division, Office of Basic Energy Science, Contract no. DE-FG02-86ER 13463 provided funds for this work.

Literature Cited

(1)   Tramontini, M.; Angliolini, L. *Tetrahedron* **1990**, *46*, 1791.
(2)   Bordunov, A. V.; Bradshaw, J. S.; Pastushok, V. N.; Izatt, R. M. *Syn. Lett.* **1996**, 933.
(3)   Pastushok, V. N.; Bradshaw, J. S.; Bordunov, A. V.; Izatt, R. M. *J. Org. Chem.* **1996**, *61*, 6888.
(4)   Bordunov, A. V.; Bradshaw, J. S.; Pastushok, V. N.; Zhang, X. X.; Kou, X.; Dalley, N. K.; Yang, Z.; Savage, P. B.; Izatt, R. M. *Tetrahedron* **1997**, in press.
(5)   Bordunov, A. V.; Bradshaw, J. S.; Zhang, X. X.; Dalley, N. K.; Kou, X.; Izatt, R. M. *Inorg. Chem.* **1996**, *35*, 7229.
(6)   Zhang, X. X.; Bordunov, A. V.; Bradshaw, J. S.; Dalley, N. K.; Kou, K.-L.; Izatt, R. M. *J. Am. Chem. Soc.* **1995**, *117*, 11507.
(7)   Zhang, X. X.; Bordunov, A. V.; Kou, X.; Dalley, N. K.; Izatt, R. M.; Mangum, J. H.; Li, D.; Bradshaw, J. S.; Hellier, P. C. *Inorg. Chem.* **1997**, *36*, 2586.
(8)   Nishida, H.; Katayama, Y.; Katsuki, H.; Nakamura, H.; Takagi, M.; Ueno, K. *Chem. Lett.* **1982**, 1853.
(9)   Katayama, Y.; Fukuda, R.; Iwasaki, T.; Nita, K.; Takagi, M. *Anal. Chim. Acta* **1988**, *204*, 113.
(10) Gatto, V. J.; Gokel, G. W. *J. Am. Chem. Soc.* **1984**, *106*, 8240.
(11) Lukyanenko, N. G.; Pastushok, V. N.; Bordunov, A. V. *Synthesis* **1991**, 241.
(12) Bogatsky, A. V.; Lukyanenko, N. G.; Pastushok, V. N.; Kostyanovsky, R. G. *Synthesis* **1983**, 992.
(13) Bogatsky, A. V.; Lukyanenko, N. G.; Pastushok, V. N.; Kostyanovsky, R. G. *Dol. Akad. Nauk SSSR* **1982**, *265*, 619; *Chem. Abstr.* **1982**, *97*, 216146c.
(14) Bordunov, A. V.; Hellier, P. C.; Bradshaw, J. S.; Dalley, N. K.; Kou, X.; Zhang, X. X.; Izatt, R. M. *J. Org. Chem.* **1995**, *60*, 6097.
(15) Bradshaw, J. S.; Krakowiak, K. E.; Izatt, R. M. In *The Chemistry of Heterocyclic Compounds*; Taylor, E. C., Ed.; Wiley: New York, 1993, Vol. 51.
(16) Habata, Y.; Akabori, S. *J. Chem. Soc., Dalton Trans.* **1996**, 3871.

(17) Czech, A.; Czech, B. P.; Bartsch, R. A.; Chang, C. A.; Ochaya, V. O. *J. Org. Chem.* **1988**, *53*, 5.
(18) Czech, A.; Czech, B. P.; Desai, D. H.; Hallman, J. L.; Phillips, J. B.; Bartsch, R. A. *J. Heterocyclic Chem.* **1986**, *23*, 1355.
(19) Chapoteau, E.; Czech, B. P.; Gebauer, C. R.; Kumar, A.; Leong, K.; Mytych, D. T.; Zazulak, W.; Desai, D. H.; Luboch, E.; Krzykawski, J.; Bartsch, R. A. *J. Org. Chem.* **1991**, *56*, 2575.
(20) Bordunov, A. V.; Lukyanenko, N. G.; Pastushok, V. N.; Krakowiak, K. E.; Bradshaw, J. S.; Dalley, N. K.; Kou, X. *J. Org. Chem.* **1995**, *60*, 4912.
(21) Lukyanenko, N. G.; Pastushok, V. N.; Bordunov, A. V.; Vetrogon, N. I.; Bradshaw, J. S. *J. Chem. Soc., Perkin Trans. 1* **1994**, 1489.
(22) Cram, D. J.; Ho, S. P.; Knobler, C. R.; Maverick, E.; Trueblood, K. N. *J. Am. Chem. Soc.* **1986**, *108*, 2989.
(23) Cram, D. J.; Ho, S. P. *J. Am. Chem. Soc.* **1986**, *108*, 2998.
(24) Helgeson, R. C.; Czech, B. P.; Chapoteau, E.; Gebauer, C. R.; Kumar, A.; Cram, D. J. *J. Am. Chem. Soc.* **1989**, *111*, 6339.
(25) Krakowiak, K. E.; Bradshaw, J. S.; Zhu, C.; Hathaway, J. K.; Dalley, N. K.; Izatt, R. M. *J. Org. Chem.* **1994**, *59*, 4082.
(26) Adam, K. R.; Lindoy, L. F.; Lip, H. C.; Rea, J. H.; Skelton, B. W.; White, A. H. *J. Chem. Soc., Dalton Trans.* **1981**, 74.
(27) Baldwin, D.; Duckworth, P. A.; Erickson, G. R.; Lindoy, L. F.; McPartlin, M.; Mockler, G. M.; Moody, W. E.; Tasker, P. A. *Aust. J. Chem.* **1987**, *40*, 1861.
(28) Bordunov, A. V.; Dalley, N. K.; Kou, X.; Bradshaw, J. S.; Pastushok, V. N. *J. Heterocyclic Chem.* **1996**, *33*, 933.
(29) Zhang, X. X.; Bradshaw, J. S.; Bordunov, A. V.; Izatt, R. M. *J. Inclusion Phenom. Molec. Recognit. Chem.* **1997**, *29*, 259.
(30) Izatt, R. M.; Pawlak, K.; Bradshaw, J. S.; Bruening, R. L. *Chem. Rev.* **1991**, *91*, 1721.

Separations Using Liquid–Liquid Systems

Chapter 9

Metal Ion Separations with Proton-Ionizable Lariat Ethers

Richard A. Bartsch

Department of Chemistry and Biochemistry, Texas Tech University, Lubbock, TX 79409–1061

A lariat ether is a crown ether to which a side arm bearing one or more potential coordination sites is attached. For a metal ion bound within the crown ether cavity, the side arm can provide axial coordination. Lariat ethers with pendant acidic functionality, such as a carboxylic acid group, are termed proton-ionizable lariat ethers. In its ionized form, a proton-ionizable lariat ether possesses both a crown ether cavity for metal ion complexation and the requisite anion for formation of an electroneutral metal ion complex. Movement of a metal ion from an aqueous solution into an organic medium in solvent extraction or transport across a liquid membrane does not require concomitant transfer of an aqueous phase anion. This factor is of immense importance to potential practical applications in which hard aqueous phase anions, such as chloride, nitrate, and sulfate, would be involved. Proton-ionizable lariat ethers with selectivity for targeted metal ions in solvent extraction and liquid membrane transport processes may be constructed by suitable structural variations within the extractant (carrier) molecule.

In the three decades since Pedersen reported the synthesis of crown ether (macrocyclic polyether) compounds and their complexation behavior toward various metal salts (*1*), interest in crown ethers as complexing agents has grown exponentially (*2-7*). With hard (*8*) donor atoms (oxygens), crown ethers are anticipated to exhibit strong association with hard metal ion species, alkali metal, alkaline earth, lanthanide and actinide cations. An important factor in determining the strength with which metal ions are complexed by a macrocyclic polyether ligand is the relationship between the diameter of the crown ether cavity and the diameters of the metal ions (*3,5*).

A potential application of crown ethers in metal ion separations is illustrated in Figure 1. Thus an aqueous stream which contains M^+ and P^+ metal ions is contacted with an organic stream which contains a crown ether compound (depicted as a circle in Figure 1). In the mixing zone, the selectivity of the crown ether for M^+ is expressed and M^+X^- is extracted into the organic phase as a crown ether complex; whereas P^+ remains in the aqueous phase. This separation process works well for lipophilic aqueous phase anions, X^-, such as perchlorate, picrate, and thiocyanate. However, for process solvent extraction of metal ions, the common anions are chloride, nitrate, and sulfate. Since such hard anions are extensively hydrated and resist transfer into an organic medium (and dehydration), distribution coefficients of the metal salt between an aqueous phase and an organic phase containing the crown ether are often too low to be useful (9-11).

Proton-Ionizable Lariat Ethers - The Concept

Attachment of one or more side arms with potential metal ion coordination sites to a crown ether framework produces complexing agents known as "lariat ethers" (12). A solution to anion transport into the organic phase is structural modification of the macrocyclic polyether ligand to include a pendent acidic function, for example a carboxylic acid group. Now when the metal ion is transferred into the organic phase (Figure 2), a proton is lost from the proton-ionizable lariat ether in an ion-exchange reaction. Thus X^- remains in the aqueous phase and an electroneutral, ionized lariat ether-metal ion complex is formed which has good solubility in the organic phase.

Marked enhancement in the efficiency of metal ion extraction for a proton-ionizable lariat ether versus a structurally related, but neutral (nonionizable) lariat ether, is shown in Figure 3 (13). Both extractants are lariat ethers based upon the same benzo-18-crown-6 unit. For the extractant shown on the right, the side arm is not ionizable; whereas the extractant on the left is a lariat ether carboxylic acid. Figure 3 shows results for competitive extraction of aqueous solutions of the five alkali metal chloride species with chloroform solutions of each of the two lariat ethers. The metals loading of the organic phase (extraction efficiency) was assessed as a function of the aqueous phase pH. For the neutral lariat ether, the extraction efficiency is seen to be independent of the aqueous phase pH and is very low overall. This is because both an alkali metal cation and a chloride ion must be transferred into the organic phase to form the neutral extraction complex. For the lariat ether carboxylic acid extractant, there is also very low extraction efficiency for an aqueous phase pH < 4. Under such conditions, the pendent acidic group does not dissociate. However for pH > 5, the extraction efficiency for K^+ and Rb^+ increases sharply as the basicity of the aqueous phase is enhanced. At pH ~11, the combined extraction percentages for the five alkali metal cation species totals 100 percent based upon formation of a 1:1 ionized lariat ether-metal ion complex.

The extraction selectivity with the lariat ether carboxylic acid is $K^+ >> Rb^+ >> Cs^+ > Na^+ > Li^+$. Strongest complexation of K^+ is predicted for a crown ether with an 18-crown-6 ring size (5).

A second important advantage for the use of proton-ionizable lariat ethers in metal ion extractions is the ease with which the extracted metal ions can be recovered (Figure 4). The separated organic phase which contains the extraction complexes can be readily stripped by treatment with 1 M hydrochloric acid (14) to release the metal ions into the acidic aqueous phase and regenerate the lariat ether carboxylic acid in the organic solvent for subsequent reuse. Such a release mechanism is not available for neutral crown ether ligands.

Strategies for Incorporation of Proton-Ionizable Groups into Macrocyclic Ligands

Three primary methods for incorporation of proton-ionizable moieties into macrocycles have evolved (15). These are:
 (1) attachment of a pendent side arm which includes or is terminated by an acidic entity (e.g. , **1** (16) in Figure 5);
 (2) inclusion within the crown ether framework of a subunit which projects an acidic entity so that both the proton and the atom which bears it are within the cavity (e.g., **2** (17) in Figure 5);
 (3) inclusion, as part of the crown ether framework, of a unit which places a heteroatom in a ligating position, with an attached proton projecting into the ring cavity (e.g., **3** (18) in Figure 5.

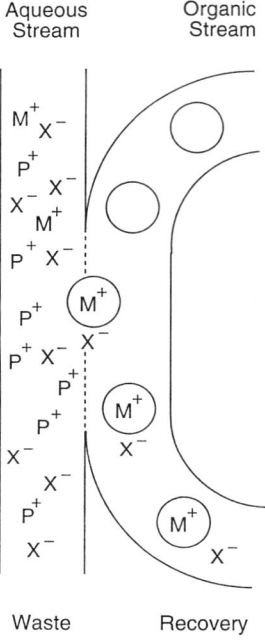

Figure 1. Metal ion separation using a crown ether in solvent extraction. (Waters of hydration are omitted for clarity.)

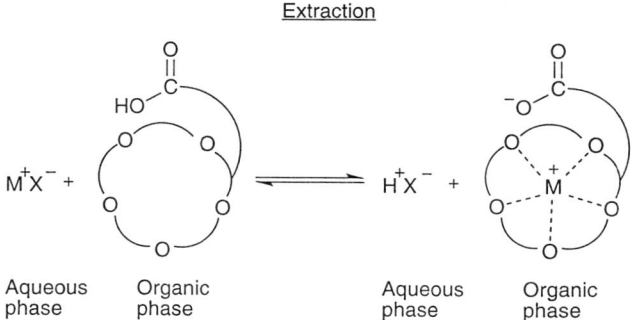

Figure 2. Metal ion extraction with a proton-ionizable lariat ether.

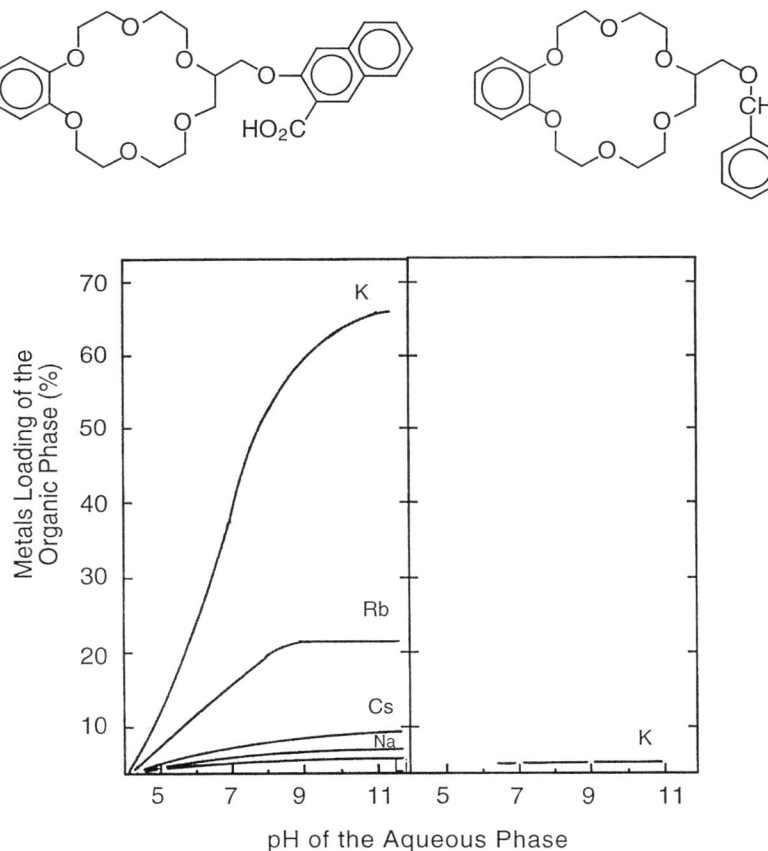

Figure 3. Results for solvent extractions of aqueous alkali metal chloride solutions with chloroform solutions of a proton-ionizable lariat ether (left) and a neutral lariat ether (right).

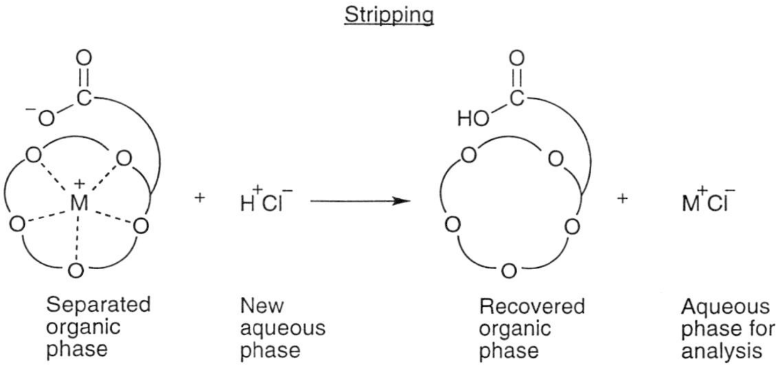

150

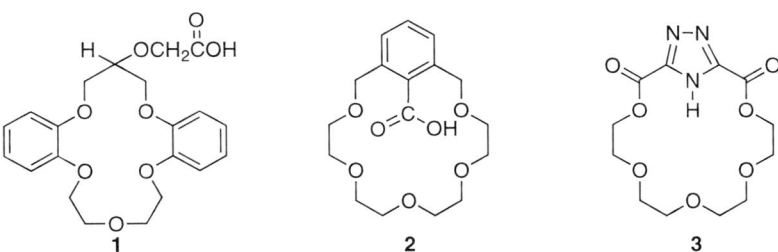

Figure 4. Release of metal ions from an ionized lariat ether-metal ion complex by treatment with aqueous acid.

Figure 5. Three different types of macrocyclic ligands with proton-ionizable moieties.

Of these three types of proton-ionizable macrocyclic ligands, only the first is a proton-ionizable lariat ether. The opportunity for axial complexation of a metal ion which is bound within the crown ether cavity is anticipated to enhance selectivity in metal ion complexation and separation processes (*12*).

Structural Variations Within Proton-Ionizable Lariat Ethers

Possible structural variations within proton-ionizable lariat ethers include the:
 (1) crown ether ring size;
 (2) length of the "arm" that connects the acidic group to the polyether ring;
 (3) attachment site(s) for lipophilic group(s) which are necessary to retain the ionized lariat ether in the organic phase;
 (4) rigidity of the polyether ring;
 (5) identity of the proton-ionizable group.

The influence of these structural variations upon the efficiency and selectivity of metal ion separations by proton-ionizable lariat ethers has been under active investigation in our laboratories for more than 15 years. Reports of the synthesis of proton-ionizable lariat ethers (*19-24*) and their applications, primarily in alkali metal cation separations by solvent extraction (*13,14,16,25-34*) and transport across various types of liquid membranes (*35-41*), have appeared in the literature.

Brief accounts of studies which probe each of the structural factors listed above in competitive solvent extraction of alkali metal cations are presented below. The procedure for competitive solvent extraction in current use is as follows. In a capped, metal-free plastic centrifuge tube is placed 2.0 mL of an aqueous solution (20 mM in each) of sodium, potassium, rubidium, and cesium chlorides and 20 mM in lithium (chloride + hydroxide). A 5.0 mM solution (2.0 mL) of the proton-ionizable lariat ether is added and the tube is shaken on a vortex mixer for four minutes and centrifuged. Of the organic phase, 1.5 mL is removed with a syringe and transferred to a new plastic centrifuge tube. After adding 1.5 mL of 1.0 N HCl, the new plastic tube is shaken on a vortex mixer and centrifuged. A portion of the aqueous phase is removed by syringe and diluted with deionized water for alkali metal cation analysis by ion chromatography.

Crown Ether Ring Size. In competitive solvent extractions of five alkali metal cation species from aqueous solution into chloroform by lariat ether carboxylic acid **4** (Figure 6), only Li^+ and Na^+ were extracted into the organic phase (*28*). The Li^+/Na^+ extraction selectivity with **4** was 20. Much lower Li^+ selectivities were noted for lariat ether carboxylic acid analogues with 12-crown-4, 13-crown-4, and 15-crown-4 rings in place of the 14-crown-4 ring in **4**. Highest Li^+ selectivity for the 14-crown-4 ring is attributed to the best fit of Li^+ within the crown ether cavity.

Length of the Arm That Connects the Acidic Group to the Polyether Ring. The influence of varying the length of the arm that links the proton-ionizable group to the crown ether ring was assessed with lariat ether phosphonic acid monoethyl esters **5** with n = 1-4 (*29*). For competitive solvent extractions of alkali metal cations from aqueous solution into chloroform, Na^+ selectivity was observed with n = 1 and 2, as would be expected for the 16-crown-5 ring. On the other hand, weak Li^+ selectivity was noted when n = 3 and 4 which indicates that the ionized group is the preferred metal ion coordination site. Examination of CPK (Corey-Pauling-Kortun) space-filling models suggests that with n = 3 and 4 the side arm is too long to allow for simultaneous coordination of a metal cation by the polyether unit and the ionized group. For extractions by **5** with n = 1 and 2, the Na^+ selectivity was much higher for the latter. The CPK models indicate that when n = 2 the side arm may be easily oriented over the crown ether cavity to provide simultaneous coordination of Na^+ by both the cyclic polyether unit and the ionized group.

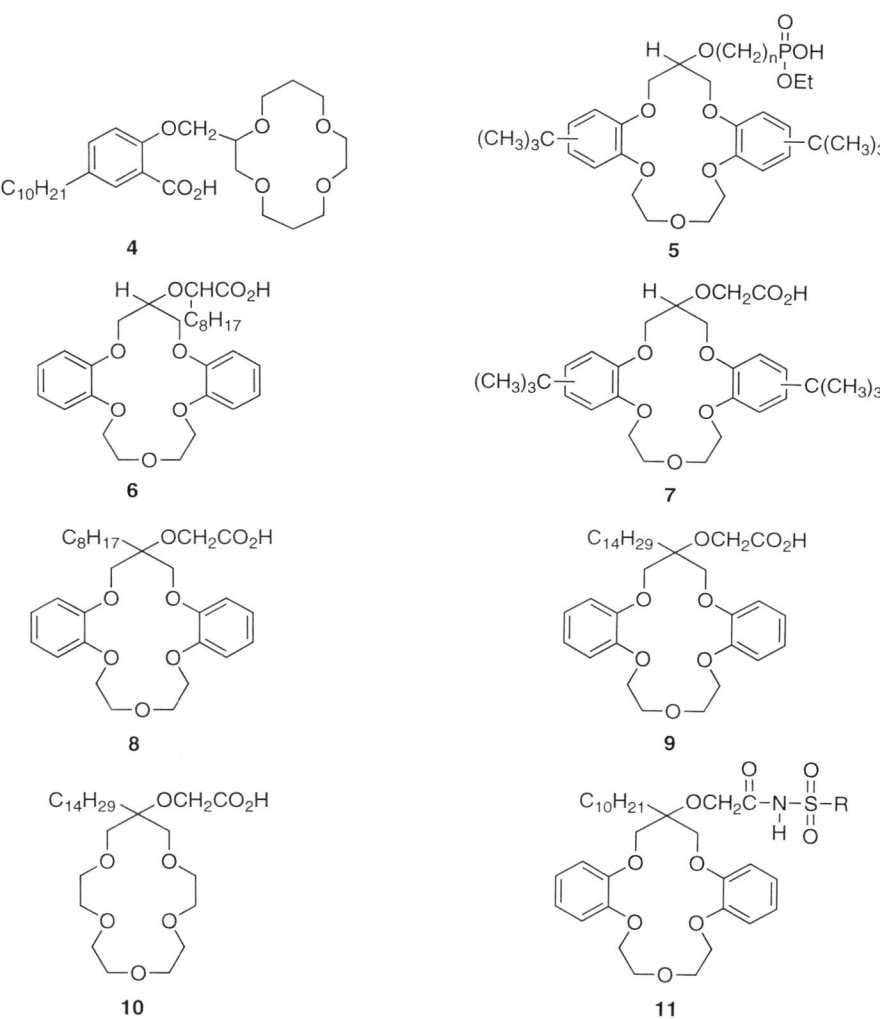

Figure 6. Structural variations within proton-ionizable lariat ethers.

Attachment Site(s) of Lipophilic Group(s). Evaluation of lariat ether carboxylic acid **1** (Figure 5), our first proton-ionizable lariat ether, in competitive solvent extraction of alkali metal cations, revealed that significant amounts of the ionized ligand were lost from the chloroform layer when it was contacted with a highly alkaline aqueous phase (*16*). In the lariat ether carboxylic acids **6-8**, eight carbons have been added to provide the requisite lipophilicity. For structural isomers **6-8**, the eight carbons are attached to the side arm, to the benzo units of the polyether ring, and to the polyether ring, respectively. For competitive solvent extraction of alkali metal cations from water into chloroform, all three isomers exhibited Na^+ selectivity, as anticipated for the 16-crown-5 ring size. However the Na^+ selectivity of **8** in which the lipophilic group is attached to the crown ether ring geminal to the functional side arm, gave much higher Na^+ selectivity (*13*). This is attributed to orientation of the carboxylic acid group over the polyether cavity when the non-polar, lipophilic group points away from the polar crown ether ring. This conformation serves to preorganize the binding site (*42*) and enhance the Na^+ selectivity. Supporting evidence for this contention is provided by the solid-state structure for an analogue of **8** with the octyl group replaced by a decyl group which shows positioning of the functional side arm over the crown ether cavity (*43*).

Rigidity of the Crown Ether Ring. Lariat ether carboxylic acids **9** and **10** have the same crown ether ring size, the same functional side arm, and the same lipophilic group attached to the crown ether ring geminal to the functional side arm. However, lariat ether carboxylic acid **9** has two benzo units in the ring in place of two ethylene units in **10**. Incorporation of two benzo groups will increase the rigidity of the polyether ring. However, this structural modification will decrease the polyether ring oxygen basicity due to replacement of four dialkyl ether oxygens with four alkyl aryl ether oxygens in which electron density on oxygen is delocalized into the aryl ring by resonance. For competitive solvent extraction of alkali metal cations from water into chloroform, the Na^+ selectivity and overall extraction efficiency of **10** was much greater than that of **9** (*13*). Apparently the enhanced rigidity of **10** helps to preorganize the binding site.

Identity of the Proton-Ionizable Group. In earlier work, we synthesized and evaluated lariat ether carboxylic acids, phosphonic acid monoethyl esters, phosphonic acids, and sulfonic acids. Within this group of ligands considerable variation of the acid strength of the proton-ionizable group is possible. However, replacement of one proton-ionizable group with another to alter the acidity may also involve other factors (*e.g.*, steric factors and the types and strengths of coordination). Very recently, we introduced a new type of proton-ionizable lariat ether **11** with pendent N-(R)sulfonyl carboxamide groups (*24*). For this type of proton-ionizable group, the acidity may be "tuned" by variation of R. For competitive solvent extraction of alkali metal cations from aqueous solutions into chloroform, compounds **11** with R = methyl, trifluoromethyl, phenyl, and *p*-nitrophenyl all exhibit quantitative metal ion loading and very high Na^+ selectivity. Only Na^+ and K^+ were extracted into the organic phase and the Na^+/K^+ selectivity was nearly 50. Such "tunable" proton-ionizable groups possess excellent potential for further development.

Summary and Future Directions

Proton-ionizable lariat ethers are shown to be effective and selective complexing agents for alkali metal cations. Results from our studies of structural variation within proton-ionizable lariat ether ligands provide valuable insight for the design of even more selective complexing agents. For the future, the new "tunable" N-(R)sulfonyl carboxamide proton-ionizable groups are an exciting area for further investigation.

Another promising area for future development is that of multi-ionizable lariat ether compounds that may be used for complexation of multivalent metal ions.

Acknowledgment

Our research program on metal ion separations by proton-ionizable lariat ethers and their polymers has received continuous support from the Division of Chemical Sciences of the Office of Basic Energy Sciences of the U.S. Department of Energy for many years. We are most grateful for this support under current grant DE-FG03-94ER14416 and from the previous grants and contracts.

Literature Cited

(1) Pedersen, C. J. *J. Am. Chem. Soc.* **1967**, *89*, 7017.
(2) Gokel, G. W.; Korzeniowski, S. H. *Macrocyclic Polyether Syntheses*; Springer-Verlag: Berlin, 1982.
(3) Hiraoka, M. *Crown Compounds. Their Characteristics and Applications*; Elsevier: New York, 1982.
(4) *Cation Binding by Macrocycles. Complexation of Cationic Species by Crown Ethers*: Inoue, Y.; Gokel, G. W., Eds.; Marcel Dekker: New York, 1990.
(5) Izatt, R. M.; Pawlak, K; Bradshaw, J. S. *Chem. Rev.* **1991**, *91*, 1721.
(6) *Crown Ethers and Analogous Compounds*; Hiraoka, M., Ed.; Elsevier: New York, 1992.
(7) *Macrocyclic Compounds in Analytical Chemistry*; Zolotov, Yu. A., Ed.; Wiley: New York, 1997.
(8) Pearson, R. G. *J. Am. Chem. Soc.* **1963**, *85*, 3533.
(9) McDowell, W. J.; Shoun, R. R. *Energy Res. Abstr.* **1978**, *3*, 4537.
(10) Gerow, I. H.; Davis, M. W. *Sep. Sci. Technol.* **1979**, *14*, 395.
(11) Kolthoff, I. M. *Anal. Chem.* **1979**, *51*, 1-22R.
(12) Gokel, G. W.; Dishong, D. M.; Diamond, C. J. *J. Chem. Soc., Chem. Commun.* **1980**, 1053.
(13) Bartsch, R. A. *Solvent Extr. Ion Exch.* **1989**, *7*, 829.
(14) Bartsch, R. A. *Sep. Sci. Technol.* **1992**, *27*, 989.
(15) Brown, P. R.; Bartsch, R. A. In *Inclusion Aspects of Membrane Chemistry*; Osa, T.; Atwood, J. L., Eds.; Kluwer Academic Publishers: Dordrecht, The Netherlands, 1991, Ch. 1.
(16) Strzelbicki, J.; Bartsch, R. A. *Anal. Chem.* **1981**, *53*, 1894.
(17) Helgeson, R. C.; Timko, J. M.; Cram, D. J. *J. Am. Chem. Soc.* **1973**, *95*, 3023.
(18) Bradshaw, J. S.; Chamberlin, D. A.; Harrison, P. E.; Wilson, B. E.; Arena, G.; Dalley, N. K.; Lamb. J. D.; Izatt, R. M. *J. Org. Chem.* **1985**, *50*, 1985.
(19) Bartsch, R. A.; Heo, G. S.; Kang, S. I.; Liu, Y.; Strzelbicki, J. *J. Org. Chem.* **1982**, *47*, 457.
(20) Bartsch, R. A.; Liu, Y.; Kang, S. I.; Son, B.; Heo, G. S.; Hipes, P. G.; Bills, L. J. *J. Org. Chem.* **1983**, *48*, 4864.
(21) Koszuk, J. F.; Czech, B. P.; Walkowiak, W.; Babb, D. A.; Bartsch, R. A. *J. Chem. Soc., Chem. Commun.* **1984**, 1504.
(22) Czech, B. P.; Czech, A.; Son, B.; Lee, H. K.; Bartsch, R. A. *J. Heterocycl. Chem.* **1986**, *23*, 465.
(23) Czech, B. P.; Desai, D. H.; Koszuk, J.; Czech, A.; Babb, D. A.; Robison, T. W.; Bartsch, R. A. *J. Heterocycl. Chem.* **1992**, *29*, 867.
(24) Huber, V. J.; Ivy, S. N.; Lu, J.; Bartsch, R. A. *J. Chem. Soc., Chem. Commun.* **1997**, 1499.
(25) Strzelbicki, J.; Bartsch, R. A. *Anal. Chem.* **1981**, *53*, 2251.
(26) Strzelbicki, J.; Heo, G. S.; Bartsch, R. A. *Sep. Sci. Technol.* **1982**, *17*, 635.

(27) Charewicz, W. A.; Heo, G. S.; Bartsch, R. A. *Anal. Chem.* **1982**, *54*, 2094.
(28) Bartsch, R. A.; Czech, B. P.; Kang, S. I.; Stewart, L. E.; Walkowiak, W.; Charewicz, W. A.; Heo, G. S.; Son, B. *J. Am. Chem. Soc.* **1985**, *107*, 4997.
(29) Pugia, M. J.; Ndip, G.; Lee, H. K.; Yang, I.-W.; Bartsch, R. A. *Anal. Chem.* **1986**, *58*, 2723.
(30) Charewicz, W. A.; Walkowiak, W.; Bartsch, R. A. *Anal. Chem.* **1987**, *59*, 494.
(31) Walkowiak, W.; Charewicz, W. A.; Kang, S. I.; Yang, I.-W.; Pugia, M. J.; Bartsch, R. A. *Anal. Chem.* **1990**, *62*, 2018.
(32) Walkowiak, W.; Kang, S. I.; Stewart, L. E.; Ndip, G.; Bartsch, R. A. *Anal. Chem.* **1990**, *62*, 2022.
(33) Walkowiak, W.; Jeon, E.-G.; Huh, H.; Bartsch, R. A. *J. Inclusion Phenom. Mol. Recog. Chem.* **1992**, *12*, 213.
(34) Bartsch, R. A.; Hayashita, T.; Lee, J. H.; Kim, J. S.; Hankins, M. G. *Supramolecular Chem.* **1993**, *1*, 305.
(35) Strzelbicki, J.; Bartsch, R. A. *J. Membr. Sci.* **1982**, *10*, 35.
(36) Charewicz, W. A.; Bartsch, R. A. *Anal. Chem.* **1982**, *54*, 2300.
(37) Charewicz, W. A.; Bartsch, R. A. *J. Membr. Sci.* **1983**, *12*, 323.
(38) Bartsch, R. A.; Charewicz, W. A.; Kang, S. I. *J. Membr. Sci.* **1984**, *17*, 97.
(39) Walkowiak, W.; Brown, P. R.; Shukla;, J. P.; Bartsch, R. A. *J. Membr. Sci.* **1987**, *32*, 59.
(40) Bartsch, R. A.; Charewicz, W. A.; Kang, S. I.; Walkowiak, W. In *Liquid Membranes. Theory and Applications*, ACS Symposium Series 347, Noble, R. D.; Way, J. D., Eds.; American Chemical Society: Washington, DC, 1987, Ch. 6.
(41) Brown, P. R.; Hallman, J. L.; Whaley, L. W.; Desai, D. H.; Pugia, M. J.; Bartsch, R. A. *J. Membr. Sci.* **1991**, *56*, 195.
(42) Cram. D. J. *Angew. Chem., Int. Ed. Engl.* **1986**, *25*, 1039.
(43) Bartsch, R. A.; Kim, J. S.; Olsher, U.; Purkiss, D. W.; Ramesh, V.; Dalley, N. K.; Hayashita, T. *Pure Appl. Chem.* **1993**, *65*, 399.

Chapter 10

Redox-Recyclable Extraction and Recovery of Heavy Metal Ions and Radionuclides from Aqueous Media

Steven H. Strauss

Department of Chemistry, Colorado State University, Fort Collins, CO 80523

Scientists involved in the separation of ionic pollutants such as radionuclides or toxic heavy metal ions from water have designed extractants with high selectivities and large capacities. Although there is still room for improvement in these parameters, there is a more urgent need to develop processes that allow the target pollutants to be recovered in a minimal volume of secondary waste and that allow the extractants to be reused (recycled). We have studied redox-active transition-metal-containing extractants which undergo reversible electron-transfer activation and deactivation as the target ions are extracted and recovered. The "redox-recyclable" extractants investigated so far include molecular organometallic complexes such as substituted ferrocenes and layered metal chalcogenides such as MoS_2. The molecular complexes can be dissolved in water-immiscible organic solvents for solvent-extraction processes or can be immobilized on inert supports for ion-exchange chromatography. The bulk metal chalcogenides themselves function as redox-recyclable ion-exchange materials. Several extractants have been tested for the repeatable extraction and recovery of aqueous $^{99}TcO_4^-$, ReO_4^-, $^{137}Cs^+$, $^{90}Sr^{2+}$, Hg^{2+}, Cd^{2+}, Pb^{2+}, and Ag^+. Using a substituted ferrocene, greater than 99% of the radioactivity was extracted from an aqueous phase containing 1 M NaOH, 1.5 M $NaNO_3$, 10^{-5} M $^{99}TcO_4^-$, and 10^{-10} M $^{95m}TcO_4^-$, and greater than 99% of the extracted radioactivity was recovered in a solid secondary waste. Using MoS_2, greater than 99% of dissolved Hg^{2+} was selectively extracted from an aqueous phase containing 1 M HNO_3 and 10^{-3} M Hg^{2+}. The final concentration of aqueous mercury was ≤ 0.033 µM in some cases (6.5 ppb), and greater than 94% of the extracted Hg^{2+} was recovered as elemental mercury after stripping. In both cases, the volume of secondary waste containing the target pollutant was a small fraction of the volume of the primary aqueous waste, and in both cases the stripped (deactivated) extractants were reusable.

Solvent extraction and ion-exchange chromatography are mature technologies (*1–6*). When they are used for the remediation of toxic and/or radioactive aqueous waste streams, however, minimization of the volume of secondary waste (e.g., the strip solution) destined for permanent disposal is still a significant challenge. This is especially problematic where hazardous but dilute waste streams are concerned. In general, the volume of secondary waste per mole of contaminant recovered is inversely proportional to the concentration of contaminant in the primary waste (*7*). The cost of permanent disposal is directly related to the volume of the secondary waste, and, in some solvent extraction and ion-exchange systems, the volume of the strip solution is equal to the volume of the primary waste. Furthermore, in recent years tougher environmental regulations and the high initial cost of new, more effective, and more selective extractants have made the reuse of the extractant an increasingly important issue. For these reasons, the reuse of extractant materials and the minimization of secondary waste volume must become the focus of scientific efforts in chemical separation in the near future. In summary, not only must a modern and effective extractant have (1) a large *capacity* and (2) a high *selectivity* for the target pollutant, it must also simultaneously satisfy two other design criteria: (3) it must allow for the recovery of the target species in a *minimal volume* of secondary waste and (4) it must be reuseable (*recyclable*).

Design Criteria for Redox-Recyclable Extractants

We are investigating and developing a relatively unexplored strategy in waste remediation, the use of redox-active transition-metal containing extractants for the separation and recovery of specific pollutant cations or anions (*8–14*). We have named the strategy, which is represented in Figure 1, Redox-Recyclable Extraction and Recovery (R^2ER). Our investigations are based on the seminal electrochemical-switching work of Porter et al. (*15*), Martin et al. (*16*), Fabbrizzi et al. (*17*),

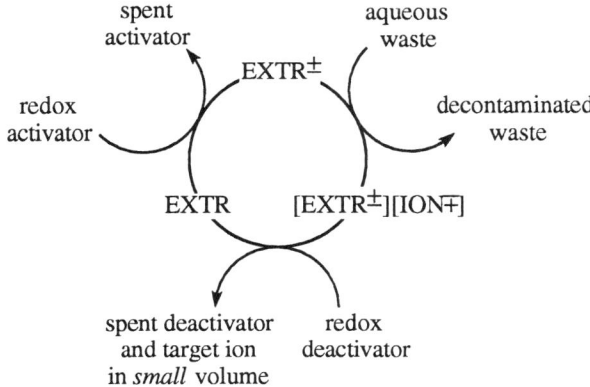

Figure 1. The complete, repeatable cycle of extraction-deactivation/recovery-reactivation with a redox-recyclable extractant, EXTR.

Echegoyen, Gokel et al. (*18*), Shinkai et al. (*19*), and Beer et al. (*20*), *but with an added emphasis on recovering the target pollutant in a minimal volume of secondary waste*. These groups, as well as others, have shown that the binding of ions can be enhanced by electrochemically switching (or redox switching) an extractant molecule. However, when one considers practical factors such as duty-cycle time, extractant stability under harsh conditions, extractant effectiveness over many cycles, extractant cost, and secondary-waste volume, much work remains to be done before useful R^2ER schemes can be developed and reduced to practice.

The key features of R^2ER are as follows. The neutral, deactivated extractant (EXTR) has little or no affinity for an ionic pollutant. When it is activated by either one-electron oxidation or reduction, it becomes cationic or anionic, respectively ($EXTR^{\pm}$), and therefore develops an electrostatic affinity for an ion of opposite charge. If an $EXTR^{\pm}$ salt dissolved in a water-immiscible phase is shaken with an aqueous phase containing an ionic pollutant, the pollutant may migrate to the extractant phase (i.e., the pollutant ion will undergo exchange with the original counterion of $EXTR^{\pm}$). The degree of ion-exchange (pollutant migration) will depend on factors that normally control ion migration between phases, such as relative hydration free energies and shape selectivity. When the cationic or anionic extractant is subsequently deactivated by either reduction or oxidation, respectively, it becomes neutral again (EXTR) and loses the electrostatic affinity it formerly had for the ionic pollutant. Depending on the choice of redox deactivator, the ionic pollutant can be recovered in a relatively small volume of secondary waste.

The design criteria for molecular redox-recyclable extractants are as follows. They should be transition-metal complexes that are very stable as neutral complexes and as one-electron oxidized or reduced cations or anions, respectively. Ideal complexes will be kinetically inert to substitution in both redox states. This criterion immediately suggests the use of polydentate ligands. The complexes should not contain acid- or base-labile functional groups. They should have redox potentials that allow the use of simple, inexpensive oxidants or reductants. They should undergo rapid one-electron oxidation or reduction, but should not undergo over-oxidation or over-reduction in the presence of an excess of oxidant or reductant. The complexes should be relatively nontoxic (e.g., iron complexes would be preferable to chromium complexes). In addition, the complexes should be relatively inexpensive (e.g., iron complexes would be preferable to ruthenium complexes). Finally, the complexes must have negligible water solubility in both working oxidation states, whether the process they will be used for is solvent extraction or stationary-phase physisorbed ion exchange.

The design criteria for solid-state redox-recyclable extractants are similar. First, the solid-state layered or channeled extractants must be stable in contact with a wide variety of aqueous phases in both active (reduced) and inactive (oxidized) forms. Second, the solid-state extractants must have suitable electrochemical potentials in the presence of simple intercalant ions such as Li^+ or Na^+, so that they can be shuttled between their active and inactive oxidation states using relatively mild and inexpensive reductants and oxidants. Third, solid-state extractant activation and deactivation redox reactions must be sufficiently rapid to allow for reasonably short duty times for complete extraction-deactivation/recovery-reactivation cycles. Fourth, the solid-state extractants must not dissolve in prolonged contact with the aqueous phase to be treated.

They must not undergo any other form of decomposition, including irreversible over-reduction or over-oxidation, during many cycles of extraction, deactivation/recovery, and reactivation. Radiolytic stability is also important for nuclear waste treatment. Finally, practical solid-state extractants should have low toxicity, low cost, and should be relatively easy to prepare and handle.

HEP. A Molecular Redox-Recyclable Extractant for Solvent Extraction and Ion-Exchange Chromatography

Liquid-Liquid Extraction with HEP. Our most complete study of redox-recyclable extraction and recovery to date involves TcO_4^- and its non-radioactive surrogate ReO_4^- (8,9,12,13). (Work in our laboratory confirmed earlier reports that the solvent extraction behavior of TcO_4^- and ReO_4^- are similar enough that ReO_4^- extraction experiments can be used to test the potential effectiveness of TcO_4^- extractants (21).) The key component in our studies is the stable, lipophilic complex HEP, a tetraalkylated ferrocene (HEP = 1,1',3,3'-tetrakis(2-methyl-2-hexyl)ferrocene).

HEP =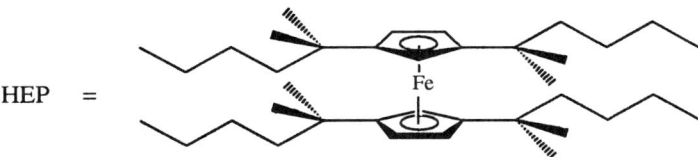

The cycle of extraction–deactivation/recovery–reactivation is given by the following three simplified chemical reactions (simplified in that $HEP^+TcO_4^-(org)$ is really a mixture of ca. 10^{-1}–10^{-3} M $HEP^+NO_3^-(org)$ and only ca. 10^{-5} M $HEP^+TcO_4^-(org)$ and that the solid secondary waste contains $Fe(NO_3)_3 \cdot xH_2O$ as well as $Fe(TcO_4)_3$ (8,13):

$$HEP^+NO_3^-(org) + TcO_4^-(aq\ waste) \rightarrow HEP^+TcO_4^-(org) + NO_3^-(aq\ waste)$$

$$HEP^+TcO_4^-(org) + Fe(s) \rightarrow HEP(org) + Fe(TcO_4)_3(s) + TcO_2(s)$$

$$HEP(org) + Ce^{4+}(aq) + NO_3^-(aq) \rightarrow HEP^+NO_3^-(org) + Ce^{3+}(aq)$$

Note that the middle reaction (deactivation/recovery) is not balanced. Some data for the extraction step are listed in Table I (for these experimental results, $TcO_4^- = {}^{99}TcO_4^-$ and ${}^{95m}TcO_4^-$). The figures of merit for these extractions are $D(TcO_4^-)$ and $D(ReO_4^-)$, which were calculated using the following equation:

$$D(MO_4^-) = \frac{[MO_4^-]_{f,org}}{[MO_4^-]_{f,aq}} = \frac{[MO_4^-]_{i,aq} - [MO_4^-]_{f,aq}}{[MO_4^-]_{f,aq}}$$

Note that the greater selectivity of HEP^+ for TcO_4^- and ReO_4^- relative to NO_3^- is the result of a natural bias based on the free energies of hydration of these anions (this has been discussed at length in an important review by Moyer and Bonneson (22)). In other words, this is an example of Hofmeister separation (23–26). Our new extractant $HEP^+NO_3^-$ is more selective than Aliquat-336$^+NO_3^-$ for ReO_4^- relative to NO_3^- (14).

Table I. Extraction of TcO_4^- and ReO_4^- Using $HEP^+NO_3^-$ [a]

aqueous waste simulant	organic phase	$D(TcO_4^-)$	$D(ReO_4^-)$
1 M NaOH/1.5 M $NaNO_3$	toluene	430(20)	230(10)
1 M NaOH/1.5 M $NaNO_3$	o-$C_6H_4Cl_2$	270(10)	120(10)
1 M NaOH/1.5 M $NaNO_3$	2-nonanone	470(20)	60(10)
1 M HNO_3	toluene	270(10)	190(10)
1 M HNO_3	o-$C_6H_4Cl_2$	210(10)	230(10)
1 M HNO_3	2-nonanone	190(10)	180(10)

[a] The aq. waste simulant contained either 10^{-2} M $KReO_4$ or ~10^{-5} M $Li^{99}TcO_4$ plus ~10^{-10} M $Li^{95m}TcO_4$. The organic phase contained 0.1 M [HEP][NO_3]. Estimated standard deviations are given in parentheses. $D(MO_4^-)$ values are for a single contact of equal volumes of the two phases.

For the most part, however, the fact that HEP^+ gives $D(TcO_4^-)$ values well over 100 is not a significant breakthrough except to show that HEP^+ is a lipophilic cation that is reasonably stable in the presence of aqueous 1 M NaOH and 1 M HNO_3 and *can* be used for $^{99}TcO_4^-$ extraction from relevant nuclear waste simulants.

The real advantage of using $HEP^+NO_3^-$ for $^{99}TcO_4^-$ remediation is that *technetium-99 can be recovered as a solid of minimal volume*. The recovery step consists of deactivating HEP^+ by one-electron reduction to neutral HEP. Several bulk metals can serve as the reducing agent, including iron, which is inexpensive and nontoxic. The plots in Figure 2 show the time course of reduction by metallic iron of HEP^+ to HEP in three organic solvents with the concomitant formation and precipitation of $Fe(TcO_4)_3 \cdot 3H_2O$ and $TcO_2(s)$ (*8,13*). Note that $HEP^+NO_3^-$ and $HEP^+TcO_4^-$ dissolved in toluene were converted to HEP and Tc-containing solids in only 10-20 minutes.

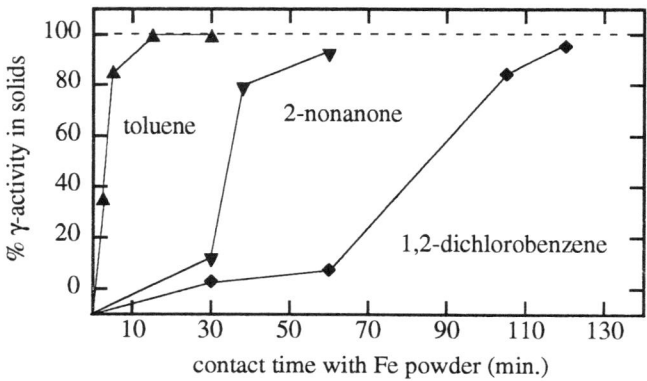

Figure 2. Time course of reactions of iron powder with ~0.01 M [HEP^+][NO_3^-], ~10^{-5} M [HEP^+][$^{99}TcO_4^-$], and ~10^{-10} M [HEP^+][$^{95m}TcO_4^-$] dissolved in water-saturated toluene, 2-nonanone, or 1,2-dichlorobenzene. The γ activity of ^{95m}Tc in the organic phase and/or the solid phase was monitored.

The following analysis shows the extent of the secondary-waste volume reduction that may be possible. We will make the following assumptions: (1) the concentration of $^{99}TcO_4^-$ in an aqueous waste stream of concern is 5×10^{-5} M; (2) the recovered solid is primarily $Fe(NO_3)_3 \cdot 9H_2O$ (density = 1.68 g cm^{-3}); (3) $HEP^+NO_3^-$ dissolved in toluene is the extractant; (4) 100 L of aqueous waste is treated. Therefore: for one extraction-deactivation-reactivation cycle with 0.1 M $HEP^+NO_3^-$, $D_{total}(TcO_4^-)$ = 430 (8), the mole ratio of NO_3^-/TcO_4^- in the precipitate ($[NO_3^-/TcO_4^-]_s$) = 2,000, and the volume of the solid precipitate (V_s) = 0.79 L; for two R^2ER cycles with 0.01 M $HEP^+NO_3^-$, $D_{total}(TcO_4^-)$ ~ 2,100 (46 × 46) (13), $[NO_3^-/TcO_4^-]_s$ = 400, and V_s = 0.16 L; for four R^2ER cycles with 0.001 M $HEP^+NO_3^-$, $D_{total}(TcO_4^-)$ ~ 256 (4 × 4 × 4 × 4) (13), $[NO_3^-/TcO_4^-]_s$ = 80, and V_s = 0.032 L, a volume reduction factor of 3,000. Although the aqueous waste simulant we used for these experiments contained only 1.0 M NaOH and 1.5 NaNO$_3$, this analysis is relevant because prior experiments with Aliquat-336$^+$NO$_3^-$ have shown that $D(TcO_4^-)$ is not affected by more than a factor of two by the addition of 0.86 M NaNO$_2$, 0.49 M NaAl(OH)$_4$, 0.39 M Na$_2$CO$_3$, 0.11 M Na$_3$PO$_4$, 0.093 M NaCl, 0.073 M Na$_3$(citrate), 0.031 M Na$_2$SO$_4$, and 0.002 M Ca(NO$_3$)$_2$ to a waste simulant containing 1.0 M NaOH and 1.5 M NaNO$_3$ (21).

Ion-Exchange Chromatography with Physisorbed $HEP^+NO_3^-$. We have also prepared R^2ER ion-exchange "resins" from $HEP^+NO_3^-$ physisorbed onto different materials ranging from silica gels to organic polymers (9,12). A complete R^2ER cycle for $HEP^+NO_3^-$ loaded onto 100 Å pore size SiO_2 (surface area = 300 m^2 g^{-1}) is shown in Figure 3. In this case, $K_d(ReO_4^-)$ from 1 M HNO$_3$ containing 0.32 mM KReO$_4$ was found to be approximately 100 mL g^{-1} (9), which can be compared with 289 mL g^{-1} for Reillex-HPQ (27). Although it seems as if K_d for our new material is lower than K_d for Reillex-HPQ, we have been limited so far by how much $HEP^+NO_3^-$ can be loaded onto the resins. When $K_d(ReO_4^-)$ is expressed in mL per mmol cation exchange group, the values are 450 and 87 for $HEP^+NO_3^-/SiO_2$ and Reillex-HPQ, respectively, demonstrating that the redox-recyclable ion-exchange material $HEP^+NO_3^-/SiO_2$ is at least four times more selective than Reillex-HPQ for extraction of ReO_4^- from

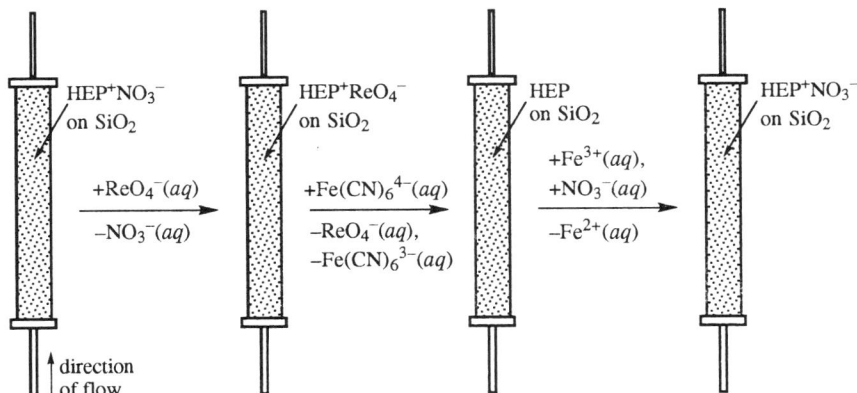

Figure 3. R^2ER cycle for ReO_4^- extraction and recovery using $HEP^+NO_3^-/SiO_2$.

1 M HNO_3. Furthermore, our ReO_4^--loaded material can be recycled by passing a 25 mM aqueous solution of $Fe(CN)_6^{4-}$ through the column, which reduces physisorbed HEP^+ to physisorbed HEP and strips all of the bound ReO_4^- in only 10 minutes (9). We found that K_d decreased by about 5% per complete cycle over five cycles (9). The decrease in K_d could be due to desorption of physisorbed extractant and/or decomposition of the physisorbed exctractant. Improving the hydrophobicity and stability of R^2ER extractants over many cycles are important goals of our ongoing research.

Other Molecular Redox-Recyclable Extractants for Solvent Extraction

Fe(Tp')$_2$. We have investigated substituted bis(hydridotris(pyrazolylborate)) iron complexes, abbreviated Fe(Tp')$_2$, as potential R^2ER extractants for anions such as ReO_4^- and $^{99}TcO_4^-$ (14). In one series of experiments, the two aqueous waste simulants contained 0.010 M $KReO_4$ and either 3 M HNO_3 or 3 M $NaNO_3$. The two organic phases examined contained 0.05 M [Fe(Tp')$_2$][NO_3] or 0.05 M Aliquat-336$^+$NO$_3^-$ dissolved in dichloromethane. The D(ReO_4^-) values for [Fe(Tp')$_2$][NO_3] were 0.89(1) for 3 M HNO_3 and 7.8(1) for 3 M $NaNO_3$. The corresponding values for Aliquat-336$^+$NO$_3^-$ were 0.77(1) and 4.4(1), respectively. Preliminary results indicated that greater than 88% of the extracted ReO_4^- could be recovered in a solid precipitate from the [Fe(Tp')$_2$][ReO_4]-containing dichloromethane phase using iron powder as the deactivating reagent (14).

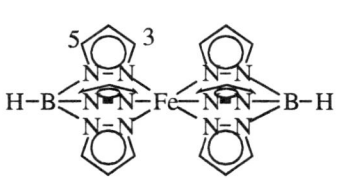

bis(hydridotris(pyrazolylborate))iron(II), Fe(Tp)$_2$

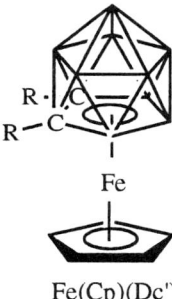

Fe(Cp)(Dc')

Fe(Cp)(Dc'). We have also synthesized organometallic extractants that cycle between their uncharged and anionic forms and can extract $^{90}Sr^{2+}$ and $^{137}Cs^+$ from a nuclear waste simulant containing 1 M NaOH and 1.5 M $NaNO_3$ (11). These are complexes of iron containing a cyclopentadienyl ligand (Cp) and a substituted dicarbollide ligand (Dc'). When a 0.05 M toluene solution of the salt [Na][Fe(Cp)(Dc')] was used, D(Sr^{2+}) and D(Cs^+) were 2 and 23, respectively (R = $C_{12}H_{25}$) (11). These values are higher than for the recently developed and very promising extractant sodium bis(dicarbollide)cobalt(1–) (28). These results are especially important because Fe(Cp)(Dc')$^-$ complexes, like HEP^+ and Fe(Tp')$_2^+$, are redox-recyclable, whereas Co(Dc)$_2^-$ is not redox-recyclable (the $E_{1/2}$ value for the Fe(Cp)(Dc')$^{0/1-}$ couple is 0.17 V vs. SCE). We are currently developing complete R^2ER cycles using Fe(Cp)(Dc')$^-$ extractants to recover $^{90}Sr^{2+}$ and $^{137}Cs^+$ in a minimal volume of secondary waste.

MoS_2. A Redox-Recyclable Material for Heavy-Metal Ion-Exchange

R²ER of Aqueous Hg^{2+} Using MoS_2. We have found that lithium-intercalated metal chalcogenides such as Li_xMoS_2 and Li_xWS_2 will rapidly undergo cation-exchange with Hg^{2+} in aqueous solution ($0.25 \leq x \leq 1.9$) (10). The capacity of Li_xMoS_2 was as high as 580 mg mercury per gram of extractant. Most importantly, greater than 94% of the ion-exchanged mercury in Hg_yMoS_2 was efficiently recovered as metallic mercury in a cold trap when Hg_yMoS_2 was heated under vacuum at 425°C. Thus, MoS_2 and WS_2 represent redox-recyclable extractants for aqueous Hg^{2+}. Note that metallic mercury, which represents the secondary waste recovered, is only 0.015% of the volume of the primary waste simulant, 10 mM aqueous Hg^{2+}. Furthermore, the molybdenum-containing material recovered after the heat-induced recovery step was essentially MoS_2, which was reactivated (recycled) to Li_xMoS_2 with n-BuLi in hexanes for subsequent R²ER cycles.

Extraction and recovery data for three sucessive cycles with the same sample of MoS_2 are listed in Table II, which also includes a scheme for the complete cycle (10).

Table II. Use of a Sample of Li_xMoS_2 for Three Consecutive Mercury Extraction-Deactivation/Recovery-Reactivation Cycles[a]

$$Li_xMoS_2 \xrightarrow[0.1 \text{ M } HNO_3(aq)]{Hg^{2+}(aq)} Hg_yMoS_2 \xrightarrow[-Hg^0]{\text{heat, vacuum}} Hg_zMoS_2$$
$$\underset{n\text{-BuLi}}{\longleftarrow}$$

cycle	x value[b]		y value[c]		z value[d]
	n-BuLi titration	Li/Mo ratio from digestion	digestion	filtrate	
1	1.4	1.3	0.35	0.32	0.02
2	1.7	1.6	0.39	0.50	<0.01
3	2.0	1.8	0.50	0.50	<0.01

[a] Samples of Li_xMoS_2 were prepared by treating weighed samples of MoS_2 with a five-fold excess of 2.5 M n-BuLi in hexane. Extraction consisted of: (i) vigorously stirring a weighed quantity of solid Li_xMoS_2 with 0.5 molar equiv. (based on Mo) of 10.0 mM $Hg(NO_3)_2$ in 0.1 M aqueous HNO_3 for 2 h at 25°C, (ii) filtering the extraction mixtures, and (iii) analyzing the filtrate and the solid for lithium, mercury, and molybdenum by ICP-AES. Deactivation (recovery of mercury) consisted of heating the solid samples of Hg_yMoS_2 under vacuum at 425°C for 6 h. Reactivation consisted of treating Hg_zMoS_2 with a five-fold molar excess (based on Mo) of n-BuLi as above. [b] The values of x were determined by: (i) titrating the n-BuLi that remained after activation of MoS_2 or Hg_zMoS_2 and (ii) digesting samples of Li_xMoS_2 in aqua regia and analyzing the resulting solutions for lithium and molybdenum by ICP-AES. [c] The values of y were determined by: (i) digesting samples of Hg_yMoS_2 in aqua regia and analyzing the resulting solutions for mercury and molybdenum by ICP-AES and (ii) determining the difference in initial and final concentrations of mercury for the extraction step. [d] The values of z were determined by digesting samples of Hg_zMoS_2 in aqua regia and analyzing the resulting solutions for mercury and molybdenum by ICP-AES.

Note that the values of x and y for Li_xMoS_2 and Hg_yMoS_2 increased from the first complete cycle to the third complete cycle. This is probably due to a decrease in MoS_2 particle size over successive cycles, which resulted in a higher Li/Mo mole ratio in Li_xMoS_2 during a fixed amount of time over successive cycles. It is sensible that the MoS_2 particles would decrease in size after several R^2ER cycles involving exfoliation and reflocculation of the material.

Two other significant papers reporting the extraction and recovery of $Hg^{2+}(aq)$ have appeared recently (29,30). In one paper, a thioalkylated montmorillonite clay (thiomont), was found to have a capacity of 65 mg mercury per gram of thiomont (29). Treatment of mercury-loaded thiomont with 0.1 M HCl resulted in exchange of protons for Hg^{2+} and regeneration of thiomont. In the other paper, a mesoporous silica with thioalkyl groups grafted to the surface had a capacity of 505 mg mercury per gram of extractant, but only for the first extraction (30). After regeneration of the material with concentrated aqueous HCl, the capacity dropped to 40% of its original value.

Extraction of Other Soft Metal Ions from Aqueous Media Using MoS_2.
In a series of extraction experiments, samples of $Li_{1.3}MoS_2$ were treated with aliquots of 0.1 M aqueous HNO_3 containing 0.20 equiv of $M(NO_3)_2$ (the initial Mo/M^{2+} mole ratio was 5.0; M^{2+} = Hg^{2+}, Pb^{2+}, Cd^{2+}, and Zn^{2+}; $[M^{2+}(aq)]_{initial}$ = 1.0 mM). After separating the supernatant from the flocculant ion-exchanged M_yMoS_2 solid by filtration, the final concentration of $M^{2+}(aq)$ in the filtrates was determined by ICP-AES analysis. The ratios of M^{2+} extracted to M^{2+} initial were 1.00(1) for Hg^{2+}, 0.74(1) for Pb^{2+}, 0.41(1) for Cd^{2+}, and only 0.13(1) for Zn^{2+} (10). The driving force for the Li^+/M^{2+} ion exchange is undoubtedly the greater affinity of soft Lewis acids such as Hg^{2+} for the solid-state soft Lewis base MoS_2^{x-} relative to the hard Lewis acids Li^+ or H_3O^+. Therefore, the observed selectivity trend is sensible since mercury is softer than cadmium, which itself is softer than zinc. Preliminary experiments demonstrated that Li_xMoS_2 also extracts Ag^+ from various aqueous media, including an aqueous medium containing a high concentration of thiosulfate ion (31). Experiments are in progress to recover the extracted lead, cadmium, and silver from Pb_yMoS_2, Cd_yMoS_2, and Ag_yMoS_2 in relatively small volumes of secondary waste.

Acknowledgment

I would like to thank all of my students, postdocs, and collaborators, whose names can be found in the references, for their insightful and dedicated work as the concepts of R^2ER were conceived, developed, and reduced to practice. Our work in this area has been generously supported by the United States DOE and NSF.

Literature Cited

(1) Thornton, J. D. *Science and Practice of Liquid-Liquid Extraction*; Clarendon Press: Oxford, 1992; Vols. 1 and 2.
(2) Baird, M. H. I. *Can. J. Chem. Eng.* **1991**, *69*, 1287.
(3) Dasgupta, P. K. *Anal. Chem.* **1992**, *64*, 775A.
(4) Peters, R. W.; Shem, L. In *Emerging Sep. Technol. Met. Fuels, Proc. Symp.*; Lakshmanan, V. I.; Bautista, R. G.; Somasundaran, P., Eds.; Miner. Met. Mater. Soc.: Warrendale, PA, 1993, pp 3-64.

(5) Abe, M. *Ion Exch. Solvent Extr.* **1995**, *12*, 381.
(6) Streat, M. *Ind. Eng. Chem. Res.* **1995**, *34*, 2841.
(7) Tedder, D. W. *Sep. Purif. Methods* **1992**, *21*, 23.
(8) Clark, J. F.; Clark, D. L.; Whitener, G. D.; Schroeder, N. C.; Strauss, S. H. *Environ. Sci. Technol.* **1996**, *30*, 3124.
(9) Chambliss, C. K.; Odom, M. A.; Morales, C. M. L.; Martin, C. R.; Strauss, S. H. *Anal. Chem.* **1998**, *70*, 757.
(10) Gash, A. E.; Spain, A. L.; Dysleski, L. M.; Flaschenriem, C. J.; Kalaveshi, A.; Dorhout, P. K.; Strauss, S. H. *Environ. Sci. Technol.* **1998**, *32*, 1007.
(11) Clark, J. F.; Chamberlin, R. E.; Strauss, S. H. **1998**, manuscript in preparation.
(12) Chambliss, C. K.; Odom, M. A.; Moyer, B. A.; Martin, C. R.; Strauss, S. H. **1998**, manuscript in preparation.
(13) Clark, J. F.; Gansle, K. M.; Whitener, G. D.; Schroeder, N. C.; Strauss, S. H. **1998**, manuscript in preparation.
(14) Gansle, K. M. **1997**, unpublished results, Colorado State University.
(15) Deinhammer, R. S.; Porter, M. D.; Shimazu, K. *J. Electroanalytical Chem.* **1995**, *387*, 35.
(16) Ghatak-Roy, A. R.; Martin, C. R. *Anal. Chem.* **1986**, *58*, 1574.
(17) De Santis, G.; Fabbrizzi, L.; Licchelli, M.; Monichino, A.; Pallavicini, P. *J. Chem. Soc., Dalton Trans.* **1992**, 2219.
(18) (a) Chen. Z.; Echegoyen, L. In *Crown Compounds*; Cooper, S. R., Ed.; VCH: New York, 1992, pp 27-39. (b) Medina, J. C.; Gay, I.; Chen. Z.; Echegoyen, L.; Gokel, G. W. *J. Am. Chem. Soc.* **1991**, *113*, 365.
(19) Shinkai, S. In *Comprehensive Supramolecular Chemistry*; Gokel, G. W., Ed.; Elsevier: Oxford, 1996, Vol. 1, pp 671-700.
(20) Beer, P. D.; Smith, D. K. *Prog. Inorg. Chem.* **1997**, *46*, 1.
(21) Rohal, K. M.; Van Seggen, D. M.; Clark, J. F.; Van Egeren, M. K.; Chambliss, C. K.; Strauss, S. H.; Schroeder, N. C. *Solvent Extr. Ion Exch.* **1996**, *14*, 401.
(22) Moyer, B. A.; Bonneson, P. V. "Physical Factors in Anion Separation," In *Supramolecular Chemistry of Anions;* Bianchi, A.; Bowman-James, K.; Garcia-España, E., Eds.; VCH: New York, 1997, p 1.
(23) Bucher, J.; Diamond, R. M.; Chu, B. *J. Phys. Chem.* **1972**, *76*, 2459.
(24) Collins, K. D.; Washabaugh, M. W. *Q. Rev. Biophys.* **1985**, *18*, 323.
(25) Marcus, Y.; Kamlet, M. J.; Taft, R. W. *J. Phys. Chem.* **1988**, *92*, 3613.
(26) Hara, H.; Ohkubo, H.; Sawai, K. *Analyst* **1993**, *118*, 549.
(27) Ashley, K. R.; Cobb, S. L.; Ball, J. R. *Solvent Extr. Ion Exch.* **1995**, *13*, 353.
(28) Miller, R. L.; Pinkerton, A. B.; Hurlburt, P. K.; Abney, K. D. *Solvent Extr. Ion Exch.* **1995**, *13*, 813.
(29) Mercier, L.; Detellier, C. *Environ Sci. Technol.* **1995**, *29*, 1318.
(30) Feng, X.; Fryxell, G. E.; Wang, L.-Q.; Kim, A. Y.; Liu, J.; Kemner, K. M. *Science* **1997**, *276*, 923.
(31) Gash, A. E.; Dysleski, L. M. **1997**, unpublished results, Colorado State University.

SEPARATIONS USING SOLID–LIQUID SYSTEMS

Chapter 11

Structural Basis of Selectivity in Tunnel Type Inorganic Ion Exchangers

Abraham Clearfield, Damodara M. Poojary, Elizabeth A. Behrens, Roy A. Cahill, Anatoly I. Bortun, and Lyudmila N. Bortun

Department of Chemistry, Texas A&M University, College Station, TX 77843-3255

The crystal structures of two tunnel type inorganic ion exchangers are described. A knowledge of structure is necessary to understand the ion exchange properties of these compounds. The titanosilicate of composition $Na_{1.64}H_{0.36}Ti_2O_3(SiO_4) \cdot 1.8H_2O$ has a square framework structure outlining a tunnel parallel to the c-axis. In addition, the faces outlining the tunnel have cavities in which Na^+ fits snugly but alkali metal ions larger than Na^+ are excluded. Cs^+ fits within the tunnels forming eight bonds with oxygen atoms of the silicate having distances of 3.183(5) Å and 3.057(6) Å. Because of its large diameter, Cs^+ can only occupy half of the tunnel sites for a maximum uptake of 25% of the total exchange capacity. The remaining charge is satisfied by Na^+ and protons within the tunnel. The affinity for Cs^+ is much greater than for Na^+ in the tunnel sites so that small amounts of Cs^+ may be removed from concentrated sodium nitrate solutions making this exchanger useful for nuclear waste remediation. The second exchanger, $K_3H(TiO)_4(SiO_4) \cdot 4H_2O$, has a structure similar to the first but with a shorter c-axis. Thus, ions occupy the face centers but not the tunnels so that the selectivity series for alkali metal ions depends upon ion size.

During the past 50 years, nuclear defense activities have produced large quantities of nuclear waste that now require safe and permanent disposal (1). The general procedure to be implemented involves the removal of ^{137}Cs and ^{90}Sr from the waste solutions for disposal in permanently vitrified media (2). These ions are present in the waste solutions at concentrations of 10^{-3}-10^{-7} M admixed with many other species and at high salt concentrations. Therefore, highly selective sorbents or ion exchangers are required. Further, at the high radiation doses present in the solution, organic exchangers or sequestrants are likely to decompose over time. Inorganic ion exchangers are resistant to radiation damage and can exhibit remarkably high selectivities (3).

A promising group of compounds are those with tunnel structures in which the tunnel space is fit to the size of the ion of interest. There are literally hundreds of inorganic compounds with tunnel structures or a combination of tunnels and cavities. However, for our purposes, the atoms constituting the framework must not be affected by either strong alkali or acid. Therefore, zeolites are unsuitable as are many compounds of amphoteric metals. Phosphates may be useable in acid solution but many of them hydrolyse in basic solution. For our purposes, we will choose certain compounds of titanium and zirconium as silicates and germanates for use in nuclear waste treatments. However, the principles we develop will be quite general and can be applied to situations other than nuclear waste systems. For use in mild acid, alkali or neutral solutions a much wider range of elements may be suitable for construction of the framework and thus many more compounds can be utilized.

Compounds With Tunnel Structures

Sodium Titanium Silicate.

A titanium silicate of ideal formula $Na_2Ti_2O_3(SiO_4) \cdot 2H_2O$ was discovered through a collaboration between Sandia National Laboratory and Texas A&M University (4). This compound was reported to be highly selective for Cs^+ and Sr^{2+} in the presence of large amounts of sodium ion and NaOH and also for Cs^+ in moderately strong acid solution. No structural details were given. We were able to synthesize this compound as a highly crystalline powder from a mixture of titanium isopropoxide, tetraethylorthosilicate and NaOH under hydrothermal conditions (5). The procedure is quite specific as small changes in experimental conditions yield related compounds such as ETS-4 (6). The structure was solved ab initio from X-ray powder data (5). This compound has a tetragonal unit cell with the arrangement of atoms forming a framework that encloses tunnels. The framework is formed from TiO_6 octahedra and SiO_4 tetrahedra. There are four octahedra at the corners of a square with 4 symmetry, that is, two octahedra up and two below turned 90° from the first two. These octahedra are bridged by silicate tetrahedra in the a- and b-axis directions (Figure 1). In the c-axis direction, the octahedra are connected by bridging oxo groups. The unit cell dimensions are approximately a = b = 7.8Å, c = 12.0Å. Half the sodium ions are situated in the ac and bc planes bonded by four oxygen atoms from silicate groups above and below the Na^+. The coordination sphere of the Na^+ is completed by two water molecules in the axial positions (Figure 2).

The remaining sodium ions are located in the tunnels but at an occupancy of 64%. The reason for this lowered occupancy is the limited space available within the tunnels. The water molecules bonded to the framework sodium ions also lie within the tunnels accounting for one of the two waters in the formula. An additional 0.8 H_2O also resides in the tunnels bonded to the sodium ions within the tunnels but since there are four formula units per unit cell there are 7.2 H_2O in the tunnel of one unit cell. The charge balance is made up by protons so the real, as opposed to the ideal, formula is $Na_{1.64}H_{0.36}Ti_2O_3(SiO_4) \cdot 1.8H_2O$.

The sodium ions in the face-centers are bonded to four silicate oxygens at a distance of 2.414(5) Å. The two bonds to the water molecules are somewhat longer 2.765(1) Å. For the sodium ions within the tunnel the Na-O bonds are longer, 2.74(2) Å to silicate oxygens and 2.79-3.17 Å to water, indicating much weaker bonding. Exhaustive treatment of the sodium titanosilicate with a 0.2 M $CsNO_3$ solution led to only partial exchange of the Na^+. The final composition was $Na_{1.49}Cs_{0.2}H_{0.31}Ti_2O_3(SiO_4) \cdot H_2O$. Solution of the crystal structure of this phase (5) revealed the reason for the low uptake. Cesium ion cannot fit in the space occupied by Na^+ in the framework. This was shown conclusively by treating the acid form of the exchanger $H_2Ti_2O_3(SiO_4) \cdot nH_2O$ with Cs^+. In both the sodium and proton phases, the Cs^+ only occupied positions within the tunnel (Figure 3).

There are two positions within the tunnels for Cs^+. In both sites the Cs^+ sits exactly in the center of the square but at either 1/4c, 3/4c or 0.13c, 0.63c. The c-axis is approximately 12Å long so a Cs^+ at 1/4c is approximately 6Å away from its mate at 3/4c. However, the second Cs^+ at 0.13c would be less than 1.5Å away from the one at 1/4c. This is smaller than the radius of the Cs^+ so only one of the two sites can be occupied in any one unit cell. Thus, the maximum capacity for Cs^+ is only 25% of the total (theoretical) capacity. As a result, some Na^+ and H_2O fill the remaining space in the tunnel. The Cs^+ at 1/4 is eight coordinate with Cs-O bond distances of 3.183(5) Å whereas the Cs^+ in the second site at 0.13c is six coordinate with four bonds of length

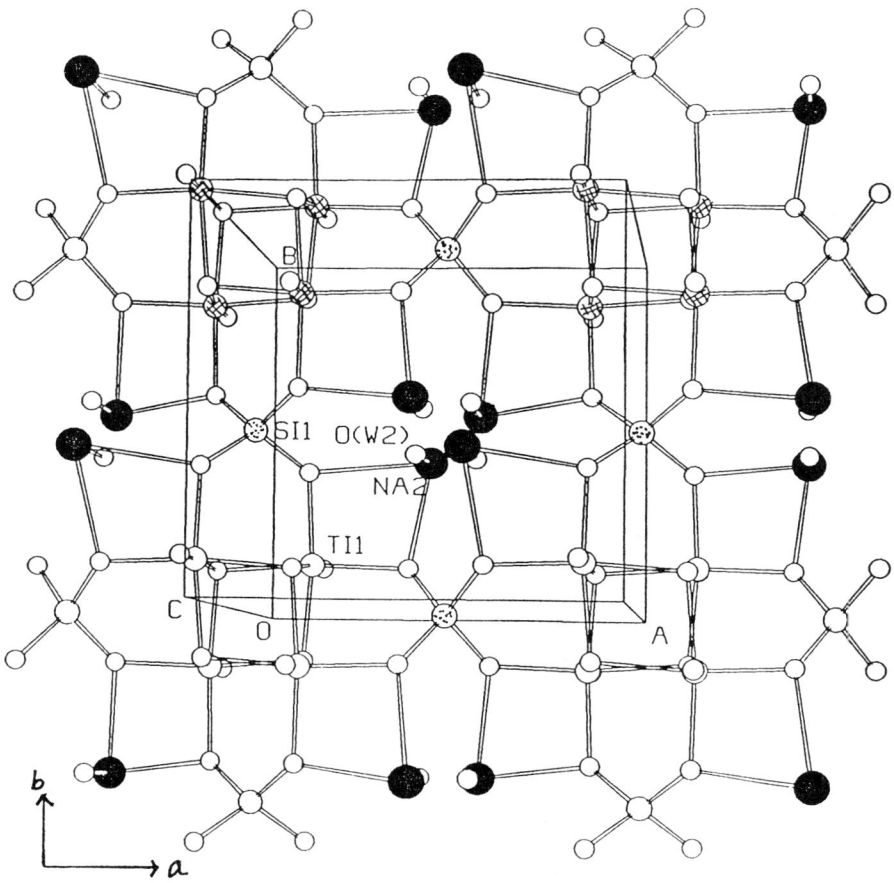

Figure 1. Schematic structure of sodium titanosilicate as viewed down the c-axis. The disordered sodium ions in the tunnel are represented by filled circles and the water molecules by open circles.

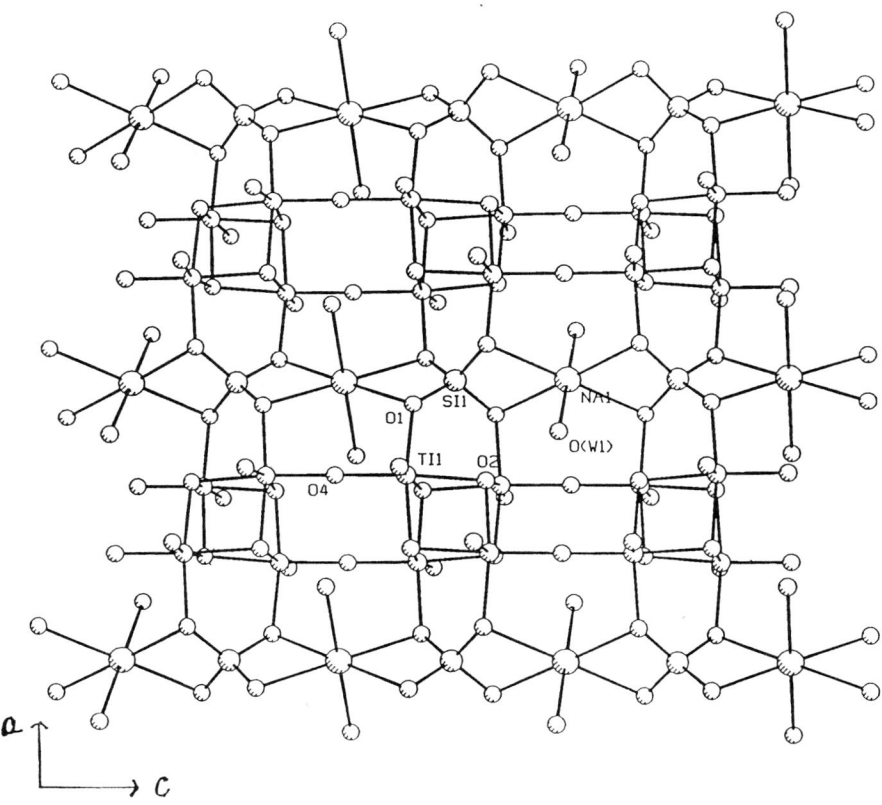

Figure 2. Representation of the titanosilicate structure as viewed down the b-axis showing the arrangement of the atoms in the ac and bc faces and the six coordinate sodium ions in the framework sites.

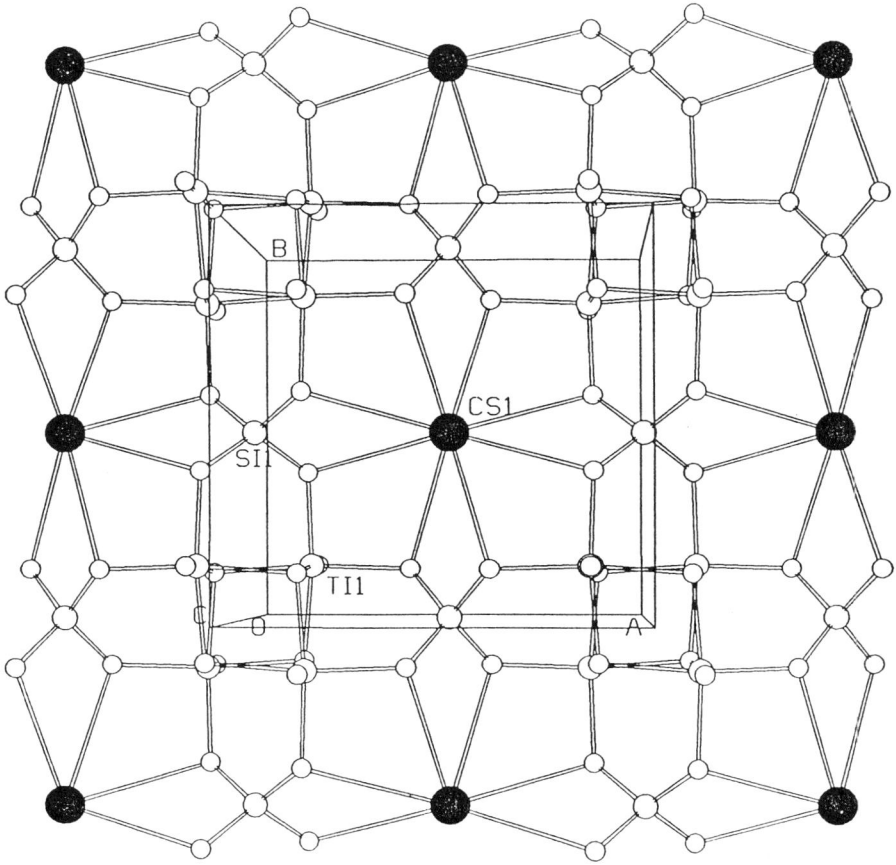

Figure 3. Schematic representation of the Cs^+ exchanged phase of the titanosilicate as viewed down the c-axis. The tunnel also contains sodium ions and water molecules.

3.057(6) Å and two bonds to water molecules of length 2.95(2) Å. These bond distances are very close to the sum of the ionic radii for the observed coordination numbers of the cesium ion and are largely responsible for the high affinity for this ion.

The crystal structure data explains very well the observed ion exchange behavior of the titanosilicate. This is illustrated by the titration curves (Figure 4) for Na^+ and Cs^+ using the acid form of the exchanger, $H_2Ti_2O_3SiO_4 \cdot H_2O$. The theoretical capacity is 7.81 meq/g. Cesium ion is exchanged in acid solution below pH 2 and attains a total uptake of about 25% of the theoretical capacity. This uptake is in keeping with the X-ray data showing that no Cs^+ resides in the framework and only half the sites in the tunnel can be occupied by Cs^+. The remainder of the sites are occupied by water and hydronium ions. In contrast, sodium ion begins to exchange at a pH near 2 and attains a total uptake of 6.1 meq/g, which is very close to that required by the sodium phase formula.

Rietveld refinement of the structure at different levels of exchange showed that the sodium ion fills the preferred sites in the framework first and then begins to fill the tunnel sites (8). However, if an equimolar mixture of sodium and cesium ions is used as the titrant only about 0.5 meq/g of Cs^+ is exchanged over the entire pH range but more than 4 meq/g of Na^+ is exchanged. We interpret this as a competition for the several ion exchange sites. The sodium ion can diffuse into both the framework sites and the tunnel sites. The larger Cs^+ ions can only move down the tunnels while the sodium ions can enter the tunnels from all three directions, down the tunnels and through the framework sites. Diffusion of Cs^+ into the lattice is further retarded by the strong bonding it experiences within the tunnels. Thus, while the affinity of the exchanger for Cs^+ is very high, kinetic and mainly steric factors determine that more Na^+ will be exchanged. By the reverse token Cs^+ is strongly retained by the exchanger and thus is easily separated from Na^+ by mild acid treatment.

Another factor that may be responsible for the slow diffusion of cesium ion may result from the significant electrostatic repulsive forces it must encounter when sodium is present in the framework sites, presenting a barrier to its diffusion. In contrast, the smaller sodium ion moving through the tunnels and faces could diffuse more rapidly, encountering less electrostatic repulsion, and, in competition with Cs^+, not only fills up the framework sites but is also found in the tunnels. This further reduces the Cs^+ occupancy. Thus, while Cs^+ is thermodynamically favored over Na^+ as shown from K_d values, we attribute its low uptake to less favorable kinetic and steric factors. Additional studies to further develop these concepts are in progress.

Many nuclear waste streams are alkaline and contain high levels of Na^+, lesser but significant amounts of K^+, and small amounts of Cs^+ (10^{-3} -10^{-8}M). Thus, in spite of the low uptake of Cs^+, the high K_d values allow Cs^+ to be removed from such solutions by the titanosilicate, albeit with a very low capacity.

Figure 5 shows the potentiometric titration curves including those for K^+ and Li^+ ions. We note that initially K^+ is sorbed from acid solution to a somewhat greater extent than Cs^+. The potassium uptake curve parallels that for Cs^+ up to pH 7 where there is a change of slope in the curve. The amount of K^+ exchanged at this pH is ~2.5 meq/g. The X-ray diffraction results show (8) that K^+ preferentially occupies the tunnel sites at 1/4c, 3/4c. There are four K-O bond distances (8-fold coordination) at 3.18(1) Å and four at 2.82(1) Å. The former value is somewhat large, but the latter is the expected value for the sum of the ionic radii for eight coordination.

Potassium ion is too large to fit into the framework sites. Instead, additional loading of K^+ results in occupation of a second site close to the framework but halfway between the K^+ ions at 1/4 c, 3/4 c (Figure 6). Because of the volume limitations of the tunnel, the maximum uptake of K^+ is less than Na^+, and the formula derived by chemical analysis is $K_{1.35}H_{0.65}Ti_2O_3(SiO_4) \cdot H_2O$. This value for K^+ exchange compares favorably to $K_{1.38}H_{0.62}$ as derived from the Rietveld refinement. Because of the positioning of the K^+ ions, their distribution is 0.5 moles within the tunnels at the centers of the ab plane and 1/4c, 3/4c and about 0.85-0.88 moles in the second site near the framework at 0 and 1/2c.

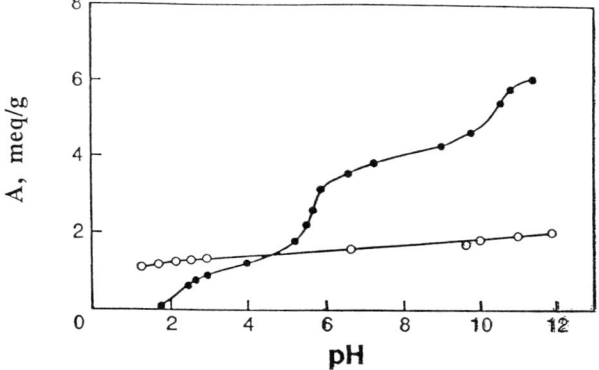

Titration curves for H₂Ti₂O₃SiO₄

● 0.1 M (NaCl + NaOH)

O 0.1 M (CsCl + CsOH)

Figure 4. Potentiometric titration curves for sodium and cesium ions obtained with the acid form of the titanosilicate, $H_2Ti_2O_3SiO_4 \cdot 1.5H_2O$. Titrant: 0.1 M(NaCl + NaOH), 0; 0.1M (CsCl + CsOH), O.

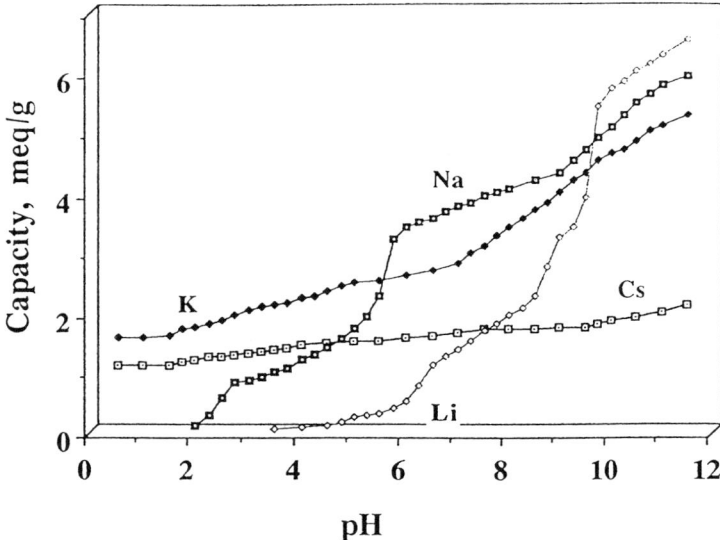

Figure 5. Potentiometric titration curves for alkali metal cations by the static equilibrium method showing the contrast in behavior of the alkali cations. Titrant: 0.1 (MCl + MOH).

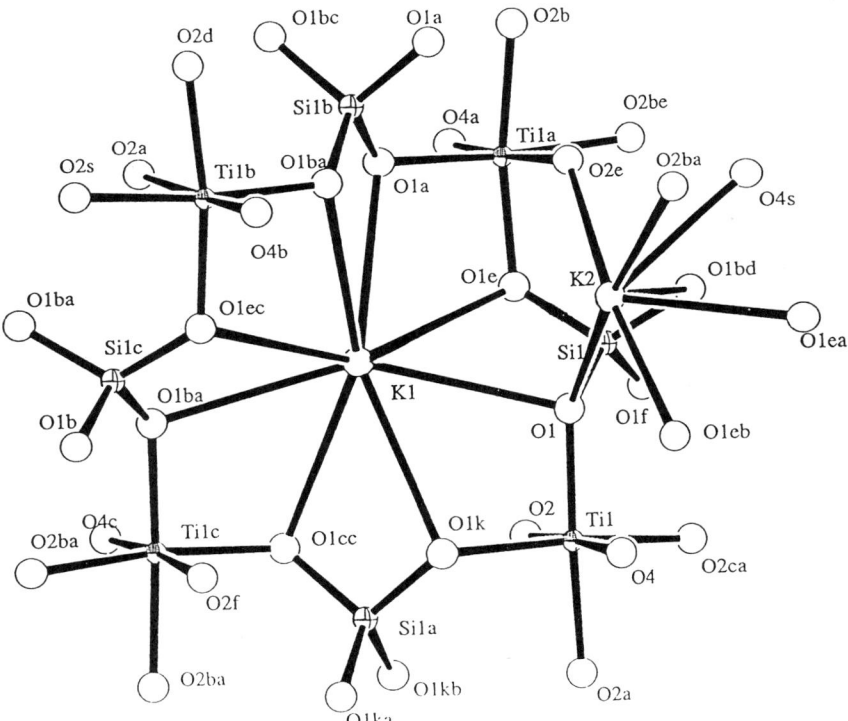

Figure 6. Schematic representation of the coordination of K⁺ in the titanosilicate phase $K_{1.38}H_{0.62}Ti_2O_3(SiO_4) \cdot H_2O$. K1 is sited in the center of the tunnel at $1/4$ c and $3/4$ c. K2 is located near the framework at 0 c and $1/2$ c with an occupancy of 0.88.

The lithium ion curve (Figure 5) is interesting in that no exchange occurs until pH 4 and then only a very small amount is taken up to pH 6. At high pH, the uptake rises rapidly with the maximum value being greater than that for Na^+. The formula obtained from elemental analysis is $Li_{1.7}H_{0.3}Ti_2O_3(SiO_4)\cdot 2.3H_2O$. Because of the small size of the lithium ion it would form long bonds if it were located in the center tunnel sites or the face center framework sites. X-ray data shows that a site similar to the one for K^+ close to the framework is favored in order to form bonds suitable to lithium's smaller ionic radius.

It is clear from this structural analysis that the affinity of the titanosilicate for alkali metal ions is controlled by the lateral dimensions of the tunnel and the cavities within the framework. The tunnel sites are highly selective for Cs^+ and K^+ whereas the framework sites prefer Na^+. In the absence of competing ions, Li^+ can be exchanged but only in neutral or alkaline solutions. In considering divalent ions, only half the sites will be occupied and this will lessen the crowding. Therefore other factors than size may be operative and investigations designed to uncover these factors are in progress (9).

Pharmacosiderites. Closely related to the titanosilicates described in the previous section are a group of compounds having structures related to the mineral pharmacosiderite (10). This mineral has the composition $KFe_4(OH)_4(AsO_4)_3$. We will describe the structure of the titanium silicate analogue first prepared by Chapman and Roe (11) by a mild hydrothermal procedure. The ideal composition is $HM_3Ti_4O_4(SiO_4)_3\cdot 4H_2O$ where M is a univalent ion. The unit cell is cubic with a=7.821(1) Å for both the proton and Cs^+ phases (12). The titanium atoms are octahedrally coordinated and form clusters of $(TiO)_4$ cubes at the corners of the cubic unit cell as shown in Figure 7. These clusters are bridged to each other by the silicate groups along all three crystallographic directions (Figure 8). This arrangement creates equivalent tunnels parallel to the three cubic unit cell axes running through the face centers. The formula of the acid form is $H_4Ti_4O_4(SiO_4)_3\cdot 8H_2O$. Hydronium ions are present in the face centers and the body center for a total of $4H_3O^+$ and the remaining water molecules are tetrahedrally disposed about the one in the body center as shown in Figure 8.

In the potassium phase (12), the K^+ ions are located precisely in the face centers of the cube and are 12-coordinate as shown in Figure 9. Eight of the bonds are to silicate oxygens with a bond length of 3.23(1) Å and the four remaining bonds are to water molecules (K-O_w = 3.17(1) Å) located within the tunnels. The ionic radius for 12-coordinate K^+ ion is not known with certainty but is approximately 1.60Å (13). Using this value, the sum of the ionic radii is 2.96Å ($r_{O^{2-}}$ =1.36Å). Thus, the bonds are somewhat longer than optimal.

If the Cs^+ ions occupied the face centers, they would have eight coordination sites at a uniform bond distance of 3.2Å. The ideal bond length for 12-coordinate Cs^+ is 3.24Å (ionic radius r_{Cs^+} =1.88Å) (13). Thus, the face center site may be ideal, but instead the cesiums locate off center into the channel a distance of 0.46Å. This positioning creates four short bonds (Cs-O = 3.14Å) and four longer ones (Cs-O = 3.41Å) to the silicate oxygens and two short (2.82(1) Å) and two long bonds (3.62(1) Å) to water molecules. These cesium ion sites are disordered with the ions dispersed randomly on one or the other side of the face center. Harrison et al. (14) were able to synthesize single crystals of $HCs_3Ti_4O_4(SiO_4)\cdot 4H_2O$ under forcing hydrothermal conditions (750°C, 30,000 psi) and found the same type of disorder with Cs^+, confirming the validity of the powder structural study.

It is now clear why only three of the four protons of the acid form are exchanged by the cesium and potassium ions. These alkali metal ions are either located in the face centers or close to these positions accommodating three cations per unit cell. There is no room for a fourth cation of the same size within the tunnels. It may be possible to place a fourth ion within the tunnels provided the cation diameter is small enough. But the smaller the cation, the less may be the tendency to fit the cations in the

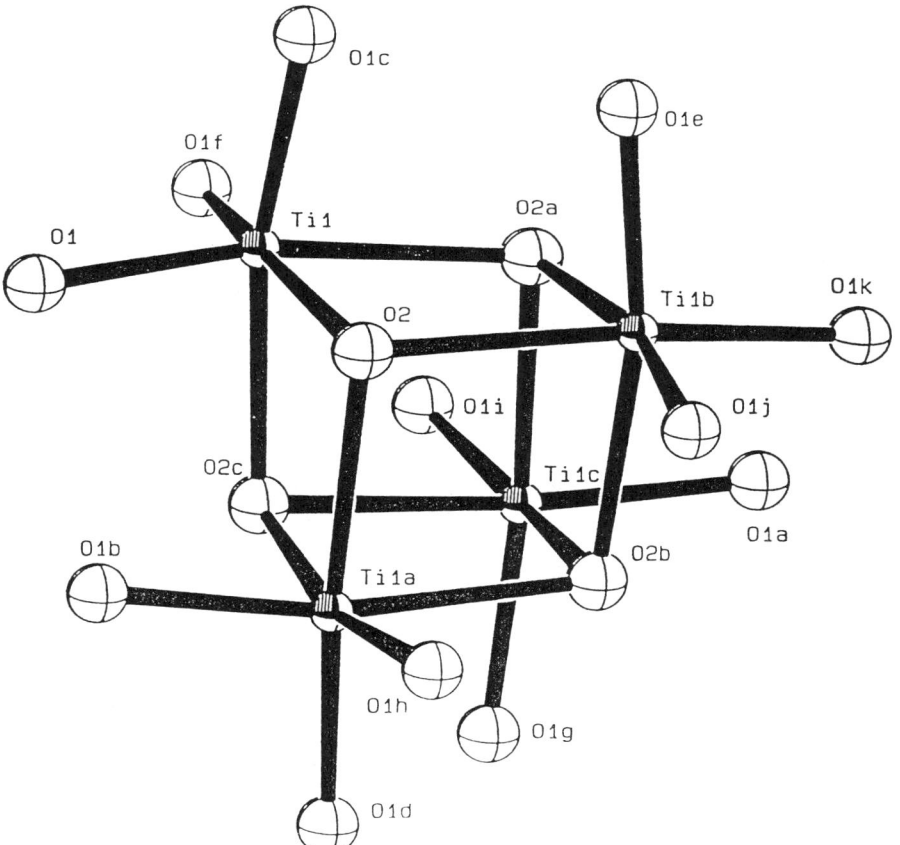

Figure 7. A section of the titanosilicate with pharmacosiderite structure showing the cluster of four titanium-oxygen octahedra at the corners of the cubic unit cell.

Figure 8. Schematic representation of the acid phase of the pharmacosiderite titanosilicate, $H_4(TiO)_4(SiO_4)_3 \cdot 8H_2O$. The stippled pattern represents the titanium octahedral clusters, the striped polyhedra represent the silicate groups and the striped squares in the center of the titania clusters are silicate groups directed down the cube.

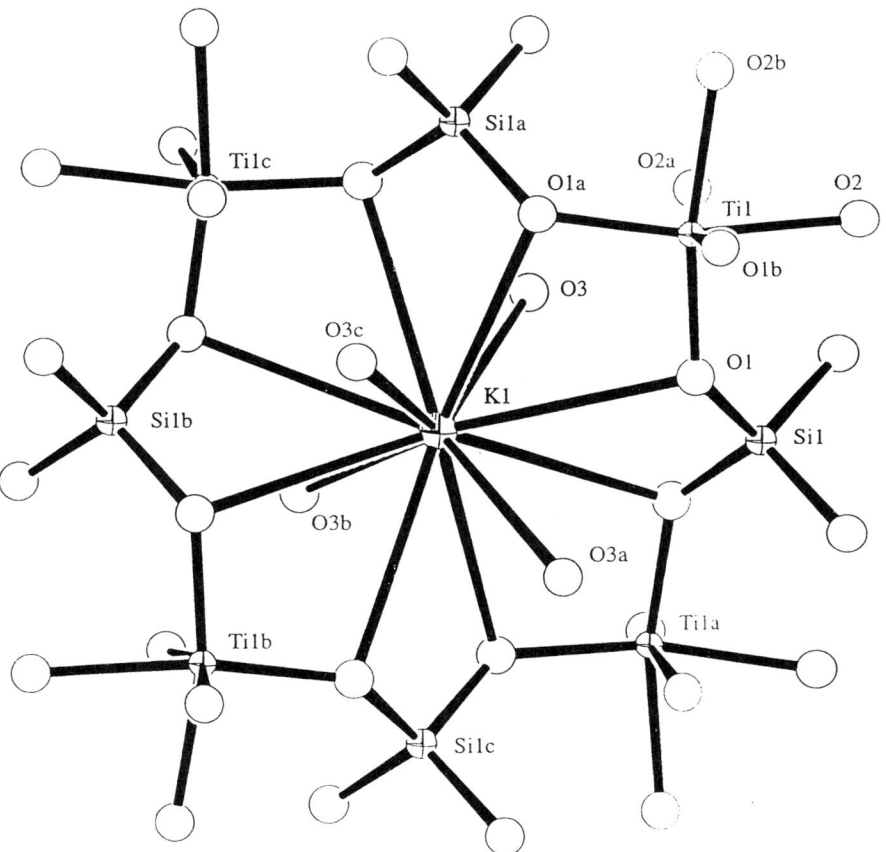

Figure 9. A portion of the pharmacosiderite titanosilicate structure showing the siting of the K$^+$ in the face center of the cube. O3 represents water molecules. With the six cubic faces filled, no potassium can occupy the cube center so the formula is $K_3H(TiO)_4(SiO_4)_3 \cdot 4H_2O$.

face centers. On the assumption that all the alkali metal cations fill similar sites, we might expect the selectivity series to be $Cs^+>K^+>Na^+>Li^+$ based on the suitability of the bond lengths. This is in fact the case for the acid phase K_d values in ml/g at pH 2.5-3 were Li, <1: Na, 180; K^+ 1800; Cs, 11,000 (*11*).

In addition to the silicate form of pharmacosiderite, there are germanate forms (*15-17*) and partially Ge substituted phases. These substitutions change the unit cell size and the size of the cavities within the framework. As a result, the selectivity for a particular ion changes as the level of framework substitution changes (*18*). This is illustrated by the data in Table I. We note that germanium may substitute for titanium as well as for silicon. By controlling the level of substitution and the size of the ion incorporated into the framework it should be possible to change selectivities drastically but over a narrow range of ion sizes. The ions most affected are those that form bonds that are slightly too long or too short for optimal affinity by the exchanger.

Conclusions

In the two tunnel structures described here the affinity of the exchanger for alkali metal cations depends upon the number and strength of the bonds formed by the framework oxygens and water molecules within the tunnels. We use as a measure of bond strengths the observed bond lengths. The exchanger framework is relatively rigid and symmetrical. Thus, the ions most preferred are those that can fit into the tunnels or faces to form preferred coordination polyhedra. This behavior is quite unlike that of organic sulfonic acid ion exchange resins where the framework is flexible, can swell and the cavities fill with water. For such resins the affinity of the resin for a series of ions of the same charge but increasing size changes modestly and separation factors are low. The opposite is true for the tunnel structures. K_d values change rapidly as the ion size approaches that required for strong bond formation. However, these tunnel exchangers may have multiple exchange sites so that one site may favor one ion and another site favor a different sized ion as is the case for the titanosilicate.

Because the crystal structures of these tunnel exchangers in both the proton and cationic phases can be determined with some degree of precision, it would appear that the affinity for ions can be determined by lattice energy type calculations. Such calculations could be used to predict feasible separations or predict the best structure type for difficult ion separations as encountered in nuclear waste remediation problems.

Acknowledgement

The authors acknowledge with thanks the generaous support of the Department of Energy under the Efficient Separations and Crosscutting Program Contract No. 206015-A-N4 and the Basic Energy Sciences Division, grant No. DE-FG07-96ER 14689 under the sponsorship of the Environmental Management Program.

Literature Cited

(1) Proceedings of the First Hanford Separation Science Workshop, Battelle PNNL, PNL-Sa-231775, Richland, WA, May 1993.
(2) Physics Today, Special Issue: *Radioactive Waste*, June, 1997.
(3) Clearfield, A.; Nancollas, G. H.; Blessing, R. H. In *Ion Exchange and Solvent Extraction*, Marinsky, J. A.; Marcus, Y., Eds.; Marcel Dekker: New York, **1973**; Vol. 5, Ch. 1.
(4) Anthony, R. G.; Dosch, R. G.; Gu, D.; Philip, C. V. Ind. Eng., *Chem. Res.* **1994**, *33*, 2702-2705.
(5) Poojary, D. M.; Cahill, R. A.; Clearfield, A., *Chem. Mater.* **1994**, *6*, 2364-2368.
(6) Kuznicki, S. M.; Thrush, K. A.; Allen, F. M.; Levine, S. M.; Hamel, M. M.; Hayhurst, D. T.; Maknaud, M. In *Synthesis of Microporous Materials*, Ocelli,

Table I. ^{89}Sr K_ds for Various Pharmacosiderites in Groundwater and Nuclear Waste Simulants

SAMPLE	K_d (mL/g) Groundwater Simulant	K_d (mL/g) Nuclear Waste Simulant
$HNa_3(TiO)_4(SiO_4)_3 \cdot nH_2O$	3.85×10^4	N/A
$HK_3(TiO)_4(SiO_4)_3 \cdot 4H_2O$	5.24×10^4	2.61×10^4
$HK_3(TiO)_4(GeO_4)_3 \cdot 4H_2O$	2.18×10^5	475
$HK_3(TiO)_{3.5}(GeO)_{0.5}(GeO_4)_{2.5} \cdot 4H_2O$	4.39×10^4	79

Volume:mass=200:1
24 hours contact time
Groundwater simulant contains: 100 ppm Ca, 3.4×10^{-4} ppm Cs, 0.4 ppm Fe(III), 1 ppm K, 8 ppm Mg, 15 ppm Na, and 0.05 µCr ^{89}Sr.

Nuclear waste simulant: 5 M $NaNO_3$ and 1 M NaOH spiked with 0.06 µCi ^{89}Sr.
*(no Sr carrier was used for either simulant)

M.; Robson, H., Eds.; Van Nostrand Reinhold, New York, **1992**; Vol. I, p. 427.
(7) Bortun, A. I.; Bortun, L. N.; Clearfield, A., *Solvent Extr. Ion Exch.* **1996**, *14*, 341.
(8) Poojary, D. M.; Bortun, A. I.; Bortun, L. N.; Clearfield, A., *Inorg. Chem.* **1996**, *35*, 6131.
(9) Clearfield, A.; Bortun, A. I.; Bortun, L. N., Proceedings of IEX '96, *Royal Society of Chem.*, Cambridge **1996**, *182*, 338-345.
(10) Buerger, M. J.; Dollase, W. A.; Garaycochea, I. *Kristallogr.* **1967**, 92-125.
(11) Chapman, D. M.; Roe, A. L., *Zeolites*, **1990**, *10*, 730.
(12) Behrens, E. A.; Poojary, D. M.; Clearfield, A. *Chem. Mater.* **1996**, *8*, 1236.
(13) Prewitt, C. T.; Shannon, R. D. *Trans .Amer. Crystallogr. Assoc.* **1969**, *5*, 57.
(14) Harrison, W. T. A.; Gier, T. E.; Stucky, G. D. *Zeolites* **1995**, *15*, 408.
(15) Nenoff, T. M.; Harrison, W. T. A.; Stucky, G. D. *Chem. Mater.* **1994**, *6*, 525.
(16) Mutter, G.; Eysel, W.; Greis, O.; Schmetz, K. N. *Jb. Miner. Mh.* **1984**, *4*, 183.
(17) Feng, S.; Greenblatt, M. *Chem. Mater.* **1992**, *4*, 462-467.
(18) Behrens, E. A.; Poojary, D. M.; Clearfield, A. *Chem. Mater.*, in press.

Chapter 12

Metal Ion Separations with Lariat Ether Ion-Exchange Resins

Richard A. Bartsch and Takashi Hayashita[1]

Department of Chemistry and Biochemistry, Texas Tech University, Lubbock, TX 79409–1061

Condensation polymerization of proton-ionizable dibenzo lariat ethers with formaldehyde in formic acid produces lariat ether ion-exchange resins. These novel, dual-function, cation-exchange resins have both cyclic polyether units and ion-exchange sites for metal ion complexation. This combination provides metal ion sorption selectivities which are unattainable with ordinary ion-exchange resins. Structural variations within the proton-ionizable lariat ether monomers influence both the selectivity and efficiency of metal ion sorption.

Three decades ago, Pedersen reported the first practical syntheses of a variety of macrocyclic polyethers, such as **1** and **2** (Figure 1), as well as the results of initial investigations of their metal salt complexation behavior (*1,2*). Pedersen proposed that an appropriately sized metal cation could be complexed within the central cavity of the macrocycle. The class name of "crown ethers" was advanced due to the resemblance of such complexes to crowns worn on the heads of royalty.

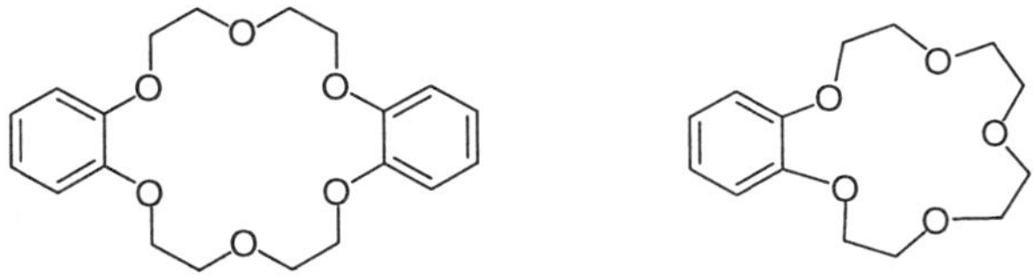

Figure 1. Structures of the crown ethers dibenzo-18-crown-6 (**1**) and benzo-15-crown-5 (**2**).

Pedersen also proposed a trivial nomenclature system for crown ether compounds (*1*) which is in widespread use today. The system consists of naming in order: (a) the number and kind of substituents on the polyether ring; (b) the total

---

[1]Current address: Department of Chemistry, Graduate School of Science, Tohoku University, Aramaki, Aoba-ku, Sendai 980–77, Japan.

number of atoms in the polyether ring; (c) the class name "crown"; and (d) the number of oxygens in the polyether ring. Thus compounds **1** and **2** are designated dibenzo-18-crown-6 and benzo-15-crown-5, respectively.

Crown Ether Polymers

Due to the strong metal ion binding behavior of crown ethers, considerable attention has been focused upon their incorporation into polymers (*3*). Immobilization of crown ethers in polymers prevents loss of these relatively expensive compounds to mobile phases during separation processes and also alleviates their potential physiological activity (*4*).

Formaldehyde-type condensation polymers of benzocrown and dibenzocrown ethers were studied by Blasius and co-workers in the mid-1970's to the early 1980's (*5-9*). The resins were usually synthesized by condensation of dibenzocrown ethers with formaldehyde in formic acid or of benzocrown ethers with formaldehyde and a crosslinking agent, such as phenol, resorcinol, or xylol, in a mixture of formic and sulfuric acids. Simplified structures for examples of such polymers are represented by **3** and **4** (Figure 2). The simplification is to represent the resin as a linear polymer, even though it is crosslinked to some degree. The formaldehyde-type crown ether polymers were found to be easy to synthesize and to have excellent resistance to heat and to acidic and basic environments. (Copolymer **3** is listed in current Aldrich and Fluka catalogs as poly(dibenzo-18-crown-6)-*co*-formaldehyde and poly(dibenzo-18-crown-6), respectively.)

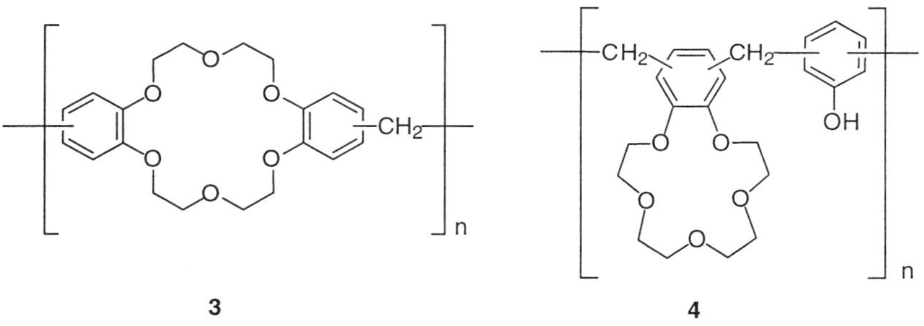

Figure 2. Simplified structures for formaldehyde condensation polymers of dibenzo- and monobenzocrown ethers.

Blasius and co-workers also investigated applications of the crown ether polymers in metal ion separation processes (*5-9*). For their use in chromatographic separations, sorption of a metal ion from solution onto the resin must be accompanied by concomitant transfer of an anion. Therefore, the selectivity and efficiency of metal ion sorption from solution by a crown ether polymer is strongly influenced by the identity of the anion(s) present in the solution.

Proton-Ionizable Lariat Ethers

Attachment of one or more side arms with potential metal ion coordination sites to a crown ether framework provides complexing agents known as "lariat ethers" (*10*). In 1981, we reported the synthesis of the lariat ether carboxylic acid **5** (Figure 3) with a hydrogen attached geminal to the functional side arm on the polyether ring and its application in the solvent extraction of alkali metal cations from aqueous solutions into chloroform (*11*). In such proton-ionizable lariat ethers, a side arm with an acidic

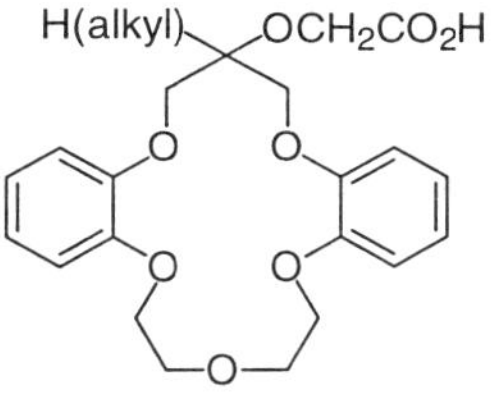

Figure 3. Structures of lariat ether carboxylic acid (**5**) and Amberlite™ CG-50 (**6**), a commercially available, cation-exchange resin.

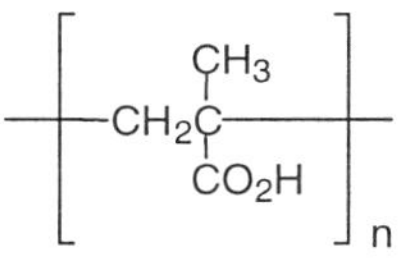

group is attached to a crown ether ring (*12*) and metal ion complexation involves the ion exchange of a metal ion for the proton of the acidic function. Eliminating the need for concomitant transfer of an aqueous phase anion into the organic medium is of immense importance for applications of crown and lariat ether ligands as the next generation of selective metal ion extractants. For process solvent extraction of metal ions, the anions normally encountered are chloride, nitrate, and sulfate, which are very hydrophilic ions. The efficiency of metal ion extraction is markedly enhanced for proton-ionizable lariat ethers compared with analogous compounds which have non-ionizable side arms (*11,13*).

From examination of Corey-Pauling-Kortun (CPK) space-filling models, the cavity diameter of dibenzo-16-crown-5 is estimated to be 2.0-2.4 Å. For the alkali metal cations, the ionic diameters are: Li^+, 1.48 Å; Na^+, 2.04 Å; K^+, 2.76 Å; Rb^+, 2.98 Å; and Cs^+, 3.40 Å (*14*). Based upon the relationship between the relative diameters of the alkali metal cations and the polyether cavities in dibenzo-16-crown-5 compounds, such as **5**, Na^+ selectivity would be predicted.

Attachment of a lipophilic alkyl group geminal to the functional side arm in **5** is beneficial for two reasons. First, introduction of the alkyl group enhances the overall lipophilicity of the extractant which decreases loss of the ionized lariat ether from the organic phase into a contacting basic aqueous phase during solvent extraction of metal ions (*11,15*). Second, the presence of an alkyl group geminal to the oxyacetic acid side arm enhances the Na^+ selectivity in competitive solvent extraction of alkali metal cations (*15*). The enhancement in Na^+ selectivity is proposed to arise from orientation of the alkyl group away from the polar polyether ring which positions the carboxylic acid group of the functional side arm directly over the crown ether cavity. Such preorganization of the binding site enhances the selectivity in metal ion complexation by macrocyclic ligands (*16*). In agreement, solid-state structures of lariat ether carboxylic acids **5** with a geminal hydrogen and a geminal decyl group have the oxyacetic acid group directed away from the polyether cavity in the former and over the cavity in the latter (*17*).

Lariat Ether Ion-Exchange Resins

It was envisioned that formaldehyde-type condensation resins could be prepared from proton-ionizable dibenzo lariat ethers, such as **5**. The resultant lariat ether ion-exchange resins would have both ion-exchange and cyclic polyether binding sites for metal ion complexation. It was anticipated that such dual function cation-exchange resins (*17*) would provide metal ion sorption selectivities which are different from those attainable with ordinary ion-exchange resins.

Lariat Ether Carboxylic Acid Resins. Reaction of appropriate acyclic and cyclic dibenzo polyether carboxylic acid monomers with formaldehyde in formic acid at reflux produced the new polyether ion-exchange resins **7-14** (Figure 4) (*19,20*).

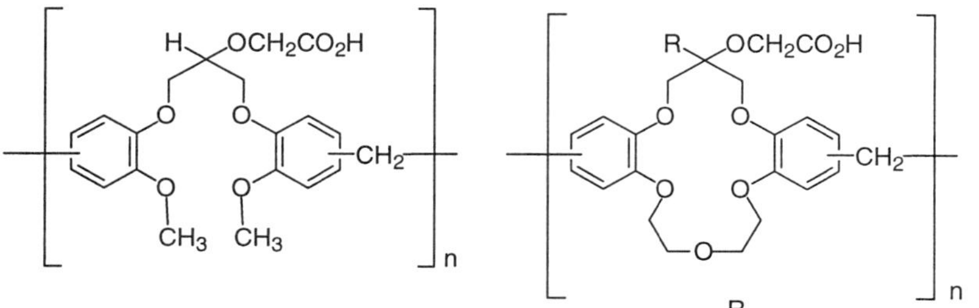

Figure 4. Dibenzo acyclic polyether and lariat ether carboxylic acid resins.

Alkali Metal Cation Sorption. The behavior of these resins was evaluated in competitive alkali metal cation sorption from aqueous solutions which were 0.10 M in each of the five alkali metal cations with a mixture of chloride and hydroxide counterions (*19,20*). In control experiments, it was shown that alkali metal cation sorption from such solutions was completed in a matter of minutes and that the sorbed metal ions could be readily stripped from the resins by washing with aqueous hydrochloric acid.

For comparison with an ion-exchange resin which contained the same acidic function but no polyether unit, competitive alkali metal cation sorption was also performed with Amberlite™ CG-50 (**6**, Figure 3), a commercially available poly(methacrylic acid) resin. For CG-50, the selectivity for alkali metal cation sorption from neutral and alkaline solutions was $Li^+ > Na^+ > K^+ \approx Rb^+ \approx Cs^+$. For acyclic polyether carboxylic acid resin **7**, the sorption selectivity was $Li^+ > Na^+ > K^+ > Rb^+ \approx Cs^+$ (Figure 5a) which shows that the introduction of ether linkages does not significantly alter the sorption selectivity from that found with CG-50 (*19*). On the other hand, the sorption selectivities for the lariat ether ion-exchange resins **8** and **11** were $Na^+ > Li^+ \approx K^+ > Cs^+ > Rb^+$ (Figure 5b) and $Na^+ >> Li^+ > K^+ \approx Cs^+ > Rb^+$ (Figure 5c), respectively. Thus the Li^+ sorption selectivity exhibited by CG-50 (**6**) and the acyclic polyether carboxylic acid resin **7** changed to Na^+ sorption selectivity for the lariat ether carboxylic acid resins **8** and **11**.

Although both lariat ether carboxylic acid resins **8** and **11** exhibit Na^+ selectivity, that for resin **11**, in which a propyl group is attached to the same crown ether ring carbon as the oxyacetic acid side arm, is much higher than that for resin **8**. Examination of CPK space-filling models reveals that when the propyl chain in the binding unit of resin **11** points away from the polar crown ether ring, the carboxylic acid group is positioned directly over the crown ether cavity. This preorganizes the binding sites in resin **11** compared with those in resin **8**. Such preorganization of the binding sites enhances complexation of that metal ion that best fits the cavity (*16*). This was the first instance in which the conformational positioning of the ion-

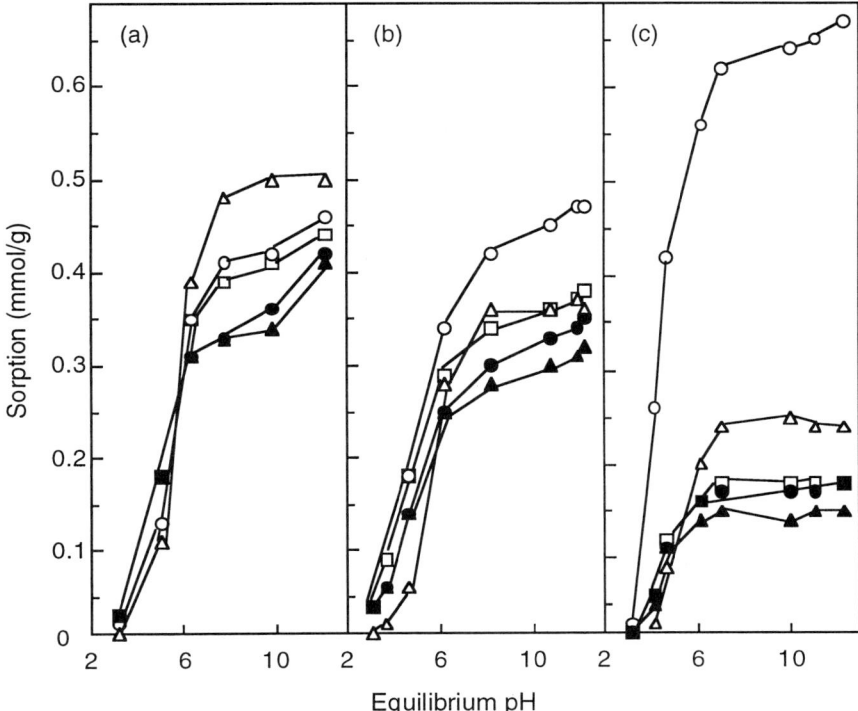

Figure 5. Competitive alkali metal cation sorption by lariat ether carboxylic resins (a) **7**, (b) **8**, and (c) **11**: Li$^+$ (Δ), Na$^+$ (O), K$^+$ ($\square$), Rb$^+$ ($\blacktriangle$), Cs$^+$ ($\bullet$).

exchange group in an ion-exchange resin has shown an important influence upon metal ion recognition.

Lariat ether carboxylic acid resin **11** also was utilized for the selective column concentration of alkali metal cations from dilute, basic aqueous solutions (20). Due to a stronger interaction of Na$^+$ with the resin, the elution peak for Na$^+$ in the acidic stripping solution was retarded relative to those for the other alkali metal cations. With gradient stripping, the concentration factor for Na$^+$ from a basic aqueous sample solution, which was 6.0 X 10^{-5} M in each of the five alkali metal cations, reached 1030 with an 84% purity (20).

The influence of the geminal alkyl group on selectivity and efficiency of competitive alkali metal cation sorption was further probed with the series of lariat ether carboxylic acid resins **9-14** (21). In this series, there is systematic structural variation of the geminal n-alkyl group from one to ten carbons. The highest metal ion loading and Na$^+$ selectivity were obtained when the geminal alkyl group was methyl, ethyl, or propyl. Longer alkyl groups were found to be detrimental to both the sorption efficiency and selectivity.

The effect of medium polarity on competitive sorption of alkali metal cations from aqueous and aqueous methanolic solutions by polyether carboxylic acid resins **7-14** was investigated (22). For the lariat ether carboxylic acid resins **8-14**, the Na$^+$ selectivity was enhanced as the percentage of methanol in the medium increased. This was attributed to strengthened metal ion-crown ether interactions as the solvent polarity decreased.

The next structural variation was to determine the influence of the crown ether ring size upon the selectivity of alkali metal cation sorption for the series of dibenzo polyether carboxylic acid resins **7** and **15-17** (Figure 6) (*23*). Once again, the lariat ether carboxylic acid resins were found to exhibit enhanced sorption selectivity over the corresponding acyclic polyether carboxylic acid resins. Good sorption selectivity for Li^+ and Na^+ over K^+, Rb^+, and Cs^+ was observed for the dibenzo-14-crown-4 carboxylic acid resin **17** which has a geminal propyl group. This change from the very good Na^+ sorption selectivity observed earlier with the dibenzo-16-crown-5 carboxylic acid resin **11** to sorption selectivity for Li^+ and Na^+ with the corresponding dibenzo-14-crown-4 resin **17** is attributed to the smaller crown ether ring size in the latter. For column concentration of alkali metal cations from dilute, basic aqueous solutions, gradient elution of the sorbed metal ions from resin **17** with aqueous acid gave selective column concentration of Li^+ and Na^+ from a solution of the five alkali metal cation species.

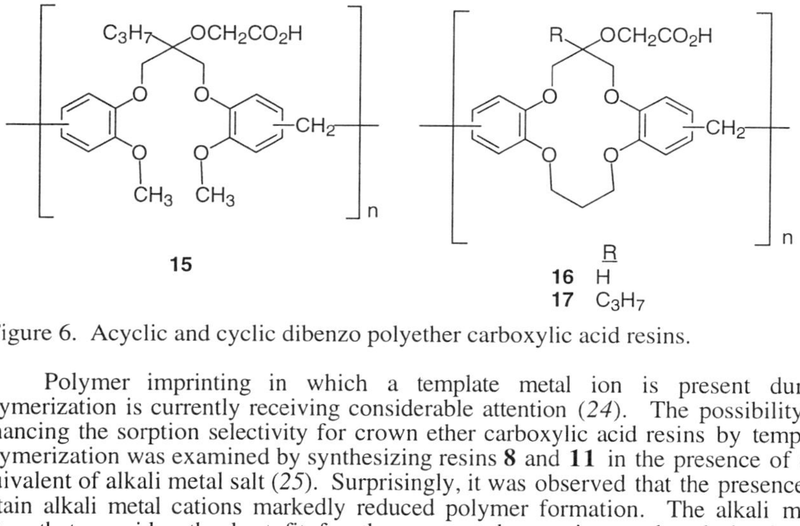

Figure 6. Acyclic and cyclic dibenzo polyether carboxylic acid resins.

Polymer imprinting in which a template metal ion is present during polymerization is currently receiving considerable attention (*24*). The possibility of enhancing the sorption selectivity for crown ether carboxylic acid resins by template polymerization was examined by synthesizing resins **8** and **11** in the presence of one equivalent of alkali metal salt (*25*). Surprisingly, it was observed that the presence of certain alkali metal cations markedly reduced polymer formation. The alkali metal cation that provides the best fit for the crown ether cavity produced the largest decrease in polymer yield. It was proposed that metal ion complexation rendered the dibenzocrown ether monomer inert to polymerization under the reaction conditions and only uncomplexed monomer was involved in polymer formation. In agreement, no template effect for alkali metal cation sorption was noted for the polymers which were produced (*25*).

Alkali and Alkaline Earth Metal Cation Sorption. Batch sorption of the four physiological metal ions Na^+, K^+, Mg^{2+}, and Ca^{2+} by CG-50 (**6**), acyclic dibenzo polyether carboxylic acid resin **7**, and lariat ether carboxylic acid resins **8** and **11** was also investigated (*26*). The aqueous solutions were 0.10 M in Na^+ and K^+ and 0.050 M in Mg^{2+} and Ca^{2+}. The observed sorption selectivity order for CG-50 was $Ca^{2+} > Mg^{2+} \gg Na^+ \approx K^+$. (Preferential sorption of the divalent metal ions results from enhanced electrostatic interaction with the carboxylate groups of the resin.) When the aqueous solution pH was 6-8, the sorption selectivity order for acyclic polyether carboxylic acid resin **7** and lariat ether carboxylic acid resin **8**

changed slightly to $Ca^{2+} \gg Mg^{2+} > Na^+ \approx K^+$. For lariat ether carboxylic acid resin **11**, which has a geminal propyl group, the sorption selectivity was found to be pH dependent. At pH = 6, the sorption selectivity order was $Na^+ > Ca^{2+} > Mg^{2+} \approx K^+$; whereas at pH = 8, it was $Ca^{2+} > Na^+ > Mg^{2+} > K^+$. Constraint of the ion-exchange group in resin **11** to a position which is highly favorable for Na^+ complexation allows sorption of this alkali metal cation to compete favorably with that of the alkaline earth metal cations.

Chromatographic Separation of Y^{3+} from Sr^{2+}. Separation of Y^{3+} and Sr^{2+} is important for the determination in environmental samples of the fission product ^{90}Sr and its daughter ^{90}Y (*27*). Extraction of pure ^{90}Y from its precursor ^{90}Sr is also important for applications in nuclear medicine (*28*). Column chromatographic separation of Y^{3+} and Sr^{2+} (each 2.5 X 10^{-3} M) from aqueous solutions by sorption on lariat ether carboxylic acid resins **8** and **11** has been reported (*29*). With both resins, Sr^{2+} was cleanly separated from the mixture of Y^{3+} and Sr^{2+}. The separation of Sr^{2+} was more effective with resin **8** which shows that in this case the presence of a geminal propyl group was detrimental.

Lariat Ether Phosphonic Acid Monoethyl Ester and Lariat Ether Sulfonic Acid Resins. To enhance the acidity of the ion-exchange site in the lariat ether ion-exchange resins, acyclic and cyclic dibenzo polyether phosphonic acid monoethyl ester monomers and analogous sulfonic acid compounds were polymerized with formaldehyde in formic acid to provide lariat ether phosphonic acid monoethyl ester resins **18-23** and **27** and lariat ether sulfonic acid resins **24-26** (Figures 7 and 8) (Laney, E. E.; Lee, J. H.; Kim, J. S.; Huang, X.; Jang, Y.; Hwang, H.-S.; Hayashita, T.; Bartsch, R. A. *React. Funct. Polym.*, **1998**, in press).

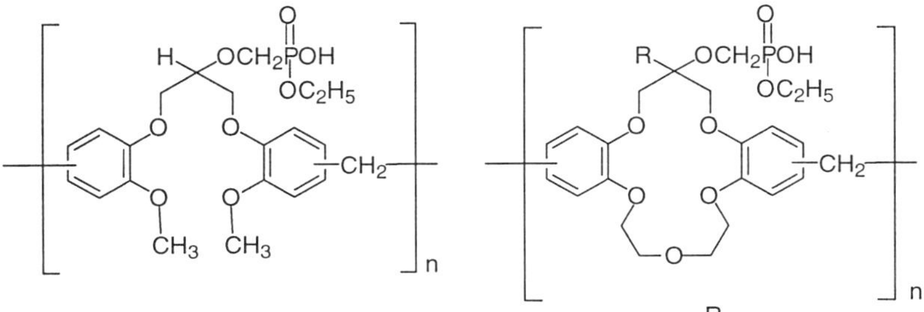

Figure 7. Acyclic and cyclic dibenzo polyether phosphonic acid monoethyl ester resins.

Alkali Metal Cation Sorption. Competitive sorption from aqueous solutions which were 0.10 M in each of the five alkali metal cations by acyclic polyether phosphonic acid monoethyl ester resin **18** and the dibenzo-16-crown-5 phosphonic acid monoethyl ester resins **19** and **20** was investigated (*30*). The sorption selectivity order for the acyclic polyether resin **18** was $Li^+ > K^+ \approx Rb^+ \approx Cs^+ > Na^+$. For the lariat ether phosphonic acid monoethyl ester **19**, the ordering changed to $Li^+ > Na^+ > K^+ > Rb^+ \approx Cs^+$. The change in the position for Na^+ in the sorption selectivity ordering from last with resin **18** to second with resin **19** is consistent with

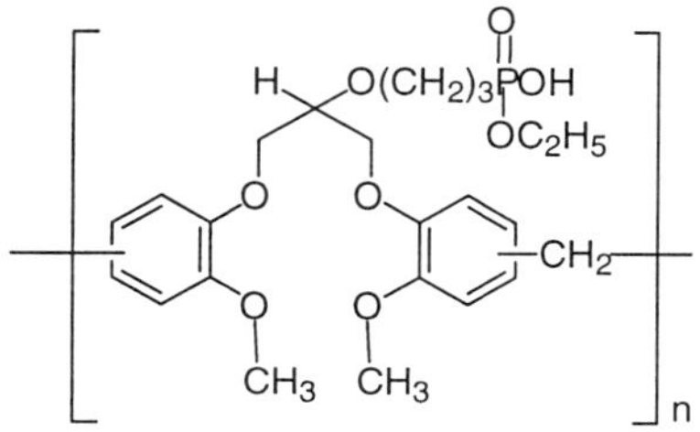

Figure 8. Acyclic and cyclic dibenzo polyether phosphonic acid monoethyl ester resins **22**, **23** and **27** and acyclic and cyclic dibenzo polyether sulfonic acid resins **24-26**.

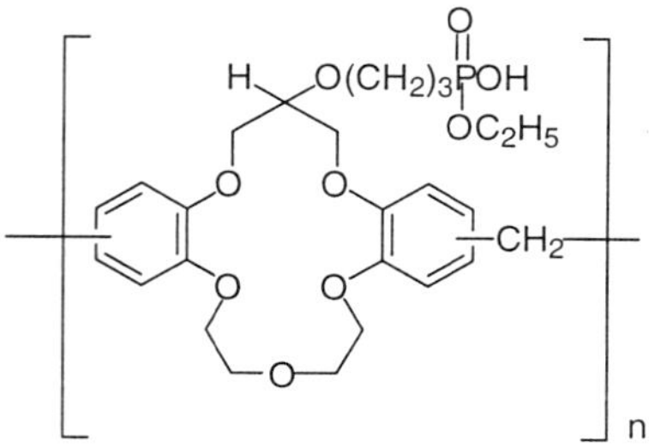

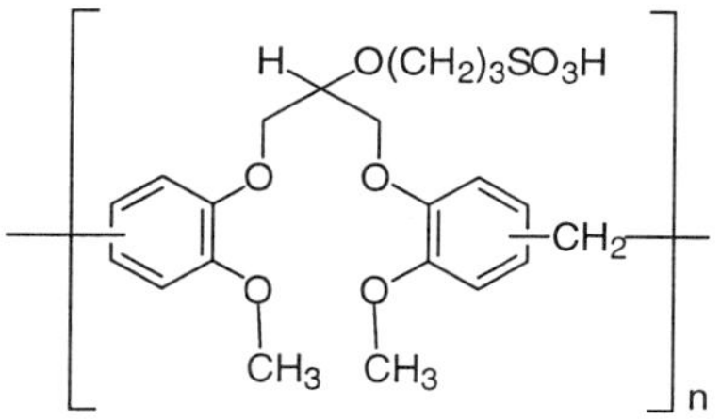

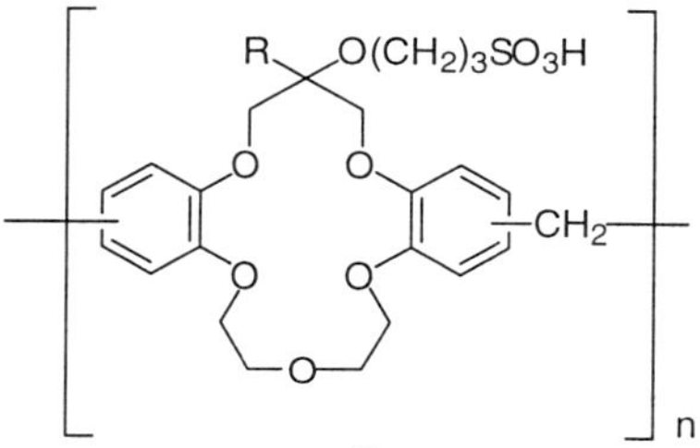

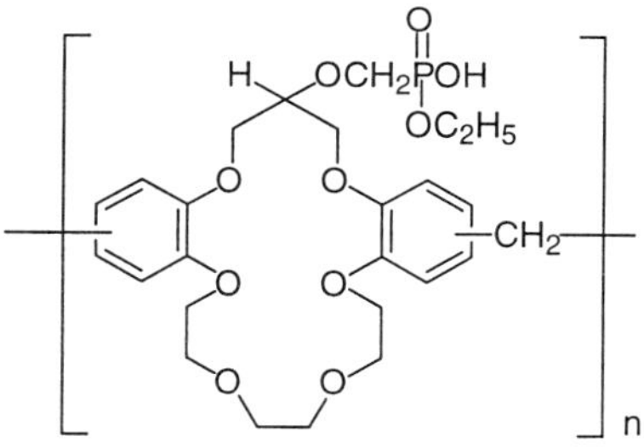

an enhanced role of the cyclic polyether unit in metal ion binding by the latter. For lariat ether resin **20**, which has a geminal methyl group, the sorption selectivity was found to depend upon the pH of the aqueous solution from which the alkali metal cations were sorbed. For pH 3-6, the sorption selectivity order for resin **20** was Na^+ > Li^+ > $K^+ \approx Rb^+ \approx Cs^+$; and for pH > 8, the ordering became Li^+ > Na^+ > $K^+ \approx Rb^+ \approx Cs^+$.

Although Na^+ sorption was enhanced by the presence of the geminal methyl group in resin **20** compared with resin **19**, resin **20** remained Li^+ selective over much of the pH region. Thus the sorption selectivity for resin **20** is found to be quite different from the very good Na^+ sorption selectivity observed for the analogous dibenzo-16-crown-5-oxyacetic acid resin **9** and suggests stronger interactions of Li^+ with a phosphonic acid monoethyl ester group than a carboxylic acid function.

Sorption of Divalent Heavy and Transition Metal Cations. The ability of proton-ionizable lariat ether resins to sorb heavy and transition metal cations from acidic aqueous solutions has been investigated (Laney, E. E.; Lee, J. H.; Kim, J. S.; Huang, X.; Jang, Y.; Hwang, H.-S., Hayashita, T.; Bartsch, R. A. *React. Funct. Polym.*, **1998**, in press). In the initial screening study, competitive batch sorptions of 1.0 mM Pb^{2+} and Zn^{2+} from aqueous solutions of pH 0-3 with a shaking time of four hours was examined for three series of acyclic and cyclic polyether resins with proton-ionizable groups. Under these conditions, CG-50 (**6**), a poly(methacrylic acid) resin, and Rexyn 101(H), a poly(vinylsulfonic acid) resin, exhibited no selectivity in competitive sorption of Pb^{2+} and Zn^{2+}.

The three series of resins included the polyether carboxylic acid resins **7**, **8**, and **11**, polyether phosphonic acid monoethyl ester resins **18**, **19**, and **21-23**, and the polyether sulfonic acid resins **24-26**. The polyether sulfonic acid resins **24-26** have longer spacers between the acidic function and the polyether unit than do the lariat ether carboxylic acid resins **7**, **8**, and **11** and the lariat ether phophonic acid monoethyl ester resins **19** and **21**. Therefore, the lariat ether phosphonic acid monoethyl ester resins **22** and **23**, which have the same spacer unit as those in the lariat ether sulfonic acids, were also examined.

For competitive sorption of Pb^{2+} and Zn^{2+} at pH = 2 or lower, the three polyether carboxylic acid resins **7**, **8**, and **11** exhibited poor metal ion loading. With the three polyether sulfonic acid resins **24-26**, both Pb^{2+} and Zn^{2+} were completely sorbed at pH = 2 or less. With pH < 2, only very slight selectivity for Pb^{2+} sorption was noted. In the region of pH = 1.0-2.5, both acyclic polyether phosphonic acid monoethyl ester resin **18** and the lariat ether phosphonic acid monoethyl ester resin **19** gave good selectivity for Pb^{2+} over Zn^{2+} with complete sorption of the Pb^{2+} at pH = 2.5. In contrast, lariat ether phosphonic acid monoethyl ester resin **21** gave poor metal ion loading, but with some selectivity for sorption of Pb^{2+}. For the phosphonic acid monoethyl ester resins **22** and **23**, which have longer spacer units in the side arm than do resins **18** and **19**, sorption efficiency was intermediate with good selectivity for sorption of Pb^{2+} over Zn^{2+}. From this screening study, it was concluded that the structural features that promote selective Pb^{2+} binding by proton-ionizable dibenzo-16-crown-5 resins are: (a) a cyclic polyether unit; (b) a proton-ionizable group of intermediate acidity; (c) a short spacer unit connecting the acidic group to the polyether framework; and (d) the absence of a geminal alkyl group.

Based upon these results, dibenzo-16-crown-5 and dibenzo-19-crown-6 phosphonic acid monoethyl ester resins **19** and **27**, respectively, were selected for more intensive study (*31*). For single species Pb^{2+} sorption by resin **19**, it was determined that the sorption complex involved one polyether unit, a Pb^{2+} cation, and a monovalent anion from the aqueous solution. Single species sorption experiments showed that Pb^{2+} binding by the dibenzo-19-crown-6 phosphonic acid monoethyl ester resin **27** was somewhat stronger than that that for the dibenzo-16-crown-5 phosphonic acid monoethyl ester resin **19** (Figure 9). However, the monomer precursor to resin **19** is
much easier to prepare than that for resin **27** which is an important compensating factor.

For the expanded studies of competitive metal cation sorption by lariat ether phosphonic acid monoethyl ester resins **19** and **27**, a broader pH region of 0-6 was utilized. For competitive batch sorption of Pb^{2+} and a second multivalent metal ion species, both resins **19** and **27** were found to exhibit good sorption selectivity for Pb^{2+} over Cd^{2+}, Ni^{2+}, Zn^{2+}, and Fe^{3+} in certain acidic pH regions. The influence of the presence of large excesses of alkali metal cations and two of the alkaline earth metal cations upon the Pb^{2+} sorption efficiency was also examined. Efficient sorption of 1.0 mM Pb^{2+} by resins **19** and **27** was maintained even for aqueous solutions which contained 0.20 M alkali metal cations. For resin **27**, efficient Pb^{2+} sorption from solutions containing 0.10 M Mg^{2+} and Ca^{2+} was observed.

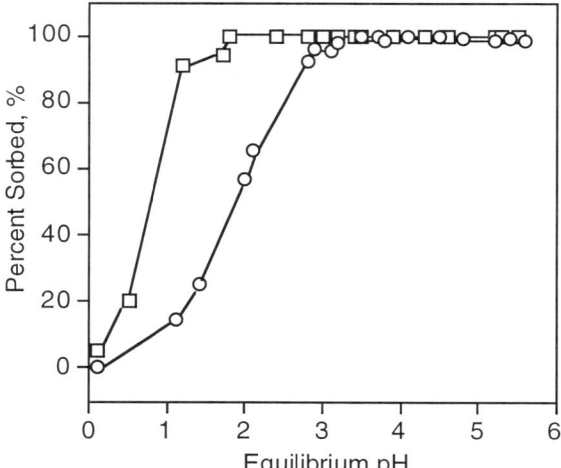

Figure 9. Non-competitive Pb^{2+} sorption by lariat ether phosphonic acid monoethyl esters **19** (O) and **27** (□).

Summary

Condensation polymerization of proton-ionizable dibenzo lariat ethers with formaldehyde in formic acid produces novel ion-exchange resins. These lariat ether ion-exchange resins have both ion-exchange and cyclic polyether binding sites for metal ion complexation and provide sorption selectivities which cannot be obtained with ordinary ion-exchange resins. From *sym*-(alkyl)dibenzo-16-crown-5-oxyacetic acid monomers, new resins have been prepared which exhibit very good Na$^+$ selectivity in alkali metal cation separations by both batch sorption and concentrator column methods. Conformational positioning of the ion-exchange group over the polyether unit is shown to have an important influence on the recognition of alkali metal cations. A resin prepared from *sym*-dibenzo-16-crown-5-oxyacetic acid monomer has been used to separate Y^{3+} from Sr^{2+}. Proton-ionizable lariat ether resins with phosphonic acid monoethyl ester groups are found to exhibit good sorption selectivity for Pb^{2+} from acidic aqueous solutions over a variety of multivalent heavy and transition metal cations, as well as large excesses of alkali and alkaline earth metal cations.

Acknowledgment

This research was supported by the Division of Chemical Sciences of the Office of Basic Energy Sciences of the U.S. Department of Energy (Grant DE-FG03-9414416) and the Texas Higher Education Coordinating Board Advanced Research Program.

Literature Cited

(1) Pedersen, C. J. *J. Am. Chem. Soc.* **1967**, *89*, 2495.
(2) Pedersen, C. J. *J. Am. Chem. Soc.* **1967**, *89*, 7017.
(3) Smid, J.; Sinta, R. *Top. Curr. Chem.* **1984**, *121*, 105.
(4) Hiraoka, M. *Crown Compounds: Their Characteristics and Applications*; Elsevier: New York, NY, 1982; Ch. 7.

(5) Blasius, E.; Adrian, W.; Janzen, K. P.; Klauthe, G. *J. Chromatogr.* **1974**, *96*, 89.
(6) Blasius, E.; Janzen, K. P. *Top. Curr. Chem.* **1981**, *98*, 165.
(7) Blasius, E.; Janzen, K. P. *Pure Appl. Chem.* **1982**, *54*, 2115.
(8) Blasius, E.; Janzen, K. P.; Keller, M.; Lander, H.; Nguyen-Tien, T.; Scholten, G. *Talanta* **1980**, *27*, 107.
(9) Blasius, E.; Maurer, P. *J. Chromatogr.* **1976**, *125*, 511.
(10) Gokel, G. W.; Dishong, D. M.; Diamond, C. J. *J. Chem. Soc., Chem. Commun.* **1980**, 1053.
(11) Strzelbicki, J.; Bartsch, R. A. *Anal. Chem.* **1981**, *53*, 1894.
(12) Bartsch, R. A. *Solvent Ext. Ion Exch.* **1989**, *7*, 829.
(13) Bartsch, R. A.; Hayashita, T.; Lee, J. H.; Kim, J. S.; Hankins, M. G. *Supramol. Chem.* **1993**, *1*, 305
(14) McBride, D. W.; Izatt, R. M.; Lamb, J. D.; Christensen, J. J. In *Inclusion Compounds*; Atwood, J. L.; Davies, J. E. D.; McNicol, D. D. Eds.; Academic Press: New York, NY, Vol. 3; p 586.
(15) Walkowiak, W.; Charewicz, W. A.; Kang, S. I.; Yang, I.-W.; Pugia, M. J.; Bartsch, R. A. *Anal. Chem.* **1990**, *62*, 2018.
(16) Cram, D. J. *Angew. Chem., Int. Ed. Engl.* **1986**, *25*, 1039.
(17) Bartsch, R. A.; Kim, J. S.; Olsher, U.; Purkiss, D. W.; Ramesh, V.; Dalley, N. K.; Hayashita, T. *Pure Appl. Chem.* **1993**, *65*, 399.
(18) Alexandratos, S. D.; Wilson, D. L. *Macrocycles* **1986**, *19*, 280.
(19) Hayashita, T.; Goo, M.-J.; Lee, J. C.; Kim, J. S.; Krzykawski, J.; Bartsch, R. A. *Anal. Chem.* **1990**, *62*, 2283.
(20) Hayashita, T.; Lee, J. H.; Chen, S.; Bartsch, R. A. *Anal. Chem.* **1991**, *63*, 1844.
(21) Hayashita, T.; Goo, M.-J.; Kim, J. S.; Bartsch, R. A. *Talanta* **1991**, *38*, 1453.
(22) Hayashita, T.; Lee, J. H.; Lee, J. C.; Krzykawski, J.; Bartsch, R. A. *Talanta* **1992**, *39*, 857.
(23) Hayashita, T.; Lee, J. H.; Hankins, M. G.; Lee, J. C.; Kim, J. S.; Knobeloch, J. M.; Bartsch, R. A. *Anal. Chem.* **1992**, *64*, 815.
(24) Wulff, G. *Angew. Chem., Int. Ed. Engl.* **1995**, *34*, 1812.
(25) Zhao, Q.; Bartsch, R. A. *J. Polym. Sci., Polym. Chem.* **1995**, *33*, 2267.
(26) Hayashita, T.; Bartsch, R. A. *Anal. Chem.* **1991**, *63*, 1847.
(27) Wilken, R. F.; Joshi, S. R. *Radioact. Radiochem.* **1991**, *2*, 95
(28) Srivistava, S. C. *Radiolabeled Monoclonal Antibodies for Imaging and Therapy*; Plenum: New York, NY, 1988.
(29) Wood, D. J.; Elshani, S.; Wai, C. M.; Bartsch, R. A.; Huntley, M.; Hartenstein, S. *Anal. Chim. Acta* **1993**, *284*, 37.
(30) Hankins, M. G., Doctoral Dissertation, Texas Tech University, 1994, pp. 41-59
(31) Laney, E. E., Doctoral Dissertation, Texas Tech University, 1994, pp. 73-173

Chapter 13

Design of Novel Polymer-Supported Reagents for Metal Ion Separations

Spiro D. Alexandratos and Latiff A. Hussain

Department of Chemistry, University of Tennessee, Knoxville, TN 37996

> Polymer-supported reagents are prepared by the immobilization of ligands onto macromolecular matrices. Appropriately chosen ligands with high affinities toward targeted metal ions allow for the application of these reagents to separations science. Ion exchange and chelating resins are the two types of reagents traditionally used in metal ion separations. The latter are more selective than the former but with significantly slower rates of complexation. Examples of both types of resins are presented. Current research has shown that combining both mechanisms within a single polymer support yields selective complexation at rapid rates. Examples will be discussed.

The complexation and separation of metal ions is an area where functionalized polymers have been used extensively. It has been estimated that over 25,000 hazardous waste sites exist in the United States alone and the cost of remediation will exceed one trillion dollars (1). For example, at the Hanford nuclear facility, nearly 1.4 billion m^3 of hazardous waste, most of it high level radioactive waste, spread over 560 mi^2 of land needs to be treated (2). Making the problem more difficult is the fact that much of the waste exists in complex mixtures which must be separated prior to transfer to a safe storage area. Radioactive metal ions may be present in low pH solutions which contain high levels of dissolved solids and/or organic compounds, as is the case with uranium (3), as well as in basic solutions, as found with cesium (4). The complexity of such mixtures poses a formidable challenge to existing separation methods. A need exists for the development of improved technologies in order to make remediation of these sites both chemically and economically feasible.

Ion Exchange Resins

Ion exchange is a technique that has found widespread use in both remediation and pollution prevention. Ligands are covalently bound to an insoluble organic or inorganic polymer (5) and the ion-containing aqueous phase is passed through a column containing the polymer, usually in the form of beads. An ion exchange resin removes the metal by replacing it with an ion such as Na$^+$ (if the resin has cation exchange sites with ionically bound sodium ions) or Cl$^-$ (if the resin has anion exchange sites with ionically bound chloride ions) while a chelating resin removes the metal through complex formation (6). After the polymer has been loaded, the metal is eluted by passing an appropriate solution through the column to regenerate the polymer and produce a solution containing the extracted metal. The regenerability of the polymer and the recoverability of the metal ion are important factors to consider when deciding whether a resin can be used under process conditions. In the case of cation exchange resins, a strongly acidic solution may be used to recover the metal ion, but a more selective resin may require costly regenerants such as EDTA or 1-hydroxyethane-1,1-diphosphonic acid. In general, ion exchange resins can be loaded and regenerated many times without a significant loss of capacity (7). The initial price of ion exchange systems can be high due to the cost of the resin, but the recyclability of the polymer and the recovery of precious metals can make the process economically viable. The kinetic performance of the polymers can also be enhanced by varying the physical properties of the polymer such as the degree of cross-linking and porosity.

The organic polymers used as supports for ion exchange resins can be synthetic or naturally occurring. Cellulose has been the most extensively studied natural polymer (8, 9). Unmodified cellulose has a low ion exchange capacity; in most cases, cellulose is used after it has been modified by oxidation, esterification or etherification. Because degradation is possible in very acidic solutions, the application of cellulose is limited to solutions with an acidity no greater than 10^{-4} M (10).

The synthetic polymer supports can be either step-growth or chain-growth polymers. The most common crosslinked step-growth polymer used as an ion exchange support is a phenolic made by the condensation of phenol and formaldehyde (11, 12). Chain-growth polymers used as ion exchange supports are prepared by free radical polymerization of vinyl monomers with divinyl monomers added as cross-linking agents. The most important example of this type of support is a copolymer of styrene and divinylbenzene (DVB) (12-14) due to its display of the following properties:

- stable to a wide range of pH with alkaline and acidic reagents;
- resists hydrolytic cleavage;
- thermally stable at normal use temperatures;
- mechanical stability can be controlled by the degree of cross-linking;
- easily prepared at different porosities;
- readily functionalized via electrophilic and nucleophilic aromatic substitution.

Polystyrene has become the most widely used support not only for ion exchange applications but also for other purposes such as immobilized catalysts. Ion exchange resins with various degrees of cross-linking have different physical and mechanical properties (13, 15). Lightly cross-linked beads can be swollen by contact with good solvents but the resin can become mechanically unstable. The resin volume can change considerably in different solutions and is a disadvantage when column work is considered. Resins with high degrees of cross-linking do not swell as much as lightly cross-linked resins and their volume does not undergo a significant change in different solutions. The disadvantages of the highly cross-linked resins are their low exchange capacity and restriction of ionic diffusion. Generally, the selectivity of the resins towards metal ions can be improved by increasing the degree of cross-linking (5).

Ion exchange resins can be classified by their functional groups (5, 10, 15):

- Cation exchange resins - contain acidic functional groups such as -SO_3H and -COOH.
- Anion exchange resins - contain basic functional groups such as -NR_2 and -$N^+R_3Cl^-$.
- Amphoteric ion exchange resins - contain both acidic and basic functional groups.
- Chelating resins - contain functional groups that chelate metal ions.

At this time, the sulfonic acid and carboxylic acid resins are the two most common cation exchangers (16-21). However, they have disadvantages that limit their application. The sulfonic acid resin is a strong cation exchanger; its most serious drawback is its lack of selectivity towards targeted metal ions when other cations are present in solution. For example, it cannot effectively recover uranium from leach liquor when iron, aluminum, and other metal ions are present along with the UO_2^{2+} (10). The weakly acidic carboxylic acid resin is more selective than the sulfonic acid resin, especially for the alkaline earth metal ions over monovalent ions (22, 23). However, the exchange capacity of the carboxylic resin is strongly dependent upon the solution pH. The effective pH range is 6 to 14 because of the high affinity of the carboxylic acid resin for H^+ (given dissociation constants between 10^{-5} and 10^{-9}) (10). This property obviates its application to the recovery of metal ions from highly acidic solutions.

In order to solve problems for which sulfonic and carboxylic acid resins are not applicable, chelating resins, which can selectively recover specific metal ions from dilute solutions, have been developed (6, 24). Their selectivity is due to the ability of the ligand to form a highly stable complex with a particular metal ion and less stable complexes with other ions. Early work in this field was published by Kennedy (25), who observed that thorium(IV), iron(III) and uranyl cations formed relatively strong inner chelate complexes with partially esterified phosphates and phosphonates as the acids or their corresponding sodium salts, whereas alkaline earth, divalent transition metal ions, and lanthanides formed weak complexes. Moreover, the chelate complexes usually show much higher stability constants than the corresponding monomer complexes (25). For example, the stability constants of the phosphonic acid resin for UO_2^{2+} and Fe(III) are greater by 10^6 or more over that of the corresponding monomers. The increased stability of the polymer-supported complexes was attributed to a combination of factors such as the polymer entropy effect and a lower dielectric constant of the resin matrix as compared to an aqueous medium.

Chelating Resins

In order to understand the interaction between polymer-supported ligands and metal ions, numerous chelating resins have been synthesized that contain nitrogen and/or oxygen as the donor atoms. The iminodiacetic acid resin (Figure 1), which is commercially available as Dowex A-1 (Dow Chemical Co.) and Chelex-100 (Bio-Rad Laboratories), is probably the most extensively studied chelating resin and has been used for selective separations of transition metal ions from alkali and alkaline-earth metal ions in different pH and ionic strength solutions (*26-28*). The selectivity of Dowex A-1 has been determined to be $Mg^{2+}<Mn^{2+}<Co^{2+}<Zn^{2+}<Ni^{2+}<<Cu^{2+}$ at near-neutral pH (*29*). The resin containing 8-quinolinol (Figure 2) as the chelating group has been synthesized in different ways. Spheron-1000 (Koch-Light Laboratories) consists of a copolymer matrix of hydroxyethyl methacrylate / 2-propenyl phenyl ketone and the functional group is bound to the polymer via an azo group (*30*). The affinity of Spheron-1000 is remarkably high for heavy metal ions and is influenced slightly by ionic strength (*30*). It does not interact with alkali, alkaline earth or ammonium ions.

Chelating resins with macrocyclic units have also been prepared. A variety of macrocyclic rings such as crown ethers (Figure 3) (*31*), calixarenes (Figure 4) (*32*) and cryptands (Figure 5) (*33*) have been incorporated onto different types of polymeric supports. The selectivity of macrocyclic units, usually for alkali and alkaline earth metal ions, is dependent mainly upon the size of the cations and the cavity of the macrocyclic rings (*34*). Polymeric pseudocrown ethers based on a one-step cyclization reaction between chloromethylated styrene-DVB copolymer and polyoxyethylene were synthesized by Warshawsky and had a high affinity for anionic transition metal halide complexes such as AuX_4^-, FeX_4^-, ZnX_4^{2-} and CuX_4^{2-} (*35*).

The highly selective chelating resins can have a number of disadvantages. The synthesis of a desired ligand on a polymer support without formation of other functional groups is difficult and the cost of most of the ligands is high. The major drawback of the chelating resins is their low ionic accessibility which is due to the lack of hydrophilic groups within the matrices.

Immobilized Phosphorus Ligands

The moderate acidity of polymer-supported phosphorus acid ligands and, more importantly, the coordinating ability of the phosphoryl oxygen makes them preferable to sulfonic acid, carboxylic acid, and chelating ion exchange resins in many applications. Phosphorus acid resins are more selective than sulfonic acid resins (*36*) while retaining that selectivity in lower pH solutions than carboxylic acid resins (*10*). The presence of ionic phosphorus acid ligands in the crosslinked polymer matrix increases resin hydrophilicity relative to chelating resins and thus ionic accessibility.

The synthesis of polymers with covalently bonded phosphonic acid ligands has been accomplished by polymerization of phosphorus-containing monomers, either by chain- or step-growth polymerization (*37*), and post-functionalization of polymers (*38*). Phosphorylation of polystyrene-DVB copolymers and their chloromethylated derivatives, the most commonly utilized method for the synthesis of phosphorus acid resins, was first reported by Jones (*39*). Subsequently, Kennedy and co-workers (*37*) synthesized several resins by phosphorylating polystyrene or chloromethylated polystyrene-DVB copolymers for the recovery of uranyl ions. They selected phosphorus-containing ligands because the electron donating phosphoryl oxygen has strong complexing properties for inorganic salts, such as uranyl nitrate and ferric chloride. Ion exchange resins with phosphorus ligands have been prepared by first treating chloromethylated polystyrene with a cyanide salt solution in an organic solvent and then reacting the product with a mixture of H_3PO_3 or a carboxylic acid and a phosphorus-containing halide (*40*). It has been found that the

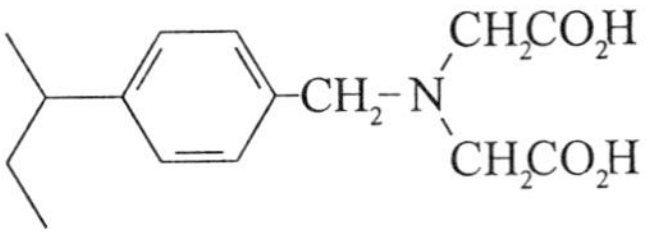

Figure 1. Iminodiacetic acid resin.

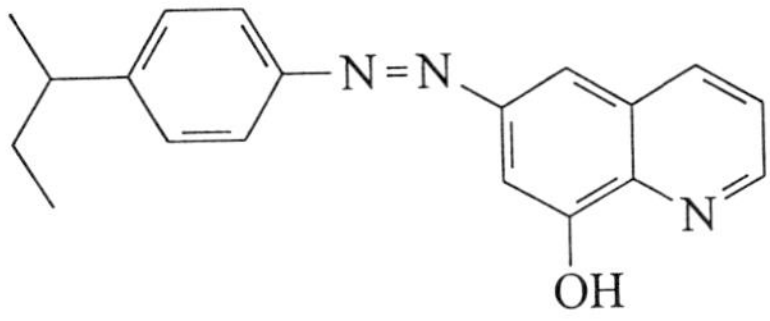

Figure 2. Quinolinol resin.

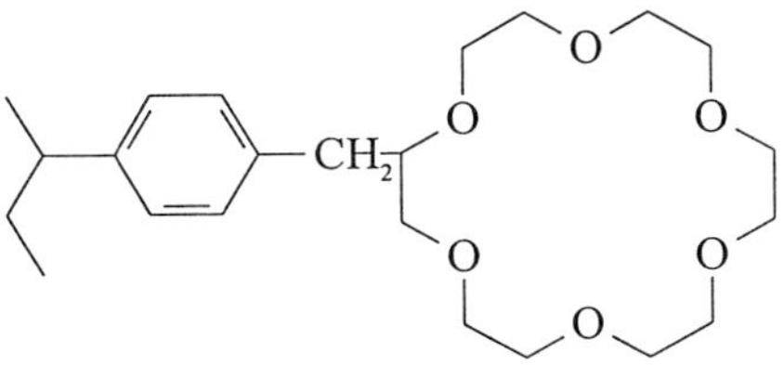

Figure 3. Polymer-supported crown ether.

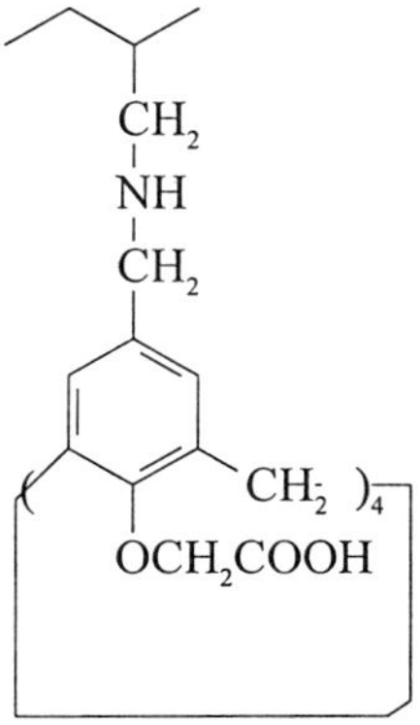

Figure 4. Polymer-supported calixarene.

phosphonic acid ligand on a polymer-support shows the same selectivity order toward various cations as does the monomer (*41*): thorium(IV)>uranium(VI)>uranyl>iron(III)> lanthanides(III)>hydrogen(I)>copper(II)>cobalt(II)> barium(II)>sodium(I).

Reactive Ion Exchange

The concept of reactive ion exchange (RIEX) was introduced by Helfferich (*42*) and extended by Janauer (*43*) as a means of enhancing the separation abilities of traditional ion exchange resins. In conventional ion exchange, the overall process is exclusively a redistribution of counterions and the exchanging ions retain their identity in the resins. However, a reactive ion exchange procedure, by definition, contains at least one ion exchange step and one chemical reaction such as reduction, precipitation, neutralization or complex formation. In general, the chemical reaction involves transformation of an initially present species to a new species. An example of RIEX is the titration of a strong acid cation exchange resin with sodium hydroxide which involves an ion exchange reaction (exchange of the resin proton with the sodium ion) accompanied by a neutralization reaction (water formed by the consumption of hydroxide ions from solution and protons from the resin).

Because the ΔG of ion exchange is only a few kilocalories per mole (*44*), the values among different metal ions are too small to impart selectivity to the separation. The advantage of employing RIEX to achieve selective separation is to utilize favorable free energy changes of accompanying chemical reactions which yield an overall negative free energy of sufficient magnitude for the desired separation (*45,46*). Since ion exchange can be coupled with many suitable reactions, RIEX is a very useful tool in the design of new extractants for selectively recovering metal ions.

Dual Mechanism Bifunctional Polymers

A new category of phosphorus-based metal ion complexing agents, termed dual mechanism bifunctional polymers (DMBPs), has been developed for the selective complexation and recovery of metal ions from aqueous solutions (*47*). These polymers are synthesized with two different functional groups, each displaying either an access mechanism or a recognition mechanism. The access mechanism is provided by an aspecific ion exchange ligand which mainly enhances ion accessibility and mobility within the polymer network. The recognition mechanism is responsible for the specificity through reduction, coordination or precipitation. During the separation process, the ion exchange group first serves to bring the metal ions into the polymer matrix and the highly specific reaction occurs subsequently between the metal ion and the recognition group.

DMBPs are divided into three classes based upon the type of recognition mechanism. The Class I resin is an ion exchange/redox resin which combines ion exchange with a recognition group capable of reducing certain metal ions. This class of resins is exemplified by the phosphinic acid resin (Figure 6). The P-O-H moiety is capable of ion exchange with the metal ions through the acidic hydrogen while the P-H group is capable of metal ion reduction. The ligand is thus oxidized to the phosphonic acid. Both Ag(I) and Hg(II) are rapidly reduced to the zerovalent state (*48*). This resin also shows a significant ability to coordinate non-reducible metal ions through the phosphoryl oxygen in highly acidic solutions wherein ion exchange is not expected to occur (*49*).

Finding that the Class I resins had a strong coordinating ability led directly to the development of Class II DMBPs: the ion exchange / coordination resins. The application of phosphoryl-containing extractants to metal ion recovery processes is well known. For example, tributyl phosphate and trioctylphosphine oxide have been used in the recovery of actinides and numerous other metal ions (*50*). Purely coordinating polymer-supported extractants have been prepared and most of these resins display excellent selectivity for different metal ions (*51*). However, the polymeric coordinating extractants usually show very low metal ion loading capacity due to their limited hydrophilicity and accessibility. The Class II resins studied most extensively consist of a phosphonic acid ligand as the ion exchange group for enhanced accessibility and either a phosphorus ester or tertiary amine group as the coordinating ligand for enhanced selectivity (Figure 7) (*52*). In one example, the distribution coefficient (the ratio of milliequivalents M^{n+} per g (dry weight) polymer to milliequivalents M^{n+} per mL solution) for Ag(I) ions was 2900 in 4 N HNO_3 for the polymer with monoethyl and diethylphosphonate groups bound to the polymer. Under the same conditions, the monofunctional polymer with diethylphosphonate ligands had a distribution coefficient of 490 and the polymer with monoethylphosphonate ligands had a distribution coefficient of 440 (*36*).

The third class of DMBPs are the ion exchange/precipitation resins (*53*). In many cases, if the metal ions cannot be reduced by the first class or coordinated by the second class of DMBPs, removal of metal ions from aqueous solutions by the formation of insoluble salts is an important alternative. The Class III resins combine the phosphonic acid ligand for enhanced ionic accessibility with a quaternary amine group (Figure 8) whose associated anion can react with the targeted metal ion to form an insoluble metal salt. The salt precipitates within the beads and can be recovered through resolubilization into a concentrated solution. For example, this class of resins is useful for the recovery of barium ions which readily form precipitates with certain anions: barium has a reduction potential too low to be reduced by the Class I resin and does not readily coordinate with the phosphoryl ligands of the Class II resins.

The DMBPs operate through the **inter-ligand cooperation** of two different groups on neighboring sites in complexing a given metal ion. The importance of **intra-ligand cooperation** for metal ion chelate formation was quantified through the synthesis of an immobilized *gem*-diphosphonic acid ligand (Figure 9) (*54*). In one example, the resulting ion exchange resin (now available as Diphonix® from Eichrom Industries, Inc.) had a distribution coefficient of 7×10^5 for UO_2^{2+} in 1 N HNO_3 as compared to a value of 900 for the analogous monophosphonic acid and 200 for the sulfonic acid resin.

Our current research emphasizes the synthesis of other polymers capable of selective intra-ligand cooperation such as the ketophosphonates and will be the subject of a future report.

Conclusion

Ion exchange and chelating resins will continue to play a pivotal role in many applications involving water treatment, the mining industry, and environmental

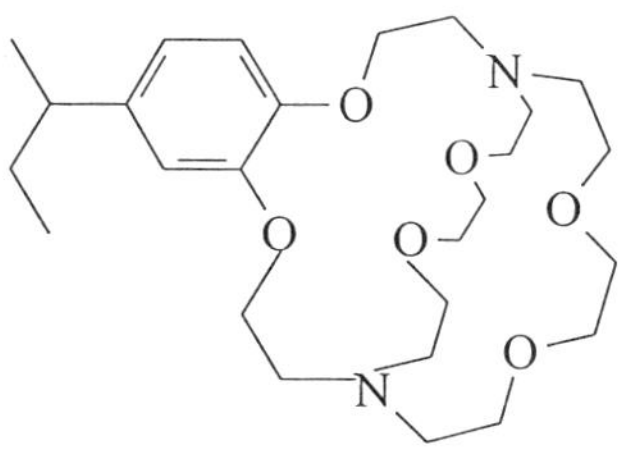

Figure 5. Polymer-supported cryptand.

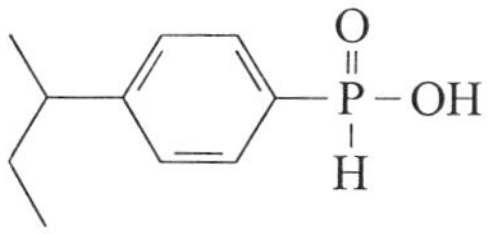

Figure 6. Class I Dual Mechanism Bifunctional Polymer.

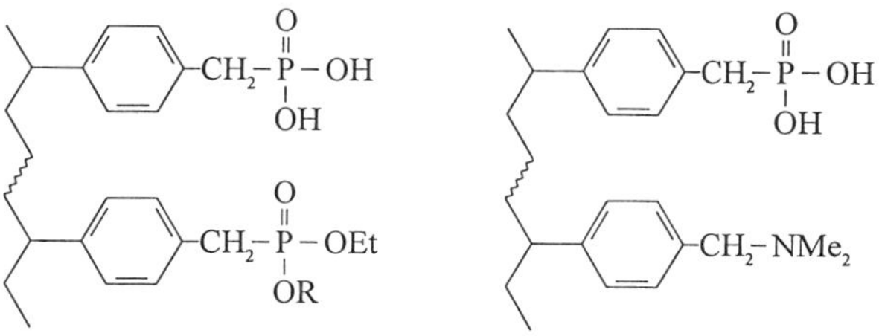

R = H, Et

Figure 7. Class II Dual Mechanism Bifunctional Polymer.

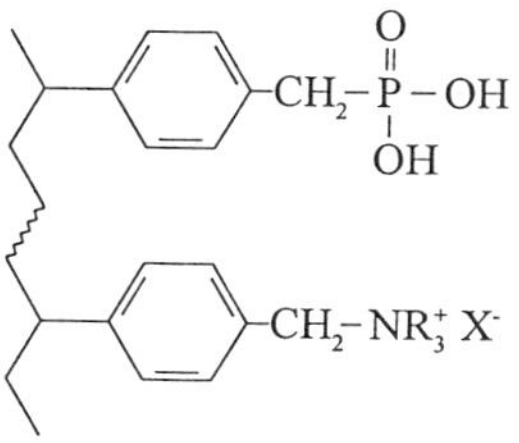

Figure 8. Class III Dual Mechanism Bifunctional Polymer.

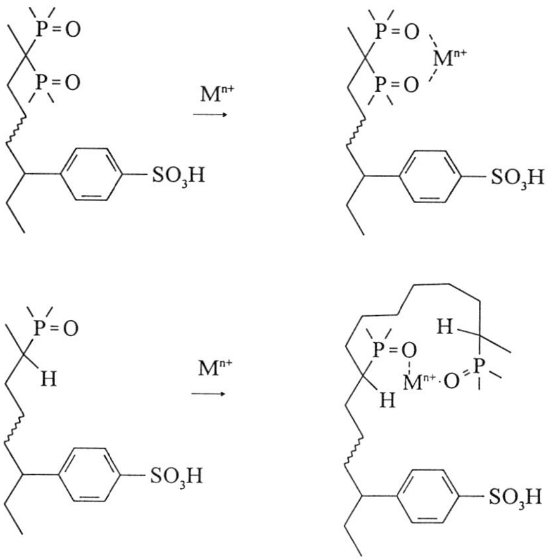

Figure 9. Intra-ligand (top) vs. inter-ligand (bottom) cooperation.

remediation for the foreseeable future. Their use within well-engineered continuous processes is an important advantage. The key issue in separations science has been selectivity coupled with rapid kinetics. It is now understood that this can be achieved with bifunctional polymers: an access ligand can greatly enhance complexation kinetics without diminishing the selectivity of the recognition ligand. Further progress is thus expected in the preparation of ion-selective polymer-supported reagents.

Acknowledgment

It is a pleasure to acknowledge the Department of Energy, Office of Energy Research, Division of Chemical Sciences, Office of Basic Energy Sciences, for their continued support of this research through grant DE-FG05-86ER13591.

Literature Cited

(1) Reed, D.T.; Tasker, I.R.; Cunnane, J.C.; Vandegrift, G.F. *Environmental Restoration and Separation Science*. In *Environmental Remediation*; Vandegrift, G.F., Reed, D.T., Tasker, I.R., Eds.; *ACS Symposium Series 509*; American Chemical Society: Washington, DC, 1992; Ch. 1.
(2) Illman, D.I., *Chem. Eng. News* **1993**, *71*, 9.
(3) Majee, S.; Ray, C.; Das, J. *Ind. J. Chem.* **1989**, 28A.
(4) Kaczvinsky, J.R., Jr.; Fritz, J.S.; Walker, D.D.; Ebra, M.A. *J. Radioanal. Nucl. Chem.* **1985**, *91*, 349.
(5) Helfferich, F. *Ion Exchange*; McGraw-Hill: New York, 1962.
(6) Sahni, S.K.; Reedijk, J. *Coord. Chem. Rev.* **1984**, *59*, 1.
(7) Tavlarides, L.L.; Bae, J.H.; Lee, C.K. *Sep. Sci. Technol.* **1987**, *22*, 581.
(8) Lieser, K.H. *Pure Appl. Chem.* **1979**, *51*, 1503.
(9) Wagscheider, W.; Knapp, G. *CRC Crit. Rev. Anal. Chem.* **1981**, *79*, 11.
(10) Dorfner, K. *Ion Exchangers: Properties and Application*; Ann Arbor Science Publisher, Inc.: Ann Arbor, MI, **1972**.
(11) Morgan, P.W. *Condensation Polymers, Polymer Reviews*; Interscience: New York, 1965, Vol. 10.
(12) Smith, D.A. *Addition Polymers*; Butterworth: London, 1968.
(13) Sherrington, D.C.; Hodge, P. *Synthesis and Separations Using Functional Polymers*; John Wiley: New York, 1987.
(14) Frechet, J.M.J.; Farrel, M.J. In *Chemistry and Properties of Crosslinked Polymers*, Labana, S.S., Ed., Academic Press: New York, 1977.
(15) Korkisch, J. *Handbook of Ion Exchange Resins: Their Application to Inorganic Analytical Chemistry*, CRC Press, Inc.: Boca Raton, 1989, Vol. 1.
(16) Streat, M.; Naden, D. In *Ion Exchange and Sorption Processes in Hydrometallurgy*; Streat, M.; Naden D., Eds.; John Wiley and Sons, Ltd.: New York, 1987, p. 1.
(17) Kunin, R. *Ion Exchange Resins*; John Wiley and Sons: New York, **1950**.
(18) Walton, H.F.; Navratil, J.D. In *Recent Developments in Separation Science*; Li, N. N., Ed.; CRC Press: Boca Raton, 1981, Vol. 6, Ch. 5.

(19) Morrison, W.S. In *Ion Exchange Technology*; Nachod, F.C.; Schubert, J., Eds.; Academic Press: New York, 1956, Ch. 13.
(20) Gerstner, F. In *Ion Exchange Technology*; Nachod, F.C.; Schubert, J., Eds.; Academic Press: New York, 1956, Ch. 14.
(21) Tompkins, E.R.; Khym, J.X.; Cohn, W.E. *J. Am. Chem. Soc.* **1947**, *69*, 2769.
(22) Kazantsev, E.I.; Denison, A.N. *Russ. J. Inorg. Chem.* **1963**, *8*, 1149.
(23) Brajter, K.; Miazek, I. *Talanta*, **1981**, *28*, 759.
(24) Warshawsky, A. In *Ion Exchange and Sorption Processes in Hydrometallurgy*; Streat, M.; Naden, D., Eds.; John Wiley and Sons, Ltd.: New York, **1987**, 166.
(25) Kennedy, J. *Chem. Ind.* **1956**, 378.
(26) Oslen, R.L.; Diehl, H.; Collins, P.E.; Ellestad, R.B. *Talanta*, **1961**, *7*, 187.
(27) Reyden, A.J.; Lingen, R.C.M. *Fresenius Z. Anal. Chem.* **1962**, *187*, 241.
(28) Turse, R.; Rieman, W. III *Anal. Chim. Acta* **1961**, 24, 202.
(29) Rosset, R. *Bull. Soc. Chim. Fr.* **1966**, 59.
(30) Slovak, Z.; Toman, J. *Fresenius, Z. Anal.Chem.* **1976**, *278*, 115.
(31) Manecke, G.; Kramer, A. *Makromol. Chem.* **1981**, *182*, 3017.
(32) Ohto, K.; Tanaka, Y.; Inoue, K. *Chem. Lett.* **1997**, 647.
(33) Manecke, G.; Reuter, P. *Makromol. Chem.* **1981**, *182*, 3017.
(34) Blasius, E.; Janzen, K.P. *Pure Appl. Chem.* **1982**, *54*, 2115.
(35) Warshawsky, A.; Kalir, R.; Berkovitz, H.; Patchornik, A. *J. Am. Chem. Soc.* **1979**, *101*, 4249.
(36) Alexandratos, S.D.; Crick, D.W.; Quillen, D.R. *Ind. Eng. Chem. Res.* **1991**, *30*, 772.
(37) Kennedy, J.; Wheeler, V. *Anal. Chim. Acta.* **1959**, *20*, 412.
(38) *Duolite Ion Exchange Manual*; Chemical Process Co.: Redwood City, CA, 1960.
(39) Thomas, S.L.S.; Jones, J.I. (Natl. Research Development. Corp.) Brit. Patent 762,085, **1956**.
(40) Rogozhin, S.V.; Davankov, V.A.; Belihich, L.; Kabachnik, M.I.; Medved, T.Y.; Palikarpov, Y.M. USSR Pat. 316,704 (1971); *Chem. Abstr.* **1972**, *76*, 86585y.
(41) Persoz, J.; Rosset, R. *Bull. Soc. Chim. Fr.* **1964**, 2197.
(42) Helfferich, F. In *Ion Exchange*; Marinsky, J.A., Ed.; Marcel Dekker: New York, 1966, Vol. 1, Ch. 2.
(43) Janauer, G.E.; Gibbons, R.E., Jr.; Bernier, W.E. In *Ion Exchange and Solvent Extraction*; Marinsky, J.A.; Marcus, Y., Eds.; Marcel Dekker: New York, 1985, Vol. 9, Ch. 2.
(44) Boyd, G.E.; Vaslow, F.; Lindenbaum, S. *J. Phys. Chem.* **1967**, *71*, 2214.
(45) Eisenman, G. *Biophys. J. Suppl.* **1962**, *2*, 259.
(46) Reichenberg, D. In *Ion Exchange*; Marinsky, J.A., Ed.; Marcel Dekker: New York, 1966; Vol. 1, Ch. 7.
(47) Alexandratos, S.D. *Sep. Purif. Meth.* **1988**, *17*, 67.
(48) Alexandratos, S.D.; Wilson, D.L. *Macromolecules* **1986**, *19*, 280.

(49) Alexandratos, S.D.; Quillen, D.R.; McDowell, W.J. *Sep. Sci. Technol.* **1987**, *22*, 983.
(50) Weaver, B. In *Ion Exchange and Solvent Extraction*; Marinsky, J.A.; Marcus Y., Eds.; Marcel Dekker: New York, 1974; Vol. 6, Ch. 4.
(51) Kantipuly, C.; Katragadda, S.; Chow, A.; Gesser, H.D. *Talanta* **1990**, *37*, 491.
(52) Alexandratos, S.D.; Quillen, D.R.; Bates, M.E. *Macromolecules* **1987**, *20*, 1911.
(53) Alexandratos, S.D.; Bates, M.E. *Macromolecules* **1988**, *21*, 2905.
(54) Alexandratos, S.D.; Trochimczuk, A.W.; Crick, D.W.; Horwitz, E.P.; Gatrone, R.C.; Chiarizia, R. *Macromolecules* **1996**, *29*, 1021.

Chapter 14

Recent Advances in the Chemistry and Applications of the Diphonix Resins

E. Philip Horwitz[1], Renato Chiarizia[1], Spiro D. Alexandratos[2], and Michael Gula[3]

[1]Chemistry Division, Argonne National Laboratory, 9700 South Cass Avenue, Argonne, IL 60439–4831
[2]Department of Chemistry, University of Tennessee, Knoxville, TN 37996
[3]Eichrom Industries, Inc., Darien, IL 60561

>The Diphonix® class of ion exchange resins is characterized by the presence of geminally substituted diphosphonic acid groups chemically bonded to a polymer matrix. Regular Diphonix contains gem-diphosphonic groups chemically bonded to a sulfonated styrene-divinylbenzene matrix. Modification of the properties of Regular Diphonix are achieved by the introduction of additional functional groups such as anion exchange groups in Diphonix-A and phenolic groups in Diphonix-CS. Diphosil has a silica matrix in which the gem-diphosphonic groups are chemically bonded to an organic polymer graft that surrounds the silica particles. Applications of the Diphonix resins range from treatment of a variety of radioactive waste to iron control in hydrometallurgy and semiconductor manufacture.

The Diphonix Class of Ion Exchange Resins

Diphonix® resin is a new bifunctional chelating ion exchange material that contains geminally substituted diphosphonic groups and sulfonic acid groups chemically bonded to a styrene-divinylbenzene polymeric network (1). The rationale for the introduction of bifunctionality into the resin has been discussed as a coupling of an access mechanism that allows all ions into the matrix rapidly and a recognition mechanism where a second ligand in the matrix selectively complexes a targeted metal ion (1). In the case of Diphonix, the sulfonic acid groups are the access functionality and the gem-diphosphonic acid groups are the recognition functionality. Hereinafter this resin will be referred to as Regular Diphonix resin. The preparation and properties of Regular Diphonix resin have been described in a series of publications (1-9). Regular Diphonix resin is now available commercially from Eichrom Industries, Darien, IL.

During the last few years, three new Diphonix-type resins have been synthesized and characterized. Each new resin retains the basic gem-diphosphonic acid functional group, but in one case a silica-based matrix is used in place of the styrene-divinylbenzene matrix and in the other two cases additional functional groups are introduced to enhance the selectivity of the resin for specific ions or groups of ions. The silica-based Diphonix, called Diphosil, is prepared by grafting silica gel with an organic polymer and then chemically bonding the gem-diphosphonic acid groups to the organic graft (10). The other two Diphonix resins consist of modified organic polymer networks. In one of the modified Diphonix resins, called Diphonix-A, anion exchange functionalities are introduced by replacing styrene with 4-vinylpyridine and then quaternizing the pyridinium nitrogen (11). In the other modified Diphonix resin, called Diphonix-CS, phenolic groups are introduced by polymerizing phenol/formaldehyde within a polymer network slightly modified from Regular Diphonix (12). Figure 1 depicts the structures of the four different types of Diphonix resin.

Properties of Diphonix Resins. The unique feature of Regular Diphonix, Diphosil, and Diphonix-A resins is their ability to rapidly sorb a wide number of metal ions from highly acidic media, even in the presence of complexing anions. Figure 2 compares the uptake of selected metal ions, as measured by their distribution ratios, for Regular Diphonix and two of the modified Diphonix resins from 0.1 to 10 M HNO_3. The data in Figure 2 show that the modified Diphonix resins are very similar to Regular Diphonix with regard to their uptake of cations but, as expected, Diphonix-A shows a much higher uptake of Tc as the pertechnetate ion. The kinetic properties of these three resins are depicted in Figure 3 using U(VI). All three show very rapid uptake of U(VI) from 0.1 M HNO_3. Rapid exchange kinetics are one of the major features of the Diphonix resins. Another noteworthy characteristic of Diphonix resins is the insensitivity of the sorption of certain metal ions, such as the tetra- and hexavalent actinides and iron (III), to different matrices. Figure 4 shows the effect of increasing concentrations of selected complexing anions on the uptake of Th(IV) on Regular Diphonix (with HF, Np(IV) was utilized). Tetravalent actinides are known to form strong complexes with complexing anions such as fluoride and oxalate. As the data in Figure 4 show, only when the concentration of these acids approaches 1 M, does the distribution ratio begin to decline significantly. Figure 5 shows the influence of commonly occurring cations on the D_{Th}. Only Fe(III), which is also strongly complexed by the diphosphonic ligand, shows any significant effect, and it must be present in macroscopic amounts before D_{Th} is depressed.

Diphonix-CS is a new ion exchange resin specifically designed for the removal of Cs^+ and Sr^{2+} from highly alkaline media (12). Figure 6 shows the distribution ratios of Cs and Sr as a function of sodium hydroxide concentration. Data points for Regular Diphonix are shown for comparison. As one can see, the uptake of Cs is significantly improved with Diphonix-CS. The improved uptake of Cs is no doubt due to the presence of the phenolic groups. In alkaline media, Diphonix-CS rapidly sorbs both Cs and Sr and shows very good radiation stability at least up to an absorbed dose of 200 Mrads (12).

Applications of Diphonix Resins

Because of the diversified nature of the four different Diphonix resins, there are a variety of applications of these unique ion exchange materials. Some of these applications are strictly potential and others are already in use in facilities ranging from pilot-scale to full-scale operations.

Radioactive Waste Treatment. The first application of Regular Diphonix resin was the removal of actinides from mixed-waste at Argonne National Laboratory-East. These wastes were generated in-house as a result of ongoing programs in the nuclear fuel cycle and research in actinide chemistry. Tables I and II show typical compositions of mixed-waste solutions processed in a special facility designed to handle at one time volumes of wastes up to 55 gallons (13). Generally, two one-half liter Diphonix columns were used in tandem to ensure high-levels of decontamination.

Another application of Regular Diphonix resin for waste treatment took place at a southeastern United States nuclear fuel fabricator that faced the need to significantly lower the concentration of alpha-emitting species in its wastewater to comply with new

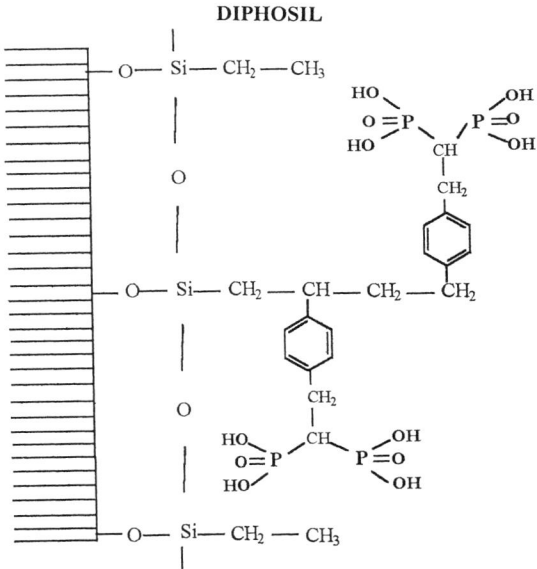

Figure 1. Structures of the four different types of Diphonix ion exchange resins. (Adapted from refs. 1, 10, 11, 12.)

Figure 1. *Continued.*

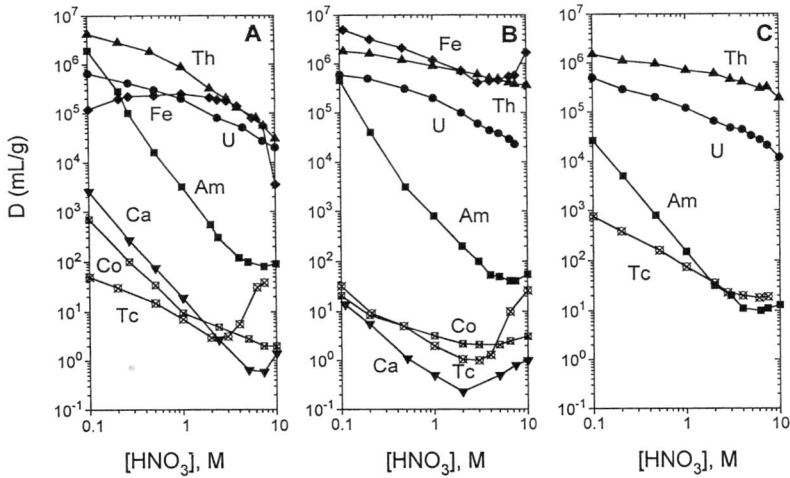

Figure 2. Distribution ratios of selected metal ions as a function of nitric acid concentration in the aqueous phase. A. Regular Diphonix, B. Diphosil, C. Diphonix-A. (Adapted from refs. 1, 2, 10, 12.)

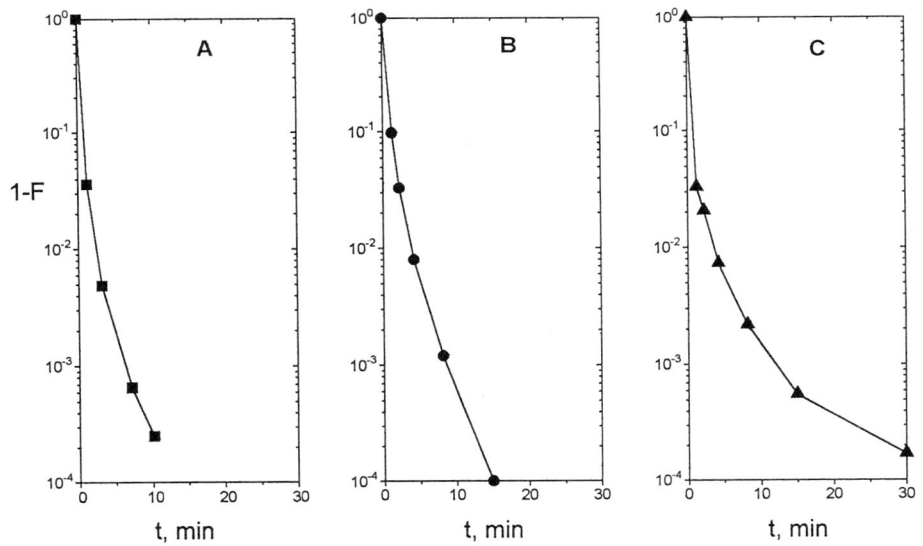

Figure 3. Fractional attainment of equilibrium for uranium(VI) uptake from 0.1 M nitric acid. A. Regular Diphonix, B. Diphosil, C. Diphonix-A. (Adapted from ref. 4.) The fractional attainment of equilibrium, F, is defined as $F = [M]_{res,t} / [M]_{res,eq}$.

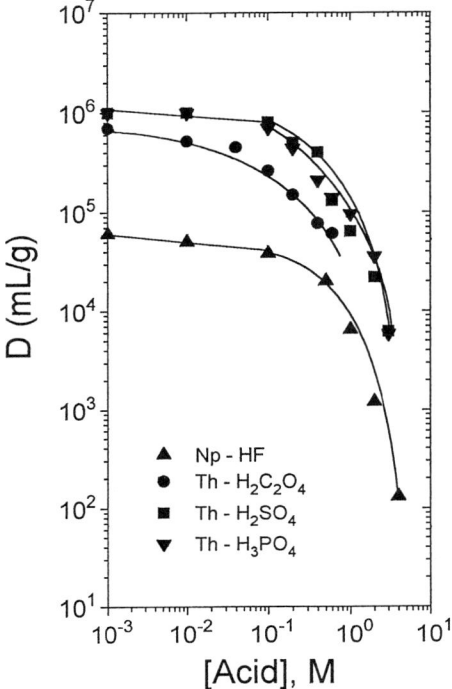

Figure 4. Distribution ratio of thorium(IV) (neptunium(IV) for HF) on Regular Diphonix as a function of the aqueous concentration of selected complexing acids. Nitric acid concentration was 1 M throughout. (Adapted from ref. 5.)

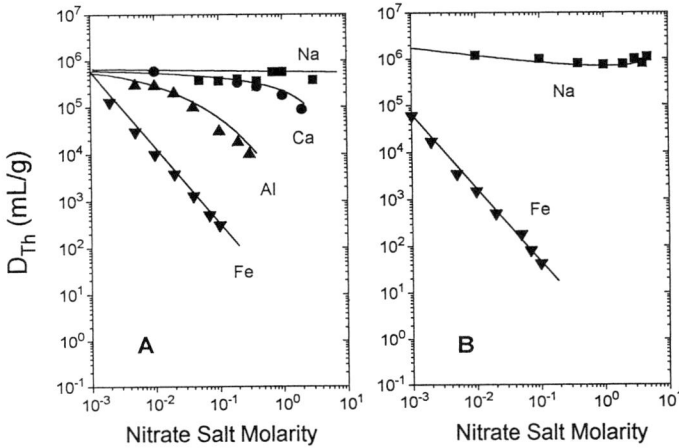

Figure 5. Distribution ratio of thorium(IV) as a function of the aqueous concentration of selected nitrate salts. Nitric acid concentration was 1 M throughout. A. Regular Diphonix, B. Diphosil. (Adapted from refs. 5 and 10.)

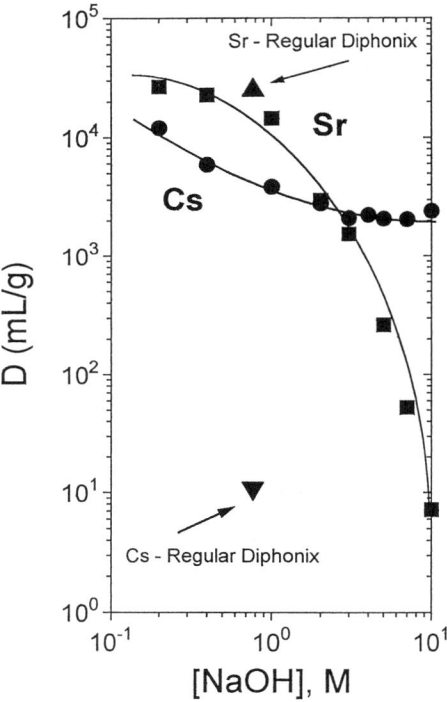

Figure 6. Distribution ratios of Sr^{2+} and Cs^+ on Diphonix-CS and Regular Diphonix as a function of the aqueous concentration of sodium hydroxide. (Adapted from ref. 12.)

Table I. Analytical Chemistry Laboratory Generated Mixed Waste from ICP/AES

Constituents:
Al, Ba, Ca, Co, Cr, Cu, Fe, Cd, Mg, Mn, Ni, Pb, Sr, Ti, V, Zn, Zr

Acids:
HNO_3, HCl, H_2SO_4

Major Hazardous Constituents:
Cu 10^{-2} M , Cd 5×10^{-2} M

Actinides
U (10^{-4} to 10^{-5} M) , ^{237}Np, ^{239}Pu, ^{241}Am (~ 1 µCi/5 gal)

Decontaminations in the range of 10^4 to 10^5 were obtained using two columns in series each containing 0.5 liters of Regular Diphonix.

Table II. Typical Composition of Actinide Containing Acidic Mixed-Waste Solutions

Constituent	M
HNO_3	0.01 - 0.1
$NaNO_3$	0.80 - 1.0
Na_2SO_4	0.14 - 0.20
NaCl	0.17 - 0.20
Mg	0.07
Al	5×10^{-4}
Ca	0.10
Fe	1×10^{-3}
Mo	3×10^{-6}
Zr	4×10^{-6}
Total Actinides in 208 L (55 gal)	
^{238}Pu	6×10^{-6} g
^{239}Pu	2×10^{-2} g
^{241}Pu	7×10^{-5} g
^{241}Am	1.3×10^{-4} g
Total alpha	4×10^9 d/m

Decontamination of ~10^4 was achieved using two columns in series each containing 0.5 L of Regular Diphonix ion exchange resin.

environmental restrictions (*14*). Regular Diphonix was tested on a pilot-scale and found to afford the necessary decontamination for compliance. Table III summarizes

Table III. Results of Testing Uptake of Alpha-Emitting Species from Wastewater System Neutralization Tank Contents using Regular Diphonix Resin

Alpha Concentration in Feed (pCi/mL)	Influent pH	Alpha Concentration in Effluent (pCi/mL)
1.2	6.7	0.043
0.78	7.0	0.022
0.47	8.2	0.043
0.16	11	<0.03
0.49	13	0.054

Source : Adapted from ref. 14.

some of the test results. A full-scale unit will soon be in place to treat much larger volumes of waste.

Regular Diphonix resin has also been tested successfully for the treatment of low-level radwaste (LLRW) generated at the Northeast Utilities, Millstone nuclear power plant (14). In this application, Regular Diphonix resin removes radioactive Co and Zn from an effluent stream originating from floor and equipment drains. Effluent from a Regular Diphonix column followed by a conventional anion exchange column in series showed no detectable Co. Decontamination factors (DFs) of >1000 for Co were achieved in pilot-test runs. In general, the Diphonix resin showed consistently superior DFs over conventional cation exchange resin for Co and Zn with throughputs at least three times greater than conventional cation exchange resin. No breakthrough of Co and Zn were detected even after passing the equivalent of 45,000 gallons of waste per cubic foot of resin bed. Although Regular Diphonix resin is more expensive than standard strong acid cation exchange resin, the much greater waste minimization achieved with Diphonix more than offsets its higher cost. The utility spends approximately $750 per cubic foot to dispose of spent ion-exchange resin (14). Conservative estimates show that the Diphonix resin system will produce roughly one-fifth the volume of spent resin than a standard cation resin-based system.

Applications of Diphosil, the silica-based diphosphonic resin, and scale-up studies for its commercial production are currently in progress at Eichrom Industries. Many of the applications of Regular Diphonix in radioactive waste treatment could be performed by Diphosil. Diphosil is, however, 85% inorganic and, therefore, has the possible advantage of being a more desirable waste form. A potential application of Diphonix-A and Diphosil is the purification of water used in steam turbines and as a coolant in nuclear power plants. The absence of sulfonic groups and their ability to sorb a wide range of cations and, in the case of Diphonix-A, anions such as chloride, sulfate, and silicate should give them major advantages in producing ultra-pure water. Conventional sulfonic acid-based cation exchange resins have the undesirable property of producing low concentrations of sulfuric acid due to the hydrolysis of sulfonic acid groups attached directly to phenyl groups. Unlike Regular Diphonix, Diphonix-A and Diphosil do not contain sulfonic acid groups because their hydrophilicity is achieved by other means.

As mentioned above, Diphonix-CS was specifically designed to remove both Sr and Cs from the alkaline supernate present in the high-level waste tanks at the Hanford and Savannah River sites. (Regular Diphonix only sorbs Sr. See Figure 6.) But unlike inorganic materials, such as the crystalline silicotitanates currently undergoing testing, both Cs and Sr can be eluted from Diphonix-CS with 1 M HNO_3 (12). To date, Diphonix-CS has only been tested with alkaline waste simulants (12). Column test runs have been favorable for Cs removal, but improvements in the stripping of Sr need to be made.

Iron Control. One of the most useful properties of Diphonix resins is their ability to strongly retain trivalent iron even from highly acidic solutions (*2,4*). Iron is a ubiquitous element in hydrometallurgy. In the production of copper by solvent extraction and electrowinning, the presence of Fe is particularly troublesome in the electrowinning operation. High concentrations of iron decrease current efficiency because of the oxidation of Fe(II) at the anode and the reduction of Fe(III) at the cathode. Iron accumulates in the electrowinning cells because the sulfuric acid in the spent electrolyte is continuously recycled to strip copper from the loaded organic extractant.

The traditional solution to iron-buildup involves bleeding out some electrolyte, which results in the loss of an expensive additive used to prevent corrosion of the lead anode; namely cobalt. Regular Diphonix resin has been shown to effectively remove Fe(III) from the spent electrolyte solution thus improving the operating efficiency of the electrowinning plant (*15*). Copper(II) and cobalt(II) are not bound by the resin because of the high (1.5 M) concentration of sulfuric acid. Regeneration of the Diphonix column is achieved by a novel stripping procedure using a 0.3 to 0.8 M sulfurous acid solution in 2 M H_2SO_4 containing 1 to 2 g/L of Cu(II). Stripping is generally carried out at a temperature of 65 to 75 °C. Under these conditions Fe(III) is reduced to Fe(II). The mechanism of the iron reduction is believed to be due to the formation of a small concentration of Cu^+ formed by reduction of Cu^{2+} by sulfurous acid, and a small displacement of Fe(III) from the diphosphonic groups by hydrogen ion. Copper(I) then rapidly reduces Fe(III) to Fe(II) by a one electron transfer. The overall stripping reaction is described by the following equation:

$$2\ \overline{Fe(III)R} + 4H^+ + H_2SO_3 + H_2O \rightarrow 2Fe^{+2} + H_2SO_4 + 2\ \overline{H_3R}$$

where $\overline{Fe(III)R}$ and $\overline{H_3R}$ denote Fe^{3+} and H^+ bound to Diphonix. Note that copper, the key reducing agent, acts as a catalyst and does not appear in the equation. Figure 7 depicts the entire iron removal cycle. The effluent from the strip cycle is added to the heap leaching solution.

Eichrom Industries successfully completed pilot studies at several solvent extraction-electrowinning sites and a full-scale unit is now in operation at the Mexicana de Cananea mine in Sonora, Mexico. Figure 8 shows a photo of the iron control system at Mexicana de Cananea. This unit has a design capacity to remove approximately one metric ton of iron per day from the Cananea copper electrowinning circuit. Eichrom is also pursuing the application of its iron control process at other copper producing sites and is developing flowsheets utilizing Regular Diphonix resin for the control of iron in the production of other metals, specifically Zn, Ni, and Co.

Acknowledgments

Diphonix resin development and initial characterization were funded by the Office of Computational and Technology Research, Division of Advanced Energy Projects. The underlying basic studies were performed under the auspices of the Office of Basic Energy Sciences, Division of Chemical Sciences.

Literature Cited

1. Horwitz, E. P.; Chiarizia, R.; Diamond, H.; Gatrone, R. C.; Alexandratos, S. D.; Trochimczuk, A. W.; Crick, D. W. *Solvent Extr. Ion Exch.* **1993**, *11*, 943-966.

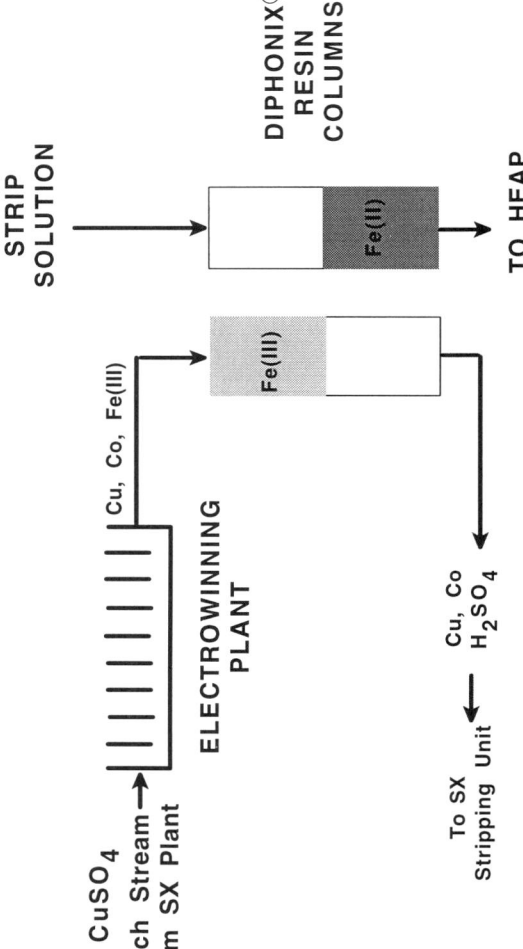

Figure 7. Schematic drawing depicting Eichrom's Iron Control System applied to the electrowinning of copper.

Figure 8. Commercial installation of Eichrom's Iron Control System at the Mexicana de Cananea plant site.

2. Chiarizia, R.; Horwitz, E. P.; Gatrone, R. C.; Alexandratos, S. D.; Trochimczuk, A. W.; Crick, D. W. *Solvent Extr. Ion Exch.* **1993**, *11*, 967-985.
3. Nash, K. L.; Rickert, P. G.; Muntean, J. V.; Alexandratos, S. D. *Solvent Extr. Ion Exch.* **1994**, *12*, 193-209.
4. Chiarizia, R.; Horwitz, E. P.; Alexandratos, S. D. *Solvent Extr. Ion Exch.* **1994**, *12*, 211-237.
5. Horwitz, E. P.; Chiarizia, R.; Alexandratos, S. D. *Solvent Extr. Ion Exch.* **1994**, *12*, 831-845.
6. Chiarizia, R.; Horwitz, E. P. *Solvent Extr. Ion Exch.* **1994**, *12*, 847-871.
7. Chiarizia, R.; Ferraro, J. R.; D'Arcy, K. A.; Horwitz, E. P. *Solvent Extr. Ion Exch.* **1995**, *13*, 1063-1082.
8. Alexandratos, S. D.; Trochimczuk, A. W.; Crick, D. W.; Horwitz, E. P.; Gatrone, R. C.; Chiarizia, R. *Macromolecules* **1996**, *29*, 1021-1026.
9. Alexandratos, S. D.; Trochimczuk, A. W.; Horwitz, E. P.; Gatrone, R. C. *J. Appl. Polymer Sci.* **1996**, *61*, 273-278.
10. Chiarizia, R.; Horwitz, E. P.; D'Arcy, K. A., Alexandratos, S. D.; Trochimczuk, A. W. *Solvent Extr. Ion Exch.* **1996**, *14*, 1077-1100.
11. Chiarizia, R.; D'Arcy, K. A.; Horwitz, E. P.; Alexandratos, S. D.; Trochimczuk, A. W. *Solvent Extr. Ion Exch.* **1996**, *14*, 519-542.
12. Chiarizia, R.; Horwitz, E. P.; Beauvais, R. A.; Alexandratos, S. D. *Solvent Extr. Ion Exch.* **1998**, *16*, in press.
13. Hines, J. J.; Diamond, H.; Young, J. E.; Mulac, W.; Chiarizia, R.; Horwitz, E. P. *Sep. Sci. Technol.* **1995**, *30*, 1373-1384.
14. Gula, M. J.; Totura, G. T.; Jassin, L. *J. Met.* **1995**, *47*, 54-57.
15. Gula, M. J.; Dreisinger, D. B.; Corona, J.; Young, S. K. In *Iron Control and Disposal*; Dutrizae, J. E.; Harris, G. B., Eds; Canadian Institute of Mining, Metallurgy, and Petroleum: Montreal, Quebec, **1996**, 315-327.

Chapter 15

Reillex-HPQ Anion Exchange Column Chromatography: Removal of Pertechnetate Ion from DSSF-5 Simulant

Norman C. Schroeder[1], Susan D. Radzinski[1], Jason R. Ball[1], Kenneth R. Ashley[2], and Glenn D. Whitener[3]

[1]Chemical Science Technology (CST-11), Los Alamos National Laboratory, Los Alamos, NM 87545
[2]Department of Chemistry, Texas A&M University at Commerce, Commerce, TX 75429
[3]Department of Chemistry, Macalester College, St. Paul, MN 55105

The repetitive loading of pertechnetate anion (TcO_4^-) from the Hanford tank waste simulant, DSSF-5, onto a 2.54 x 50 cm Reillex-HPQ anion exchange column has established the viability of this resin for pertechnetate anion removal. The column was loaded at a linear flow rate of 3.00 cm/min (15 mL/min) until at least 1% breakthrough occurred, and then up-flow eluted at the same flow rate with 0.005 M Sn^{2+}/1.0 M ethylenediamine/1.0 M NaOH. A total of 11 cycles were run which kept the column in service and in contact with DSSF-5 simulant or caustic solution for 94 days. Recovery of technetium over this period of service was nearly quantitative (>98%).

An early attempt to separate pertechnetate anion (TcO_4^-) from neutralized Hanford waste was performed by Wheelwright's group during the early 1960's (1). They used Dowex-1 anion exchange resin to recover pertechnetate anion from freshly generated waste. This work clearly demonstrated anion exchange as the baseline technology for this separation. An additional thirty years of anion exchange resin development has produced resins with improved performance compared to Dowex-1. These new resins may be able to exceed the decontamination factor of five obtained by Wheelwright's group. F. Marsh at Los Alamos National Laboratory (LANL), in collaboration with Reilly Industries, has developed a new resin, Reillex-HPQ, to separate plutonium nitrate (i.e., $Pu(NO_3)_6^{2-}$) from 7-8 M nitric acid solutions (2). This resin is a copolymer of divinylbenzene and 4-vinylpyridine that has been subsequently methylated at the pyridine nitrogen to give pyridinium [$-C_5H_4N(CH_3)^+$] strong base anion exchange sites. The pyridinium functionality of Reillex-HPQ is unique; most other strong base anion resins like Dowex-1 are alkyl quaternary amine resins. Reillex-HPQ, compared to other resins, has superior stability to radiolysis and nitric acid (3, 4). At LANL's plutonium facility, the lifetime of the resin is at least four times that of conventional resins under 7-8 M nitric acid processing conditions.

More specifically for pertechnetate anion, this resin was included in a thesis project that evaluated 22 anion exchange resins for their ability to separate pertechnetate anion from acidic, neutral, and high salt solutions (5). Reillex-HPQ was the highest ranked resin in this study. On the basis of these data, we chose to examine the qualities of Reillex-HPQ to partition pertechnetate anion from caustic Hanford

Double Shell Slurry (DSS) and Double Shell Slurry Feed (DSSF) tank waste simulants (6-9). The Tank Waste Remediation Program requested that we study these simulants because they represented wastes being considered for pretreatment at the time of these studies.

We have previously reported the results of a flow study which measured the breakthrough volumes for DSS simulant using 1.00 x 5.67 cm Reillex-HPQ columns (10). These preliminary results were encouraging with respect to using Reillex-HPQ resin for pertechnetate anion partitioning. We extended these studies to a 1.00 x 20.0 cm column which was down-flow loaded with DSS simulant containing 5.0 x 10^{-5} M pertechnetate anion (8). The column was down-flow eluted with a reducing/complexing eluent containing 0.005 M Sn^{2+}, 1.0 M $NH_2C_2H_4NH_2$ (en), and 1.0 M NaOH. The column was loaded and eluted for 11 cycles over a 52 day period. The 1% breakthrough volume decreased from an initial 57.7 bed volumes (BV) to 39.8 BV at the eighth loading to 24.0 BV at the eleventh loading. A 1% breakthrough volume is defined as the volume of effluent, measured in geometrical resin bed volumes (BV), that has a technetium activity equal to 1% of the column feed activity. The column elutions removed an average of 97% of the loaded activity. Column bleed was a problem during loadings 2-11; an average of 0.42% of the activity being loaded bled during the runs. Overall, these results were taken as positive and encouraging.

This report describes research using 2.54 x 50.0 cm columns with DSSF type simulants. The 1% breakthrough volumes are reported for 11 loading and eluting cycles for DSSF-5 simulant. Between cycles the resin column was slowly washed with a 6.0 M $NaNO_3$/2.0 M NaOH solution; thus the column was in contact with a caustic solution for 94 days. This work continues our prior studies to test the viability of using Reillex-HPQ to separate technetium, as pertechnetate anion, from Hanford waste streams by applying the baseline technology of anion exchange. It assumes that an efficient organic destruction process will precede technetium partitioning and that all the technetium has been brought to the pertechnetate anion state prior to processing. These problems are not trivial and are addressed in our other works (9, 11). (Schroeder, N. C.; Ashley, K. R.; Radzinski, S. D.; Truong, A. P.; Szczepaniak, P. A.; Whitener, G. D. *Science and Technology for Disposal of Radioactive Tank Waste*, submitted.) Lastly, this report attempts to explain the decreasing column performance and column bleed problems observed in this work and our earlier studies.

Experimental

Reagents. All water used was 18-MΩ water (Millipore, Bedford, MA). Reillex-HPQ resin, chloride form, 30-60 mesh, was obtained from Reilly Industries (Indianapolis, IN). Standard solutions of HNO_3 were prepared from concentrated HNO_3 (J. T. Baker) and standardized against P. S. sodium carbonate (J. T. Baker). Sodium hydroxide solutions were standardized against 0.1600 or 8.000 N H_2SO_4 solutions (Hach Chemical, Ames, IA). All other reagents were analytical grade, except where noted, and obtained from either Aldrich Chemical Co., J. T. Baker, EM Science (Merck), or Mallinckrodt.

Pertechnetate Anion Preparation and Assay Techniques. Lithium pertechnetate ($Li^{99}TcO_4$) was added to simulants to produce a macro concentration of $\approx 5 \times 10^{-5}$ M. The $Li^{99}TcO_4$ stock solution (0.12 M) was metathesized from $NH_4^{99}TcO_4$ (Oak Ridge National Laboratory) by a published procedure (10, 12). The lithium salt was initially prepared for accelerator transmutation of technetium studies; we deemed it unnecessary to convert it to the sodium salt for this work. A tracer technetium isotope, ^{95m}Tc ($t_{1/2}$ = 61 days), was also added to the simulants to give a $Na^{95m}TcO_4$ concentration of < 10^{-9} M. This gamma emitting isotope was readily obtained from the Medical Radioisotopes Group at LANL. Both macro and tracer technetium were adjusted to the same chemical form (i.e., TcO_4^-) before adding them to the simulants by taking each isotope solution to incipient dryness, three times, in HNO_3 under mild heating conditions; the surface of the hot plate never exceeded 200° C. Each technetium isotope, as TcO_4^-, was tested with a standard 60 minute batch contact experiment between DSS simulant and Reillex-HPQ resin; a batch K_d value of (330 ± 30) mL simulant/g dry resin qualified the isotope (8).

Gamma counting of ^{95m}Tc was performed using a Packard Auto-Gamma Model 5530 instrument (Packard Instrument Company, Downers Grove, IL). The detector is a three inch diameter thallium activated sodium iodide (NaI(Tl)) crystal that has a through-hole design. Technetium-95m decays by isomeric transition (4%), positron emission (0.4%), and electron capture with gamma emission (>95%). Seventy percent of all decay events produce a gamma photon of energy 204.2 keV (13). These are the photons that are counted (the lower and upper limits of photon energies counted by the instrument were set as 165 and 245 keV, respectively). Counting efficiency for the 204.2 keV gamma peak of ^{95m}Tc is 45%. Samples were counted in 25 mL polyethylene scintillation vials. The number of counts per minute (cpm) for each sample and for several blanks (background, bkg) were recorded. Corrections for the decay of the ^{95m}Tc were automatically made by the instrument. The macro ^{99}Tc concentration does not interfere with the gamma counting of ^{95m}Tc since it decays 100% by ß⁻ emission without producing any gamma photons. The instrument is equipped with a QA program to correct for background change and position of the calibration peak (^{137}Cs). An auto-calibration program adjusts the detector voltage to compensate for temperature and detector drift.

Simulants. DSSF-7 simulant (7.00 M Na^+) was prepared; the composition is shown in Table I. Weights of the reagents in the table were added to water with complete dissolution of each reagent prior to the addition of the next reagent (8). The simulant after preparation was a clear, slightly yellow solution, which yielded a small amount of white precipitate upon cooling. DSSF-5 (5.00 M Na^+) simulant was prepared from filtered (medium fritted disc glass funnel) DSSF-7 simulant by diluting 5.00 L of DSSF-7 simulant with 2.00 L of water to give a total volume of 7.00 L. Other dilutions of DSSF-7 were prepared similarly. There was less than 1% deviation from the volumes being additive. Simulants were characterized by some of the qualification methods (density, percentage water, and [OH^-] titer) described previously (8). The proper amount of $Li^{99}TcO_4$ and $Na^{95m}TcO_4$ to give a macro concentration of 3.5 x 10^{-5} M and ≈ 6000 cpm/4 mL, respectively, were added to the simulants. The simulants were then mixed vigorously and stirred for 24 hours prior to the column chromatography experiments.

Batch Distribution Coefficients. Reillex-HPQ resin was converted to the nitrate form as previously described and was stored in water (9, 14). A weighed portion of 60 °C dried resin was placed into a polyethylene vial (caps with Teflon liners were used) with the desired amount of simulant solution; the solution-to-resin ratio was 10:1. The sample was then contacted for the desired length of time using a Burrell Wrist-Action Shaker or Lab-Line Constant Temperature Shaker Bath at 25 °C. At the end of the contact period, the resin/solution sample was poured into a Bio-Rad Econo-Column and the filtrate collected. Appropriate aliquots of the initial solution and filtrate were counted and the technetium batch distribution coefficient (K_d) calculated as described previously (8). Equation 1 defines K_d with ($TcO_4^-)_{total}$ as the total mmols of pertechnetate anion in the solution before contact with the resin, $[TcO_4^-]_{sol}$ as the concentration of pertechnetate anion in V_s mL of solution in contact with the mass of dry resin (g), cpm as the counts per minute of the uncontacted solution, and cpm_{sol} as the total counts per minute of the contacted solution.

Table I. Double-Shell Slurry Feed (DSSF-7) Tank Waste Simulant Formulation

Material	Molarity	Material	Molarity
$NaNO_3$	1.162	$NaOH$	3.885
KNO_3	0.196	$Al(NO_3)_3 \cdot 9H_2O$	0.721
KOH	0.749	Na_2CO_3	0.147
$CsNO_3$	7.00×10^{-5}	$NaCl$	0.102
Na_2SO_4	0.008	$NaNO_2$	1.512
$Na_2HPO_4 \cdot 7H_2O$	0.014		
$Li^{99}TcO_4$	5×10^{-5}	$Na^{95m}TcO_4$	$<10^{-9}$
[cation] = 7.95 M[a]		[anion] = 7.76 M[a]	ionic strength (m) = 8.14 M[a]

[a] Aluminum assumed to be $Al(OH)_4^-$ and phosphate to be PO_4^{3-}.

$$K_d = \left\{ \frac{\left(TcO_4^-\right)_{total} - \left(\left[TcO_4^-\right]_{sol} \times V_{sol}\right)}{\left[TcO_4^-\right]_{sol} \times V_{sol}} \right\} \times \left(\frac{V_{sol}}{g}\right)$$

$$= \left\{\frac{cpm_t - cpm_{sol}}{cpm_{sol}}\right\} \times \left(\frac{V_{sol}}{g}\right) \quad (1)$$

Elution Reagents. The stripping eluent was a solution containing 0.0050 M $SnCl_2 \cdot 0.5H_2O$ or $SnCl_2 \cdot 2H_2O$, 1.00 M ethylenediamine ($C_2H_8N_2$, en), and 1.00 M NaOH. The NaOH pellets, $SnCl_2 \cdot xH_2O$ crystals, and en liquid were added to 700 mL of N_2-purged water in a 1000 mL volumetric flask. After dissolution, the solution was diluted to the mark with N_2-purged water. A piece of a mossy tin was added to the solution to maintain the stannous ion; solutions were used the same day as their preparation.

General Experimental Column Set Up. Chromaflex (Kontes) glass chromatography columns (2.54 x 60 cm) with clear plastic safety/water jackets and accessories were used. The columns were connected to a Masterflex (Cole-Parmer) peristaltic pump. All tubing and fittings were PTFE. The two- and three-way valves were aluminum housing with PTFE plugs. The valving arrangements were such that various solutions could be pumped into the top or bottom of the column and the effluent from the bottom or top of the column could be collected.

Approximately 250 mL of Reillex-HPQ nitrate form resin was added to the column as a water slurry. After the resin had settled, it was backwashed several times with water. All of the resin was completely lifted (floated) by the backwash current of water. A packing reservoir attached to the top of the column was used to ensure enough volume for the resin and backwash liquid. The resin was well classified using this procedure. After backwashing and settling (2.54 x 50.8 cm resin bed), a flow adapter was placed on the top of the column and adjusted so that there was no space between it and the top of the resin bed. Approximately two bed volumes (500 mL) of a 6.0 M $NaNO_3$/2.0 M NaOH shrinking solution were then pumped down through the column. It was necessary to restrain the bed with the flow adapter in order to prevent it from floating and to form a solid resin bed. As the resin shrank, the flow adapter was readjusted to fit against the top of the resin. This process shrinks the length of the bed 3% (2.54 x 49.3 cm resin bed).

DSSF-5 simulant, containing pertechnetate anion, was pumped down through the column. Generally, 240 mL of effluent was collected in a 250 mL graduated cylinder. Then 10 mL was collected in a 10 mL graduated cylinder; 4.00 mL of this was taken and counted. The simulant was pumped through the column at a rate of 3.0 cm/min (15 mL/min) until the desired percent breakthrough had been achieved.

Stripping. An up-flow wash of the column with 2-4 BV of 1.0 M en/1.0 M NaOH solution preceded the elutions. Technetium was up-flow eluted from the column at various flow rates, usually 3.0 cm/min. The first breakthrough experiment (BT-1) used 0.010 M Sn^{2+}/1.0 M en/1.0 M NaOH as the eluent solution. BT-2 through BT-11 used 0.0050 M Sn^{2+}/1.0 M en/1.0 M NaOH as the eluent solution. The volume of the fractions collected varied; however, 4.00 mL from each fraction were taken and γ-counted.

The column was regenerated by washing it, in an up-flow mode, with 2 BV of the 6.0 M $NaNO_3$/2.0 M NaOH shrinking solution. The column was maintained between each loading and elution cycle, 5-7 days, with an up-flow wash with the shrinking solution at 0.1 mL/min. The flow adapter was adjusted prior to the start of the next loading to constrain the resin bed. A total of 11 cycles were run which kept the column in service and in contact with a caustic solution for 94 days.

These sustainability studies used the same resin for eleven cycles of loading and elution. The resin bed was not disturbed through the initial six experiments. Mechanical problems occurred during the sixth (BT-6) and seventh (BT-7) experiments. Hence, the resin bed was reclassified with a backwash of water and then down-flow treated with several bed volumes of shrinking solution before placing the column back into service. No further mechanical problems were encountered for the remaining four runs.

Results and Discussion

DSSF-7 supernatant from Hanford tank AP-105 is 60.3% water with a density of 1.333 g/mL (*15*). Five different DSSF-7 simulant preparations had an average density of 1.351 ± 0.007 g/mL with 58.0 ± 0.5% water and a OH^- titer of 1.16 ± 0.02 M. The DSSF-5 simulant had a calculated density of 1.251 g/mL and a measured density of 1.266 g/mL for one sample. The calculated concentrations are $[Na^+]$ = 5.00 M, [cation] = 5.68 M, [anion] = 5.54 M, and ionic strength (μ) = 5.86 M. The calculated OH^- titer of 0.826 M was lower than the measured value of 1.13 M. The DSSF-2.33 simulant had a calculated density of 1.117 g/mL and a calculated OH^- titer of 0.384 M. The calculated concentrations are $[Na^+]$ = 2.33 M, [cation] = 2.65 M, [anion] = 2.59 M, and ionic strength (μ) = 2.71 M.

The batch pertechnetate anion K_d values between Reillex-HPQ and the DSSF-7, DSSF-5, and DSSF-2.33 simulants, measured as a function of time, are shown in Figure 1. The first-order approach to equilibrium are depicted by the lines; they are the least squares fit of the data to the equation $K_d = {}^fK_d - {}^fK_d e^{-kt}$ where ${}^fK_d$ is the final pertechnetate anion K_d value and k is the time constant for pertechnetate anion sorption (*16*). The corresponding values for the three DSSF simulants are: ${}^fK_d$ = (210 ± 10) mL/g and k = (0.11 ± 0.02) min^{-1} for DSSF-7, ${}^fK_d$ = (245 ± 6) mL/g and k = (0.12 ± 0.01) min^{-1} for DSSF-5, and ${}^fK_d$ = (300 ± 20) mL/g and k = (0.26 ± 0.08)min^{-1} for DSSF-2.33. The first order rates decrease as the ionic strength or density of the solutions increase. These characteristics are related to the viscosity of the simulants, thus the rates may correlate with the diffusion rate of the simulant solutions, containing pertechnetate, into the resin pores. However, lower pertechnetate anion diffusion rates may occur in solutions that have high concentrations of water-structuring anions; these are ions with high negative Gibb's free energies of hydration, ΔG_{hyd}, such as OH^- and $Al(OH)_4^-$ which are present in high concentrations in Hanford simulants and actual wastes (*17*). We have measured the k values for a broad range of Hanford simulants (Ashley, K. R.; Whitener, G. D.; Schroeder, N. C.;

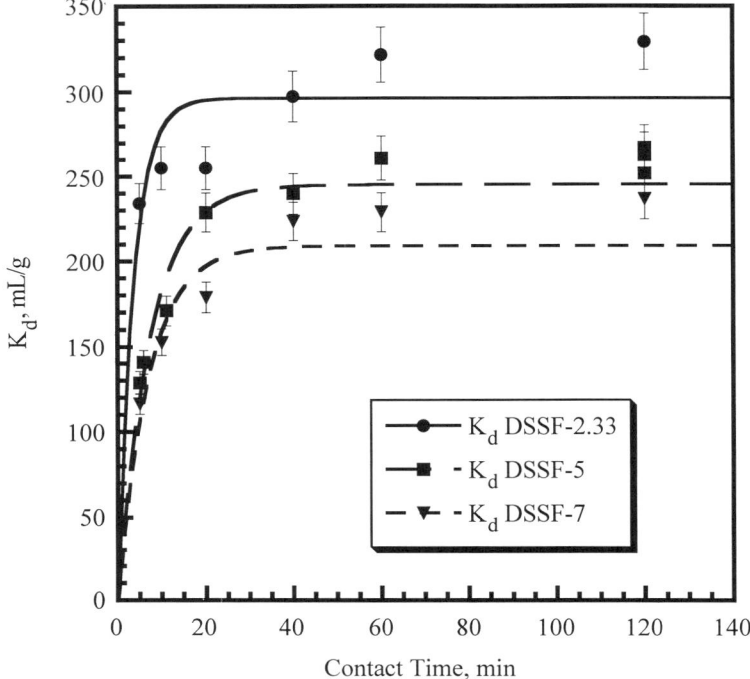

Figure 1. Batch K_d values for pertechnetate anion between DSSF-7, DSSF-5, and DSSF-2.33 simulants and Reillex-HPQ resin as a function of time. The lines are the least-squares fits of the data to the equation $K_d = {}^fK_d - {}^fK_d e^{-kt}$ (see text).

Ball, J. R; Radzinski, S. D. *Solvent Extr. Ion Exch.*, in press). These k values show a strong inverse correlation with simulant total hydroxide concentrations and a very poor correlation with the simulant densities (Schroeder, N. C.; Los Alamos National Laboratory, unpublished data).

The rapid increase and the magnitude of the pertechnetate anion K_d values with time indicates that good column performance would be expected at a 3 cm/min (15 mL/min) simulant flow rate. This flow corresponds to a ~17 minute residence time in a 50 cm column; the estimated K_d values for the DSSF-7, DSSF-5, and DSSF-2.33 simulants at 17 minutes are 190, 220, and 280 mL/g, respectively. The column distribution ratio (λ) can be defined by the equation, $\lambda = K_d \rho_b$ (*18*). The term ρ_b is the bed density, g dry resin/mL of settled resin bed in the feed matrix. Lambda is the number of geometric bed volumes of feed that has passed through the column by the 50% breakthrough point. For Reillex-HPQ, the value of ρ_b is (0.361 ± 0.003) g/mL. Thus the expected 50% breakthrough are 68, 79, and 101 BV for the DSSF-7, DSSF-5, and DSSF 2.33, respectively. However, the physical behavior of these solutions in the anion exchange columns must be considered, *vide infra*.

Our previous 1 x 20 cm column experiments with DSS simulant columns were trouble free (*8*). In contrast, experiments with DSSF-7 simulant with 1 x 20 cm columns were wrought with problems. This simulant was loaded directly onto a resin bed that had been packed in water and conditioned with 1 M NaOH; the same procedure that was used with DSS simulant. Abnormal breakthrough curves, characterized by earlier than expected breakthrough volumes (< 7 BV), and curves that had a decreasing slope shortly after achieving breakthrough were obtained. Column performance improved as the runs proceeded or as the feed flow rates increased. What was striking about these initial runs was the 8% shrinkage of the resin bed length after contacting the DSSF-7 simulant. Conditioning the column with a 4.0 M $NaNO_3$ solution pre-shrunk the resin bed slightly and improved the column performance. However, it was apparent that resin bed instability was going to continue to be a problem. At this point, it was decided to switch to 2.54 cm diameter columns with the expectation that the wider column would not support an unstable resin bed; the resin bed would naturally settle to a well packed state. In addition, this column diameter is 42 to 85 times greater than the range of resin bead diameters (300 to 600 mm). This should negate wall effects and allow the results from these experiments to be scaled to any larger size column (19). The 50 cm resin bed height gives a geometric bed volume of ≈ 250 mL.

The initial 2.54 cm column run with DSSF-7 was as poor as any of the 1.00 cm runs. We observed that this simulant floats the 50 cm resin bed. Constraining the bed with a flow adapter did not improve performance. Constraining and preshrinking the resin bed with DSSF-7 simulant containing no pertechnetate anion improved the column performance. However, preconditioning columns with solutions as complex as simulants is not a remedy for a potential processing operation. Two options were available to solve this problem; dilute the simulant enough to give good column performance and/or preshrink the resin with a simple solution. Two dilutions of DSSF-7, DSSF-1.17 (d = 1.07 g/mL) and DSSF-3.5 (d = 1.18 g/mL), gave good column performance. This is not unexpected since these dilutions have densities less than the DSS simulant (d = 1.24 g/mL) which never gave us these types of problems. However, Hanford tank waste processing restricts the allowed dilutions to 5.0 M Na^+; thus, we were limited to a 1:1.4 dilution of the DSSF-7 to give a DSSF-5 (d = 1.266 g/mL). Fortunately we were able to find a shrinking solution consisting of 6.00 M $NaNO_3$ and 2.00 M NaOH that could precondition the resin bed to match the expected shrinkage by the DSSF-5 solution. Concentrations greater than that of DSSF-5 will be difficult to process because shrinking solutions with higher nitrate and hydroxide concentrations will exceed the limit of solubility.

Figure 2 shows the breakthrough curves for the first, seventh, and eleventh runs (BT-1, BT-7, and BT-11) with DSSF-5. The 1% breakthrough for the BT-1 run occurred at 59.1 geometric bed volumes. The breakthrough curves of BT-2 through BT-11 were affected by "activity bleeding" from the column. Thus, the 1% breakthrough volumes are reported at 1% above the bleed activity. For example, the 1% breakthrough point for BT-7 is at 1.55% because activity bleeding from the column creates a baseline at 0.55%. The breakthrough curves for the other runs are similar.

Table II summarizes the column performance for eleven consecutive DSSF-5 experiments with a single bed of Reillex-HPQ resin. The 1% breakthrough volumes decreased from 59.1 BV in the initial run to 40.0 BV in BT-8 to 39.3 BV in BT-11, a 33% decrease. These breakthrough volumes are for the geometric bed volume of that particular run. Hence, they have been corrected for the 14% decrease in volume of the resin bed. In processing terms, the volume of waste or simulant that could be processed to the 1% breakthrough point has decreased by 43% over the 11 runs. These numbers reflect only the efficiency of the column as measured by the 1% breakthrough volumes; they do not reflect the total pertechnetate anion capacity of the column as a function of time.

The decrease in column performance is consistent with the results of the Reillex-HPQ resin stability study performed in 2.0 M NaOH at 50 °C (Ashley, K. R.; Cobb, S. L.; Cuttrell B.; Ball, J. R.; Schroeder, N. C.; Radzinski, S. D. *Solvent Extr. Ion Exch.*, to be submitted). That study also showed that after an initially stable period of 2 days, a 30% decrease occurred in the resin's capacity over the next 3 days, with a

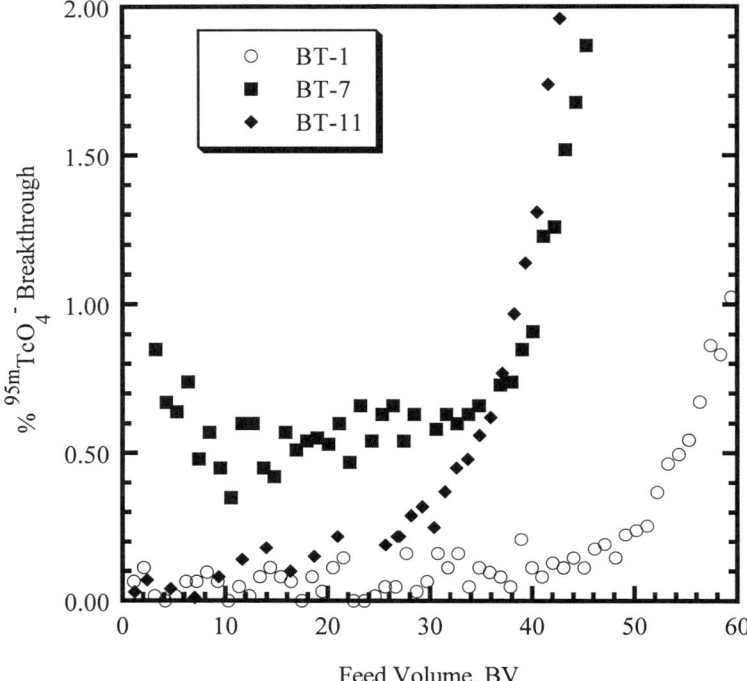

Figure 2. Breakthrough Curves for the BT-1, BT-7, and BT-11 Column Experiments with DSSF-5.

Table II. Column Performance Experiments[a]

Run	Elapsed time, days[b]	Total volume processed, L	Resin bed volume, mL	1% Breakthrough volume, BV[c]	Volume @ 1%, L
BT-1[a]	2	14.75	249.8	59.1	14.76
BT-2	7	14.45	237.1	58.7	13.92
BT-3	15	14.50	233.6	57.9	13.53
BT-4	21	14.00	232.7	58.4	13.59
BT-5	37	13.75	230.6	53.2	12.27
BT-6	43	12.53	230.6	51.0	11.76
BT-7	50	11.75	238.6	43.8	10.45
BT-8	58	10.4	230.6	40.0	9.22
BT-9	70	10.5	222.4	44.1	9.81
BT-10	80	9.50	222.4	39.8	8.85
BT-11	94	9.50	214.8	39.3	8.44

[a]The flow rate was 3.0 cm/min (15 mL/min) except for BT-1 in which the flow rate was 3.12 cm/min (15.6 mL/min). [b]The cumulative time that the resin has been in service. [c]The breakthrough volume is in units of geometric bed volumes (BV) of the bed for that run.

subsequent stable period out to at least 27 days. There is a loss in the total capacity of the resin, not just the strong base capacity. This implies that the resin is not just being demethylated. We are tentatively explaining the decrease in resin performance to be the result of solubilization of low molecular weight polymers from the Reillex-HPQ resin matrix by the caustic solutions. This would account for the observed "activity bleeding" if the low molecular weight polymers were bringing TcO_4^- with them. Solubilization could also explain why the resin bed volume decreased in our column experiments. The higher temperature in the resin stability experiment explains the accelerated rate of change as compared to the column runs.

Analysis of the 1% breakthrough volumes for the BT-1 through BT-11 runs as a function of time shows an initial plateau for the first four runs, decreasing values for runs five through seven, and then another plateau for the remaining runs. This is qualitatively similar to the results from the Reillex-HPQ resin stability studies.

Sorbing pertechnetate anion onto a resin column is the first step of the technetium partitioning process. Efficient elution of the sorbed technetium is required to recycle the column. In the past, 8.0 M HNO_3 has been used to elute pertechnetate anion from anion exchange resins; however, this presents several problems. First, large-scale processing of an organic resin with an oxidizing acid is potentially explosive (20, 21). Secondly, significant additional wastes are generated by the large volume of acid required to elute pertechnetate anion.

A second method is to reduce TcO_4^- to TcO_2 and to wash the electrostatically neutral precipitate from the resin. Our experience is that TcO_2 is a very difficult material to remove from the resin. A third method is to reduce and complex technetium to form a cationic species. This method is based on the synthesis of soluble ^{99}Tc species for use in the medical radioisotope field (22,23). The general reaction to form a reduced ^{99}Tc complex is given by equation 2.

$$^{99}TcO_4^- + \text{Reductant} + \text{Ligand} \rightleftharpoons [Tc(V) \text{ Product Complex}]^z \quad (2)$$

Stannous ion is the most common reductant for the synthesis in equation 2 because of its rapid kinetics to give Tc(V); further reduction by Sn(II) to Tc(IV) is slower. In the presence of a ligand, the Tc(V) is coordinated and stabilized. The geometry of the complex is dependent upon the ligand type; oxygen and sulfur ligands yield a square pyramid geometry with a technetyl core, TcO^{3+}, while nitrogen ligands favor an octahedral complex containing a TcO_2^+ core with *trans* oxygens (22, 23). The ligand type will determine the final charge (z) of the technetium complex.

Eluting technetium from an anion exchange resin would be most efficient if a cationic complex formed because it would be electrostatically rejected by the resin. The reaction must occur rapidly in aqueous caustic media and produce a single soluble product. In addition, the whole process should have the potential of recycling the reagents. Thus, neutral nitrogen chelate ligands are the preferred complexants. The reported species, *trans*-$[TcO_2(en)_2]^+$ appeared to be a desirable species to try to form during elution (23).

We have found that Sn(II)/en/NaOH solutions quantitatively remove technetium from a 1.00 x 5.7 cm Reillex-HPQ column that had been loaded with 7.5 x 10^{-4} mmol pertechnetate anion from a DSS simulant (8, 9). Figure 3 compares the technetium elution curves using the 0.005 M Sn(II)/1.0 M en/1.0 M NaOH and 8.0 M HNO_3 eluents. The tin elution encompasses aliquots 9-12; ≈ 97.5% of the technetium was eluted with 4.4 geometric bed volumes of this eluent. The 8.0 M HNO_3 elution starts at aliquot 9; it was necessary to pass 55.4 geometric bed volumes of the acid through the column to achieve the same recovery. These experiments clearly demonstrate that both the time for technetium removal and the amount of waste generated is substantially reduced when Sn(II)/en/NaOH is used as an eluent instead of nitric acid.

We also demonstrated that the eluent would effectively remove the technetium from these larger columns loaded from DSSF-5 simulant (8, 9). The most successful

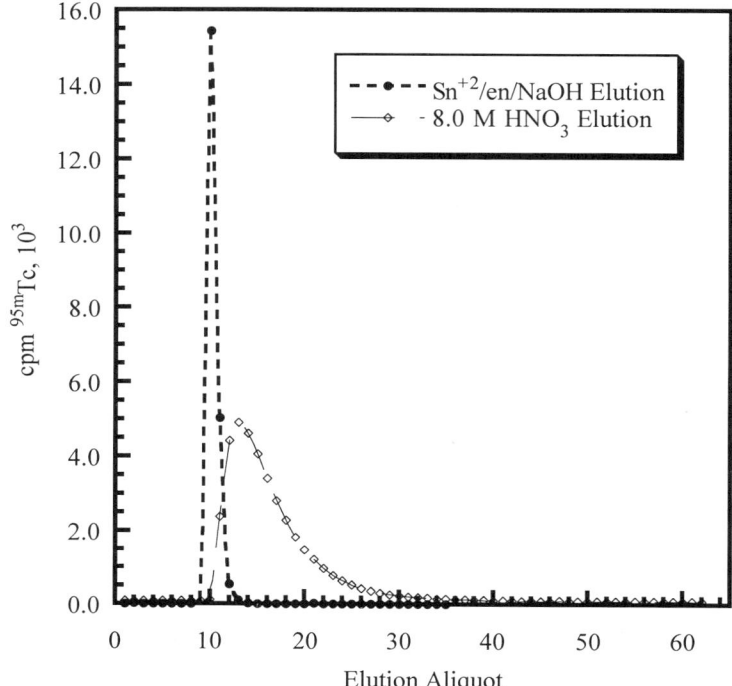

Figure 3. Comparative elution curves for technetium from a 1.00 x 5.73 cm column (4.50 mL geometric bed volume) of Reillex-HPQ using the tin and nitric acid eluents. Each elution aliquot is 5.0 mL.

eluent system found was a Sn(II)/en/NaOH solution in which the NaOH and the en concentrations ranged between 0.10 and 1.0 M and the Sn(II) concentration ranged from 0.005 to 0.016 M.

In addition, we have shown that the Sn(II)/en/NaOH eluent system is effective for removing technetium sorbed onto AG MP-1, a 20-25% crosslinked, strong base resin having a methyl quaternary amine functionality (24). (Ashley, K. R.; Whitener, G. D.; Schroeder, N. C.; Radzinski, S. D. *Solvent Extr. Ion Exch.*, to be submitted.) On the other hand, elution of pertechnetate from AG 1-X8, an 8% crosslinked resin with the same functionality, is less than 75% effective.

Table III shows the bed volumes of 0.005 M Sn(II)/1.0 M en/1.0 M NaOH eluent used to remove technetium from each of the BT experiments. The *% Elution* is based on the amount of activity recovered during the elution relative to the amount sorbed during the loading (*vide infra*, Table IV: column 4/[column 2-column 3]). The larger than 100% elutions may be just experimental error or elution of technetium from previous loadings might be occurring. The average BV used to elute 97.4% of the technetium is quite consistent with the small column experiment in Figure 3. The BT-1 elution was relatively inefficient since it left 10% of the technetium on the column. Although using only 2 BV of eluent may account for this inefficiency, there are alternative explanations. If transition metal impurities in the simulant were deposited in the resin bed during the loading phase, then during elution they may have become

Table III. Bed Volumes of Sn(II) Eluent Required to Elute Technetium from the BT Column Runs

Run	BV Eluent[a]	% Elution[b]
BT-1	2.03	91.09
BT-2	3.66	102.5
BT-3	4.08	98.60
BT-4	4.34	96.33
BT-5	3.25	83.90
BT-6	4.14	98.44
BT-7	3.77	102.0
B7-8	3.72	103.9
BT-9	4.37	102.3
BT-10	4.50	104.6
BT-11	4.50	98.99
Average	3.85 ± 0.72	98.41 ± 6.18

[a]BV is the geometrical bed volume. [b]Calculated from data in Table IV; column 4/(column 2-column 3).

reduced hydroxides and therefore capable of sorbing reduced technetium species. Reduced transition metals, especially iron, are capable of such behavior (25). Normally, transition metal ions in 1.7 M NaOH would be expected to be insoluble. However, in the high ionic strength simulants, anionic complexes might form. To prevent this, all subsequent tin elutions were preceded by a 1.0 M en/1.0 M NaOH up-flow elution of the column. Except for BT-5, this seemed to rectify this problem.

Alternatively, there may be competition between ethylenediamine and the weak base pyridine (py) sites in the Reillex-HPQ resin for Tc(V). Loading the resin with 1.0 M en/1.0 M NaOH before the tin elution increases the concentration of ethylenediamine at the reduction site. Hence, the formation of $trans$-$[TcO_2(en)_2]^+$ becomes more favorable over $trans$-$[TcO_2(py)_4]^+$ or $trans$-$[TcO_2(en)(py)_2]^+$ type complexes which are tethered to the resin.

The bed volumes of eluent listed in Table III do not reflect the real efficiency of this eluent. Figure 4 is the elution curve for the BT-8 experiment; this shows that 99.4% of the technetium is contained in the first 1.5 bed volumes of eluent. The remaining technetium was contained in an extended tail. All of the elution curves were of this form, regardless of the percentage of the total technetium eluted. When this elution is in progress, a pink band of the $trans$-$[TcO_2(en)_2]^+$ complex ($\approx 10^{-3}$ M) passes up the column; the pink solution becomes dark red upon standing 30 minutes.

Table IV summarizes the technetium inventory, based on the ^{95m}Tc activity, for the eleven experiments. The *Added activity* (column 2) is the cpm of ^{95m}TcO$_4^-$ loaded for each column run. The *Bleed activity* (column 3) is the cpm of ^{95m}TcO$_4^-$ that comes through the column as it is being loaded. The *Eluted activity* (column 4) is the cpm of ^{95m}TcO$_4^-$ that is removed with the eluent. The *Activity left on column* (column 5) is the *Added activity - Bleed activity - Eluted activity*, (column 2 - column 3 - column 4). The averages for each column are given. Notice that in each run there is a small bleed of activity from the column. Two factors could influence the magnitude of the bleed. One is incomplete elution of reduced technetium from a prior run's elution; this effect is seen in BT-2 and BT-3 which followed the inefficient BT-1 elution. The second possibility is the solubilization of low molecular weight polymers of the resin that have sorbed technetium on them. Table II shows there was a significant decrease in the 1% breakthrough after BT-6. Solubilization may explain the higher bleed levels for BT-7 and -8.

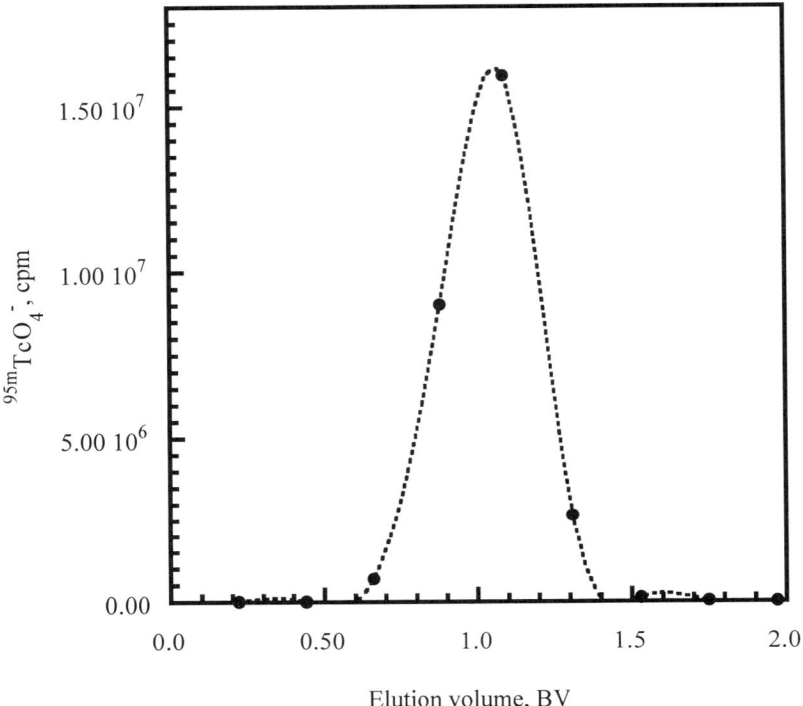

Figure 4. Elution curve for the BT-8 breakthrough experiment.

Table IV. Inventory of ^{95m}Tc for Each Experiment During the Column Performance Experiments

Run	Added activity, 10^6 cpm	Bleed activity 10^6 cpm	Eluted activity 10^6 cpm	Amount left on column 10^6 cpm
BT-1	23.05	0.0389	20.96	2.087
BT-2	22.98	0.8127	22.73	0.2459
BT-3	23.29	0.5614	22.41	0.8813
BT-4	23.18	0.1229	22.21	0.9691
BT-5	25.33	0.1339	21.14	4.191
BT-6	23.32	0.1483	22.81	0.5138
BT-7	20.12	0.2495	20.26	0.0000
BT-8	28.98	0.2352	29.87	0.0000
BT-9	21.08	0.0826	21.47	0.0000
BT-10	14.84	0.0590	15.46	0.0000
BT-11	17.34	0.0659	17.10	0.2445
Average	22.14	0.2282	21.49	0.8303
Std. Dev.	3.79	0.2432	3.64	1.282

Table V is a moving inventory of the technetium activity for the eleven experiments. Columns 2-4 have the same explanation as the previous table except that each breakthrough run is added to the previous one; each row is the sum of all of the previous rows. Column 5 is a moving summation of columns 3 and 4 and accounts for the activity that leaves the column. Column 6 is a moving percentage of column 5 relative to column 2 and illustrates that 98.11% of the activity added to the column over the 3 month life of this experiment is accounted for. Overall, 1.89% of the technetium was not recovered; unrecovered technetium may have been lost to radioactive decay or reduction and deposition on the column.

Table V. Moving Inventory of ^{95m}Tc During the Column Performance Experiments

Run	Total added activity, 10^6 cpm	Total bleed activity, 10^6 cpm	Total eluted activity, 10^6 cpm	Total recovered activity, 10^6 cpm	Total % recovery
BT-1	23.05	0.039	20.96	21.00	91.10
BT-2	46.03	0.852	43.69	44.54	96.77
BT-3	69.32	1.413	66.10	67.51	97.39
BT-4	92.49	1.536	88.29	89.83	97.12
BT-5	117.8	1.670	109.4	111.1	94.27
BT-6	141.2	1.818	132.2	134.1	94.97
BT-7	161.3	1.843	152.5	154.6	95.84
BT-8	190.3	2.078	182.4	184.7	97.06
BT-9	211.3	2.161	203.8	206.2	97.57
BT-10	226.2	2.444	219.3	221.7	98.04
BT-11	243.5	2.510	236.4	238.9	98.11

Summary

The purpose of the repetitive column loading and elution experiments was to establish the viability of recycling a 2.54 x 50 cm Reillex-HPQ column that is used to remove TcO_4^- from a DSSF-5 simulant. The viability of the column was indicated by the 1.0% breakthrough volume measured in bed volumes (BV); the larger the number the better the performance. The 2.54 x 50 cm column was packed, conditioned with a solution containing 6.0 M $NaNO_3$ and 2.0 M NaOH, loaded with pertechnetate anion from DSSF-5 simulant at a flow rate of 3.0 cm/min until at least 1% breakthrough occurred, up-flow washed with 2-4 BV of 1.0 M en/1.0 M NaOH solution, and then up-flow eluted with the reducing/complexing reagent (0.005 M Sn^{2+}/1.0 M en/1.0 M NaOH). After regeneration with a solution containing 6.0 M $NaNO_3$ and 2.0 M NaOH, the column was put into service for another loading and elution cycle. A total of 11 cycles were run which kept the column in service or in contact with a 6.0 M $NaNO_3$ and 2.0 M NaOH solution for 94 days.

The conclusions are:

- A 2.54 x 50 cm Reillex-HPQ resin column can be used repetitively over a 94 day service period to successfully remove pertechnetate from a DSSF-5 feed.

- Reillex-HPQ resin loses 30% of its efficiency as indicated by decreased 1% breakthrough volume in 2 M NaOH over this 94 day time period.

- A Sn(II)/en/NaOH solution is an effective eluent for removing technetium from Reillex-HPQ. Recovery of technetium over this period of service is nearly quantitative (> 98%).

- High ionic strength (high density) caustic wastes are difficult to process with organic resins.

Acknowledgment

This work was supported by the Department of Energy's Tank Waste Remediation Program and Efficient Separations and Processing Crosscutting Program.

Literature Cited

(1) Roberts, F. P.; Smith, F. M.; Wheelwright, E. J. *Recovery and Purification of Technetium-99 from Neutralized PUREX Waste*, Report HW-SA-2581, General Electric Company, 1962.

(2) Marsh, S. F. *Solvent Extr. Ion Exch.* **1989**, *7*, 889.

(3) Marsh, S. F. *The Effects of Ionizing Radiation on Reillex™-HPQ, a New Macroporous Polyvinylpyridine Resin, and on Four Conventional Polystyrene Anion Exchange Resins*, Report LA-11912, Los Alamos National Laboratory, 1990.

(4) Marsh, S. F. *The Effects of in Situ Alpha–Particle Irradiation on Six Strong-Base Anion Exchange Resins*, Report LA-12055, Los Alamos National Laboratory, 1991.

(5) McGinnis, D. F. *Organic Resin Anion Exchangers for the Treatment of Radioactive Waste*, Ph.D. Thesis, University of Salford, Salford, England M5 4WT, 1988.

(6) Schroeder, N. C.; Abney, K. D.; Attrep, Jr, M.; Radzinski, S. D.; Brewer, J.; Ashley, K. R.; Ball, J. R.; Stanmore, F.; LaFebre, N.; Pinkerton, A. B.; Turner, R. *Technetium Partitioning for the Hanford Tank Waste Remediation System: Adsorption and Extraction of Technetium From Double-Shell Slurry Waste Simulant*, Report LA-UR 93-4092, Los Alamos National Laboratory, 1993.

(7) Schroeder, N. C.; Abney, K. D.; Attrep, Jr, M.; Radzinski, S. D.; Brewer, J.; Ashley, K. R.; Ball, J. R.; Stanmore, F.; LaFebre, N.; Pinkerton, A. B.; Turner, R. *Technetium Partitioning for the Hanford Tank Waste Remediation System: Sorption and Extraction of Technetium from Simple Caustic Solutions*, Report LA-UR-94-62, Los Alamos National Laboratory, 1994.

(8) Schroeder, N. C.; Radzinski, S. D.; Ashley, K. R.; Ball, J. R.; Stanmore, F.; Whitener, G. D. *Technetium Partitioning for the Hanford Tank Waste Remediation System: Sorption of Technetium from DSS and DSSF-7 Waste Simulants Using Reillex™-HPQ Resin*, Report LA-UR-95-40, Los Alamos National Laboratory, 1995.

(9) Schroeder, N. C.; Radzinski, S. D.; Ball, J. R.; Ashley, K. R.; Cobb, S. L.; Cuttrell, B.; Adams, J. M.; Johnson, C.; Whitener, G. D. *Technetium Partitioning of the Hanford Tank Waste Remediation System: Anion Exchange Studies for Partitioning Technetium from Synthetic DSSF and DSS Simulants and Actual Wastes (101-SY and 103-SY) Using Reillex™-HPQ Resin*, Report LA-UR-95-4440, Los Alamos National Laboratory, 1995.

(10) Ashley, K. R.; Turner, R.; Ball, J. R.; Abney, K. D.; Schroeder, N. C. *J. Radioanal. Nucl. Chem. Articles* **1995**, *194*, 71.

(11) Schroeder, N. C.; Ashley, K. R.; Whitener, G. D.; Truong, A. P. *LANL Pretreatment: Technetium Removal Studies,* Report LA-UR-96-4470, Los Alamos National Laboratory, 1996.
(12) Ashley, K. R.; Ball, J. R.; Pinkerton, A. B.; Abney, K. D.; Schroeder, N. C. *Solvent Extr. Ion Exch.* **1994**, *12*, 239.
(13) Lederer, C. M.; Hollander, J. M.; Perlman, I. *Table of the Isotopes, 6th Ed.*; John Wiley: New York, 1967; pp 46, 232-3.
(14) Wu, Y-Y. J.; Williamson, J. M. A.; Zhang, Q.; Grissom, M. R.; Chu, I. C. *Solvent Extr. Ion Exch.* **1996**, *14*, 285.
(15) De Lorenzo, D. S.; DiCenso, A. T.; Amato, L. C.; Stephens, R. H.; Johnson, K. W.; Simpson, B. C.; Welsh, T. L. *Tank Characterization Report for Double–Shell Tank 241-AP-105*, Report WHC-SD-WM-ER-360, Westinghouse Hanford Company, 1994.
(16) The least-squares fit of the data was done using the program KaleidaGraph™ for the Macintosh® computer. The uncertainties reported with the numbers are the estimated one standard deviation of that number.
(17) Rogers R. D., Griffin, S. T., Horwitz, E. P., Diamond, H. *Solvent Extr. Ion Exch.* **1997**, *14*, 547.
(18) Helfferich, F. *Ion Exchange*, Dover Publications, Inc.: New York, NY, 1995; p 452.
(19) Helfferich, F. *Ion Exchange*, Dover Publications, Inc.: New York, NY, 1995; p 487 and references therein.
(20) Calmon, C. *Chem. Eng.* **1980**, *87*, 271.
(21) Schmidt, W. C. *Technetium Recovery and Storage at 224-B: Budget Design Criteria*, Report RL-SEP-373, 1965.
(22) Kastner, M. E.; Lindsay, M. J.; Clarke, M. J. *Inorg. Chem.* **1982**, *21*, 2037.
(23) Baldas, J. In *Advances in Inorganic Chemistry*, Sikes, A. G., Ed.; Academic Press: New York, 1994, Vol. 41; p 1.
(24) Attrep, Jr., M. *Fission Product Separation Using Ion Exchange, Solvent Extraction and Cobalt Dicarbollide*, Report LA-UR-94-3647, Los Alamos National Laboratory, 1994.
(25) Del Cul, G. D.; Bostick, W. D.; Trotter, D. R.; Osborne, P. E. *Sep. Science Technol.* **1993**, *28*, 551.

Chapter 16

Extraction Chromatography: Progress and Opportunities

Mark L. Dietz, E. Philip Horwitz, and Andrew H. Bond

[1]Chemistry Division, Argonne National Laboratory, 9700 South Cass Avenue, Argonne, IL 60439-4831

> Extraction chromatography provides a simple and effective method for the analytical and preparative-scale separation of a variety of metal ions. Recent advances in extractant design, particularly the development of extractants capable of metal ion recognition or of strong complex formation in highly acidic media, have significantly improved the utility of the technique. Advances in support design, most notably the introduction of functionalized supports to enhance metal ion retention, promise to yield further improvements. Column instability remains a significant obstacle, however, to the process-scale application of extraction chromatography.

Extraction chromatography (EXC) is a type of liquid-liquid chromatography that couples the selectivity of solvent extraction (*1*) with the multistage character of a chromatographic process and the ease of handling associated with ion-exchange resins (*2*). Typically, extraction chromatographic materials are prepared by simple adsorption of an organic extractant onto any of a wide variety of inorganic (e.g., alumina, silica) or organic (e.g., cellulose, styrene-divinylbenzene copolymers) supports. In contrast to ordinary partition chromatography, in which the partitioning solute undergoes little, if any, chemical change, the sorption of a metal ion in EXC involves the complex chemical changes associated with the conversion of a hydrated metal ion into a neutral organophilic metal complex, just as in liquid-liquid extraction. This conversion often involves a number of interactions and equilibria, manipulation of which affords opportunities for the design of systems capable of the efficient and selective separations of a variety of metal ions (*3*).

Since its introduction by Siekierski in 1959 (*4*), extraction chromatography has been studied extensively, and a number of reviews (*5,6*) and monographs (*7*) have appeared summarizing various aspects of the technique. In this chapter, we examine recent progress in this field, with particular emphasis on work directed at improving the performance of extraction chromatographic materials and broadening their range of applications.

Background

Conventional extraction chromatographic materials are prepared by the physical impregnation of an inert substrate with either an undiluted extractant or a solution of the extractant in an appropriate diluent. This impregnation can be accomplished by

©1999 American Chemical Society

any of a variety of techniques (*5-7*). Most commonly, a porous support material is contacted with a solution of the extractant or of an extractant-diluent mixture in a volatile solvent, and the solvent slowly removed by evaporation under vacuum. Alternatively, the support is contacted with a solution of the extractant in a mixture of an organic solvent and water, the support separated from the solution by filtration, and the excess organic solvent removed by water washing. For very hydrophobic extractants, the most satisfactory results (i.e., homogeneous impregnation of the support) have been obtained by contacting a solution of the extractant in a precalculated amount of solvent with a support until all of the liquid has been absorbed. Because none of these methods is particularly well-suited to the preparation of large quantities of EXC materials, procedures have been devised by which the extractant can be incorporated directly into the support during its preparation. The Levextrel resins (*8,9*), for example, are macroporous styrene-divinylbenzene copolymers containing an extractant added to the mixture of monomers during the polymerization process. By appropriate choice of reaction conditions, the amount of extractant incorporated, the extent of cross linking of the polymer, and other resin characteristics (e.g., porosity, surface area) can be varied as desired (*10*). Although in principle this approach could be applied with any extractant, in actual practice, only certain types of extractants (e.g., neutral organophosphorus extractants, aliphatic amines) have appropriate physical and chemical properties (e.g., viscosity, solubility, acidity) for inclusion, as the polymerization process is affected by the extractant properties (*9*).

Regardless of the method by which an EXC material is prepared, the retention of the extractant on the support is the result of physical interactions, not covalent bond formation between the extractant and the support. For certain EXC materials, most notably those employing a polymeric support, the nature of these interactions have been the subject of considerable interest. In a recent series of studies by Cortina et al. (*11-13*), for example, the adsorption of various acidic organophosphorus extractants (e.g., di-(2-ethylhexyl)phosphoric acid, HDEHP) on Amberlite XAD-2 (a non-polar macroporous styrenic polymer) was examined using FT IR. Comparison of the infrared spectrum of supported HDEHP to that of the free extractant in carbon tetrachloride showed that adsorption is accompanied by only small shifts in both the phosphoryl (P=O) and P-O-C stretching bands of the extractant and in the various stretching frequencies (e.g., normal modes of methyl and methylene units) associated with Amberlite XAD-2. This indicates that the interaction between the extractant and support is quite weak, consisting of only the attractive forces between alkyl chains and/or aromatic rings of the ligand and those of the support. Much the same can be said of the interactions of other extractants studied with the same or related supports (e.g., polyesters) (*13-15*). Thus, in most instances, the immobilization of an extractant on a support is the result of a combination of weak adsorption and/or physical entrapment of the extractant.

Given the absence of significant interactions between the extractant and support in conventional EXC materials, the complexation properties of a supported extractant would be expected to closely parallel those of the same extractant in a liquid-liquid system. In fact, qualitatively, this is what is generally observed. Recently, the behavior of several extractants supported on various substrates has been examined quantitatively by treating EXC materials as an extractant homogeneously dispersed in a solid matrix and applying equilibrium models developed originally for liquid-liquid systems in which an extractant is homogeneously dispersed in an organic solvent. Kimura (*16*), for example, studied the uptake of Th(IV) and U(VI) by tri-*n*-butyl phosphate (TBP)-loaded Amberlite XAD-4, a high surface area macroporous styrene-divinylbenzene copolymer. Measurements of the extractant, acid, and nitrate dependencies of the sorption of the two cations by the resin were found to be consistent with the extraction of $Th(NO_3)_4 \cdot 2TBP$ and $UO_2(NO_3)_2 \cdot 2TBP$, respectively, in agreement with solvent extraction data. In effect then, XAD-4 functions as an inert diluent for the TBP (*16*). In more recent work, Cortina et al. (*12*) examined the sorption of Zn(II), Cu(II), and Cd(II) from nitrate solutions by HDEHP-loaded Amberlite XAD-2. Treatment of the pH and extractant dependence of log D (where D is defined as $[M]_{resin}/[M]_{aq}$) using Letagrop-distr indicated that the nature of the metal species extracted is somewhat different than that reported for extraction of the same metals into toluene (Zn) (*17*) or paraffinic hydrocarbons (Cu, Cd) (*18-21*), with the extracted species in the EXC system typically less solvated than in the organic solvents. In addition, it was observed that the extractant, which exists largely as a dimer in non-polar solvents, is less associated in the support. Also, it was found that while the extractability of zinc, copper, and cadmium into organic solvents by HDEHP follows the order Zn > Cu > Cd, for the supported reagent, cadmium extraction exceeds that of copper. Other recent work by Strikovsky et al. (*22*), which examined copper extraction by the sulfur analog of HDEHP, di(2-ethylhexyl)dithiophosphoric acid (DEHTPA), supported on Amberlite XAD-2, however, found that the extraction equilibria/constant correspond well with liquid-liquid extraction results in *n*-heptane. Taken together, the results of these and other related studies (*23*) suggest that for many EXC materials, quantitative predictions of metal ion retention properties from liquid-liquid extraction data are feasible. Nonetheless, in certain instances, even a relatively inert support can have unanticipated effects on the physicochemical properties of an extractant.

Many aspects of the performance of an extraction chromatographic resin are not readily predictable from liquid-liquid extraction data, as far more factors are involved in a dynamic chromatographic process than in batch (i.e., static) liquid-liquid extraction (*24*). The performance of an extraction chromatographic material is normally defined in terms of seven parameters: retention, selectivity, efficiency, capacity, stability (physical and chemical), ease of regeneration, and reproducibility/repeatability. Like retention, the selectivity of a given EXC material is governed primarily by the nature of the extractant and the composition of the mobile phase. The selectivity of a given chromatographic system can, however, also be altered by the presence of macrolevels of retained elements, by impurities in the extractant, and by the presence of active sites on the support material.

As with all chromatographic processes, column efficiency is of considerable importance in EXC, since excessive band spreading (i.e., poor column efficiency) can render the separation of two metals difficult or impossible, even if the extractant comprising the stationary phase exhibits very high selectivity for one of the ions. Column efficiency is generally expressed in terms of height equivalent to a theoretical plate (HETP) or simply, plate height (H), and is a complex function of a number of system characteristics. In EXC systems, efficiency is determined primarily by the sum of three factors: flow phenomena, diffusion in the stationary phase, and extraction kinetics (*25*). The contribution to the plate height due to flow phenomena is given by the following equation:

$$H_{flow} = \Sigma \left(\frac{1}{2} \lambda_i d_p + \frac{D_M}{\omega_i d_p^2 v} \right)^{-1} \quad (1)$$

where λ_i and ω_i are parameters related to the bed structure and velocity inequalities of the mobile phase in the interstitial space, d_p is the diameter of the support particle, D_M is the diffusion coefficient of the ion in the mobile phase, and v is the interstitial mobile phase velocity. The contribution to the plate height arising from stationary phase diffusion is given by equation 2:

$$H_{diff} = q \bullet \frac{k'}{(1+k')^2} \bullet \frac{d_l^2 v}{D_S} \quad (2)$$

where q is a configuration factor which depends on the shape of the stationary phase, k' is the capacity factor (the number of free column volumes to the peak maximum), d_l is the depth of the stationary phase, and D_s is the diffusion coefficient of the metal ion-extractant complex in the stationary phase. Finally, the band spreading resulting from extraction kinetics is given by the equation:

$$H_{kin} = \frac{2k'}{(1 + k')^2} \bullet \frac{v}{k_{oa}} \qquad (3)$$

where k_{oa} is the first-order rate constant for the extraction of the solute ion from the stationary phase into the mobile phase. The relative contributions of each of these three factors to column efficiency depends on the specific chemical system, the particle size and porosity of the support, the operating temperature, the extractant loading, and the mobile phase flow velocity. The predominant factor for a particular EXC system can be determined by measuring the plate height as a function of mobile phase flow velocity for materials of various particle sizes and extractant loadings at several temperatures (25). This type of study, although essential in obtaining a full understanding of the parameters governing band broadening in a given EXC system, has only rarely been carried out.

Since extraction chromatography has most frequently been applied in analytical-scale separations, the capacity of EXC materials has typically been regarded as being of secondary importance. In preparative or process-scale applications, however, capacity can become an important consideration. The maximum capacity of an EXC material for a particular metal ion will depend on the total amount of extractant that can be loaded onto the support. This, in turn, is dependent on the nature of the support. Clearly, to maintain a given capacity, the extractant loaded onto the support must be satisfactorily retained; that is, the EXC material must exhibit adequate physical stability. Given the absence of strong interactions between the extractant and the support in a typical extraction chromatographic material, it is not surprising that loss of extractant into the eluent is often cited as the major limitation of EXC. Although loss of extractant can sometimes be minimized by careful control of eluent composition (e.g., pH) or by presaturation of the mobile phase with extractant, these are not always viable options.

Although most compounds employed as stationary phases in EXC have been selected on the basis of prior satisfactory application in liquid-liquid extraction, and are thus expected to have adequate chemical stability, the possibility of chemical degradation of the extractant during use must also be considered. Taken together, it is the physical and chemical stability of a particular EXC system that determines the feasibility of its regeneration, as well as the reproducibility of the results obtained upon repeated use of a given column.

Recent Developments

Extractants. Recent advances in molecular design and synthetic methodology have led to a wide array of new extractants, among them crown ethers (26), cryptands (27), calixarenes (28), and bifunctional organophosphorus reagents capable of stronger and more selective binding of a target ion (particularly in acidic media) than many of the extractants traditionally employed in EXC (e.g., TBP, HDEHP). The availability of such extractants has led to the development of extraction chromatographic sorbents exhibiting strong, and often, quite selective retention of a number of metal ions from aqueous solution. The effect of improved extractants on the performance of EXC materials is especially evident in the development of EXC methods for the separation and preconcentration of the radionuclides ^{241}Am and ^{90}Sr from environmental and biological matrices and nuclear waste samples for subsequent determination.

The development of such methods for ^{90}Sr separation and preconcentration is complicated by two factors. First, a number of procedures for the determination of radionuclides in biological or environmental samples (e.g., urine, feces (*29*), soil (*30*)) involve a preliminary digestion or leaching of the sample with acid, producing a final sample solution often several molar in nitric acid. Thus, any proposed separation method should be effective for highly acidic samples. Strontium sorption from acidic media (i.e., mineral acid solutions) is further complicated by the chemistry of the ion itself. That is, because of its large ionic radius and low charge, Sr(II) has a relatively low charge density. As a result, the energy associated with bond formation between it and the functional groups of many organic extractants is insufficient to completely dehydrate both the cation and the anions that must accompany the strontium into the organic (stationary) phase to maintain electrical neutrality. Thus, sorption of strontium from acidic media requires the transfer of a complex bearing a number of associated water molecules into an organic phase. As would be expected, the net result is generally very poor strontium extraction.

During the 1960's and 1970's, several extraction chromatographic systems were devised for the separation of strontium from other alkaline earth cations or from various other elements (*31-34*). All of these systems, however, suffered from serious shortcomings, among them insufficient selectivity over calcium, inadequate retention of strontium (particularly from solutions containing high concentrations of mineral acids), and the need for cumbersome sample treatment. Recent work by two of the authors (MLD and EPH) has shown that the use of a crown ether-based extraction chromatographic material can provide a means of overcoming all of the limitations associated with previous EXC systems (*35-37*). Crown ethers possess a macrocyclic ring whose dimensions can be tailored to provide a good fit for the cation of interest, thereby yielding selective complexation. Earlier work directed at the development of methods for the removal and recovery of strontium-90 from nuclear waste streams led to a process known as SREX (for S tRontium EXtraction), capable of selectively extracting strontium from solutions containing even high ($\geq$ 3 M) concentrations of nitric acid:

$$Sr^{2+}(aq) + NO_3^-(aq) + \overline{Crown\ Ether} \rightleftharpoons \overline{Sr(NO_3)_2(Crown\ Ether)}$$

The process solvent used, which consists of a 0.2 M solution of bis-4,4'(5')-[*tert*-butylcyclohexano*]*-18-crown-6 (Figure 1a) in 1-octanol, provides a solution to the charge density problem (noted above) by combining the crown ether with a solvent capable of dissolving a substantial amount of water (*38-40*). In such a system, extraction involves the transfer of the strontium/crown ether complex from aqueous solution into a water-like, yet water-immiscible, organic medium. The result is greatly improved extraction efficiency. In subsequent work (*35-37*), it was shown that impregnation of an inert polymeric support (e.g., Amberlite XAD-7) with a 1 M solution of this same crown ether in 1-octanol yields an extraction chromatographic resin that retains all of the favorable properties (e.g., strontium selectivity) of the corresponding liquid-liquid system.

Figure 2 shows the nitric acid dependency of the uptake of alkali and alkaline earth cations (expressed as k', the number of free column volumes to the maximum of the elution band) by the new resin (Sr resin). As can be seen, strontium is well retained at acid concentrations exceeding ~ 2 M. Under these same conditions, barium is the only other alkali or alkaline earth cation that exhibits appreciable sorption. Barium interference is easily dealt with, however, by loading the sample at sufficiently high acidity. The selectivity of the resin over calcium is particularly noteworthy. In fact, calcium concentrations of up to 0.5 M do not appreciably decrease strontium sorption. Also noteworthy is the significant decrease in strontium retention as the nitric acid content of the mobile phase is reduced. Thus, sorbed strontium can be recovered simply by rinsing the resin with dilute acid or water. The ability to recover strontium in water simplifies subsequent processing and analysis of

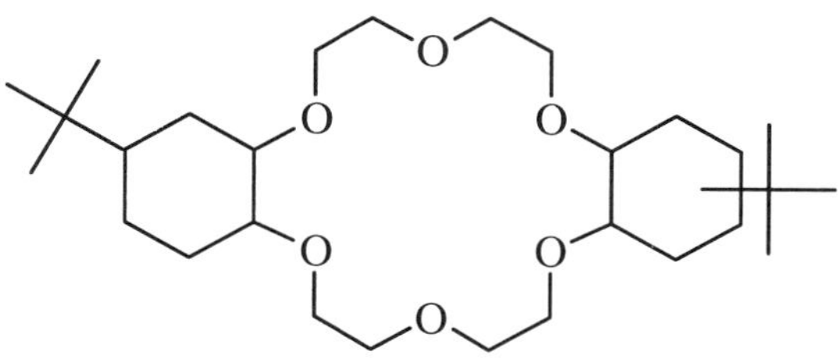

Figure 1. a. Bis-4,4'(5')-[*tert*-butylcyclohexano]-18-crown-6.

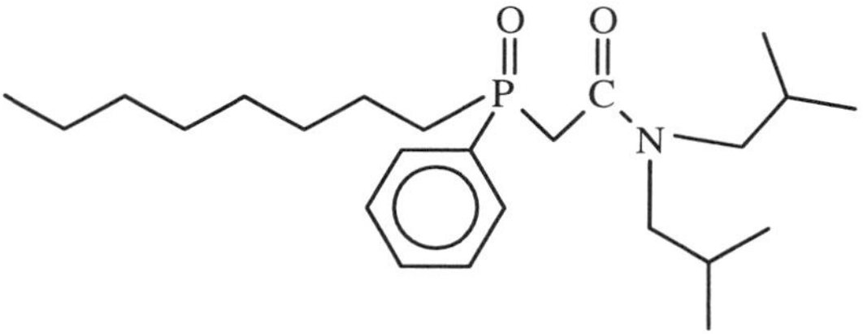

b. Octyl(phenyl)-N, N-diisobutylcarbamoylmethylphosphine oxide.

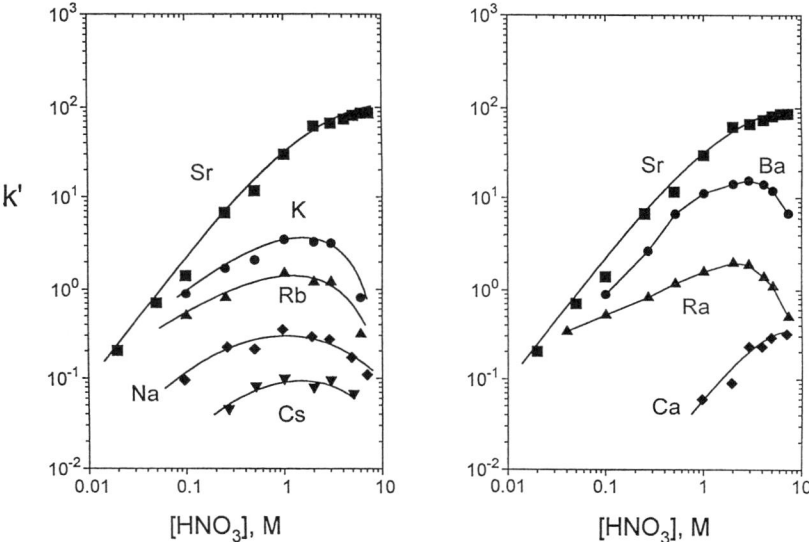

Figure 2. Nitric acid dependency of the retention of alkali and alkaline earth elements on the strontium selective resin. (Adapted from ref. 3.)

the sample and minimizes waste generation. Table I summarizes the elution behavior of 25 elements on the Sr resin. As shown, most of the elements, many of which are common constituents of environmental, biological, or geological samples, are essentially unretained by the resin. Only lead is retained as well as strontium.

As is the case for strontium, the development of EXC methods for ^{241}Am separation and preconcentration is complicated by the high concentrations of mineral acids often present in samples, either as a result of pretreatment steps (e.g., soil leaching) or the nature of the sample itself (e.g., nuclear waste solutions). Moreover, the concentration of most or all other sample constituents generally exceeds that of americium by several orders of magnitude. During the 1960's, several extraction chromatographic procedures for americium separations, particularly from other actinides (*41-43*) or rare earth elements (*44*), were described. The separation of Am from complex matrices such as environmental samples, however, involved cumbersome procedures requiring combinations of solvent extraction and ion-exchange (*45*).

By the mid-1970's, efforts to improve available processes for the removal of actinides from acidic nuclear waste solutions demonstrated that tetra- and hexavalent actinides could be extracted from wastes containing a wide range of acid concentrations by any of a variety of acidic or neutral organophosphorus reagents. In contrast, trivalent species such as ^{241}Am(III) were far less efficiently extracted. Moreover, the available extractants often showed poor selectivity over other waste constituents, required a salting-out agent to enhance extraction efficiency, or suffered from third phase formation problems (*46*). In the mid-1980's, work by Horwitz and co-workers on the fundamental chemistry of neutral bifunctional organophosphorus extractants (*47*) culminated in the development of the TRUEX Process (for TRansUranium EXtraction) (*46,48*). With this process, which employs a mixture of octyl(phenyl)-N, N-diisobutylcarbamoylmethylphosphine oxide (CMPO, Figure 1b) and TBP in a paraffinic hydrocarbon, americium (as well as tetra- and hexavalent actinides) can be extracted from aqueous solutions containing a wide range (1-6 M) of nitric acid concentrations. In addition, the process exhibits excellent selectivity for actinides over many of the other inert and fission product constituents of typical waste solutions. Subsequent work by two of the authors (MLD and EPH) has shown that TRUEX Process chemistry can provide the basis of a versatile extraction chromatographic material capable of the efficient uptake of ^{241}Am and other transuranium elements from acidic nitrate media (*49,50*).

Figure 3 shows the nitric acid dependency of the uptake of Am(III), Pu(IV), Np(IV), Th(IV), Np(V), and U(VI) by an EXC resin consisting of a 0.75 M solution of CMPO in TBP sorbed on Amberlite XAD-7. As shown, Am is reasonably well retained from solutions containing 1-6 M nitric acid. Under these same conditions, few other ions (e.g., lanthanides, Fe(III), Bi(III)) are appreciably sorbed. In fact, of the common major constituents of environmental and geological samples (Na, K, Ca, Fe, and Al), only Fe(III) significantly reduces Am sorption. Reduction of Fe(III) to Fe(II) by addition of ascorbic acid, however, eliminates this effect. Since its introduction, the TRU resin (as it has come to be called) has found application in actinide determinations in a variety of complex matrices (*51-54*). In the course of this work, however, certain limitations of the material have become evident. Americium retention, for example, while better than that seen with any previous EXC materials, is still inadequate for certain applications. That is, satisfactory Am separation generally requires acidification of the sample to ≥ 1 M nitric acid. For a large volume sample, (e.g., natural waters) such acidification is not always feasible, as it generates substantial waste. Although this problem can be partly overcome by the use of a larger column, the additional cost makes this an unattractive option.

In the early 1990's, work at Argonne National Laboratory on the use of water-soluble complexants as masking or stripping agents led to the development of a series of substituted methane diphosphonic acid derivatives capable of forming highly stable complexes in acidic media with a variety of metal ions in the tri-, tetra-, and

Table I. Elution Behavior of Selected Elements on the Sr Specific Chromatographic Resin[a]

Elution Volume[b]	Element
< 5	Li, Na, Rb, Cs, Mg, Ca, Al, Cr(VI), Mn(II), Fe(III), Co(II), Rh(III), Pd(II), Cd, La, Eu,
< 10	K, Ra, Tc(VII), Ag
< 20	Mo(VI)
< 25	Ba
< 30	Hg
> 30	Sr, Pb

[a]Column parameters: resin particle diameter = 50-100 µm; bed volume = 1 mL; bed height = 5.0 cm; 1 free column volume = 0.60 mL.
[b]Elution volume refers to the number of free column volumes required to elute ~ 99% of the indicated element. The eluent was 3 M nitric acid/0.01 M oxalic acid.

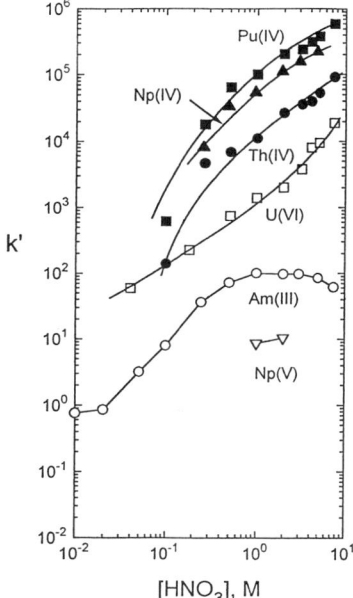

Figure 3. Nitric acid dependency of the retention of actinides on the TRU resin. (Adapted from ref. 3.)

hexavalent oxidation states (55). It has been shown that by replacing two of the four hydrogen ions with an alkyl group, lipophilic dialkyl substituted diphosphonic acids can be prepared which can serve as powerful actinide extractants from highly acidic solutions (56). Work by two of the authors (MLD and EPH) has shown that if an inert polymeric support is impregnated with one such compound, di-(2-ethylhexyl)methanediphosphonic acid (Figure 4), abbreviated $H_2DEH[MDP]$, an extraction chromatographic resin capable of extraordinarily strong retention of actinides, particularly americium, is obtained (57). Figure 5 shows the acid dependency of k' for various actinide species on an EXC resin (now known as Dipex) consisting of 40% (w/w) of the undiluted extractant supported on beads of Amberchrom CG-71ms acrylic ester resin. As shown, the retention of all of the actinides is extraordinarily high, a result consistent with prior studies in the liquid-liquid mode (56). Especially noteworthy is the strong retention of Am(III): k'_{Am} exceeds that obtained using the TRU resin under the same conditions by more than four orders of magnitude. This makes possible the preconcentration of Am from large volume samples using only a small EXC column. Figure 6, for example, shows the effect of the sample volume to bed mass ratio on the uptake of Am by the Dipex resin. As can be seen, Am sorption is quantitative until a ratio of 4000:1 is reached, meaning that a 250 mg column would be adequate for the treatment of 1 L of sample.

Supports. The extraction chromatography of metal ions, unlike the partition chromatography of organic substances, only infrequently involves the separation of substances with very similar properties. In most instances, an extraction chromatographic column is employed as a "sorption filter" on which the species of interest is retained while others are not (58). Extraction chromatographic separations, therefore, often require only that the sorption of the species of interest be much stronger than that of other sample constituents. Not surprisingly then, the majority of work to improve EXC materials has focused on improved extractants, because it is the extractant that normally governs the selectivity of the system. Despite this, there has been growing awareness of the potential importance of the physical and chemical properties of the support in determining certain aspects of the effectiveness of an EXC system.

A variety of different materials have been employed as EXC supports, among them cellulose, silica gel, diatomaceous earth, alumina, and a variety of polymers (e.g., polyethylene, polytetrafluoroethylene) (58). Several of these (e.g., silica) are generally considered unsuitable for use without preliminary treatment to render their surface hydrophobic and to reduce or eliminate functionalities capable of ion-exchange. In the early 1970's, Warshawsky (59) noted the advantages offered by a then new series of macroporous, high surface area styrene-divinylbenzene- and acrylic ester-based polymers (marketed by Rohm and Haas under the designation "XAD") as EXC supports. These supports display good wettability by most extractants, do not swell appreciably as the solution composition changes, exhibit good mechanical stability, and are relatively inexpensive. As a result, most recent research activity in extraction chromatography, including the little work directly concerning the influence of support properties on EXC performance, has involved these and other related polymeric materials impregnated with various extractants. (Note that although extractant-impregnated polymers have been referred to as "solvent impregnated resins", there does not appear to be any compelling reason why these materials should be distinguished from other extraction chromatographic materials.) Parrish (60) compared the sorption of copper ions by a series of XAD resins impregnated with Kelex 100, a commercially available chelating extractant based on a high molecular weight derivative of 8-hydroxyquinoline, to that observed for similarly impregnated silanized and unsilanized silica gel and diatomaceous earth. The rate of copper uptake on the XAD resins was somewhat slower, although differences in particle sizes and loading levels among the resins made strict comparisons difficult. The XAD resins, however, offered the advantage of not requiring silanization prior to use. Parrish noted that the rate of copper sorption did

243

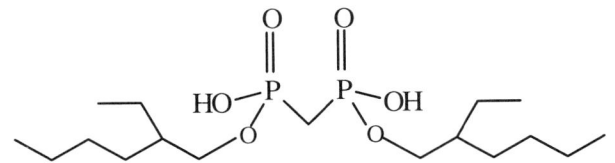

Figure 4. Di-(2-ethylhexyl)methanediphosphonic acid (H$_2$DEH[MDP]).

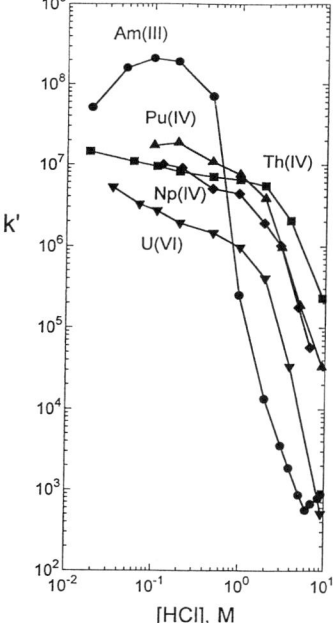

Figure 5. Hydrochloric acid dependency of actinide retention on the Dipex™ resin. (Adapted from ref. 57.)

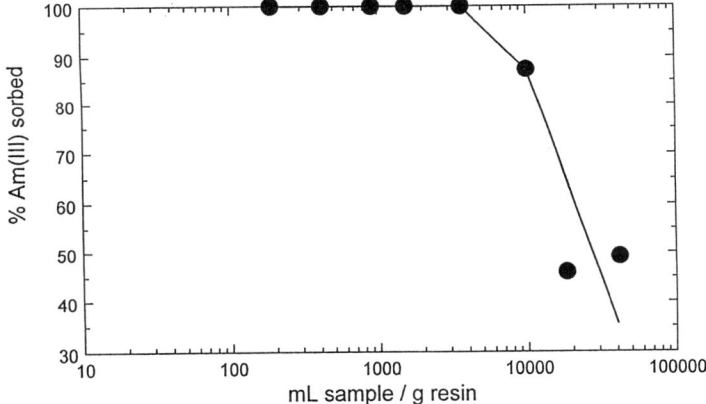

Figure 6. Effect of sample volume to resin mass ratio on the uptake of Am(III) by Dipex™ resin from Des Plaines River water. (Adapted from ref. 57.)

not correlate with either the specific surface area or the average pore diameter of the resins. Instead, this rate was found to increase with increasing water regain by the resin. These results appear to contradict those of a previous study of the effect of support geometry on the uptake of europium by HDEHP-loaded silica, which showed that the europium distribution coefficients, the total capacity of the resin for europium, and the kinetics of europium uptake were functions of median pore diameter and pore size distribution (*61,62*). Recent work by Jerabek et al. (*63*) suggests that the uptake of certain species (e.g., methanol) is indeed affected by the support morphology (e.g., pore size distribution) for XAD resins, but the relationship of these observations to the performance of the extractant-loaded resins as metal ion sorbents has yet to be determined. Additional work in this area is clearly warranted.

As already noted, an extraction chromatographic support is typically chosen to function as an inert "reservoir" which can be filled to the desired capacity with an extractant. A few studies, however, have described the use of an "active support", defined here as a substrate which interacts with the extractant through other than the weak adsorptive forces typical of EXC materials or one that actually participates in the metal ion uptake process. Several authors, for example, have examined the metal ion sorption properties of EXC materials prepared by impregnating an anion-exchange resin with an extractant containing an anionic functional group (*64-71*). Tanaka et al. (*65,67-69*) evaluated a series of resins in which the sulfonic acid derivative of dithizone (*65*), thiopyrine (azothiopyrine sulfonic acid, ATPS) (*67*), or tetraphenylporphine (*68*) was sorbed on Amberlite IRA-400, an anion-exchange resin. Similarly, Lee et al. (*66*) evaluated a resin consisting of Dowex 1-X8 impregnated with a sulfonated 8-quinolinol derivative. Subsequent work by Sarzanini et al. (*70*) employed pyrocatechol violet, a chelating agent whose structure contains a sulfonic acid functional group not involved in metal ion coordination, sorbed on AG MP-1, a macroporous anion-exchange resin. In each of these studies, the extractants were found to be strongly bound by the ion-exchange resin, most likely by a combination of simple physical sorption and ion-exchange. For ATPS on IRA-400, this sorption was sufficiently strong that the extractant was retained even in the presence of 0.5-1 M HCl or 1 M NaOH (using a volume:mass ratio of 100:1). The sorption of the metal ion(s) of interest, mercury (*64,65,67-69*), aluminum (*70*), copper (*64,66,68*), and lead (*66*) was generally found to be satisfactory, although there were indications that adsorption reduced the binding ability of certain ligands (*70*). In addition, the metal ion uptake capacity of certain of the resins was less than that expected from the amount of chelating agent loaded on the support (*66*). Despite these problems, the use of a support capable of anion-exchange, in conjunction with an extractant bearing an anionic functionality, may represent a step toward significant reduction of extractant loss from EXC materials.

Moyer et al. (*72*) have examined the use of cation-exchange resins as supports in extraction chromatography. In their work, a series of polystyrene-divinylbenzene-based cation-exchange resins (among them the commercially available Dowex 50W-X8) were impregnated with several weight percent tetrathia-14-crown-4 and the sorption of copper(II) by the resin from sulfuric acid measured. While neither unfunctionalized polystyrene-divinylbenzene resin nor the same resin loaded with the macrocycle extracted any detectable copper, impregnation of the cation-exchanger produced a 10-100-fold enhancement in the observed copper distribution ratio vs. the cation-exchanger alone. This enhancement was attributed to a synergistic effect involving coordination of the copper by the mobile macrocycle and cation-exchange by the polymer-bound sulfonic acid functional groups. Although such synergistic effects are common in liquid-liquid extraction systems involving mixtures of crown ethers and liquid cation-exchangers (*73*), Moyer's results represent the first demonstration of synergism in an EXC material involving a functionalized support. Such supported synergistic systems appear to offer a wealth of opportunities for the development of new EXC materials exhibiting enhanced metal ion uptake and selectivity.

Applications

The earliest extraction chromatographic materials employed as the stationary phase compounds originally developed for use as liquid-liquid extractants in the reprocessing of spent nuclear fuel (7). Not surprisingly then, until relatively recently, the major portion of the published applications of EXC involved nuclear/radioactive materials, particularly their small-scale separation and purification or analytical determination. During the last decade or so, there has been growing interest in the nonnuclear applications of extraction chromatography and in the possible use of the technique for industrial-scale metal ion separations. Of particular interest has been the potential of EXC in the recovery and/or purification of metals which are valuable (e.g., gold) or of some strategic importance (e.g., rare earths, platinum group metals) from both dilute solutions (e.g., dump leaching solutions) and the more concentrated solutions arising from hydrometallurgical processing. Also of interest has been the possible application of EXC materials in the removal of hazardous or toxic metals from industrial effluents (6). Table II summarizes much of this research activity in process-scale extraction chromatography. As can be seen, a variety of both supports and extractants have been investigated in an effort to develop large-scale EXC systems. Despite some success, these systems have typically been plagued by one major problem: the leakage of extractant into the column effluent and the concomitant changes in column behavior. In an examination of TBP-Levextrel resin for the removal of uranium from nuclear reprocessing solutions, for example, Krochbel and Meyer (8) observed a loss of 0.163 g of TBP per liter of solution processed, an unacceptably high loss. Similarly, in an evaluation of zinc removal from cobalt solutions by HDEHP on XAD-7 or Levextrel OC1026, Warshawsky (74) noted that the low concentration of HDEHP leaking into the column effluent created difficulty in subsequent electrowinning. A study of copper removal from dilute solution using LIX-64N-impregnated XAD-4 yielded similar results. Reagent losses, although low ($\leq$ 30 mg/L processed), were still deemed unacceptable. Given the weak interaction between extractant and support characteristic of EXC systems, extractant loss in these systems is not surprising. Until extractant losses are reduced, however, EXC is unlikely to achieve its full potential as a process-scale metal ion separation technique.

Conclusions

Extraction chromatography provides a simple and effective means of performing a wide variety of metal ion separations. Recent advances in extractant design, in particular the development of extractants capable of metal ion recognition or of strong complex formation even in acidic media, have substantially improved the utility of the method. Advances in support design, most notably the introduction of functionalized ("active") supports to enhance metal ion retention or to increase column stability, promise to yield further improvements.

Any efforts to improve extraction chromatographic materials must recognize that there are a number of measures by which EXC performance is assessed, some of which cannot be simultaneously improved. Gradual loss of column capacity due to extractant loss to the mobile phase is presently the sole remaining obstacle to large-scale application of EXC materials. Clearly then, stability is *the* aspect of EXC performance which now most warrants additional research effort.

Acknowledgments

This paper is based in part on joint publications with our colleagues in the Chemistry, Chemical Technology, and Environment, Safety, and Health Divisions at Argonne National Laboratory. Their efforts are gratefully acknowledged. This work was performed under the auspices of the Office of Basic Energy Sciences, Division of Chemical Sciences, U.S. Department of Energy, under contract number W-31-109-ENG-38.

Table II. Extraction Chromatographic Systems Evaluated for Possible Process-Scale Application

Target Ion	Matrix	Extractant/Support	Comments	Reference
U	Leachate of slimes remaining from Au extraction	Alamine-336 on silicone-treated fine clay	Extractant loss of 7.1 mg/kg resin/L processed	(5)
Au	Dilute cyanide solutions	Trialkylammonium compounds in decyl alcohol/kerosene on granulated polyethylene	8 mg/L bed capacity	(5)
U	Reprocessing solutions	TBP-Levextrel	Extractant loss of 0.163 g TBP/L processed	(8, 79)
Zn	Co electrowinning solution	HDEHP on XAD-4	Reagent leakage interferes with electrowinning	(74)
Cu	Dilute solution containing divalent and trivalent cations	LIX-64N on XAD-4	Extractant loss of ca. 30 mg/L processed	(75)
Am	Waste solutions from nuclear fuel reprocessing	di-n-hexyl-octoxymethyl phosphine oxide		(76)
Sc	Sc-bearing waste solutions	HDEHP on polystyrene		(77)
Am, Pu, U	Acidic nuclear waste solutions (Hanford)	CMPO-TBP, CMPO alone, and diamyl amyl phosphonate on Amberlite XAD-7 or XAD-16	Little change in CMPO-TBP column performance noted after 36 cycles	(78)

Literature Cited

(1) Wai, C. M. In *Preconcentration Techniques for Trace Elements*; Alfassi, J. B., Wai, C. M., Eds.; CRC Press: Boca Raton, FL, 1992; p 101.
(2) Walton, H. F.; Rocklin, R. D. *Ion Exchange in Analytical Chemistry*; CRC Press: Boca Raton, FL, 1990.
(3) Dietz, M. L.; Horwitz, E. P. *LC.GC* **1993**, *11*, 424.
(4) Siekierski, S.; Kotlinska, B. *At. Energiya* **1959**, *7*, 160.
(5) Warshawsky, A. In *Ion Exchange and Solvent Extraction-A Series of Advances*; Marinsky, J. A., Marcus, Y., Eds.; Marcel Dekker: New York, 1981; Vol. 8, p 229.
(6) Cortina, J. L.; Warshawsky, A. In *Ion Exchange and Solvent Extraction*; Marinsky, J. A., Marcus, Y., Eds.; Marcel Dekker: New York, 1997; Vol. 13, p 195.
(7) *Extraction Chromatography*; Braun, T.; Ghersini, G., Eds.; Elsevier: New York, 1975.
(8) Krochbel, R.; Meyer, A. In *Proceedings of the International Solvent Extraction Conference-1974*; Society of Chemical Industry: 1974; p 2095.
(9) Kauczor, H. U.; Meyer, A. *Hydrometallurgy* **1978**, *3*, 65.
(10) Poinescu, I.; Popescu, V.; Carpov, A. *Angew. Makromol. Chem.* **1985**, *135*, 21.
(11) Cortina, J. L.; Miralles, N.; Aguilar, M.; Sastre, A. M. *Solvent Extr. Ion Exch.* **1994**, *12*, 349.
(12) Cortina, J. L.; Miralles, N.; Aguilar, M.; Sastre, A. M. *Solvent Extr. Ion Exch.* **1994**, *12*, 371.
(13) Cortina, J. L.; Miralles, N.; Aguilar, M.; Warshawsky, A. *React. Funct. Polymers* **1995**, *27*, 61.
(14) Cote, G.; Laupretre, F.; Chessagnard, C. *React. Polymers* **1987**, *5*, 141.
(15) Bokobza, L.; Cote, G. *Polyhedron* **1985**, *4*, 1499.
(16) Kimura, T. *J. Radioanal. Nucl. Chem.* **1990**, *141*, 295.
(17) Miralles, N.; Sastre, A. M.; Aguilar, M.; Cox, M. *Solvent Extr. Ion Exch.* **1992**, *10*, 51.
(18) Sastre, A. M.; Muhammed, M. *Hydrometallurgy* **1984**, *12*, 177.
(19) Sastre, A. M.; Miralles, N.; Aguilar, M. *Chem. Scripta* **1984**, *24*, 44.
(20) Casas, I.; Miralles, N.; Sastre, A. M.; Aguilar, M. *Polyhedron* **1989**, *8*, 2535.
(21) Casas, I.; Miralles, N.; Sastre, A. M.; Aguilar, M. *Polyhedron* **1986**, *5*, 2039.
(22) Strikovsky, A. G.; Jerabek, K.; Cortina, J. L.; Sastre, A. M.; Warshawsky, A. *React. Funct. Polymers* **1996**, *28*, 149.
(23) Azaka, I. In *Extraction Chromatography*; Braun, T., Ghersini, G., Eds.; Elsevier: New York, 1975; p 17.
(24) Ghersini, G. In *Extraction Chromatography*; Braun, T., Ghersini, G., Eds.; Elsevier: New York, 1975; p 68.
(25) Horwitz, E. P.; Bloomquist, C. A. A. *J. Inorg. Nucl. Chem.* **1972**, *34*, 3851.
(26) Hiraoka, M. *Crown Compounds: Their Characteristics and Applications*; Elsevier: New York, 1982.
(27) Gokel, G. *Crown Ethers and Cryptands*; Royal Society of Chemistry: Cambridge, England, 1991.
(28) Gutsche, C. D. *Calixarenes*; Royal Society of Chemistry: Cambridge, England, 1989.
(29) Veselsky, J. C. *Mikrochim. Acta* **1978**, *1*, 79.
(30) Juznick, K.; Korun, M. *J. Radioanal. Nucl. Chem. Lett.* **1989**, *137*, 235.
(31) Akaza, I. *Bull. Chem. Soc. Jpn.* **1966**, *39*, 980.
(32) Lieser, K. H.; Bernhard, H. *Z. Anal. Chim.* **1966**, *219*, 401.
(33) Akaza, I.; Tajima, T.; Keba, T. *Bull. Chem. Soc. Jpn.* **1973**, *46*, 1199.
(34) Lada, W. A.; Smulek, U. *Radiochem. Radioanal. Lett.* **1978**, *34*, 41.
(35) Horwitz, E. P.; Dietz, M. L.; Fisher, D. E. *Anal. Chem.* **1991**, *63*, 522.

(36) Horwitz, E. P.; Chiarizia, R.; Dietz, M. L. *Solvent Extr. Ion Exch.* **1992**, *10*, 313.
(37) Chiarizia, R.; Horwitz, E. P.; Dietz, M. L. *Solvent Extr. Ion Exch.* **1992**, *10*, 337.
(38) Horwitz, E. P.; Dietz, M. L.; Fisher, D. E. *Solvent Extr. Ion Exch.* **1990**, *8*, 199.
(39) Horwitz, E. P.; Dietz, M. L.; Fisher, D. E. *Solvent Extr. Ion Exch.* **1990**, *8*, 557.
(40) Horwitz, E. P.; Dietz, M. L.; Fisher, D. E. *Solvent Extr. Ion Exch.* **1991**, *9*, 1.
(41) Moore, F. L. *Anal. Chem.* **1968**, *40*, 2130.
(42) Horwitz, E. P.; Bloomquist, C. A. A.; Orlandini, K. A.; Henderson, D. J. *Radiochim. Acta* **1967**, *8*, 127.
(43) Horwitz, E. P.; Bloomquist, C. A. A.; Henderson, D. J.; Nelson, D. E. *J. Inorg. Nucl. Chem.* **1969**, *31*, 3255.
(44) Barbano, P. G.; Rigali, L. *J. Chromatogr.* **1967**, *29*, 309.
(45) Ham, G. J. *Sci. Tot. Environ.* **1995**, *173-174*, 19.
(46) Horwitz, E. P.; Kalina, D. G.; Diamond, H.; Vandegrift, G. F.; Schulz, W. W. *Solvent Extr. Ion Exch.* **1985**, *3*, 75.
(47) Horwitz, E. P.; Diamond, H.; Kalina, D. G. In *Plutonium Chemistry*; Carnall, W. T., Choppin, G. R., Eds.; American Chemical Society: Washington, DC, 1983; p 433.
(48) Schulz, W. W.; Horwitz, E. P. *Sep. Sci. Technol.* **1988**, *23*, 1191.
(49) Horwitz, E. P.; Dietz, M. L.; Nelson, D. M.; LaRosa, J. J.; Fairman, W. D. *Anal. Chim. Acta* **1990**, *238*, 263.
(50) Horwitz, E. P.; Chiarizia, R.; Dietz, M. L.; Diamond, H.; Nelson, D. M. *Anal. Chim. Acta* **1993**, *281*, 361.
(51) Kaye, J. H.; Strebin, R. S.; Ou, R. D. *J. Radioanal. Nucl. Chem.* **1995**, *194*, 191.
(52) Crain, J. S.; Smith, L. L.; Yaeger, J. S.; Alvarado, J. A. *J. Radioanal. Nucl. Chem.* **1995**, *194*, 133.
(53) Smith, L. L.; Crain, J. S.; Yaeger, J. S.; Horwitz, E. P.; Diamond, H.; Chiarizia, R. *J. Radioanal. Nucl. Chem.* **1995**, *194*, 151.
(54) Goldstein, S. J.; Hensley, C. A.; Armenta, C. A.; Peters, R. J. *Anal. Chem.* **1997**, *69*, 809.
(55) Horwitz, E. P.; Diamond, H.; Gatrone, R. C.; Nash, K. L.; Rickert, P. G. In *Solvent Extraction 1990*; Sekine, T., Ed.; Elsevier: New York, 1992; p 357.
(56) Chiarizia, R.; Horwitz, E. P.; Rickert, P. G.; Herlinger, A. W. *Solvent Extr. Ion Exch.* **1996**, *14*, 773.
(57) Horwitz, E. P.; Chiarizia, R.; Dietz, M. L. *React. Funct. Polymers* **1997**, *33*, 25.
(58) Katykhin, G. S. In *Extraction Chromatography*; Braun, T.; Ghersini, G., Eds.; Elsevier: New York, 1975; p 134.
(59) Warshawsky, A. *Inst. Min. Metall. Trans., Sect. C* **1974**, *83*, C101.
(60) Parrish, J. R. *Anal. Chem.* **1977**, *49*, 1189.
(61) Nolte, R. F.; Specht, S.; Born, H. J. *J. Chromatogr.* **1975**, *110*, 239.
(62) Specht, S.; Nolte, R. F.; Born, H. J. *J. Chromatogr.* **1975**, *110*, 253.
(63) Jerabek, K.; Hankova, L.; Strikovsky, A. G.; Warshawsky, A. *React. Funct. Polymers* **1996**, *28*, 201.
(64) Akaiwa, H.; Kawamoto, H.; Nakata, N.; Ozeki, Y. *Chem. Lett.* **1975**, *10*, 1049.
(65) Tanaka, H.; Chikuma, M.; Harada, A.; Ueda, T.; Yube, S. *Talanta* **1976**, *23*, 489.
(66) Lee, K. S.; Lee, W.; Lee, D. W. *Anal. Chem.* **1978**, *50*, 255.
(67) Chikuma, M.; Nakayama, M.; Tanaka, T.; Tanaka, H. *Talanta* **1979**, *26*, 911.
(68) Chikuma, M.; Nakayama, M.; Itoh, T.; Tanaka, H.; Itoh, K. *Talanta* **1980**, *27*, 807.
(69) Nakayama, M.; Chikuma, M.; Tanaka, H.; Tanaka, T. *Talanta* **1982**, *29*, 503.

(70) Sarzanini, C.; Mentasti, E.; Porta, V.; Gennaro, M. C. *Anal. Chem.* **1987**, *59*, 484.
(71) Warshawsky, A.; Strikovsky, A. G.; Jerabek, K.; Cortina, J. L. *Solvent Extr. Ion Exch.* **1997**, *15*, 259.
(72) Moyer, B. A.; Case, G. N.; Alexandratos, S. D.; Kriger, A. A. *Anal. Chem.* **1993**, *65*, 3389.
(73) McDowell, W. J. *Sep. Sci. Technol.* **1988**, *23*, 1251.
(74) Warshawsky, A.; Kalir, R.; Berkovitz, H. *Inst. Min. Metall. Trans., Sect. C* **1979**, *88*, 31.
(75) Warshawsky, A.; Berkovitz, H. *Inst. Min. Metall. Trans., Sect. C* **1979**, *88*, 36.
(76) Madic, C.; Kertesz, C.; Sontag, R.; Koehly, G. *Sep. Sci. Technol.* **1980**, *15*, 745.
(77) Kim, T. K.; McClintic, R. P. "Process for Recovering Scandium from Waste Material"; U.S. Patent 4,751,061; 1988.
(78) Barney, G. S.; Cowan, R. G. In *Chemical Pretreatment of Nuclear Waste for Disposal*; Schulz, W. W., Horwitz, E. P., Eds.; Plenum Press: New York, 1994; p 51.
(79) Eschrich, H.; Ochsenfeld, W. *Sep. Sci. Technol.* **1980**, 15, 697.

Chapter 17

Metal-Ion Separations Using SuperLig or AnaLig Materials Encased in Empore Cartridges and Disks

Garold L. Goken, Ronald L. Bruening, Krzysztof E. Krakowiak, and Reed M Izatt[1]

IBC Advanced Technologies, Inc., P. O. Box 98, 856 East Utah Valley Drive, American Fork, UT 84003

By combining two state-of-the-art technologies, a promising membrane approach has been found to have broad utility in the separation of metal ions from solution. Molecular recognition technology consists of selective separation chemistries in the form of macrocycles or chelating ligands and solid phase extraction particles. The products resulting from the combination are referred to as SuperLig or Analig materials. Loaded Empore membranes are very densely packed with small high surface area SuperLig or Analig particles, yet achieve exceptional flow rates. IBC's selective chemistry ignores most non-target species, therefore membranes containing these materials achieve very high separation efficiencies. The membranes can be fabricated into filter-shaped disks or cartridges ranging from analytical-scale to process-size components. These combined characteristics have been incorporated into analytical, processing, and final polishing products that have had a major impact on current processes. This paper will present the application of these technologies to selective analytical separations involving mercury, lead, strontium, and radium.

---

[1]Current address: Department of Chemistry and Biochemistry, Brigham Young University, Provo, UT 84602.

For the past decade, IBC Advanced Technologies, Inc. (IBC) has been a provider of selective separation materials *(1-9)* in the forms of macrocycles and solid phase extraction (SPE) particles. Key to the chemistry is the development of macrocyclic and chelating ligands that selectively recognize the shape and electrical charge of a target ion, then preferentially complex with this ion. These ligands show little affinity for non-target species, even those with similar electric charges, ionic radii, molecular shapes, or other target-like attributes. These ligands are highly selective for a single ionic species, even when other similar species are present at concentrations many times (i.e., 10^4-10^6) greater than those of the target ion.

Sophisticated reactions to covalently bond ligands to solid phase supports such as silica gel *(2)* and polymeric resin particles have been developed by IBC. In these particles, the ligand and solid phase support are connected by a spacer. This spacer is attached by stable covalent bonds to the ligand and the solid phase support. The spacer has the important function of allowing the ligand to be immersed in the aqueous phase *(6)*. The solid phase in the SPE system has an important role in the separation procedure. This phase must be compatible with the solution from which the separation is to be made and should not react with or dissolve in the medium used. A variety of supports have been used allowing effective separations to be made in highly acidic, highly basic, and HF-containing solutions. The support used in the studies reported here is silica gel. This support material is one of the most effective for use in SPE separations. It can be obtained in pure form, is hydrophilic making it particularly effective for use in aqueous systems, has a high concentration of active sites for chemical bonding, and is readily converted to an SPE material by synthesis *(6)*.

Families of highly selective SPE materials for use in ion separation applications result from this procedure. Unlike ion-exchange resins, which attract almost all similarly charged species present, SPE systems based on molecular recognition technology (MRT) are highly discriminating. High selectivity is an essential criterion for many applications such as concentrating and purifying analytical samples, removing residual process catalysts or toxic compounds, removing interfering species present in other separation or analytical processes, and recovering precious metals. IBC's SPE particles are called AnaLig for analytical applications and SuperLig for the process industry.

Empore membranes were used initially in a process for preparing water samples for analysis *(10)*. These thin membranes (0.5 mm) are composed of reactive particles enmeshed in an inert, fibrous matrix without the use of binders or adhesives. The particles can compose up to 95% of the membrane's weight and are very small compared to particles used in typical SPE columns. Even though the particles are small and closely packed, flow rates can be 10 to 100 times faster than those in columns, yet equal extraction efficiencies can be achieved. Channeling or wall effects typical of columns are absent.

The combination, in the early 1990s, of IBC's selective separation particles and 3M's membrane technology produced a novel and effective separation procedure. The resulting Empore disk may be considered to be a short column that is much more compact than a normal column. The enclosed SuperLig or AnaLig material has a mesh size of less than 10μm (compared to 100 μm in the column mode) allowing a very large concentration of active sites to be present. These effects result in a system capable of much more rapid separations than are possible with a fixed bed column or with ion-exchange *(6-8)*. For example, flow rates in the Empore system can be up to 400 times those of a column system and 4000 times those of ion-exchange. The selectivity and removal ability of the ligands employed in the SPE systems are maintained at these flow rates making separations of trace amounts of metal ions from large volumes of solution feasible. This highly efficient use of particle capacity has been employed for a number of unique process and analytical applications *(4-9)*. One of the most interesting applications is a series of Rapid Analysis Products (RAP) which could bring about a fundamental change in the way sample preparation is performed. Four of these RAP products have been commercialized, Pb and Hg test kits (Hach, Loveland, Colorado) and Sr and Ra Rad Disks (3M, St. Paul, Minnesota).

RAP Membranes

The ability of these SPE materials to distinguish and separate target ions in one easy step eliminates the numerous chemical manipulations one would normally encounter in a purification process. Since sample preparation is often a purification process, streamlined procedures to improve the analytical process appeared. The name Rapid Analysis Products became associated with numerous developmental products resulting from the combination of selective particle and membrane technologies.

If the selectivity for a target species is high, purification and isolation of this species can be accomplished in a one-step process. The need for other chemical manipulations to separate the target ion, such as extraction, precipitation, filtration, or crystallization are not needed. Moreover, even though the selective ligands exhibit several orders of magnitude more affinity for a specific species, selective elution and recovery chemistries have also been developed. In most cases, the target species can be removed and retrieved in quantitative amounts by applying a variety of conditions such as pH change, water wash, or treatment with a complexing agent.

A prime example of the RAP target is the published EPA Method 905 *(11)* for radioactive strontium determination where over fifty steps are listed in the process. Traditional determinations of ^{90}Sr in aqueous matrices have involved the application of tedious, time-consuming, and labor-intensive separation schemes in order to analyze radiostrontium via its radioactive progeny, ^{90}Y *(12)*. The published Sr Rad Disk method requires only six steps and approximately twenty minutes to complete the analysis *(12,13)*. There is a similar reduction in steps for analysis using the Ra Rad Disk *(14)*. These improvements amount to considerable savings in costs and time.

Other RAP procedures are also simpler and can be accomplished in fewer steps. For example, the Hg and Pb test kits eliminate the need for advanced instrumentation, such as graphite furnace operation for lead and cold vapor atomic absorption for mercury. These test kits for Hg and Pb extraction also eliminate the need for several chemicals that are associated with health risks. Reducing the time spent in preparing samples increases the time available for method development and analysis.

An important objective for RAP is to develop a broad-based sample preparation technique that is applicable to diverse analytes including inorganic cations and anions, and even select organic molecules. IBC and 3M provide most of the materials science and technologies to develop RAP and the products introduced for Pb^{2+} and Hg^{2+} resulted from joint efforts of IBC and 3M with Hach scientists. The Rad Disk methods and results for Sr^{2+} and Ra^{2+} came from joint efforts with scientists at Argonne National Laboratory and DOE's Environmental Measurements Laboratory.

Applications

Hach Test Kits. Hach has developed test kits (15-17) for mercury and lead that fit under the RAP objectives for time, labor, and material savings. The membranes incorporate AnaLig PbO1 for lead and AnaLig HgO1 for mercury. It is necessary to convert all mercury present to Hg^{2+} prior to analysis. The 25 mm diameter membrane disk is sealed in a syringe barrel type lead or mercury extractor that can be fitted to a hand-operated vacuum pump. A measured acidified or digested sample is drawn through the appropriate extractor. The absorbed target is washed with deionized water and then eluted. A color producing porphyrin indicator is added to the eluate and the pH adjusted to above pH 13. Color measurement is done with the Hach DR/4000, or equivalent, UV-visible spectrophotometer. Color absorption is measured at 412 nm for Hg and at 477 nm for Pb. A complexing reagent is added to the eluate to produce a zero blank and color absorption is measured again. The difference in absorbances represents the amount of the target ion present.

Hach has found that the capture efficiency for lead as Pb^{2+} is 90 to 95% from several standard solutions. Estimated detection limits at the 99% confidence level were 10 $\mu g \cdot L^{-1}$ for a 10 mL sample. Precision for a 250 $\mu g \cdot L^{-1}$ standard at the 95% confidence interval was ± 5 $\mu g \cdot L^{-1}$ Pb^{2+}.

The high selectivity of the particles is evident even in the presence of numerous potential interferences. Table I summarizes the results of an interference study for the lead test kit.

The results in Table I and in the preceeding discussion demonstrate that the system is capable of accurate and precise determination of Pb^{2+} even in the presence of high concentrations of a wide range of possible interferring ions. The ions with the lowest tolerable interference levels are those where even traces (low $\mu g \cdot L^{-1}$ or less) present in the porphyrin-based spectrophotmetric analyses (following Pb^{2+} concentration and separation) can interfere with the final spectral analysis. Ba^{2+} and Sr^{2+} have chemistry similar to that of Pb^{2+}. However, the interference levels for both Ba^{2+} and Sr^{2+} are high, thereby providing evidence for the high selectivity of the AnaLig material. Mercury, as Hg^{2+}, was found to interfere with the test at a concentration of 50 $\mu g \cdot L^{-1}$ Hg^{2+}. However, sodium thiosulfate can be used to mask mercury interference up to 10 $mg \cdot L^{-1}$ Hg^{2+} in the final eluate solution. The common ubiquitous environmental ions, Na^+, Mg^{2+}, Ca^{2+}, and Fe^{3+}, are seen to offer little interference.

Table I. Interference Levels[a] for Various Ions in a Lead Test Kit[b] (16)

Species	Interference Levels, (mg·L^{-1})	Species	Interference Levels, (mg·L^{-1})
Ag$^+$	50	Mn^{2+}	500
Al^{3+}	100	Mo(VI)	100
Ba^{2+}	100	Na$^+$	20000
Br$^-$	1000	NaCl	50000
Ca^{2+}	1000	Na$_2$S$_2$O$_3$	20000
Cd^{2+}	10	NH$_4^+$	500
Cl$^-$	3000	Ni^{2+}	300
Co^{2+}	50	NO$_2^-$	500
Cr(III/VI)	10	NO$_3^-$	500
Cu^{2+}	1000	PO$_4^{3-}$	100
F$^-$	10	Sn(IV)	40
Fe^{2+}	1000	SO$_4^{2-}$	250
Fe^{3+}	100	Sr^{2+}	250
I$^-$	75	V(V)	10
K$^+$	750	W(VI)	100
Li$^+$	75	Zn^{2+}	250
Mg^{2+}	500		

[a]Defined as the levels at which the Pb recoveries are 90% or greater.
[b]Concentrations of Pb^{2+} were either 50 µg·L^{-1} or 200 µg·L^{-1}. Recoveries >90% for Pb^{2+} were found for each solution.

Strontium Rad Disks. The procedure for analysis with the Sr Rad Disk (12,13), which contains AnaLig SrO1, involves passing a sample in 2 M HNO$_3$ through a 47 mm disk positioned on a vacuum filter apparatus at a rate of 50 mL·min^{-1}. The radioactive strontium may then be quantified using a variety of techniques. For proportional counting, the disk is dried with 2 mL of an acetone wash and placed in a planchet. Counting may then be done with a low background gas proportional counter. For liquid scintillation counting, the Rad Disk is placed in a scintillation vial with liquid scintillation cocktail such as Ultima Gold (Canberra, Meridan, CT).

Some investigators prefer methods that use an added tracer to monitor the separation of the strontium. Considerable work has been done to validate results using tracers with the Sr Rad Disk. For example, ^{85}Sr tracer was added to a sample being tested for ^{90}Sr, prior to the sample being passed through a Sr Rad Disk. The disk was counted using a liquid scintillation counter capable of evaluating dual labeled samples. A 0-12 KeV window was used for ^{85}Sr and a 12-500 KeV window for ^{90}Sr measurement. Corrected tracer results showed >95% recoveries.

An alternate technique was used for gamma spectroscopy. The original sample was processed through a disk and a second sample, spiked with ^{85}Sr, was processed through a second disk. The original sample was analyzed by either gas proportional or liquid scintillation counting. The second disk with the ^{85}Sr was analyzed by gamma spectroscopy. The corrected tracer results again showed >95% recoveries.

The results of interference studies for the Sr Rad Disks are given in Table II. Quantitative strontium recoveries with 1 L samples were achieved in the presence of the indicated interferences.

Table II. Ion Interferences^a for the Strontium Rad Disk^b (13,18)

Ion	Concentration	Ion	Concentration
Mg^{2+}	10,000 mg·L^{-1}	Pb^{2+}	1 mg·L^{-1}
Ca^{2+}	500 mg·L^{-1}	Na^+	1000 mg·L^{-1}
Ba^{2+}	0.1 mg·L^{-1}	K^+	10 mg·L^{-1}
Ra^{2+}	300 pCi·L^{-1}		

^{a}Defined as the levels at which the Sr recoveries are 95% or greater.
^{b}One liter solutions were spiked with 3 mg Sr^{2+}. Recoveries >95% for Sr^{2+} were found for each solution.

The data in Table II demonstrate the effectiveness of the Sr Rad Disk in separating Sr^{2+} from the indicated ions. Some sample solutions with high Ca^{2+} concentrations can create a competition for sorbent sites between Sr^{2+} and Ca^{2+}. Solutions containing Ca^{2+} at levels as high as 500 mg·L^{-1} have been tested for the simultaneous uptake of Sr^{2+} (13-18). Quantitative recoveries were obtained from one liter solutions containing one mg Sr^{2+}. Typical environmental samples may contain an assortment of these cationic species. It is of particular interest that separations can be achieved at high Mg^{2+}, Ca^{2+}, and Na$^+$ levels since these ions are usually common in these samples. Typical potable water levels of K$^+$ are less than that shown in Table II. However, levels higher than 10 mg·L^{-1} would interfere with Sr^{2+} analysis (as may be found in some ground water samples).

Radium Rad Disks. Use of the Ra Rad Disk, which contains AnaLig RaO1, allows simultaneous quantitation of ^{226}Ra and ^{228}Ra. Due to their origin in the naturally occurring ^{238}U and ^{232}Th decay chains, these two radium isotopes, with half-lives of 1600 and 5.76 years, respectively, represent the most significant health hazards and concern for accurate detection. The sample preparation process is the same as for the Sr Rad Disk; a sample is acidified to 2 M with concentrated nitric acid and then passed through a 47 mm disk positioned on a filter apparatus. Once isolated on the disk, the investigator has a number of quantification options. Radiation can be measured directly from the disk but interpretation is difficult because of the multiple ingrowth paths occurring.

Where EPA methods are required, Smith et al. (14) describe how ^{226}Ra may be determined by eluting the radium with 15 to 20 mL of basic 0.25 M EDTA solution and then transferring this solution to a radon bubbler for determination by the radon emanation technique, EPA Method 903.1. Radium-228 may be determined by allowing the ingrowth of ^{228}Ac and the decay of ^{224}Ra to occur and then eluting actinium with 15 to 20 mL of 0.5 M HNO$_3$. The eluate solution is processed for gas flow proportional counting by evaporation on a planchet, or by coprecipitation on yttrium oxalate as described in EPA Method 904.0.

A new method described by Seely and Osterheim (*18*) demonstrates the RAP objectives by providing a simultaneous measurement of ^{226}Ra and ^{228}Ra. After the acidified sample (2 M with HNO$_3$) is drawn through the Ra Rad Disk, it is washed with 20 mL of nitric acid, dried with a 20 mL wash of acetone, and heated at 60 °C for 30 min. The dried disk is heat sealed in a 3.5 mm Mylar (polyethyleneterephthalate) envelope and set aside for 21 days to allow for equilibrium to be approached between radium and its progeny. The envelope is then placed directly on a gamma spectrometer detector endcap for counting. A multichannel analyzer is used to identify ^{214}Pb peaks at 352 KeV and 295 KeV for the quantification of ^{226}Ra and ^{228}Ac. Peaks at 338 KeV, 911 KeV, and 969 KeV are used to quantify ^{228}Ra.

Scarpitta and Miller (*19*) have reported on the viability of using Ra Rad Disks in DOE's Environmental Measurements Laboratory's (EML) Quality Assessment Program (QAP). They conclude that of twenty potential nuclide interferences tested, only ^{210}Pb, ^{90}Sr, and ^{133}Ba might interfere with ^{226}Ra determination. Since these three interferents are primarily β or γ emitters, their presence would only interfere if ^{226}Ra determination is by liquid scintillation analysis (LSA). Their presence is of little consequence if one uses an appropriate counting method that measures α-radiation. For QAP purposes, stripping with EDTA or liquid scintillation counting of the disk itself gave alpha efficiencies of 98%. Their summary states that the "Radium disk is effective at separating Ra from other elements and was found to be both a time and reagent saving product suitable for EML's QAP."

Ion interferences for the Ra Rad Disk reported by Seely and Osterheim (*18*) are given in Table III.

Table III. Ion Interferences[a] for the Radium Rad Disk[b] (*18*)

Ion	Concentration, mg·L^{-1}
Mg^{2+}	10,000
Ca^{2+}	10,000
Sr^{2+}	100
Pb^{2+}	10
Na$^+$	10,000
K$^+$	1,000
Ba^{2+}	10
NH$_4^+$	100

[a]Defined as the levels at which the Ra recoveries are 95% or greater.
[b]One liter solutions were spiked with 10.9 Bq·L^{-1} ^{226}Ra^{2+}. Recoveries >95% for ^{226}Ra^{2+} were found for each solution.

In Table III, it is seen that the Ra Rad Disk effectively removes the Ra isotopes present in a one liter sample containing cations that might be present in an environmental sample (i.e., Mg^{2+}, Ca^{2+}, Na$^+$, and K$^+$). As would be expected, selective removal is most difficult from solutions containing Sr^{2+}, Ba^{2+}, and Pb^{2+}. The properties (e.g., ionic radii) of these three cations are most similar to those of Ra^{2+}. However, as noted by Scarpitta and Miller (*19*) their presence may not be a problem in ^{226}Ra analysis since their interferents are primarily β or γ emitters.

Conclusions

In conclusion, it has been shown that the RAP concept is possible using state-of-the-art selective particle and membrane technologies. The purification and isolation of a target ion can be accomplished in one step with the appropriate solid phase extraction membranes. Elution can be employed as a secondary purification step though the particles need to be rugged and neither bleed nor alter the chemistry of the eluent. In some cases, direct quantification of the absorbed target ion from the disk is possible. Time, accuracy, and costs inherent in elemental analysis can all be improved by using RAP.

Acknowledgments

We greatly appreciate the cooperative work done in testing these materials by personnel at 3M, Hach, Argonne National Laboratory, and the DOE Environmental Measurements Laboratory.

SuperLig and AnaLig are registered trademarks of IBC Advanced Technologies, Inc. and Empore is a trademark of 3M.

Literature Cited

(1) Izatt, R. M.; Bruening, R. L.; Bruening, M. L.; Tarbet, B. J.; Krakowiak, K. E.; Bradshaw, J. S.; Christensen, J. J. *Anal. Chem.* **1988**, *60*, 1825-1826.
(2) Bradshaw, J. S.; Bruening, R. L.; Krakowiak, K. E.; Tarbet, B. J.; Bruening, M. L.; Izatt, R. M.; Christensen, J. J. *J. Chem. Soc., Chem. Commun.* **1988**, 812-814.
(3) Izatt, R. M.; Bradshaw, J. S.; Bruening R. L.; Tarbet, B. J.; Krakowiak, K. E. In *Emerging Separation Technologies for Metals and Fuels*; Lakshmanan, V. I., Bautista, R. G.; Somasundaran, P., Eds.; TMS: Warrendale, PA, 1993; pp 67-75.
(4) Izatt, N. E.; Bruening, R. L.; Anthian, L.; Griffin, L. D.; Tarbet, B. J.; Izatt, R. M.; Bradshaw, J. S. In *Proceedings: Metallurgical Processes for Early 21st Century*; Sohn, H. Y., Ed.; TMS: Warrendale, PA, 1994; pp 1001-1018.
(5) Izatt, R. M.; Bradshaw, J. S.; Bruening, R. L.; Tarbet, B. J.; Bruening, M. L. *Pure Appl. Chem.* **1995**, *67*, 1069-1074.
(6) Izatt, R. M.; Bradshaw, J. S.; Bruening, R. L. *Pure Appl. Chem.* **1996**, *68*, 1237-1241.
(7) Izatt, R. M.; Bradshaw, J. S. In *Comprehensive Supramolecular Chemistry: Supramolecular Technology;* Reinhoudt, D. N., Ed.; Vol. 10; Elsevier: New York; 1996, pp 1-11.
(8) Izatt, R. M.; Bradshaw, J. S.; Bruening, R. L. In *Perspectives in Supramolecular Chemistry*, Reinhoudt, D. N., Ed.; John Wiley & Sons: New York; 1998.
(9) Izatt, R. M.; Bradshaw, J. S.; Bruening, R. L.; Bruening, M. L. *Am. Lab.* **1994**, *26*, no. 18, 28c-28m.
(10) Hagen, D. F.; Markell, C.; Schmitt, G. *Anal. Chim. Acta* **1990**, *236*, 157-164.

(11) "Prescribed Procedures for Measurement of Radioactivity in Drinking Water, Method 905.0; EPA-600/4-80-032; U.S. Environmental Protection Agency; 1980.
(12) Smith, L. L.; Orlandini, K. A.; Alvarado, J. S.; Hoffmann, K. M.; Seely, D. C.; Shannon, R.T. *Radiochim. Acta* **1996**, *71*, 165-170.
(13) Empore™ Strontium Rad Disks, Strontium Interference Summary, 3M: St. Paul, MN; 1997.
(14) Smith, L. L.; Alvarado, J. S.; Markun, F. J.; Hoffman, K. M.; Seely, D. C.; Shannon, R. T. *Radioact. Radiochem.* **1997**, *8*, 30-37.
(15) Products for Analysis Catalog and Technical Data, Hach; Loveland, CO, 1997, p 35.
(16) Heinzig, M.W. "Lead Determination with the New AnaLig/Pb Ex Method"; Hach, Loveland, CO, 1996.
(17) "Analytical Procedures: Mercury (0 to 250 $\mu g \cdot L^{-1}$)"; Hach, Loveland, CO, 1997.
(18) Seely, D. C.; Osterheim, J. A., "Radiochemical Analyses Using Empore™ Disk Technology", *Fourth Conference on Methods and Applications of Radio-Analytical Chemistry*; Kona, HI, April 6-11, 1997.
(19) Scarpitta, S. C.; Miller, P. W., "Evaluation of 3M Empore™ Rad Disks for Radium Determination in Water", *42nd Annual Conference on BioAssay, Analytical and Environmental Radiochemistry*; San Francisco, CA, October 13-17, 1996.

Chapter 18

Biologically Generated Materials for Metal-Ion Binding: Answers to Some Fundamental Chemical Questions

Gary D. Rayson[1], Lawrence R. Drake[1,3], Hongying Xia[1], Shan Lin[1], and Paul J. Jackson[2]

[1]Department of Chemistry and Biochemistry, New Mexico State University, Box 30001, Department 3C, Las Cruces, NM 88003
[2]Environmental Molecular Biology Group, LS-7, MS M880, Los Alamos National Laboratory, Los Alamos, NM 87545

The ability of non-viable biologically generated materials to remove heavy metal ions from contaminated water has been recognized for several years. Such biogenic materials are not, however, widely used as substrates for the separation or preconcentration of metal ions. Much of the reluctance to use these potentially inexpensive materials stems from a lack of an acceptable level of predictability regarding their behavior in "real world" systems. This is primarily a result of the predominately phenomenological studies which have been reported in the past for these materials. An objective of on-going investigations within our laboratory is the elucidation of the fundamental chemical interactions governing the binding of metal ions to a plant-based biomaterial. We have selected for these studies a material comprised of fragments of cultured cells from the anther of the plant *Datura innoxia*. Because of the chemical complexity of this material, three orthogonal probes have been applied to the study of this material. These probes have included europium ion luminescence (spectrally and temporally resolved), metal nuclei nuclear magnetic resonance spectrometry (both solution and solid phases), and frontal affinity chromatographic methodologies. The chemical functionalities which have been identified include a collection of carboxylate-containing sites involved in mono-, bi-, and tridentate coordination modes where the metal ion is

---

[1]Current address: Los Alamos National Laboratory, CST-9, MS J514, Los Alamos, NM 87545.

engaged in an ion-exchange process. Additionally, sulfonate moieties have been determined to be present in separate ion-exchange sites.

Clean water can be said to be the most valuable natural resource in the world today (1). It is imperative that methods and technologies be developed and implemented for the purification of polluted waters. Specifically, the contamination of water with unacceptably high concentrations of toxic heavy metals must be addressed. The definition(s) of acceptability regarding the concentrations of these species is continually decreasing as a result of an improved understanding of the health risks associated with the ingestion of such components as lead, copper, and cadmium ions (2). The necessity of the selective removal of toxic heavy metals down to parts per billion (ppb) levels places greater demands on the technologies of remediation. Conventional methodologies for the remediation of such waters are often unable to achieve the concentration levels required for safe drinking water without considerable cost. These costs typically include the need for large quantities of material, the incorporation of complex treatment systems, or both. Commercial ion-exchange resins, including specialty chelating resins, may be effective in some instances for removing and recovering heavy metals (3-5). However, the extraction proficiency of these resins drops dramatically if the waters contain a high salt content or large amounts of calcium or magnesium ions. The costs associated with the manufacture of these organic-polymer-based extraction resins are typically passed-on to the end-user and can contribute substantially to the price of remediating a contaminated water supply or the treatment of an industrial waste stream. It is therefore necessary to identify and characterize alternate binding agents which can be inexpensively exploited to develop a cost effective process for the treatment of heavy metal polluted waters.

An alternative to these more classical approaches (ion-exchange resins, chelating resins, reverse osmosis, or electrowinning) is the application of biologically-generated materials (i.e., biomaterials). Such biogenic materials enable the reversible, selective removal of toxic heavy metals in the presence of large excesses of relatively benign metal ions (e.g., calcium, magnesium, and sodium). The characteristics of certain biomaterials for the selective removal of heavy metals (e.g., Pb, Cd, Zn) from contaminated waters have been observed to be favorable. *Datura innoxia* cell-wall fragments immobilized in a polysilicate matrix (i.e., a sol gel) have demonstrated favorable binding characteristics for ions such as zinc, lead, and cadmium. Under batch equilibrium conditions, the supernatant concentrations of separate solutions containing 10 ppm of each of these metals has been observed to decrease to 54, 150, and 70 ppb, respectively, when exposed to the immobilized *D. innoxia* biomaterial (10 µg biomaterial/mL solution). However, it has been the lack of an understanding of the chemical interactions responsible for that binding which has prohibited their widespread utilization in clean-up efforts. Without such a fundamental understanding of the parameters that influence the abilities of these materials to bind heavy metals, it is difficult, at best, to predict the response of these materials to changes in operating conditions (e.g., solution pH, ionic strength, temperature, and influent composition).

Two broad categories of binding mechanisms can be used to describe biomass-based materials: passive and active (6). Passive binding occurs in both living and non-living cells. It involves either physical adsorption or an ion-exchange interaction with the cell surface. Passive binding is typically very rapid. Conversely, active binding is characteristic of only living cells, and is often characterized by a slower metal uptake rate.

Complications often arise when living materials are used for remediation due to the onset of metal toxicity, eventually killing the organism. This limits the predictability of the metal uptake effectiveness of the biomass and thus their utility in ion selective binding. Non-living plant materials do not exhibit this limitation. Metal ion uptake for these materials is only dependent on passive binding which leads to potentially greater predictability and ease of use.

Metal uptake by non-living biomaterials has been proposed to occur through sorption processes involving the functional groups associated with the proteins, polysaccharides, lignin, and other biopolymers that are found in the cell and cell walls (7-8). Unlike conventional ion-exchange resins, which are designed with a single functionality, biomass materials can contain numerous functionalities, including amino, carboxylate, hydroxide, imidazole, sulfate, and sulfhydryl groups (7-8). These poly-functional biomass materials can exhibit unique metal adsorption abilities.

The ability of several biomaterials to bind Cu^{2+}, Al^{3+}, and Au^{3+} (as $AuCl_4^-$) at varied solution pH conditions has been reported (9). These metals were selected because they are representative of different classes of metal ions according to the hard and soft acid-base classification scheme (10). The biomaterials studied included blue-green algae, *Datura innoxia*, roots and stems of cattail plants (*Typha latifolia*), the leaves of young and mature tumbles weeds (*Salsola spp.*), spanish moss, and alfalfa sprouts (*Medicago sativa*). At pH 5, the uptake efficiency of these biomaterials for the intermediate Cu^{2+} metal ion was observed to decrease in the following order: *D. innoxia*, alfalfa sprouts > spanish moss > cattail roots > young tumble weeds > mature tumble weeds.

Most of these biomaterials exhibited pH-dependent binding profiles for the Cu^{2+} and Al^{3+} metal ions, however, Au^{3+} adsorption was reported to be independent of pH for *D. innoxia* and is approximately 100% efficient under the conditions studied. This high affinity for Au^{3+} indicates that there likely are significant concentrations of soft ligands in the cell wall fragments of *D. innoxia* (e.g., sulfur and nitrogen containing species). Al^{3+} binding was rather low for all species studied except for alfalfa sprouts, which was 100% efficient at pH 5.

In their native state, biomasses often exhibit poor mechanical characteristics. To alleviate this problem, the biomass can be encapsulated in a supporting matrix. Several methods of immobilization have been reported including the use of alginate microbeads (11), carrageenan gel (12), formaldehyde-based polymer (13), polyacrylamide (14-15), and silica-based polymers (14-16). Each of these materials has been reported to exhibit superior mechanical and physical properties with enhanced metal binding capabilities compared to the native material. Additionally, improvements in the porosity, chemical resistivity, and structural integrity of the biosorbent enable the biomaterial to be used in column or fluidized bed experiments.

In our laboratory, the silica-based immobilization scheme has been used (*14-15*). In this scheme, the resulting material consists of as much as 90% biomass by weight. The impact of pH and the immobilization strategy in a study involving several different metal ions and their uptake by *D. innoxia* have been previously described (*17*). The examination of the metal ion sorption ability of a number of different biomaterials after immobilization has been reported for both batch and flowing systems. For each biomass investigated, an enhancement in metal ion uptake after immobilization was observed.

The chemical heterogeneity of binding sites characterizes biomaterials. Such heterogeneity results from either the existence of multiple functionalities to which metal ions can bind, or the presence of specific functionalities in different chemical environments on the material's surface. For our biomaterial system, *D. innoxia* cell-wall fragments, luminescence and NMR studies have enabled the identification of carboxylates and sulfates (or sulfonates) as the major functional groups responsible for metal ion binding to this material (*18-27*). Also, it has been proposed that the complex chemical environments around these functionalities results in multiple binding sites with varied affinities in the biomaterial system. This chemical heterogeneity complicates both the description of overall sorption properties and the extraction of binding parameters from experimental isotherms.

An elucidation of biosorption processes of non-viable *D. innoxia* plant material, specifically fragments of the cultured anther cells of *D. innoxia* to metal ions, has been the focus in our research (*18-27*). These studies have included the identification of functionalities involved in metal ion binding using Eu^{3+} luminescence (*18-22*), NMR (*23*), and the characterization of metal ion binding processes in terms of binding capacities, affinities, and metal ion sorption mechanisms (*26*).

Chemical modification techniques have also been used to explore the chemistry involved in metal ion adsorption. Esterification of carboxylate functionalities contained in the cell walls of several different algae species resulted in a semi-quantitative agreement between the extent of esterification and the reduction in metal ion uptake of Cu^{2+} (*28*). These results confirmed that carboxylate groups were the dominant functional groups responsible for adsorption of this metal ion. In a similar experiment, we tested the effects of esterification on *D. innoxia*. Although carboxylate moieties were found to be partially responsible for metal ion uptake, our results indicate that other functionalities also play a significant role in metal adsorption. Interestingly, by saponifying the esterified cell fragments of *D. innoxia*, we were able to partially reverse the esterification effects and restore a portion of the metal adsorption capacity lost to esterification.

Lanthanide Luminescence Measurements

Lanthanide luminescence, particularly Eu^{3+}, has been used to aid in the identification of functionalities as well as the number of sites responsible for metal binding to biomaterials (*18-19,29*). The power of this method lies in the unique $^7F_0 \rightarrow \, ^5D_0$ transition of Eu^{3+}. This transition is singly degenerate and will exhibit a single peak for each unique binding site. In addition, the luminescence lifetimes are sensitive to ligation. Functionalities are identified by comparing the spectral peak(s) and lifetime(s) to known Eu^{3+}-ligand species. In a competitive binding experiment, a relative affinity order for metal ion uptake was classified using the decrease in the sensitivity of the Eu^{3+} luminescence (*24*). This may eventually lead to a relatively rapid classification scheme for determining a plant's utility for metal ion uptake as well as identification of the functionalities responsible for a particular biomaterial's metal adsorption capabilities.

To obtain a better understanding of the contributions of the various functional groups to the uptake of Eu^{3+}, an accurate deconvolution of the excitation spectra is required. The number of these peaks, the relative intensities, and the spacing between peaks can provide significant insight into the binding environment and the bond type of the individual sites (e.g., electrostatic bonding or chelating) (*18*). Previous attempts at deconvolving the Eu^{3+}-*D. innoxia* spectra have relied upon lifetime measurements to determine the number of sites responsible for Eu^{3+} uptake. The fitting of two Lorentzians (corresponding to the two measured lifetimes) enabled the identification of the involvement of carboxylate and sulfate (or sulfonate) functionalities (*18,19,24*). Even so, the limiting of the curve fitting algorithm to two Lorentzians resulted in an inability to accurately describe the longer-wavelength tail of the excitation spectra. More recent results obtained from the measurement of luminescence decays have indicated multiple (i.e., more than two) lifetimes involved in the decay process. Although Horrocks has suggested that the Lorentzian peak shape describes the peak shape of solid Eu^{3+} compounds (*30*), a Gaussian peak shape was found to more accurately describe the individual binding sites in *D. innoxia*. Due to the complexity of the biomaterial, it is plausible to propose that the Eu ion experiences varying molecular environments resulting in slight differences in the wavelength of light absorbed by the Eu ion and yield an inhomogeneous broadening of the individual peaks in the excitation spectra (*31*).

Accurate curve fits were therefore obtained by employing various methods to identify the individual sites responsible for Eu^{3+} uptake. These included lifetimes, a comparison of luminescence sensitivities obtained under various conditions, subtraction techniques, and comparison to model compounds. The curve fitting routine utilized for the deconvolution of the Eu^{3+} excitation spectra was based upon the Levenberg-Marquardt algorithm (*32*). The input parameters included the peak shape (e.g., Gaussian, Lorentzian, etc.), the number of peaks, and the peak parameters (i.e., width, height, or position).

An accurate curve fit for the Eu^{3+}-*D. innoxia* excitation spectra is needed to describe each of the following observations: (1) The blue shift observed for the excitation spectra with increasing Eu^{3+} concentration, (2) The blue shift observed for chemically modified cell wall fragments and the spectra obtained at lower pH, (3) Lifetime measurements requiring more than two different sites to be responsible for metal uptake under all conditions. According to Crist (*33*), the different binding sites available on these biomaterials should occur under all conditions unless sites become unavailable due to competition with H^+, other metal ions (e.g., Na^+), or through modification. This may not be a totally valid hypothesis for each of the individual sites within the excitation spectral profiles of *D. innoxia* because of the relatively poor luminescent properties of some binding sites. Lifetime measurements and the similarity of the peak shapes obtained for each of the studied conditions have indicated that the individual peaks comprising the excitation spectra should be present.

The narrow, blue-shifted excitation spectra obtained for the modified samples contacted at pH 2 (Figure 1) was subsequently subtracted from the spectral profile obtained for the native material contacted at pH 5 (Figure 2) with the lowest Eu^{3+} concentration. This resulted in a single symmetric peak. This peak exhibited a Gaussian shape with a maximum occurring at 579.42 nm. Efforts to fit Lorentzian peak-shapes to this feature yielded sufficient variability to support the earlier interpretation of this component as the result of numerous, chemically similar binding sites. To determine the number and contribution of other peaks that comprised the spectral signature, a peak was set to this fixed wavelength within the deconvolution algorithm. Peak parameters were then systematically varied until a satisfactory fit was obtained that explained the excitation spectra profiles under all conditions. The fits obtained (S1-S4) for high and low initial concentrations for the native and modified cell wall material contacted at both pH 2 and pH 5 are shown in Figures 1 and 2, respectively. The extremes of these combinations of conditions are also illustrated in these figures (modified cells at pH 2 and native cells at pH 5, respectively).

It should be noted that the peak fit returned contains the minimum number of peaks to obtain a successful fit of the excitation spectra. The saturation, blocking, and growth of the individual peaks that comprise the excitation spectra successfully described the blue shift as the concentration of the Eu^{3+}-*D. innoxia* complex increased and the blue shift observed in the spectral profiles obtained at lower pH. Site 3 (S3) has been proposed to be composed of two and possibly three fully overlapping peaks. From lifetime information, these have been postulated to involve both carboxylates forming a 1:1 complex with Eu^{3+} and a Eu^{3+}-sulfonate species. Both of these species have a peak maximum occurring at approximately 579.0 nm.

There is also strong evidence indicating that sites 1 (S1, the deconvoluted peak at the longest wavelength) and 2 (S2) appearing at 579.42 nm and 579.21 nm in Figure 2, respectively, are due to multidentate bond formation. For carboxylate groups it is well known that luminescence from these sites would occur at lower energies. An empirical equation relating the charge and transition energy for Eu^{3+}-carboxylate groups enables an estimation of the coordination involved (*35-36*). Utilizing this equation, the formation of tridentate and bidentate bonds can be

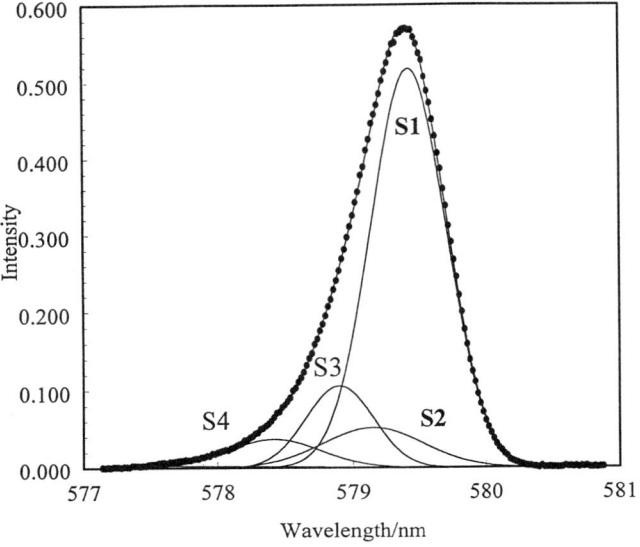

Figure 1. Eu^{3+} luminescence excitation spectrum for Eu^{3+} (0.3 mM) bound to native *D. innoxia* material at pH 5.0. (Adapted from ref. 34.)

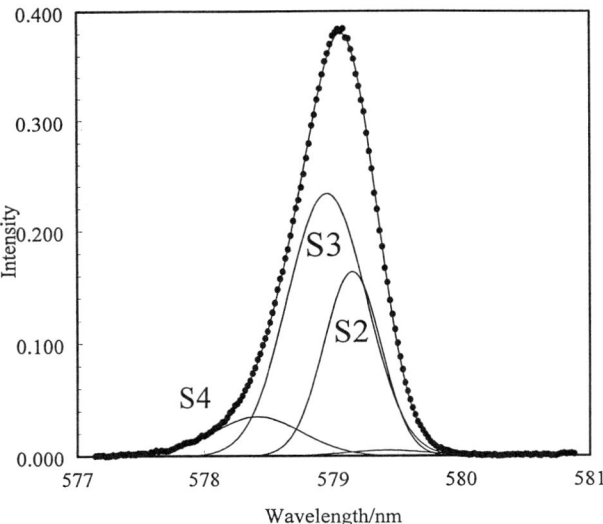

Figure 2. Eu(III) luminescence excitation spectrum for Eu^{3+} (3.0 mM) bound to chemically modified *D. innoxia* material at pH 2.0. (Adapted from ref. 34.)

surmised for sites S1 and S2, respectively. The monodentate site, S3, was observed to become more important with higher of Eu^{3+} concentrations. These assignments are also consistent with the measured luminescence lifetimes. The shorter lifetime is consistent with formation of a 1:1 complex. As a second and third carboxylate are bound to Eu^{3+}, water molecules are displaced from the inner coordination sphere resulting in an increased lifetime. The fourth site in the deconvolved spectra can be attributed to the uptake of Eu^{3+} by the polysaccharide component of the cell wall material. The luminescence obtained from Eu^{3+}-cellulose samples were observed to be weak and would be predicted to only minimally contribute to the total excitation peak obtained for the Eu^{3+}-*D. innoxia* samples contacted at pH 5. However, the spectra obtained for the cellulose model system were observed to exhibit a tail extending into shorter wavelengths providing a significant contribution to that portion of the excitation spectrum for the *D. innoxia* samples.

The peak in the excitation spectra obtained for esterified Eu^{3+}-*D. innoxia* samples contacted at pH 2 (Figure 2) appears at 579.11 nm. These spectra are considerably narrower than any of the spectra obtained under the above conditions. Comparison to the spectra obtained for Eu^{3+}-cellulose spectra suggest that this polysaccharide may be responsible for a majority of the uptake under these conditions, but the lifetime values obtained suggest that Eu^{3+}-sulfonate compounds are responsible for the spectral signature. These sites would also have been affected by the competition between Eu^{3+} and the H^+, but considering that this is the more acidic site the uptake of Eu^{3+} is expected. Although the peak positions for the esterified samples contacted at pH 2 are not dependent upon the concentration of Eu^{3+} bound, lifetime decays consisted of a multiple exponential which corresponded to the lifetimes obtained for the sulfonate resin (Dowex 50W-X8). Unfortunately, the agreement between these lifetimes is not conclusive. The deconvolved spectra indicated that the bidentate carboxylate groups are a minor contributor to the uptake of the metal ions even under these conditions and that the longer lifetime may be due to the bidentate species.

The curve fit obtained for the *D. innoxia* excitation spectra accurately described the spectral profile obtained under all conditions studied. The high affinity site, S1, saturates as the concentration of Eu^{3+} bound to the cell wall fragments increases. An upper limit for the capacity of this site is estimated to be 33 µmole/g *D. innoxia*. As the pH was decreased, this site exhibited minimal contribution. Sites 2 and 3 were populated simultaneously, but the rapid growth of S2 as a function of Eu bound suggested the carboxylates comprising this site have a greater affinity for the Eu ion. Caution in the quantitative interpretation of the data is advised because the molar absorptivity and the quantum yield for these sites are yet unknown.

Nuclear Magnetic Resonance Spectroscopic Measurements

Nuclear magnetic resonance (NMR) spectroscopy has also proven valuable for determining the chemical functionalities responsible for metal ion binding (*23,37-*

38). Solution phase ^{113}Cd NMR has been reported to be effective in determining the functionalities responsible for Cd uptake by two types of algae (37-38) and *D. innoxia* (23). The observed chemical shift covers a range of approximately 850 ppm, within which one can expect to find the Cd ion coordinated to the ligands found in biological systems. The sensitivity of Cd chemical shifts to variations in the local chemical environment make it possible for this NMR technique to reflect subtle differences in the functional groups involved in Cd ion binding. Both carboxylate functionalities and amine groups have been suggested to be responsible for metal ion uptake to *D. innoxia*. For these experiments, an extensive chemical shift table was composed for Cd^{2+} bound to functionalities that could possibly be present in the cell walls of *D. innoxia*. These compounds contained carboxylate, amine, hydroxyl, sulfhydryl, oxalate, sulfate, and sulfonate functionalities. NMR studies suggested a preference for Cd uptake by carboxyl functionalities at pH < 5. However, diamine-type groups were indicated to be involved as the pH was raised to pH 6. Unfortunately, these studies were limited by the same factor that makes them so attractive, that is, their sensitivity. Solution phase ^{113}Cd NMR spectroscopy can determine even minor changes in the chemical environment present. This technique is not only sensitive to changes in the Cd bound species, but also to changes in the surrounding matrix.

Recently, ^{113}Cd NMR has been used as a direct spectrometric probe to investigate the chemical composition of metal binding sites on a variety of biomaterials. Previous research had studied the total metal-binding behavior for a series of non-living biomaterials including four strains of algae (*Chlorella pyrenoidosa, Bryopsis, Cladophora* and *Entiomorpha*), the fragments of cultured *D. innoxia* cells, peat moss, sphagnum peat, freeze-dried roots and stems of cattail plants (*Typha latifolia*), freeze-dried leaves, roots and stems of mature and young tumble weed (*Salsola spp.*), and pecan shells. Differences in the gross metal binding capabilities and its dependence on solution conditions, such as pH, were found among these biomaterials.

Figure 3 is a summary of the Cd chemical shifts for each of these biomaterials in 0.05 M $CdSO_4$ at pH 5.0. Each of these biomaterials, as slurries, yielded NMR spectra with one narrow peak and one broad peak. The narrow resonance has been attributed to the Cd in solution while the broad peak was assigned to the Cd bound on the biomaterial. The Cd chemical shifts for solution species varied significantly between the biomaterials studied. Conversely, the Cd chemical shifts for the bound Cd varied only slightly for all biomaterials studied. These broad resonances were observed in the range between -10 to -18 ppm.

To identify the functional groups responsible for the bound Cd resonances, ^{113}Cd NMR data were again obtained for a series of model ligands (Figure 4). A comparison of the data in Figures 3 and 4 suggests the involvement of carboxylate or sulfonate groups in the binding of Cd^{2+} to each of the biomaterials in Figure 3. Because the carboxylate functionality is a weak acid, the pH of the solution would be predicted to dramatically affect the binding capacity of the material. Conversely, the strong acid characteristics of sulfonates suggest an independence of the material's binding capacity for a solution pH range of 1 to 5. All of the

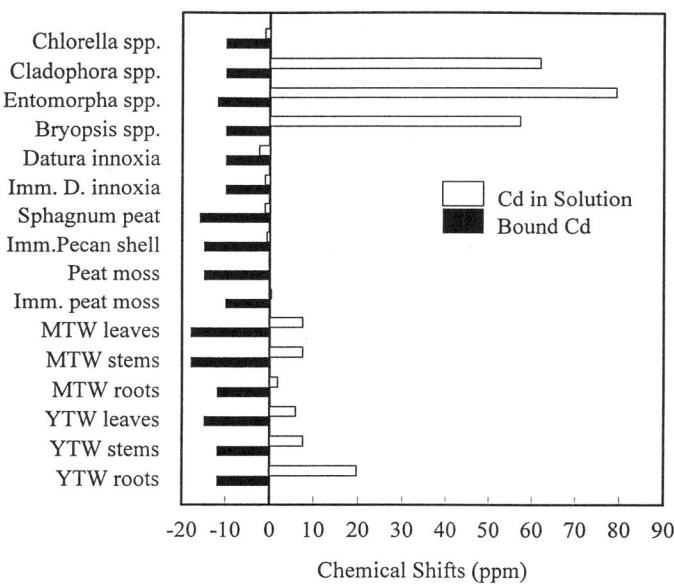

Figure 3. Chemical shifts for ^{113}Cd-NMR with various biomaterials. Imm. Indicates the material immobilized within a polysilicate matrix, MTW indicates mature tumble weed (*Salsola spp.*) tissues, and YTW indicates the young tumble weed plant materials.

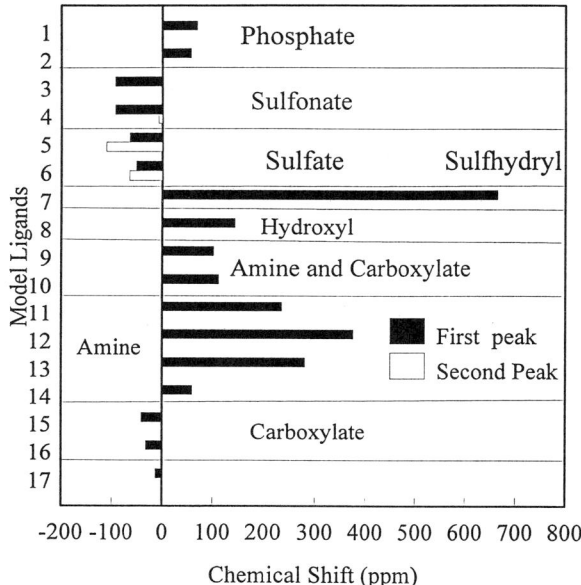

Figure 4. Chemical shifts for ^{113}Cd-NMR with various model ligands. 1) Cd-adenosine 5'-monophosphate, 2) CdHPO$_4$, 3) Cd-toluenesulfonic acid (1:3), 4) Cd-toluenesulfonic acid (1:2), 5) Cd-dodecyl sulfate, 6) 3CdSO$_4$·8H$_2$O, 7) Cd(SCH$_2$CH$_2$S), 8) Cd(OH)$_2$, 9) Na$_2$Cd(EDTA), 10) Cd(NH$_2$CH$_2$CO$_2$)$_2$·H$_2$O, 11) im$_6$Cd(NO$_3$)$_2$, 12) Cd(en)$_3$Cl$_2$·H$_2$O, 13) Cd(NH$_3$)$_6$Cl$_2$, 14) Cd(NH$_4$)$_2$(SO$_4$)$_2$·6H$_2$O, 15) Cd(O$_2$CCO$_2$), 16) Cd(O$_2$CCH$_3$)$_2$·2H$_2$O, and 17) the average chemical shift observed for the biomaterials. EDTA indicates ethylenediaminetetra acetic acid, en indicates ethylenediamine, and im indicates imidizole

biomaterials studied exhibited a significant positive dependence of Cd^{2+} binding capacity on solution pH. It was concluded that carboxylate-containing sites were again involved in the dominate mechanism of metal ion binding for each of these widely-varied biomaterials. This conclusion was also supported by the ^{113}Cd NMR results on resins with carboxylate and sulfonate groups respectively (Table I).

Among all the non-living biomaterials we studied, *D. innoxia* has been identified as the most promising biosorbent for the removal of metal ions. Advanced NMR studies have focused on *D. innoxia* to identify the functionalities responsible for metal binding. Solid state ^{113}Cd NMR with high speed magic angle spin was applied to wetted *D. innoxia* bound with enriched ^{113}Cd. The reasons for using solid state NMR are to circumvent problems possibly associated with chemical exchange and solvent effects in solution state NMR and to eliminate the anisotropy effect by using magic angle spin. The solid state NMR spectra of *D. innoxia* with bound Cd showed a very broad peak quite similar to the ^{113}Cd resonance in slurries which suggested that the Cd was bound to a distribution of slightly different carboxylate sites.

The hypothesis of the involvement of carboxylate groups in Cd binding was further investigated through chemical modification of the *D. innoxia* material. A portion of *D. innoxia* was esterified to deactivate the carboxylate groups on the biomass. The disappearance of the broad resonance of bound Cd in esterified *D. innoxia* further verified the involvement of carboxylate groups as the major binding site on *D. innoxia*.

The efficient application of biomaterials to metal ion removal from contaminated waters requires an understanding of the fundamental chemistry of the binding process for all of the metals of interest such as Zn, Cu, and Mn. The NMR information for these metals cannot be obtained directly, but the competition between these metal ions and Cd^{2+} is an indirect approach to investigate the binding behavior of metals other than Cd. When *D. innoxia* material that was already bound with Cd was soaked into Zn^{2+}, Cu^{2+}, and Mn^{2+} solutions, respectively, release of Cd from *D. innoxia* into the solution phase was observed in all cases. The replacement of Cd by Zn, Cu, and Mn suggested that these metals bind on common sites. The relative binding affinity to these common sites is: $Cu^{2+} > Zn^{2+} > Mn^{2+}$.

Solid state ^{27}Al NMR has also been used for the direct investigation of metal binding to *D. innoxia*. Aluminum NMR was chosen because of its high sensitivity (158 times higher than that of ^{113}Cd) and its short relaxation time in the solid state.

Table II is a summary of ^{27}Al chemical shifts for different materials which had been soaked in 0.05 M $AlCl_3$ (pH 3.5). From Table II, only one major Al binding site (chemical shift about 0 ppm) was observed for the non-immobilized *D. innoxia* cell-wall material. Besides this 0 ppm site, an additional binding site at 58 ppm was observed on immobilized *D. innoxia* which was shown to arise from the binding of Al to the silicate polymer. Previous ^{113}Cd NMR studies inferred that the carboxylate group was the dominant functionality responsible for the binding of Cd^{2+} to the biomaterial *D. innoxia*. To determine if carboxylates are involved in Al binding, Cd replacement experiments for *D. innoxia* bound with Al were undertaken. The disappearance of the 0 ppm Al binding site was observed as the material was exposed to increasing amounts of Cd^{2+}. Sites predominately responsible for the binding of Al to the *D. innoxia* cell-wall material can therefore be concluded to involve carboxylate functionalities. This hypothesis was further tested through ^{27}Al NMR investigations of the esterified *D. innoxia* material through the dramatic decrease in the corresponding ^{27}Al resonance.

Binding behavior of *D. innoxia* material when exposed to a pH 5.0 Al solution was investigated. Previous studies of Al binding capacity on *D. innoxia* material under different solution pH had shown a significant binding capacity increase as pH increased. At pH 5.0, Al_{13}^{7+} polymeric species become the major species rather than the Al^{3+} monomer species. The Al_{13}^{7+} polymer exhibits a characteristic 63 ppm resonance which is attributed to the central tetrahedrally coordinated Al atom in this structure. Therefore ^{27}Al NMR provides an effective approach to determining the fate of the Al_{13}^{7+} polymer in contact with *D. innoxia*. Table III is a summary of ^{27}Al chemical shifts for different materials which had been exposed to the pH 5.0 Al solution. The sorption of Al_{13}^{7+} was observed on free *D. innoxia* material and the carboxylate-containing resin. A literature review shows this is the first instance in which the Al_{13}^{7+} polymer was observed to bind to biomaterials directly. The coexistence of the 7 ppm peak for free *D. innoxia* and the 0 ppm peak for the carboxylate containing resin suggested Al_{13}^{7+} polymeric species were also bound to biomaterial through their dissociation to the monomeric species. For immobilized *D. innoxia,* the Al_{13}^{7+} resonance overlaps with the resonance arising from the binding of Al^{3+} to the silicate polymer (58 ppm peak in table I). However, magnitude of the peak area of the 59 ppm resonance for immobilized *D. innoxia* in Table II cannot be fully explained by this interaction and thus suggests the direct binding of Al_{13}^{7+} on immobilized *D. innoxia* biomaterial.

Binding Isotherms from Affinity Chromatographic Measurements

To describe the overall sorption isotherm on such chemically heterogeneous surfaces, the distribution of affinity constants is required as well as a model for a homogeneous "subsurface" -- an ensemble of "equal intrinsic affinity" sites or a group of identical sites (*39-40*). Frequently, the isotherms for this "subsurface" are assumed to exhibit Langmuirian behavior (*41-43*). For these materials, there exist two primary models for the extraction of the distributions of affinity constants. They are the discrete site and continuous site models (*44-46*). In discrete site models, it is assumed that the surface consists of sets of discrete ligand sites with distinguishable binding affinities. Typically, only a few types of sites (2-8 types) are required to adequately fit experimental binding data (*47-48*). In contrast, continuous site models assume the surface consists of heterogeneous binding sites and that their corresponding affinity constants with respect to these binding sites are characterized by continuous and smooth distribution functions. Given the great complexity of biomaterials, the discrete site models are considered only to be a

Table I. ^{113}Cd Chemical Shifts for Ion Exchange Resins

Functional group	Resin	Chemical Shifts (width in Hz)	
		Slurry	Solid
Carboxylate	Bio-Rex 70	-30 (2K) pH=5	-30(8K)
	Bio-Rex 70	-2.5 (50) pH=1	
Sulfonic acid	Dowex 50W-X8	-7.9 (50) pH=5	-17(1.2K)
	Dowex 50W-X8	-7.9 (50) pH=1	

Table II. ^{27}Al Chemical Shifts at pH 3.5

Samples	First Peak	Second Peak
Datura innoxia	0 ppm	
Immobilized *D. innoxia*	0 ppm	58 ppm
Silicate Polymer		54 ppm
Carboxylate Resin	0 ppm	

Table III. ^{27}Al Chemical Shifts at pH 5.0

Samples	First Peak (ppm)	Second Peak (ppm)
Datura innoxia	7 ppm	63 ppm
Immobilized *D. innoxia*	9 ppm	59 ppm
Silicate Polymer	4 ppm	55 ppm
Carboxylate Resin	0 ppm	63 ppm

simple way to describe the observed metal binding behaviors. Conversely, the continuous site models are a physically more realistic description of these heterogeneous systems (45).

In principle, the overall sorption isotherm for a continuous, heterogeneous surface can be considered as the superposition or integration of local sorption isotherms from all possible subsurfaces. As indicated in the following equation for the heterogeneous surface, the differences between the affinity constants K_i can be treated as infinitesimal (49):

$$q_t = \int_\Delta q(K,[M]) f(\log K) d \log K$$

Here q_t is the total sorption isotherm, $q(K,[M])$ is the local sorption isotherm describing the binding to the subsurface, and $f(\log K)$ is the distribution function of the affinity constants. The product of $f(\log K)$ and $\log K$ is the concentration of the sites, s_j, for a given set of affinity constants, K_j. The integration space, Δ, is the range of log K values.

As is well recognized (39), it is very difficult to solve the above integral (a Fredholm integral equation of the first kind) to obtain $f(\log K)$ from experimental sorption data q_t and known $q(K,[M])$. This is especially true when the experimental data are subject to substantial noise. In view of this, three main approaches have been attempted to solve the equation for $f(\log K)$. The first group of methods is associated with complicated special numerical methods. Examples of these methods are regularization (50), singular value decomposition, and heterogeneity investigated by a Loughborough distribution analysis and computed adsorptive energy distribution in the monolayer (51). The second group of methods utilizes specific simplified approximations of the local isotherm equation. These enable relatively simple analytical expressions to be obtained for $f(\log K)$ with the sacrifice of deviations with respect to the local isotherm (39,49). The third group of methods involves the use of the prior assumptions of the distribution functions (typically simple distribution functions, such as a quasi-Gaussian distribution) in combination with using Langmuir or Henderson-Hasselbalch equations to describe the local isotherm (52-53). By this algorithm, the integral adsorption equation can then be solved analytically.

Recently, Borkovec and co-workers (54) have described the use of regularized least-squares methods to obtain affinity distribution functions for the interaction systems between metal ions and the heterogeneous sorbents. Regularized least-squares methods were so employed as to resolve problems (such as instability) associated with many of the proposed approaches. The program (QUASI) developed by these researchers using this algorithm has been used for the extraction of affinity distribution of Pb^{2+} ions to immobilized *D. innoxia* cultured cell fragments. Variances of solution conditions such as ionic strength and pH and their impact on the resulting affinity distributions were studied. These parameters can provide unique insights into the nature of the interaction between Pb^{2+} and this biosorbent.

Determination of these binding sites regarding binding capacities and affinities (affinity distributions) has also been undertaken. This is important for the prediction of metal ion binding under different chemical environments. Studies of apparent affinity distributions of the immobilized *Datura innoxia* to Pb^{2+} ions at various solution compositions were conducted (55). Results show there are two classes of binding sites between the interaction of Pb ions and the biomaterial, one class having low affinities with mean affinity constant about 200 M^{-1} and another having higher affinities with mean affinity constant around 10^5 M^{-1}, depending on solution conditions (Figure 5). Investigations of the affinity distributions with varied pH revealed that the low-affinity sites involve both carboxylate and sulfonate functional groups, while that the high-affinity sites primarily involve carboxylates (Figure 6).

Conclusions

The chemical functionalities on biologically generated materials have been investigated through the application of Eu^{3+} luminescence, ^{113}Cd and ^{27}Al NMR, and affinity chromatography. These coordinated studies have indicated that carboxylates are the primary functionality involved in metal binding to a variety of biomaterials and specifically to the material derived from the plant *D. innoxia*. These moieties have been found to bind the metal ions by both electrostatic (i.e., ion-exchange) processes and through the formation of a chelate. They are also present in binding sites containing one, two, or three carboxylates as separate sites. Sulfonates have also been identified has contributing to the overall metal binding ability of this material through an electrostatic attraction.

Equally as important as the identification of those chemical functionalities involved in metal ion binding to the chemically complex materials is the determination of the minimal impact of many plausible moieties such as amines and sulfhydryl groups. This information will enable increased predictability and thus improved efficiencies in the modification or application of these biomaterials in the remediation of contaminated waters and waste streams.

Acknowledgments

The authors gratefully acknowledge the financial support of the National Science Foundation (Grant: CHE-9312219), the New Mexico Water Resources Research Institute (USGS #14-08-0001-G2035), and the New Mexico Waste-management Education and Research Consortium.

Literature Cited

(1) Peavy, H. S., Rowe, D. R., Tchobanoglous, G. *Environmental Engineering,*; McGraw-Hill: New York, 1985, 54-57.
(2) *National Interim Drinking Water Regulations;* Federal Register, Part IV, December 24, 1975.
(3) Sahni, S. S.; Reedijk, K. *Coord. Chem. Rev.*, 1984, *59*, 1.

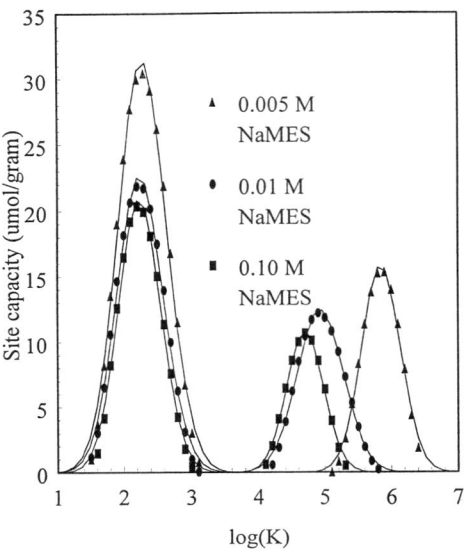

Figure 5. Affinity distributions for Pb^{2+} on immobilized *D. innoxia* as a function of sodium ion concentration (i.e., ionic strength). (Adapted from ref. 55.)

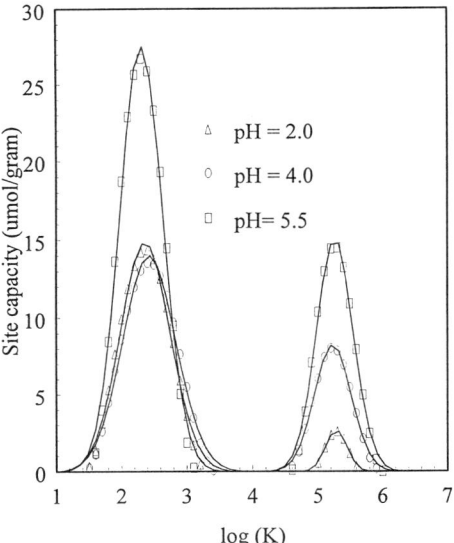

Figure 6. Affinity distributions for Pb^{2+} on immobilized *D. innoxia* as a function of influent solution pH. (Adapted from ref. 55.)

(4) *Synthesis and Separation Using Functional Polymers*, Sherrington, D. C.; Hodge, D., Eds. John Wiley: New York, 1988.
(5) Warshawsky, A.; In *Ion Exchange and Sorption Processes in Hydrometallurgy*, M. Streat; D. Naden, Eds., John Wiley: New York, 1987, p 166.
(6) Khummongkol, D.; Canterford, G. S.; Fryer, C. *Biotech. Bioeng.* 1982, *24*, 2643.
(7) Crist, R. H.; Oberholser, K.; Shank, N.; Nguyen, M. *Environ. Sci. Technol.* 1981, *15*, 1212.
(8) Greene, B.; Hosea, M.; McPherson, R.; Henzl, M.; Alexander, M. D.; Darnall, D.W. *Environ. Sci. Technol.* 1986, *20*, 627.
(9) Lujan, J. R.; Darnall, D. W.; Stark, P. C.; Rayson, G. D.; Gardea-Torresdey, J. L. *Solvent Extr. Ion Exch.* 1994, *12*, 803.
(10) Pearson, R.G. In *Hard and Soft Acids and Bases*; Pearson, R. G., Ed. Dowden, Hutchinson; Ross: Stroudsburg, P A, 1973; Part II, p. 53.
(11) Garnham, G.W.; Codd, G. A.; Gadd, G. M. *Environ. Sci. Technol.* 1992, *26*, 1764.
(12) Scott, C. D. *Biotech. Bioeng.* 1992, *39*, 1064.
(13) Nakajima, A.; Sakaguchi, T. *Biomass* 1990, 55.
(14) Darnall, D.W.; Greene, B.; Hosea, M.; McPherson, R.A.; Henzl, M.; Alexander, M. D. In *Trace Metal Removal from Aqueous Solution* Thompson, R. Ed.; Burlington House, London, 1986; pp. 1-24.
(15) Greene, B.; McPherson, R.; Darnall, D. In *Metals Speciation, Separation, And Recovery* Patterson, J. W.; Passino, R. Eds., Lewis Publishers, Inc., Chelsea, MI 1987; pp. 315.
(16) Mahan, C. A.; Holcombe, J. A. *Spectrochimica Acta*. 1992, *47B*, 1483.
(17) Darnall, D. W.; Greene, B.; Henzl, M. T.; Hosea, J. M.; McPherson, R. A.; Sneddon, J.; Alexander, M.D. *Environ. Sci. Technol.* 1986, *20*, 206.
(18) Ke, H. Y.; Birnbaum, E. R.; Darnall, D. W.; Rayson, G. D.; Jackson, P. J. *Appl. Spectrosc*. 1992, *46*, 479.
(19) Ke, H. Y.; Birnbaum, E. R.; Darnall, D. W.; Rayson, G. D.; Jackson, P. J. *Environ. Sci. Technol.* 1992, *26*, 782
(20) Ke, H. Y.; Rayson G. D. *Appl. Spectrosc.* 1992, *46*, 1168.
(21) Ke, H. Y.; Rayson G. D. *Appl. Spectrosc.* 1992, *46*, 1376.
(22) Ke, H. Y.; Rayson G. D. *Appl. Spectrosc.* 1992, *47*, 44.
(23) Ke, H .Y.; Rayson G. D. *Environ. Sci. Technol.* 1992, *26*, 1202.
(24) Ke, H. Y.; Anderson W. L.; Moncrief, R. M.; Rayson, G. D.; Jackson, P. J. *Environ. Sci. Technol.* 1994, *28*, 586.
(25) Drake, L. R.; Lin, S.; Rayson, G. D. *Environ. Sci. Technol.* 1996, *30*, 110.
(26) Lin, S.; Drake, L. R.; Rayson, G. D. *Anal. Chem.* 1996, *68,* 4087.
(27) Drake, L. R.; Rayson, G. D. *Anal. Chem.* 1996, *68*, 22A.
(28) Gardea-Torresdey, J. L.; Becker-Hapak, M. K.; Hosea, J. M.; Darnall, D. W. *Environ. Sci. Technol.* 1990, *24*, 1372.
(29) Moncrief, R. M.; Anderson, W. L.; Ke, H-Y. D.; Rayson, G. D. *Sep. Sci. Technol.* 1995, *30*, 2421.

(30) McNemar, C. W.; Horrocks, Jr., W. DeW. *Appl. Spectrosc.* 1989, *43*, 816.
(31) Steinfeld, J. I. In *Molecules and Radiation*, S. A. Rice, Ed. (Harper and Row, Publishers, Inc., New York, New York 1974) Chap. 1.7.
(32) Marquardt, D. W. *J. Soc. Ind. Appl. Math.,* 1963, *11*, 431.
(33) Crist, R. H.; Martin, J. R.; Carr, D.; Watson, J. R.; Clarke, H. J.; Crist, D. R. *Environ. Sci. Technol.* 1994, *28*, 1859.
(34) Drake, L. R.; Hensman, C. E.; Shan Lin; Rayson, G. D. *Appl. Spectrosc.* 1997 *51*, 1476.
(35) Horrocks, Jr., W. DeW. *Methods in Enzymology* 1993, *226*, 495.
(36) Albin, M.; Horrocks, Jr., W. DeW. *Inorg. Chem.* 1985, *24*, 885.
(37) Zhang, W.; Majidi, V. *Environ. Sci. Technol.* 1994, *28*, 1577.
(38) Zhang, W.; Majidi, V. *Appl. Spectrosc.* 1993, *47*, 2151.
(39) Nederlof, M. M.; van Riemsdijk, W. H.; Koopal, L. K. *J. Colloid Interface Sci.* 1990, *135,* 410.
(40) Koopal, L. K.; van Riemsdijk, W. H.; de Wit, J. C. M.; Benedetti, M. F. *J. Colloid Interface Sci.* 1994, *166,* 51.
(41) Tipping, E.; Hurley, M. A. *Geochim. Cosmochim. Acta* 1992, *56,* 3627.
(42) Milne, C. J.; Kinniburgh, D. G.; de Wit, J. C. M.; van Riemsdijk, W. H.; Koopal, L. K. *Geochim. Cosmochim. Acta* 1995, *59,* 110.
(43) Ephraim, J.; Marinsky, J. A. *Environ. Sci. Technol.* 1986, *20,* 367.
(44) Dobbs, J. C.; Susetyo, W.; Carreira, L. A. *Anal. Chem.* 1989, *61,* 1519.
(45) Dzombak, D. A.; Fish, W.; Morel, F. M. M. *Environ. Sci. Technol.* 1986, *20,* 669.
(46) de Wit, J. C. M.; van Riemsdijk, W. H.; Koopal, L. K. *Environ. Sci. Technol.* 1993, *27,* 2015.
(47) Tipping, E. *Environ. Sci. Technol.* 1993, *27,* 520.
(48) Fish, W.; Dzombak, D. A.; Morel, F. M. M. *Environ. Sci. Technol.* 1986, *20,* 676.
(49) Nederlof, M. M.; Van Riemsdijk, W. H.; Koopal, L. K. *Environ. Sci. Technol.* 1992, *26,* 763.
(50) House, W. A. *J. Colloid Interface Sci.* 1978, *67,* 166.
(51) Vos, C. H. W.; Koopal, L. K. *J. Colloid Interface Sci.* 1985, *105,* 183-196.
(52) de Wit, J. C. M.; van Riemsdijk, W. H.; Koopal, L. K. *Environ. Sci. Technol.* 1993, *27,* 2015.
(53) Benedetti, M. F.; Milne, C. J.; Kinniburgh, D. G.; Van Riemsdijk, W. H.; Koopal, L. K. *Environ. Sci. Technol.* 1995, *29,* 446.
(54) Cernik, M.; Borkovec, M.; Westall, J. C. *Environ. Sci. Technol.* 1995, *29,* 413.
(55) Lin, S.; Rayson, G. D. *Environ. Sci. Technol.* (Submitted, 1997).

SEPARATIONS USING MEMBRANES

Chapter 19

Use of Ligand-Modified Micellar-Enhanced Ultrafiltration to Selectively Remove Copper from Water

Susan B. Shadizadeh[1,2], Richard W. Taylor[3,4], John F. Scamehorn[1,2], Annette L. Schovanec[3], and Sherril D. Christian[2,3]

[1]School of Chemical Engineering and Materials Science, [2]Institute for Applied Surfactant Research, and [3]Department of Chemistry, University of Oklahoma, Norman, OK 73019

> The effectiveness of the ligand 4-hexadecyloxybenzyliminodiacetic acid ($C_{16}BIDA$) for the selective removal of Cu^{2+} from aqueous solutions using ligand-modified micellar-enhanced ultrafiltration (LM-MEUF) has been investigated. The cationic surfactant N-hexadecylpyridinium chloride (CPC) was used as the added colloid. Stirred cell ultrafiltration (UF) and semi-equilibrium dialysis (SED) methods were used to determine the effects of solution composition on the rejection of copper ions. The parameters studies include the concentrations of Cu^{2+}, ligand, Ca^{2+}, surfactant, and NaCl in the initial (feed) solution for UF and SED experiments and the applied pressure in the UF experiments. In solutions containing 100 mM CPC at pH 5.5, predominately 1:2 metal-ligand complexes are formed. Rejections of copper of up to 99.8% are observed, with almost no rejection of calcium, demonstrating the excellent selectivity and separation efficiency of $C_{16}BIDA$ in LM-MEUF. SED experiments showed that the ligand can be recovered for recycling by acid-stripping at pH 2.

Surfactant-based separation processes are a class of chemical engineering separations that have great potential for removal and/or recovery of heavy metals and organic pollutants from water streams (1,2). Surfactant-based separation processes can require less energy, be less expensive, and have less environmental impact than traditional separation methods.

Micellar-enhanced ultrafiltration (MEUF) is a surfactant-based separation technique that can be used to remove metal ions and/or dissolved organics from aqueous streams. In MEUF, the surfactant is present at a concentration well above its critical micelle concentration (CMC), so most of the surfactant is present as micelles. The micelles are roughly spherical aggregates containing about 50 to 150 surfactant molecules (3). The hydrocarbon chains of the surfactant fill the micelle interior making this core hydrophobic, and the hydrophilic portions of the surfactant are situated at the micelle surface. Cationic multivalent metal ions adsorb or bind to the surface of negatively charged micelles of an anionic surfactant while organic solutes tend to

[4]Corresponding author.

solubilize or dissolve within the micelles (*1,2,4-12*). A disadvantage of MEUF in removing dissolved metals from aqueous solutions is that ions of the same charge will be removed with approximately the same efficiency, that is, there is a lack of selectivity. For example, using an anionic surfactant, Ca^{2+}, Cu^{2+}, Ni^{2+}, and Zn^{2+} were all removed with nearly the same rejection (*2,10*).

Ligand-modified micellar-enhanced ultrafiltration (LM-MEUF) is a modification of MEUF that can provide selectivity in the removal of cations (*13-17*). In LM-MEUF, a surfactant and a ligand are added to the contaminated solution. A suitable ligand for LM-MEUF needs to have a group which can selectively complex the target metal ion. Furthermore, the ligand should have a large hydrophobic group in order to have a high tendency to solubilize in the micelles so the ligand and complexed metal ion are attached to the micelle to a large extent. This solution containing surfactant/ligand/ion is then treated by an ultrafiltration process, shown schematically in Figure 1, using a membrane with pore sizes small enough to reject or block the micelles. As micelles are rejected, the solubilized ligand and its associated ions will also be rejected. The unsolubilized ligand, uncomplexed ions, and surfactant monomers pass through the ultrafiltration membrane to the permeate side (*13-16*). The effectiveness of the separation process in retaining the target species in the retentate can be defined in terms of a rejection value. The retentate-based rejection (in %) of species X is defined as:

$$R_X (\%) = \{ 1 - ([X]_{per} / [X]_{ret}) \} \times 100 \qquad (1)$$

where $[X]_{per}$ and $[X]_{ret}$ are the concentrations of the target species in the permeate and retentate, respectively.

The resulting permeate stream from LM-MEUF contains very low concentrations of the surfactant, target metal, and the ligand and, hence, high rejection values for each of these species. By using a cationic surfactant in the LM-MEUF process, it is possible to expel the cations, which are not specifically complexed with the solubilized ligand, into the permeate by a process called ion-expulsion ultrafiltration (IEUF) (*11,18,19*). A Donnan-equilibrium effect causes the uncomplexed cations to become concentrated in the permeate; in this way, extremely large selectivities can be achieved with LM-MEUF (*13,16*). Consequently, a dilute process stream with a fairly large volume can be separated into a small volume concentrated retentate stream containing a large percentage of the surfactant/ligand/target ion, and a large volume permeate that contains toxins in low concentrations. Staging the ultrafiltration units (*6*) can permit the ultimate permeate to have any desired degree of purity so it can be discarded or reused. The retentate stream is considerably smaller in volume than the original process stream, and therefore, further treatment or disposal of this retentate can be less expensive than treatment of the original process stream. Schemes have been developed to recover the surfactant and ligand from the retentate for reuse (*17*). Micellar extraction of metal ions using an ultrafiltration technique has also been reported by Tondre and co-workers, including studies on ligand-metal complexation and kinetically controlled separation of metal ions (*20-23*).

In previous LM-MEUF studies, selective removal of Cu^{2+} from a solution containing a Cu^{2+}/Ca^{2+} mixture has been accomplished using the ligand *N*-(*n*-dodecyl)iminodiacetic acid with the cationic surfactant *N*-hexadecylpyridinium chloride (CPC). Rejections of Cu^{2+} of greater than 99% were reported with no rejection of Ca^{2+} (*13,14*). Similar results were observed using a commercially available alkyl-β-diketone ligand for the Cu^{2+}/Ca^{2+} mixture with *N*-hexadecyltrimethylammonium chloride as the surfactant (*17*). A regeneration scheme was developed and demonstrated for this latter system which allows reuse of the surfactant and ligand (*17*). In the current study, LM-MEUF is demonstrated for

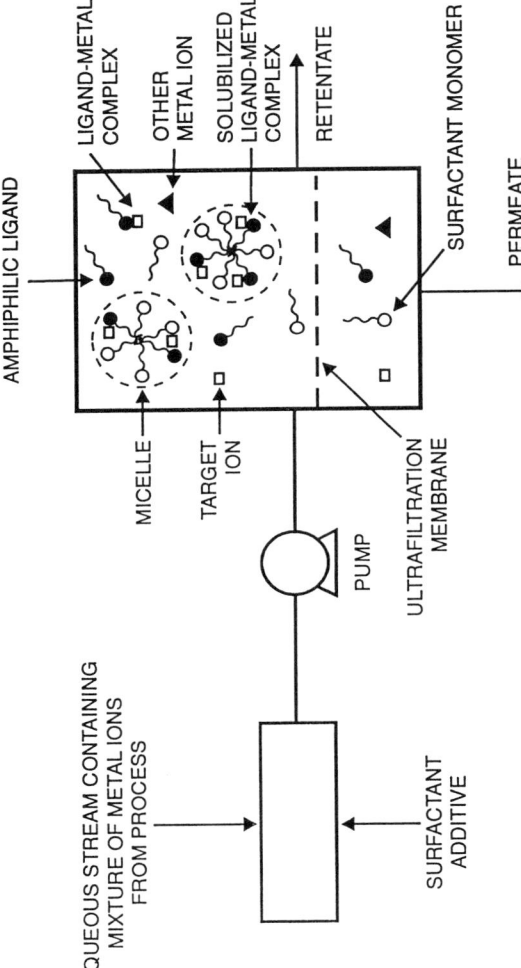

Figure 1. Diagram of an LM-MEUF unit for the removal of cationic metal species from an aqueous stream.

removal of Cu^{2+} from aqueous solutions containing Cu^{2+} and Ca^{2+} using the ligand 4-hexadecyloxybenzyliminodiacetic acid ($C_{16}BIDA$) and the cationic surfactant CPC. The effects of solute concentrations and ultrafiltration operating parameters on the separation efficiency and selectivity are investigated.

$$CH_3(CH_2)_{14}CH_2O-\underset{}{\bigcirc}-CH_2N\begin{matrix}CH_2COOH\\CH_2COOH\end{matrix}$$

$C_{16}BIDA$

Most of the data discussed here were obtained by stirred cell ultrafiltration. However, we also use the semi-equilibrium dialysis (SED) technique which has been developed in our laboratory as a simple experimental method for investigating both solubilization of organic species in surfactant micelles and the binding or expulsion of ions by micelles. Since MEUF processes are equilibrium-controlled, rather than kinetically-controlled; the results of simple SED experiments can be used to predict the effectiveness of ultrafiltration (UF) purification processes for rejected species (8,11,14-16,24-29).

Experimental Section

Materials. $CuCl_2 \cdot 2H_2O$, $CaCl_2 \cdot 2H_2O$, and NaCl (Certified ACS grade) were purchased from Fisher and dried at 240 °C for 24 hours. Compounds used to synthesize the ligand were obtained from Aldrich Chemical Co. and used as provided. Standard buffer solutions for pH 4 and pH 7, copper and calcium atomic absorption reference solutions, HCl and NaOH (Fisher Chemical Co.), and methanol (HPLC grade, J. T. Baker) were also used. N-Hexadecylpyridinium chloride (CPC) was obtained from Hexcel Chemical Co. and used as provided. The purity of CPC (pharmaceutical grade) was confirmed by surface tension and HPLC analysis. N-Hexadecyltrimethylammonium bromide (CTAB) was obtained from Fluka and used as provided. All test solutions were prepared using water that was purified by two stages of ion exchange with subsequent carbon filtering. Air used to pressurize ultrafiltration cells was passed through several filters and water removal traps.

The ligand 4-hexadecyloxybenzyliminodiacetic acid, $C_{16}BIDA$, was synthesized following literature procedures. 4-Hydroxybenzaldehyde was alkylated with 1-bromohexadecane (30) and the aldehyde converted to an oxime (31). The oxime was reduced to an amine (32), followed by N-alkylation of the benzylamine with 1-bromoacetic acid (33). The final product was purified by recrystallization twice from ethanol. NMR and mass spectra of the intermediates and final product were consistent with the compounds sought (16).

Procedures. All test solutions were adjusted to pH 5.5 ± 0.5 using HCl and/or NaOH. pH measurements were made using a Markson pH meter (model 6102) equipped with a Markson Q-830 epoxy body combination electrode. The pH meter was calibrated using standard buffer solutions (pH 4.00 and 7.00). UV-Vis absorbance measurements were made using an HP8452A diode-array spectrometer with 10 second signal-averaging or with a Hitachi 100-80 double-beam spectrophotometer. A holmium oxide filter was used to confirm wavelength accuracy.

Ultrafiltration experiments were performed using a Nuclepore 400 mL stirred cell (Model Number S76-400, Spectrum) thermostatted at 25 °C using a circulating water bath. Unless otherwise noted, the pressure on the retentate side was 60 psig

(414 kPa) in all the experiments. For the stirred cell experiments, anisotropic cellulose acetate membranes (Spectrum) with a molecular weight cut-off (MWCO) of 10,000 and a diameter of 76 mm were used. The ultrafiltration membranes were soaked in the feed solution for 24 hours before the run was started. The cell was initially filled with 300 mL of surfactant solution and the contents stirred at 840 rpm to ensure adequate mixing within the cell (7,12). For each run, a total of 200 mL of permeate was collected in 25 mL aliquots, along with samples of feed and retentate solutions. The rejection (R_X) was calculated using equation 1 from permeate and retentate concentrations measured at the midpoint of each run; that is, after 100 mL of permeate was collected.

The semi-equilibrium dialysis (SED) method has been used as described previously (8,11,14-16,24-29). Regenerated cellulose acetate membranes (Fisher) with an average MWCO of 6000 were soaked in deionized water for 24 hours prior to use. The test solution (~5 mL) was placed in the retentate side of the cell and the permeate compartment was filled with deionized water (or NaCl solution). The cells were equilibrated for 24 hours at 25 °C and then the concentrations in permeate and retentate solutions were analyzed. All ultrafiltration and SED experiments were performed in duplicate and the results reported are the average of two runs.

Sample Analysis. The concentrations of copper and calcium in the feed, permeate, and retentate solutions were determined by atomic absorption spectrometry using a Varian SpectrAA 20 atomic absorption spectrometer equipped with a GTA-96 graphite tube atomizer. The feed and retentate solutions were diluted to the concentration range of the calibration curve with doubly deionized water. Matrix effects due to the presence of surfactant were minimized by adding sufficient surfactant to the calibration standards to match the average found in the feed, retentate, and permeate solutions after dilution (14-17). Each copper and calcium sample was analyzed three times and the results averaged.

HPLC with UV detection was used to determine the CPC concentrations in the feed, permeate, and retentate. The columns (4.6 mm, I.D. x 115 mm long) employed Whatman Partisil ODS II (Alltech Assoc., Inc.) with ten micron particle size. Peak heights were measured at 260 nm. Water was used initially as the solvent to wash any salts from the column, followed by a methanol-water mixture to elute the CPC.

Results and Discussion

Ligand Properties. In pure water, the solubility of the ligand as a function of solution pH was measured using UV spectroscopy. The concentrations of the ligand at saturation vary in the following manner; pH 5.5-6.5, 14 µM; pH 7.5-8.5; 45 µM and pH 9.5-10.5, 90 µM. The increase in solubility as the pH increases is expected since the carboxylic acid moieties undergo ionization forming mono- and dianionic species; that is, at pH 1.5, 6.0, and 10.0 the predominant ligand species are H_2L, HL^-, and L^{2-}, respectively (16,34). The solubility of the ligand in the presence of a micellar surfactant (i.e., above the CMC) is important since it gives a measure of the metal complexing (loading) capacity of the LM-MEUF system. For a solution containing 100 mM CPC at pH 5.5, the nominal (total) solubility is greater than or equal to 2 mM (16). In another study utilizing an alkyl-β-diketone ligand in CTAB, this level of ligand concentration (about 2% of total surfactant) causes a decrease in the CMC of about 20-30% compared to the pure surfactant (17).

The partition of the ligand in the absence of metal ions was studied using the SED technique. Because CPC interferes in the wavelength region where the ligand absorbs, the partition experiments were carried out using the cationic surfactant CTAB which is transparent at wavelengths greater than 250 nm. The concentrations of ligand in the SED permeate and retentate solutions were measured at different pH values by absorbance measurements at 270 nm. For solutions containing 50 mM CTAB, the ligand rejection values, R_L, were calculated using equation 1 and found to be 95.3% (pH 1.5), 98.0% (pH 5.5), and 99.6% (pH 12). The increase in the R_L values parallels the progressive deprotonation of the carboxylic acid moieties as the pH rises from 1.5 to 12.0. As the charge on the ligand becomes more negative, the electrostatic interaction with the cationic micelle increases, causing a larger fraction of the ligand to be retained by the micelle. The results of these preliminary studies show that $C_{16}BIDA$ has very low solubility in water but is soluble in cationic CPC micelles at a level of at least 2 mM. The strong partition of the ligand ($\geq$ 99%) into micellar surfactant at pH $\geq$ 5.5 is important to minimize ligand losses during the ultrafiltration process.

Ultrafiltration Studies. A series of ultrafiltration experiments was carried out with $C_{16}BIDA$ in micellar CPC at pH 5.5 (feed) to determine how differing experimental conditions affect the rejection of copper, R_{Cu}. The feed concentrations of Cu^{2+}, Ca^{2+}, $C_{16}BIDA$, CPC, and NaCl and the applied pressure were varied individually while the other parameters were held constant. For each set of experimental conditions, the feed, permeate, and retentate concentrations of Cu and Ca, along with the rejection values for copper, are given in Tables I, II, and III. Measurements of the CPC concentrations in the permeate and retentate were carried out for all ultrafiltration runs. Based on HPLC analysis and using equation 1, the average value for the rejection, R_{CPC}, of CPC was 99.63%. Figures 2 to 4 show plots of $[Cu]_{per}$ and $[Ca]_{per}$ versus $[X]_{ret}$, where X represents the parameter varied. The insets in these figures give the average values of the concentration of the solution components that were held constant. Although the feed concentration of a given solute was the same for the separate runs in each set of experiments, differences in separation efficiency caused small variations ($< \pm 10\%$) in the values at the midpoint of the individual ultrafiltration runs (*13*).

Metal-Ligand Stoichiometry in Surfactant Micelles. The feed concentrations of Cu^{2+} and $C_{16}BIDA$ were varied independently and the effects on $[Cu^{2+}]_{per}$, $[Ca^{2+}]_{per}$, and R_{Cu} are shown Figures 2 and 3 and in Table I. The permeate concentrations of copper are relatively low and copper rejection values greater than 99.7% are observed as long as $[C_{16}BIDA]_{feed}:[Cu^{2+}]_{feed} \geq 2.0$. The abrupt change in R_{Cu} values as the [ligand]:[metal] ratio exceeds 2.0 is seen more clearly in the results from the SED experiments (see Table IV). These results suggest that $Cu(C_{16}BIDA)_2^{2-}$ is the predominant complex species in micellar CPC. This finding differs from that calculated (*34,35*) for comparable conditions in aqueous solution where the 1:1 $Cu(C_{16}BIDA)$ complex was the major species ($\geq$ 90%) for pH $\leq$~5.5. In the micellar system, formation of $Cu(C_{16}BIDA)_2^{2-}$ by disproportionation of the 1:1 complex (equation 2) would be favored due to electrostatic interaction of the anionic product

$$2 \, Cu(C_{16}BIDA) \rightleftharpoons Cu(C_{16}BIDA)_2^{2-} + Cu^{2+} \qquad (2)$$

with the cationic micelles. Similar changes in the stoichiometry of the predominant metal-ligand complex have been observed for other systems in cationic colloids, particularly those where the number of donor atoms in the ligand is three or less and where the resulting complex has increased formal negative charge (*15,28*). Also, the local environment of the micelle-bound ligand will result in changes in the protonation constants and local $[H^+]$ that favor deprotonation and, as a result, the formation of higher order (ML_2) complexes. The practical result of this behavior is to diminish the loading capacity of the system from the preferred case where formation of 1:1 complexes predominates.

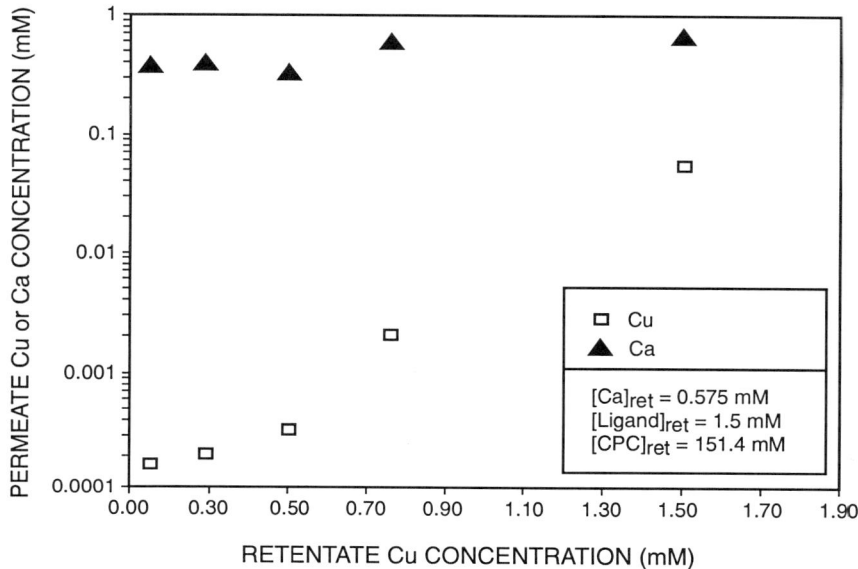

Figure 2. Effect of retentate copper concentration on separation.

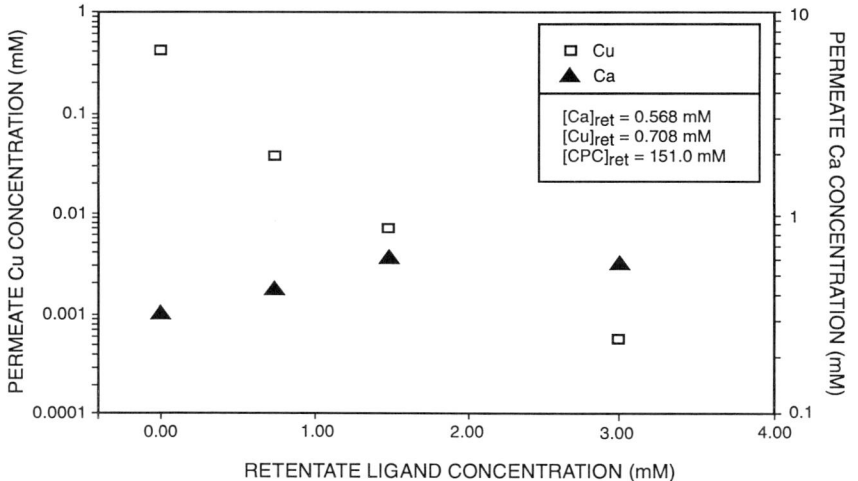

Figure 3. Effect of retentate ligand concentration on separation.

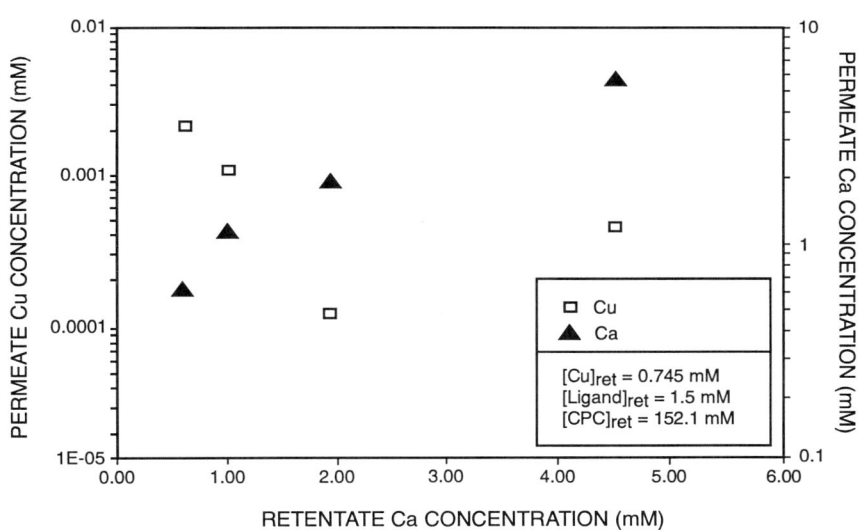

Figure 4. Effect of retentate calcium concentration on separation.

Table I. Ultrafiltration Results for $C_{16}BIDA$ in CPC[a]

Feed (mM)		Retentate (mM)		Permeate (mM)		R_{Cu} (%)[b]
[Cu]	[C_{16}BIDA]	[Cu]	[Ca]	[Cu]	[Ca]	
0.10	1.00	0.150	0.540	0.000165	0.382	99.89
0.20	1.00	0.291	0.564	0.000200	0.417	99.93
0.325	1.00	0.502	0.588	0.000322	0.327	99.93
0.50	1.00	0.763	0.604	0.00200	0.585	99.74
0.75	1.00	1.508	0.579	0.0494	0.614	96.79
0.50	0.50	0.742	0.582	0.0154	0.403	97.92
0.50	2.00	0.751	0.524	0.0000769	0.487	99.98
0.50	0.0	0.575	0.563	0.331	0.331	42.39
0.50	0.0[c]	0.574	0.000	0.545	0.000	4.90

[a]Composition of feed solution: $[CPC]_0$ = 100 mM; $[Ca^{2+}]_0$ = 0.50 mM; $[NaCl]_0$ = 0.0 mM; Feed pH = 5.50; Temp. = 25 °C; P = 60 psig (414 KPa).
[b]R_{Cu} (%) given by equation 1.
[c]Feed solution contained only copper and CPC; no Ca^{2+}.

Effect of Applied Pressure UF Separations. Table I shows that in most cases $[Ca^{2+}]_{ret} > [Ca^{2+}]_{per}$, giving rejection values for calcium as large as 45%. Furthermore, rejection values for both Cu (and Ca) are greater than zero even when no ligand is present, indicating that rejection is not due to complex formation. Table II lists data showing the effect of pressure on metal rejection for solutions of 100 mM CPC with no ligand present. As the pressure is lowered, the values of R_{Cu} and R_{Ca} decrease with negligible rejection (Cu) or expulsion (Ca) observed at 20 psig, the lowest pressure tested. Previous UF studies with stirred cells showed that the extent of ion expulsion into the permeate decreases as the applied pressure increases and ultimately reverses giving rejection at pressures of ~ 60-70 psig (19). This behavior and that observed in the current study can be attributed to concentration polarization (flux limitation) which increases with pressure (4,19). The results of SED experiments, where the applied pressure is zero and Ca^{2+} is concentrated in the permeate by ion expulsion, are discussed in a subsequent section. An important consequence of concentration polarization in UF is a reduction in the separation selectivity for Cu vs. Ca, unless low pressures are used.

Table II. Effect of Applied Pressure on Ultrafiltration Experiments[a]

P (psig)	Retentate (mM)		Permeate (mM)		R_{Cu} (%)[b]	R_{Ca} (%)[b]
	[Cu]	[Ca]	[Cu]	[Ca]		
20	0.588	0.481	0.560	0.585	4.77	-----
40	0.595	0.582	0.334	0.363	43.81	37.69
60	0.524	0.585	0.272	0.272	48.06	53.50

[a]Composition of feed solution: $[CPC]_0$ = 100 mM; $[C_{16}BIDA]_0$ = 1.0 mM; $[Cu^{2+}]_0$ = 0.50 mM; $[Ca^{2+}]_0$ = 0.50 mM; $[NaCl]_0$ = 0.0 mM; Feed pH = 5.50; Temp. = 25 °C.
[b]R_{Cu} (%) given by equation 1.

Effect of Concentration of Ca^{2+}. Figure 4 shows a plot of permeate Cu^{2+} and Ca^{2+} concentrations vs. retentate Ca^{2+} concentration (Table III, rows 1 to 5). As the retentate Ca^{2+} concentration increases, the permeate calcium concentration increases while the permeate copper concentration shows a general decrease. This general decrease may be attributed to a reduced ion expulsion effect from the increased $[Ca^{2+}]$. The slight increase in $[Cu^{2+}]_{per}$ at the highest $[Ca^{2+}]_{ret}$ is not due to competition by Ca^{2+} for the ligand, because SED studies with $C_{16}BIDA$ and Ca^{2+}, with no Cu^{2+} present, showed no evidence of Ca^{2+} complexation in the pH range 5.5 to 12 (*16*).

Table III. Ultrafiltration Results for $C_{16}BIDA$ in CPC[a]

Feed (mM)		Retentate (mM)		Permeate (mM)		R_{Cu} (%)[b]
[Ca]	[CPC]	[Cu]	[Ca]	[Cu]	[Ca]	
0.0	100	0.740	0.000	0.000178	0.000	99.97
0.50	100	0.763	0.604	0.00200	0.585	99.74
1.0	100	0.768	1.002	0.00100	1.086	99.87
2.0	100	0.679	1.928	0.000103	1.884	99.98
5.0	100	0.768	4.495	0.000391	5.748	99.95
0.50	150	0.721	0.514	0.000195	0.470	99.97
0.50	200	0.791	0.434	0.0000707	0.524	99.99
0.50[c]	100	0.744	0.552	0.000963	0.308	99.87
0.50[d]	100	0.783	0.616	0.000602	0.237	99.92

[a]Composition of feed solution: $[C_{16}BIDA]_0 = 1.0$ mM; $[Cu^{2+}]_0 = 0.50$ mM; $[NaCl]_0 = 0.0$ mM; Feed pH = 5.50; Temp = 25 °C; P = 60 psig (414 KPa).
[b]R_{Cu} (%) given by equation 1.
[c]$[NaCl]_0 = 100$ mM.
[d]$[NaCl]_0 = 200$ mM.

Effect of Concentration of CPC. The data in Table III (rows 2, 6, and 7) show that the permeate copper concentration decreases while the calcium concentration in the permeate shows no distinct trend as the retentate CPC concentration increases. Increasing the concentration of the surfactant in the retentate should result in greater solubilization of the ligand and the copper-ligand complexes and hence, higher copper rejection. On the other hand, increasing the concentration of the cationic surfactant would be expected to increase the extent of ion expulsion for uncomplexed cations such as Ca^{2+}. The data in Table III also show that the ratio of $[Ca^{2+}]_{per}:[Ca^{2+}]_{ret}$ increases as CPC goes from 100 mM to 200 mM with a positive, albeit small, ion expulsion effect at the highest CPC concentration. The effect of CPC concentration on flux is of interest because it is the predominant component and thus controls concentration polarization behavior (*4*). Ultrafiltration experiments showed that the absolute (L/m²·hr) and relative flux through the membrane decreased only slightly as the retentate CPC concentration increased to ~300 mM. At concentrations higher than this, the flux decreased dramatically, reaching zero flux at CPC levels of about 500-600 mM.

Effect of Electrolyte (NaCl) on the Separation Process. Ultrafiltration experiments were performed in the presence of added electrolyte in order to study the effects of increased ionic strength on the separation process. The data in Table III (rows 2, 8, and 9) show that the concentrations of copper and calcium in the permeate decrease with increasing concentration of the electrolyte (NaCl). This behavior can be attributed to a diminution of the ion expulsion effect as the total ionic strength increases (14). Although the copper rejection improves as the concentration of electrolyte increases, the selectivity of the separation remains almost constant because the calcium concentration in the permeate decreases as well.

Determination of the Krafft Temperature of the Retentate. The Krafft temperature is the temperature below which an ionic surfactant precipitates from water at concentrations above the CMC (36). Practical separations based on the water soluble colloids must operate at temperatures above the Krafft temperature. In order to assess the effect of the added solutes on surfactant behavior, the Krafft temperature was measured for each of the retentate solutions from the ultrafiltration runs. In all cases the Krafft temperatures observed were lower than 23 °C; therefore, surfactant precipitation did not occur in these systems under the conditions studied.

Separation Behavior at Equilibrium - SED Results. The efficiency of the separation process with $C_{16}BIDA$ and CPC (at equilibrium) has been studied using the SED technique with initial conditions comparable to those used in the ultrafiltration experiments. The results of the SED studies are presented in Table IV. Variation of the initial concentrations of the reactants, $[Cu^{2+}]_{init}$ and $[C_{16}BIDA]_{init}$, shows that rejection of copper is high ($R_{Cu} \geq 99.6\%$) when $[Cu^{2+}]_{init}:[C_{16}BIDA]_{init} \geq 2.0$.

Table IV. Results for SED Experiments of $C_{16}BIDA$ in CPC[a]

	Initial (mM)		Retentate(mM)[b]		Permeate(mM)[b]		R_{Cu} (%)[c]
[Cu]	[C_{16}BIDA]	[Ca]	[Cu]	[Ca]	[Cu]	[Ca]	
0.113	1.02	0.734	0.0955	0.0545	0.00038	0.800	99.6
0.207	1.02	0.778	0.150	0.0405	0.00057	0.899	99.6
0.355	1.02	0.746	0.288	0.0594	0.00010	0.836	99.9
0.467	1.02	0.799	0.340	0.0974	0.000170	0.836	99.9
0.547	1.02	0.750	0.414	0.0435	0.00431	0.871	99.0
0.720	1.02	0.707	0.552	0.0776	0.0549	0.879	90.1
0.897	1.02	0.827	0.635	0.118	0.0732	1.04	88.5
1.05	1.02	0.872	0.670	0.129	0.172	1.03	74.3
0.418	0.0	0.648	0.0153	0.0836	0.349	0.774	------
0.434	0.52	0.535	0.291	0.0633	0.0916	0.686	68.5
0.535	2.03	0.898	0.447	0.130	0.00112	0.951	99.7
0.543[d]	1.02	0.961	0.262	0.248	0.265	1.30	------
0.509[e]	1.02	0.820	0.344	0.0693	0.0126	1.09	96.3
0.542[f]	1.02	0.900	0.374	0.256	0.00104	1.11	99.7

[a]Composition of feed solution: $[CPC]_0 = 100$ mM, $[NaCl]_0 = 0.0$, $pH_0 = 5.50$.
[b]All permeate and retentate results reflect an average of two SED runs; results have a relative average deviation < 3.5%.
[c]R_{Cu} (%) given by equation 1.
[d]$pH_0 = 2.0$.
[e]$pH_0 = 4.0$.
[f]$pH_0 = 7.0$.

The decrease in R_{Cu} for stoichiometric ratios less than 2.0 is more pronounced in the SED studies because uncomplexed Cu^{2+} is more readily expelled into the permeate due to minimal concentration polarization effects relative to those in the UF experiments. For the same reason, uncomplexed Ca^{2+} is concentrated in the permeate solutions, with ion expulsion ratios ($[Ca^{2+}]_{per}$:$[Ca^{2+}]_{ret}$) in the range 7-22 observed. This behavior is comparable to that observed for other colloid systems with SED or UF at low pressure (0.5 atm) (18,19). Under these conditions (low or no applied pressure) separation factors ($\{[Ca^{2+}]$:$[Cu^{2+}]\}_{per}$) as large as 200-1000 are obtained for cases where $([L]:[Cu])_{init} \geq 2.0$. The data in rows 5, 12, 13, and 14 in Table IV show the effect of solution pH on the separation process. As the pH decreases, R_{Cu} becomes smaller until, at pH 2.0, there is no net rejection of copper. Increasing acid concentration results in the release of Cu^{2+} according to the reaction:

$$nH^+ + CuL_2^{2-} \text{(micellar)} \rightleftharpoons Cu^{2+} + 2 H_nL^{n-2} \text{(micellar)} \qquad (3)$$

This type of acid stripping approach can be used to regenerate the ligand/surfactant from the retentate solution for recycling. Acid stripping/ultrafiltration has been shown to be effective in recovering both ligand and surfactant using an alkyl-β-diketone as the ligand and N-hexadecyltrimethylammonium chloride as the surfactant (17).

The selectivity of $C_{16}BIDA$ for Cu^{2+} relative to other divalent cations was investigated using the SED method. Studies were carried out with Cd^{2+} and Zn^{2+}, alone and in three-component mixtures containing Cu^{2+}, under conditions similar to those employed for other SED and UF experiments. The results, listed in Table V, show only modest rejection of Zn^{2+} and no rejection of Cd^{2+}, when the cations are

Table V. Results of SED Experiments for Mixtures Containing Zn^{2+}, Cd^{2+}, and Cu^{2+} with $C_{16}BIDA$ in CPC[a]

Initial concentration (mM)			$R_M(\%)$[b]		
$[Zn^{2+}]$	$[Cd^{2+}]$	$[Cu^{2+}]$	Zn	Cd	Cu
0.519	0.0	0.0	33.2	-----	-----
0.0	0.546	0.0	-----	n.r[c]	-----
0.602	0.507	0.488	n.r[c]	n.r[c]	99.1
1.10	0.496	0.469	n.r[c]	n.r[c]	99.5

[a]Composition of feed solution: $[CPC]_0 = 100$ mM; $[C_{16}BIDA]_0 = 1.08$ mM; $[NaCl]_0 = 0.0$ mM; feed pH = 5.50; Temp. = 25 °C.
[b]R_M (%) given by equation 1. Permeate and retentate concentrations used to calculate rejection are the average of two SED runs, with relative average deviation < 3.0%.
[c]No net rejection; that is, $[M^{2+}]_{per} \geq [M^{2+}]_{ret}$.

tested individually. For the solutions containing all three cations, high rejection values are obtained for Cu^{2+}, with no rejection of either Zn^{2+} or Cd^{2+}. In fact, for both of the latter cations, there is net expulsion into the permeate, with $[M^{2+}]_{per}$:$[M^{2+}]_{ret} \sim 2$, giving separation factors ($\{[M^{2+}]$:$[Cu^{2+}]\}_{per}$) of about 100. For initial conditions where $[Cu^{2+}]$:$[M^{2+}]$:$[L] = 1$:1:2, the fraction (f_{Cu}) of Cu^{2+} present as CuL_2 provides an estimate of the rejection ($R_{Cu} = 100 f_{Cu}$), where $f = (10^{Kex/2}) / (10^{Kex/2} + 1)$ and

K_{ex} (= β_{CuL2}/β_{ML2}) is the overall equilibrium constant for the exchange reaction, $Cu^{2+} + ML_2 \rightleftharpoons CuL_2 + M^{2+}$. Using N-methyliminodiacetic acid (34) as a model for $C_{16}BIDA$, the calculated estimates of R_{Cu} are 98.7% for Zn^{2+} and 99.8% for Cd^{2+}. These values are consistent with results obtained with the three-component mixtures.

Conclusions

The amphiphilic ligand, $C_{16}BIDA$, has sufficient solubility (≥ 2 mM) in CPC to provide excellent metal binding capacity. Partition of the $C_{16}BIDA$ and its Cu^{2+}-complexes into the micelles is high ($R_L \geq 98\%$) thus minimizing losses of the ligand. The predominate form of the copper complex in CPC at pH 5.5 is $Cu(C_{16}BIDA)_2^{2-}$. For stirred cell ultrafiltration of mixtures containing Cu^{2+} and Ca^{2+}, rejection values of Cu^{2+} are equal to or greater than 99.7% provided that $([C_{16}BIDA]:[Cu^{2+}])_{feed} \geq 2.0$. For feed solutions initially equimolar in copper and calcium, the $[Cu^{2+}]:[Ca^{2+}]$ ratios in the permeate vary from 300-6500, demonstrating excellent selectivity for copper in LM-MEUF separations. However, concentration polarization effects in the ultrafiltration cell decrease expulsion of uncomplexed Ca^{2+} into the permeate, causing a slight reduction in the overall separation efficiency. Semi-equilibrium dialysis studies show the feasibility of acid stripping at pH 2 for removal of Cu^{2+} and regeneration of the ligand-surfactant mixture. Future studies will investigate the use of pH to control selectivity during the separation and recycling stages, and the design of ligands with selectivity for other cations, particularly toxic metals such as Hg^{2+}, Pb^{2+} and Cd^{2+}.

Acknowledgments

Financial support for this work was provided by the National Science Foundation Grant No's. CBT-8814147 and CTS-9123388, and an Applied Research Grant from the Oklahoma Center for the Advancement of Science and Technology. In addition, support was received from sponsors of the Institute for Applied Surfactant Research including Akzo Nobel, Amway Colgate-Palmolive, Dow, DowElanco, DuPont, Henkel, ICI, Kerr-McGee, Lever, Lubrizol, Nikko Chemical, Phillips Petroleum, Pilot Chemical, Reckitt and Coleman, Shell, Sun, and Witco. Dr. Scamehorn holds the Asahi Glass Chair in Chemical Engineering at the University of Oklahoma.

Literature Cited

(1) Scamehorn, J. F.; Christian, S. D. In *Surfactant-Based Separation Processes*; Scamehorn, J. F.; Harwell, J. H., Eds.; Dekker: New York, 1989; Ch 1.
(2) Scamehorn, J. F.; Christian, S. D.; Ellington, R. T. In *Surfactant-Based Separation Processes*; Scamehorn, J. F.; Harwell, J. H., Eds.; Dekker: New York, 1989; Ch 2.
(3) Rosen, M. J. *Surfactant and Interfacial Phenomena*; Wiley: New York, 1989; Ch 1, 3.
(4) Smith, G. A.; Christian, S. D.; Tucker, E. E.; Scamehorn, J. F. In *Use of Ordered Media in Chemical Separations*; Hinze, W. L.; Armstrong, D. W., Eds.; ACS Symposium Series 342; American Chemical Society: Washington, DC, 1987; pp 184-198.
(5) Dunn, R. O.; Scamehorn, J. F.; Christian S. D. *Sep. Sci. Technol.* **1985**, *20*, 257-284.
(6) Roberts, B. L. Ph.D. Dissertation, University of Oklahoma, 1993.
(7) Scamehorn, J. F.; Ellington, R. T.; Christian, S. D.; Penney, B. W.; Dunn, R. O.; Bhat, S. N. In *Recent Advances in Separation Techniques - III*; Li, N. N., Ed.; AIChE Symposium Series 250; AIChE: New York, 1986; pp 48-58.

(8) Christian, S. D.; Bhat, S. N.; Tucker, E. E.; Scamehorn, J. F.; El-Sayed, D. A. *AIChE J.* **1988**, *34*, 189-194.
(9) Dunn, R. O.; Scamehorn, J. F.; Christian, S. D. *Colloids Surf.* **1989**, *35*, 49-56.
(10) Scamehorn, J. F.; Christian, S. D.; El-Sayed, D. A.; Uchiyama, H. J.; Younis, S. S. *Sep. Sci. Technol.* **1994**, *29*, 809-830.
(11) Christian, S. D.; Tucker, E. E.; Scamehorn, J. F. *Am. Envir. Lab.* **1990**, *2*, 13-20.
(12) Dunn, R. O.; Scamehorn, J. F.; Christian, S. D. *Sep. Sci. Technol.* **1987**, *22*, 763-789.
(13) Klepac, J.; Simmons, D. L.; Taylor, R. W.; Scamehorn, J. F.; Christian, S. D. *Sep. Sci. Technol.* **1991**, *26*, 165-173.
(14) Dharmawardana, U. R.; Christian, S. D.; Taylor, R. W.; Scamehorn, J. F. *Langmuir* **1992**, *8*, 414-419.
(15) Simmons, D. L.; Schovanec, A. L.; Scamehorn, J. F.; Christian, S. D.; Taylor, R. W. In *Environmental Remediation: Removing Organic and Metal Ion Pollutants*; Vandegrift, G. F.; Reed, D. T.; Tasker, I. R., Eds.; ACS Symposium Series 509; American Chemical Society: Washington, DC, 1992; pp 180-193.
(16) Schovanec, A. L. Ph.D. Dissertation, University of Oklahoma, 1991.
(17) Fillipi, B. R.; Scamehorn, J. F.; Taylor, R. W.; Christian, S. D. *Sep. Sci. Technol.* **1997**, *32*, 2401-2424.
(18) Christian, S. D.; Tucker, E. E.; Scamehorn, J. F.; Lee, B. H.; Sasaki, K. J. *Langmuir* **1989**, *5*, 876-879.
(19) Krehbiel, D. K.; Scamehorn, J. F.; Ritter, R.; Christian, S. D.; Tucker, E. E. *Sep. Sci. Technol.* **1992**, *27*, 1775-1787.
(20) Tondre, C.; Son, S. G.; Hebrant, M. *Langmuir* **1993**, *9*, 950-955.
(21) Tondre, C.; Boumezioud, M. J. *J. Phys. Chem.* **1989**, *93*, 846-854.
(22) Ismael, M.; Tondre, C. *Sep. Sci. Technol.* **1994**, *29*, 651-662.
(23) Ismael, M.; Tondre, C. *Langmuir* **1992**, *8*, 1039-1041.
(24) Sasaki, K. J.; Burnett, S. L.; Christian, S. D.; Tucker, E. E.; Scamehorn, J. F. *Langmuir* **1989**, *5*, 363-369.
(25) Christian, S. D.; Smith, G. A.; Tucker, E. E.; Scamehorn, J. F. *Langmuir* **1985**, *1*, 564-567.
(26) Smith, G. A.; Christian, S. D.; Tucker, E. E.; Scamehorn, J. F. *J. Soln. Chem.* **1986**, *15*, 519-529.
(27) Mahmoud, F. Z.; Christian, S. D.; Tucker, E. E.; Scamehorn, J. F. *J. Phys. Chem.* **1989**, *20*, 5903-5906.
(28) Tuncay, M.; Christian, S. D.; Tucker, E. E.; Taylor, R. W.; Scamehorn, J. F. *Langmuir* **1994**, *10*, 4688-4692.
(29) Christian, S. D.; Tucker, E. E.; Scamehorn, J. F.; Uchiyama, H. *Colloid Polym. Sci.* **1994**, *271*, 745-754.
(30) March, J. *Advanced Organic Chemistry: Reactions, Mechanisms, and Structure*, 3rd ed.; Wiley: New York, 1985; pp 386-397.
(31) Nerdel, F.; Huldschinsky, I. *Chem. Ber.* **1953**, *86*, 1005-1011.
(32) Hartung, W. H. *J. Am. Chem. Soc.* **1928**, *50*, 3370-3375.
(33) Stein, A. S.; Gregor, A. P.; Spoerri, P. E. *J. Am. Chem. Soc.* **1955**, *77*, 191-192.
(34) Martell, A. E.; Smith, R. M. *Critical Stability Constants*; Plenum: New York, 1974; Vol. 1, pp 124-126.
(35) Perrin, D. D.; Sayce, I. G. *Talanta* **1967**, *14*, 833-842.
(36) Scamehorn, J. F.; Harwell, J. H., In *Mixed Surfactant Systems*; Ogino, K.; Abe, M., Eds.; Dekker: New York, 1993; pp 283-315.

Chapter 20

Water-Soluble Metal-Binding Polymers with Ultrafiltration

A Technology for the Removal, Concentration, and Recovery of Metal Ions from Aqueous Streams

Barbara F. Smith[1], Thomas W. Robison[1], and Gordon D. Jarvinen[2]

[1]Chemical Science and Technology Division and [2]Nuclear Materials and Technology Division, Los Alamos National Laboratory, Los Alamos, NM 87544

The use of water-soluble metal-binding polymers coupled with ultrafiltration (UF) is a technology under development to selectively concentrate and recover valuable or regulated metal-ions from dilute process or waste waters. The polymers have a sufficiently large molecular size that they can be separated and concentrated using commercially available UF technology. The polymers can then be reused by changing the solution conditions to release the metal-ions, which are recovered in a concentrated form for recycle or disposal. Pilot-scale demonstrations have been completed for a variety of waste streams containing low concentrations of metal ions including electroplating wastes (zinc and nickel) and nuclear waste streams (plutonium and americium). Many other potential commercial applications exist including remediation of contaminated solids. An overview of both the pilot-scale demonstrated applications and small scale testing of this technology are presented.

Regulatory limits for discharge of radioactive metal-ions from the United States Department of Energy (DOE) nuclear facilities have become markedly lower in recent years, and older technologies for treatment of waste streams such as carrier precipitation have become much less efficient from an overall systems engineering perspective. In the late 1980s our separations team at Los Alamos National Laboratory began evaluating the use of water-soluble metal-binding polymers in combination with UF as a more cost-effective way of meeting the increasingly stringent regulatory requirements for removal of actinides from waste waters.

The concept of using water-soluble metal-binding polymers with UF as a process was first proposed in the late 1960s by Michaels (1). Relevant actinide work had been done by Bayer and Geckeler (2) who, in collaboration with

Myasoedov's group (*3*), tested actinide binding with a water-soluble metal-binding polymer containing the 8-hydroxyquinoline ligand. We tested the concept with the same laboratory-prepared polymer along with a number of other commercially available polymers (*4*) and found that though some polymers did indeed concentrate americium (III) and plutonium (III)/(IV) (the major alpha-active contaminants in our waste waters), our goal of reaching ultra-low discharge levels could not be met with these polymer systems.

Consequently, we began designing polymers that would have higher binding constants for the actinides, particularly americium and plutonium, and would have overall better physical properties for use in the UF process. This approach coincided with the development in our laboratory of rapid survey techniques for evaluation of new polymers. The concept was developed into a preconcentration procedure for analysis of actinides in very dilute solutions (*5*). From the analytical-scale the process evolved to the bench- and pilot-scale for actinide waste water treatment.

Shortly after our work with actinides began, we had the opportunity to collaborate with the Boeing Space and Defense Group, Seattle, WA, on a joint project for electroplating waste minimization. A key issue for the electroplating industry was the removal of valuable or hazardous metal-ions from dilute waste streams without generating sludge that requires disposal in landfills. The goal became recycling in a near-closed loop process. After considering a number of technologies in a best available technology review (*6*), it was decided that water-soluble metal-binding polymers with UF had the potential to meet the needed goals of this project for dilute rinse water treatment and metal-ion recycling. From the analytical-scale studies a process was developed and taken to the bench- and pilot-scale for an electroplating rinse water recycling process (*7*). The first metal recovery systems targeted for commercialization for the electroplating industry were nickel (bright nickel, nickel strike), zinc, copper (copper strike), zinc/nickel alloy, and nickel/tungsten alloy (*8*).

The need in the electroplating and the nuclear industries for the recovery and removal of metals that exist as oxyanions has led to many studies using water-soluble polymers for the removal of a variety of oxyanions. These oxyanions include chromate, tungstate, molybdate, selenate, arsenate, and pertechnetate. The removal of some oxyanions from aqueous solutions using water-soluble metal-binding polymers with UF has been reported (*9*).

A logical extension of this technology, once it became commercially available, was further evaluation for removal and recovery applications for other transition and main-group metals (toxic, valuable, or nuisance) from other aqueous process and waste solutions such as acid mine drainage (*10*) and from solid surfaces. We have been studying the separations chemistry of the elements highlighted in periodic chart format in Figure 1.

The Concept of Water-Soluble Metal-Binding Polymers with UF (Polymer Filtration)

Polymer Filtration (PF) technology uses water-soluble polymers prepared with chelating or ion-exchange sites to sequester metal-ions in dilute aqueous solutions. The water-soluble polymers have a sufficiently large molecular size that they can be separated and concentrated using commercial UF technology. Water and smaller unbound components of the solution pass freely through the UF membrane allowing for the concentration of the polymer/metal complex. By adjusting the solution conditions, the metal-ions are released and are recovered in a concentrated form for recycling or disposal using a diafiltration process. The water-soluble polymer can be regenerated for further waste-stream processing.

The relative efficiency with which an UF membrane retains or rejects a metal species can be determined experimentally with each species assigned a numerical value between 0 and 1 called the rejection coefficient (σ). A rejection coefficient of 0 means that the species freely passes through an UF membrane (permeate) while species with a rejection coefficient of 1 are completely retained (retentate). Small metal-ions will pass freely through the membrane ($\sigma = 0$) unless the effective size is temporarily increased by binding to the polymer ($\sigma = 1$). In the case of a polymer/metal-ion complex in which the polymer (P) is physically too large to pass through the UF membrane, the rejection coefficient of the metal-ions (M^{n+}) in the presence of a complexing polymer (P) is a reflection of the equilibrium or stability constant (K_s) of the complex, which is a measure of the affinity of the polymer for a metal-ion.

$$P + M^{n+} \rightleftharpoons PM^{n+} \tag{1}$$

$$K_s = \frac{[PM^{n+}]}{[P][M^{n+}]} \tag{2}$$

Concentration Mode. Generally, there are two modes of operation in PF (*11*). The first is the concentration mode, schematically shown in Figure 2, where the volume in the retentate is reduced by simple UF. The final concentration of any species in solution can be determined by:

$$C_f = C_0 \cdot \left(\frac{V_0}{V_f}\right)^\sigma \tag{3}$$

where C_f is the final concentration of the species, C_0 the initial concentration, V_0 the initial volume of solution, V_f the final volume, and V_p the permeate volume. If the rejection coefficient of the species is 1, as would be the case for the water-soluble metal-binding polymers, then equation 3 simplifies to:

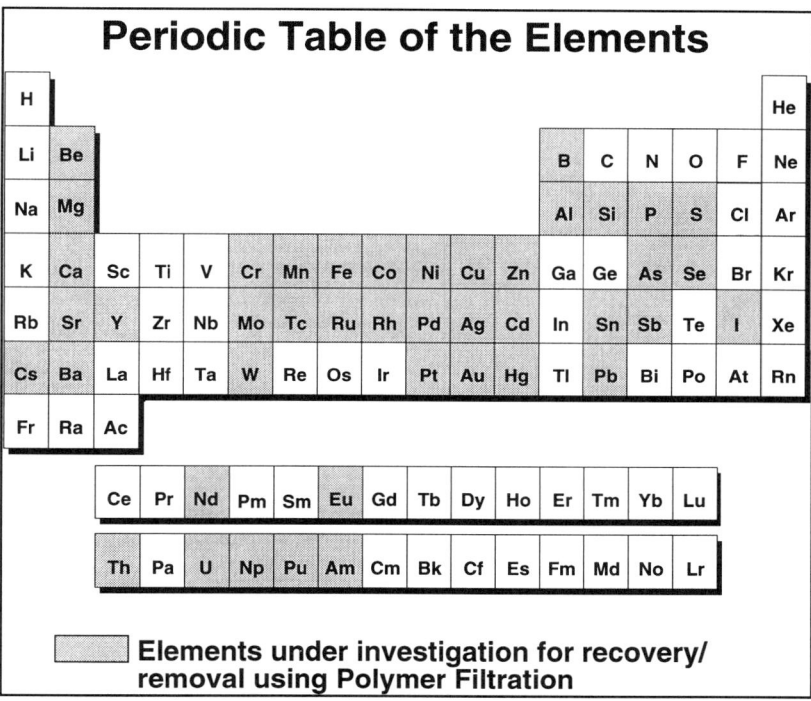

Figure 1. A Periodic Table Summarizing Elements Under Study for Application of Polymer Filtration Technology.

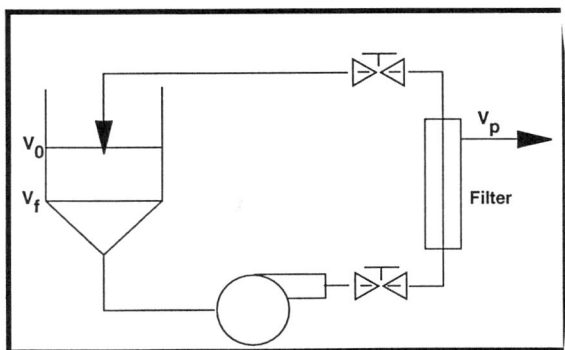

Figure 2. Schematic of Concentration Process by UF.

$$C_f = C_0 \frac{V_0}{V_f} \tag{4}$$

For two species in solution, a polymeric/metal-ion species (PM) and a molecular impurity (A), where $\sigma PM \gg \sigma A$, the UF of the solution should result in the concentration and enrichment of P based on:

$$\left(\frac{C_A}{C_{PM}}\right)_f = \left(\frac{C_A}{C_{PM}}\right)_0 \cdot \left(\frac{V_0}{V_f}\right)^{-(\sigma_a - \sigma_{PM})} \tag{5}$$

UF is the basis for a significant degree of purification during concentration of polymer/metal-ion complexes in solution.

Diafiltration Mode. The second mode of operation in PF is diafiltration (see Figure 3). Wash water (V_w) is added to the retentate at the same rate that the permeate is generated so as to maintain a constant retentate volume. In the diafiltration mode, the lower molecular weight species in solution are removed at a maximum rate when the rejection coefficient equals 0. The retentate is, in effect, washed free of smaller solute. Theoretically, the percent solute (any dissolved species) remaining in the retentate can be calculated by using equation 6:

$$C_f = C_0 \cdot e^{-\frac{V_w}{V_0}(1-\sigma)} \tag{6}$$

where V_w is the volume of solute free liquid (volume-equivalents) added, which also equals the amount of permeate produced (V_p).

The effects of various rejection coefficients on the percent solute retained during a diafiltration process are shown in Figure 4. It can be seen that theoretically, after 5 volume equivalents of processed solution, >99% of the lower molecular weight species with a rejection coefficient of 0.0 should be removed. Experimentally, however, rejection coefficients of 0.0 are not ordinarily observed. Even weak interactions between the solute and the water-soluble polymer or the UF membrane can yield a small retention value. The curves for low retention coefficients follow an exponential decay with each additional volume equivalent giving diminishing returns in percent solute removed, while higher rejection coefficients approach a linear response to solute removal.

Methods of Metal Release. The polymer-bound metal-ion can be released from the polymer by a variety of processes including those shown in the following equations:

$$M^{n+}(P) + nH^+ \rightleftharpoons H_nP + M^{n+} \tag{7}$$

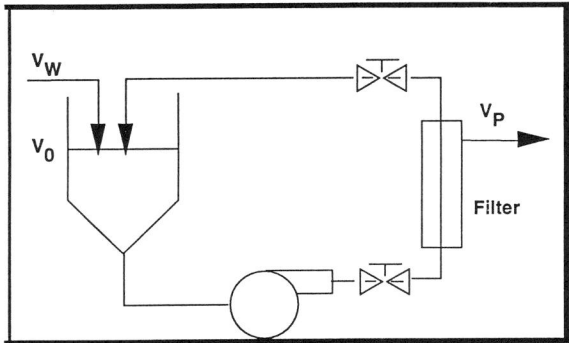

Figure 3. Schematic of the Diafiltration Process.

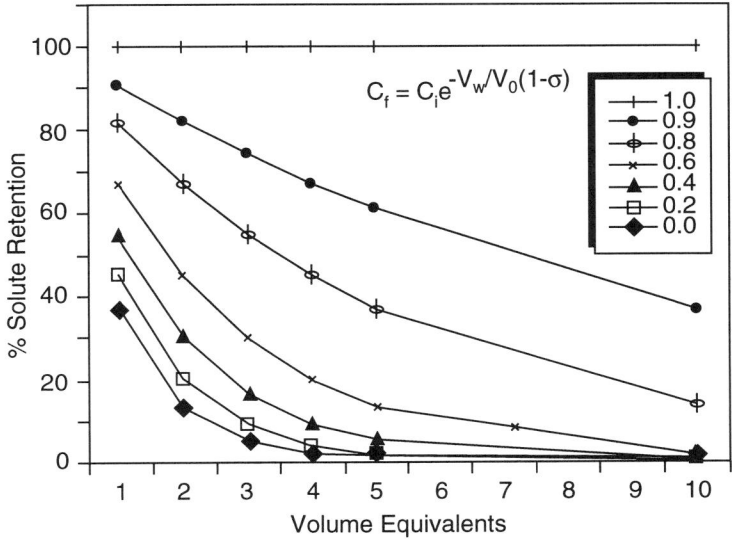

$$C_f = C_i e^{-V_w/V_0(1-\sigma)}$$

Figure 4. Plot of Solute Retained as a Function of Volume Equivalents for Various Rejection Coefficients During Diafiltration

$$M(P)+L \rightleftharpoons ML+(P) \tag{8}$$

$$M^n(P)+e^- \rightleftharpoons M^{(n-1)^+}+(P) \tag{9}$$

where M is the metal-ion, (P) is the water-soluble polymer, L is a competing molecular complexant, n is the oxidation state of the metal-ion, and the reduction reaction can be either chemically or electrically driven. When the metal is released by a proton (equation 7) or by a complexant (equation 8), the polymer-free metal-ion is recovered by a diafiltration process. In some unusual instances, the metal-ion may be so tightly bound to the polymer that destruction of the polymer (incineration, hot acid digestion, smelting, etc.) is required to recover the metal. Optionally, for waste management purposes it may be most economic to solidify the polymer-bound metal, for example, in a grout or cement material, such that it passes Environmental Protection Agency (EPA) toxicity characteristic leaching procedure (TCLP).

General Process Conditions. Generally, the concentration range for the water-soluble polymer in solution ranges from 0.001 weight/volume percent to 20 weight/volume percent of final concentrated solution. It is sufficient, and in some cases desirable, to have only enough polymer in solution such that the polymer's metal-ion loading approaches 90 to 100%. Using higher concentrations of the water-soluble polymer will result in lower flux rates through the membrane during the concentration stage. The use of a high initial polymer concentration can sometimes cause aggregation of the polymer and reduced metal-ion binding capability. In this case, operation at lower initial polymer concentrations can allow more complete metal binding and the polymer can then be concentrated to higher final concentrations with overall improved performance.

During the concentration stage for analytical applications, the polymer and metal-loaded polymer concentration can often become quite high and, in the case where the solution goes to near dryness, it can approach 90% of the weight of the concentrate. For a semi-continuous process it is necessary to work at low polymer concentrations to maintain high permeate flux across the membrane during the concentration stage of the process. During the diafiltration stage the polymer concentrations will always be higher, but at this juncture the volumes being treated are small. The flux is dependent on the transmembrane pressure which is commonly in the range of 25 to 50 psi. However, the increase in flux with transmembrane pressure is limited by concentration polarization and the flux gains are often small beyond 50 psi for typical tangential-flow, hollow-fiber UF units (*11*).

Polymer Leakage Through the Membrane. Ideally, there is no polymer permeating through the UF membrane. If there is any polymer breakthrough ($\sigma < 1.0$), it will ultimately be lost from the system (*12*). This result is unacceptable from a number of process perspectives. First, the polymer must remain in the system to maintain its working concentration. Second, polymer contamination in the permeate can create further problems downstream. Third, loss of metal-loaded polymer that would carry bound metal-ions into the permeate can result in failure to achieve target discharge limits. If there were only 1 ppm loss of polymer from a system that contains 1000 ppm polymer (which might be environmentally acceptable), a 50% loss of material in approximately a million volume equivalents would occur. At ≈2% polymer breakthrough, as has been reported for one experimental system (*13*), 50% polymer loss would occur in approximately 35 volume equivalents. This amount of polymer loss is unacceptable for a viable process.

Measurement of polymer breakthrough has required the use of a variety of methods because each polymer has different functional groups that require different techniques to determine their presence in low concentration. For example, we have used UV-Vis absorption spectra of highly colored metal-ion complexes and total organic carbon analysis to detect polymer in the permeate. However, the absorption spectra are generally limited to levels of 1 ppm or higher and the presence of other organic compounds in the feed solution can interfere with the carbon analysis.
We are developing tags for the polymer backbones that will detect breakthrough at the ppb level or less and will be useful for a wide variety of polymers.

Concentration Factors and Low Level Metal Ion Removal. When the goal of a process involves the removal of metal-ions to very low levels, it can be useful to think in terms of the resulting metal-ion concentration in the permeate solution as opposed to percent metal-ion retained in the retentate. At 99% retention of metal-ions from a solution with an initial concentration of 1000 ppm, 10 ppm remain in the permeate. For most RCRA or toxic metals this amount still represents an unacceptable discharge level and the resulting aqueous stream could not be discharged to any Publicly Owned Treatment Works (POTW). In aqueous streams containing radioactive metal-ions the decontamination factors must often be even greater. A solution containing 1×10^6 pCi/L of plutonium-239 will require 99.999% removal to reach 10 pCi/L. Thus, in evaluating a process it may be more practical to take into account the final waste stream concentration than to consider the percent of metal removed.

Calculation of a concentration factor (CF) for a PF process cycle will require knowledge of the process system which includes: the initial feed metal-ion concentration and volume, the final metal-ion concentrate and volume, the size of the reactor, and the initial polymer concentration (*12*). The polymer concentrations do not exceed 20% in a continuous process because of the reduction in flux rates. Thus, if the reactor size is 20 L (this includes the holdup volume of the system) and the initial polymer concentration is 1% w/v, we can concentrate the reactor volume from 20 L to 1 L at the end of the concentration phase. If the metal-ion concentration in the feed going to the reactor is 100 ppm (e.g., Cu, AW 63.5) and a 1% w/v polymer solution has a capacity of 0.25 g Cu per g of polymer and a 20 L reactor has 200 g of polymer, we can bind 50 g of copper which represents 500 L of feed. Thus, the concentration of 500 L to 1 L is a CF of 500 (CF = V_o/V_f).

The majority of this process will thus be run at a polymer concentration of 1% w/v or less to maintain the high flux rates. Only at the end of the batch will flux rates be reduced substantially as the polymer concentration is increased from 1 to 20% w/v. In actual single-stage practice <100% of the polymer capacity is used to avoid metal-ion breakthrough that would exceed the discharge limit. When the metal-ion is released from the polymer by diafiltration it will take three volume equivalents (one liter is one volume equivalent in this case) or three liters to collect 95% of the metal. Thus the actual CF will be 500/3 = 166.

If the feed metal-ion concentration were ten times less, then the CF would be 10 times greater, or 5000/3 = 1666. For very dilute metal-ion solutions the CF value can be very large.

Rapid Survey Techniques of New Polymers. Various UF equipment has been used to evaluate a polymer's ability to both bind and release selected metal-ions. The most common bench-scale units are stirred cells (*3*). These units are driven by gas pressure and are too laborious and time consuming to assemble for rapid survey of large numbers of polymers under a variety of reaction conditions. We have adapted centrifugation driven UF units developed for protein purification to a rapid survey technique of our polymers under various conditions (*5*). These units are commercially available through companies such as Amicon, Fisher, Millipore, etc. They use both a dead-end filtration as in the Centricon-10 units (Amicon) and a reverse dead-end for the Centraprep units (Amicon) and have sample volumes from 0.5 to 20 mL. Photos of both a stirred cell (200 mL volume) and of a Centricon-10 (2 mL volume) unit appear in Figure 5. The membranes we use typically have a molecular weight cut off (MWCO) of 10,000 to 100,000 Daltons and are composed of a variety of membrane materials such as cellulose acetate, polysulfone, and fluoropolymers.

These small centrifugal units are particularly useful for waste minimization purposes when evaluating polymers for actinides and other radionuclide separations because of the small amount of sample required for testing and because there are no transfer losses. Both the top and the bottom compartments of the unit are placed in separate scintillation vials and the whole unit can be measured (*4, 5*). The data is usually reported as ppm metal or pCi/L remaining in the permeate as a function of the particular parameter being tested. Data for actinide removal is often reported as distribution coefficients (D) as a function of the parameter under study. The D value calculation was adapted from the D determinations for ion exchange resins where D = (Total Bound Metal/Total Unbound Metal) X Phase Ratio, where the Phase Ratio is (Initial Solution Volume in mL/Initial Polymer Weight in gm). Since this equation includes a phase ratio and very small amounts of soluble polymer can have large effects, we can realize some very large D values. For example, a 0.1% w/v polymer solution that retains 99% of a metal-ion gives a log D value of 5.0.

303

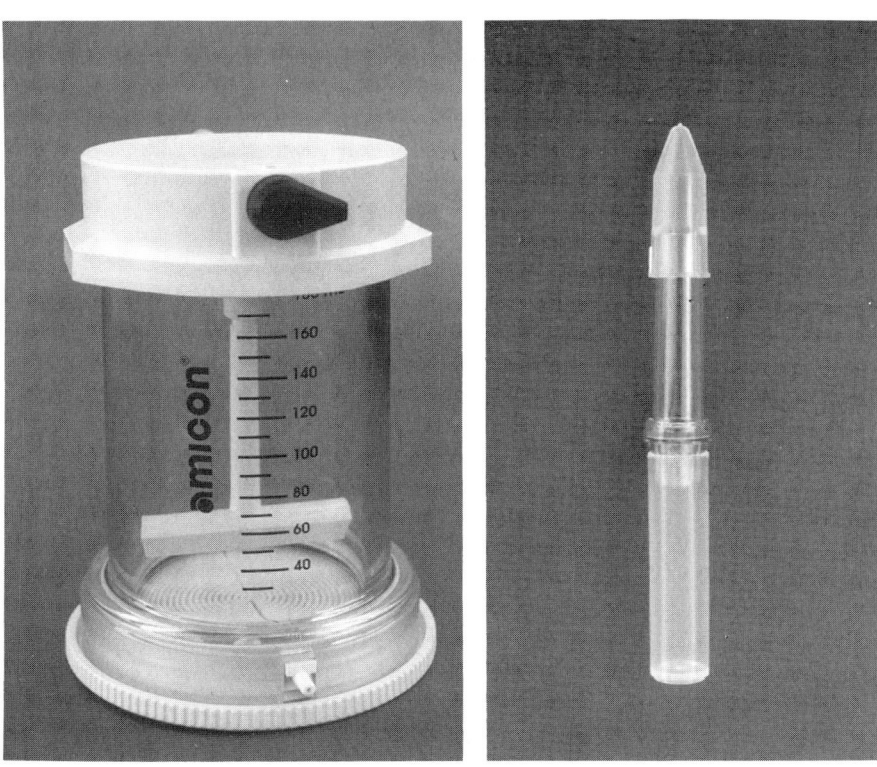

Figure 5. Two Hundred mL Stirred Cell (left) and a two mL Centricon-10 unit (right).

Blank measurements performed in the absence of polymer to determine the behavior of the metal-ion under the experimental conditions used can be useful information. With metals that form hydroxide precipitates or other polymeric inorganic species under the solution conditions, it can be difficult to sort out which species are being removed in the UF step. The presence of the polymer can influence these reactions in a variety of ways. The complexing functionality on the polymer can suppress precipitation reactions, but water-soluble polymers are also commonly used as flocculation agents. Clearly, the order and timing of reagent addition can be crucial to the PF process. Figure 6 gives an example of results of UF with a 10,000 MWCO ultrafilter for a number of divalent and trivalent metal-ions after addition of base to an acid solution in the absence of polymer. An UF operation on a solution of this composition using a polymer that was very selective for Hg(II) at pH 4 would remove most of the Fe(III) as a precipitate. Hydrolyzed metal-ions can bind with the polymer, but their presence in solution influences the binding constants as with any metal/ligand complex system.

Past Versions of Polymer Filtration

After Michaels' (*1*) first proposal of the process concept in 1968, a number of researchers have developed and evaluated the concept under a variety of process names. French researchers worked on the concept in the early 1970s (*14*). A Japanese patent was issued in the late 1970s (*15*) and a German team, Bayer and Geckeler, reported their work in the 1980s (*3*), calling the process Liquid-Phase Polymer-Based Retention. An American team reported work in the early 1990s (*13*) and called the technology Polyelectrolyte-Enhanced UF. An excellent review article by Geckeler and Volchek appeared in 1996 (*16*) which gives the current status of the technology from those authors' perspective. We started using the name Polymer Filtration (PF) when our commercialization activities began as the previous terms were considered too cumbersome by our industrial collaborators (*17*).

Comparison of Polymer Filtration with Other Commonly Used Separations Technologies

PF is a technology for the concentration, removal, and recovery of metal-ions from dilute aqueous solutions. In general, we have applied this technology to feed concentrations of ≤ 1000 ppm metal content. Though higher metal-ion concentrations can be treated, the concentration factors become small. Other processes for metal removal/recovery from dilute solutions include Precipitation (PPT), biphasic Liquid-Liquid EXtraction (LLEX), Ion eXchange (IX), Chelating Ion eXchange (CIX), Reverse Osmosis (RO), Evaporation (EV), filtration (carbon, sand, etc.), ElectroDeposition (ED), and ElectroRecovery (ER). Aqueous chelating ion exchange is the technology most closely aligned with PF because the metal-ion binding chemistry is similar and the chelators can have high metal-ion selectivity.

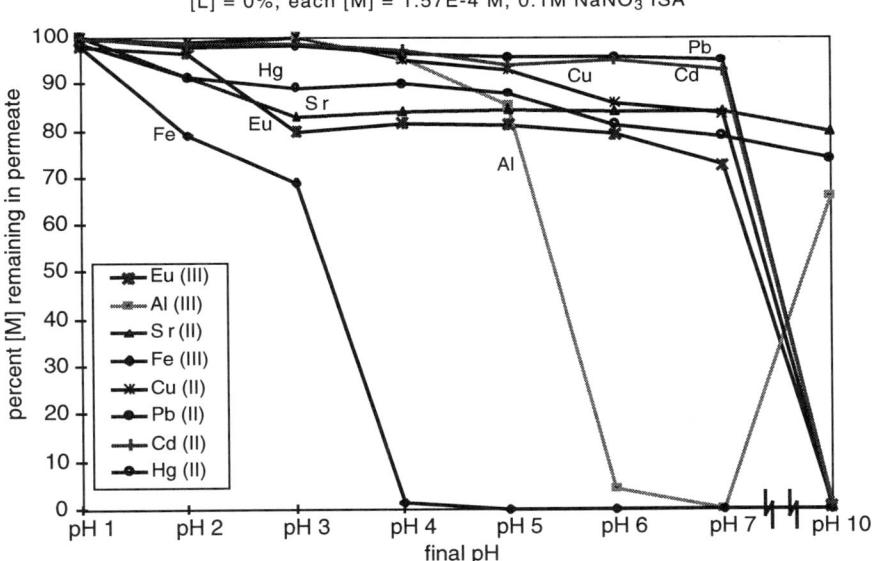

Figure 6. Plot of % Metal Ions Remaining In Solution after UF as a Function of pH in the Absence of Water-Soluble Polymer.

LLEX can also employ very selective chelators, but uses two immiscible liquid phases rather than a solid and liquid phase as in CIX.

Binding kinetics are very rapid with PF because of the homogeneity of the system. With CIX phase transfer between the aqueous solution and the solid resin must occur. This process can be relatively slow in both the metal uptake and release. For example, 90% loading can be attained in PF within seconds, while it may require hours to attain the same level of loading with some resins. This difference makes the kinetics of PF in the range of 10^4 times faster than CIX. Thus, in CIX, column flow rates and column material amounts have to be optimized to allow for slower metal binding and release, and the amount of regeneration solution required to recover the metal-ions can be large. PF can significantly reduce processing times and process volumes relative to CIX. A useful aspect of PF is its ability to recover metal-ion concentrate in a small volume and potentially recycle it directly to the original process all in a single unit. This ability can translate into smaller equipment and fewer polymer requirements for PF technology.

A water-soluble chelating polymer can have metal-ion loading capacities considerably greater than that of similar chelating ion exchange resins because of the greater density of binding sites. For example, Amberlite IRC-718 has a loading capacity of approximately 0.025 g Ni/gram of dry resin, while a water-soluble analogue called Metal-Set-Z has about 0.25 g of Ni/gram of dry polymer.

Metal-binding groups can be built into the water-soluble polymer structure to select specific metal-ions and reject benign impurities such as calcium, potassium, and other salts. Unlike LLEX, no organic solvents are required. In addition, cooperative effects between ligands on soluble polyelectrolytes can give higher binding affinity than the monomer ligands. For example, polyacrylic acid has a 10^4 greater binding constant than the monomer ligand, glutaric acid (18). PF systems can potentially take advantage of such cooperative effects to obtain higher metal binding relative to monomeric extractants.

The PF system can have advantages over other conventional metal recovery processes depending on the application. By contrast with RO, PF is carried out at low pressure (commonly ≤ 25 psi). PF is a relatively low energy process compared to EV and will not damage heat-sensitive solutes. RO and EV, as compared to PF, are unselective processes for solutes, concentrating all waste stream salts and materials, including metal-ions that may be impurities. PPT is often unspecific, generates large amounts of secondary waste, and is limited by solubility products. PF functions well, perhaps even better, at low metal-ion concentration (19), whereas some technologies like PPT have limited applicability. ED can recover metal-ions selectively as pure solids, but not as ions in solution. This process does not allow for efficient recycling in some applications. ED/ER tends to be inefficient at low metal-ion concentrations. The choice of a particular technology is dictated by the required end result and the total system cost. For dilute solution and waste polishing requirements, PF is a cost-effective option.

A very useful aspect of PF is the possibility of developing formulations (mixtures) of polymers with different chelators to recover suites of metal-ions and of separating the concentrated metal-ions from each other with different stripping chemistry. The polymers can also lend themselves to having multiple ligand groups on one polymer. We have over 30 different polymers under development with a variety of functionalities and many polymeric structures are already reported in the literature (*16*). The applications described in more detail below use water-soluble polymers reported in the literature as well as some new proprietary formulations.

The combination of concentration and diafiltration UF processes provides an effective method for the recovery, concentration, and purification of metal-ions in solution. Permeate streams 'free' (in a regulatory sense) of hazardous metal-ions will result. A number of industries successfully use simple UF processes for various applications, including water purification, waste treatment, pharmaceuticals, and the food and beverage industries. Consequently, UF is an accepted technology in industry.

Process Applications

We have been able to apply PF successfully to a variety of actual waste streams and have been involved with the commercialization of the technology for certain applications by working closely with industry. Further research and development work is ongoing for additional applications. The following paragraphs describe some of the developments leading toward commercialization.

Actinide Removal from Aqueous Streams. Because of the nature of LANL's mission and our involvement in actinide separations work for many years, our first application developed for PF has been in the area of ultra-low level analysis of actinides and removal of actinides from process and waste streams.

Development of an Analytical Procedure for Preconcentration of Actinides. The analysis of trace elements in environmental and industrial processes has become very important. Though modern instrumentation can measure increasingly lower concentrations, elements are often still present at levels near or below the detection limit. Routine radiochemical counting methods and inductively coupled plasma-mass spectrometry (ICP-MS) cannot easily measure directly such concentration levels as low as 30 pCi/L (1.1 Bq/L) total alpha, the new DOE Derived Concentration Guideline (DCG) for process waters containing alpha-emitting radionuclides or the 0.05 pCi/L limit for americium or plutonium in groundwaters near the Rocky Flats site in Colorado. Analyses can be further complicated by high concentrations of alkali and alkaline earth salts and silicates in waste waters. These ions may interfere with the analysis, making preconcentration of samples by evaporation ineffective. This has necessitated the development of rapid, reliable, and robust analytical techniques for measuring low concentrations of actinide ions.

PF had been shown to be useful in preconcentrating actinide ions from aqueous systems by Bayer and co-workers (*3*). We have been able to use water-soluble metal-binding polymers combined with UF as an effective method for selectively removing dilute actinide ions from high salt solutions on an analytical-scale (*4,5*). For some of the preliminary studies we used a waste water simulant that is typical of the Radioactive Liquid Waste Water Treatment Facility (RLWTF, TA-50)(Technical Area 50) at LANL (*4*). Development studies were performed using ^{241}Am spiked simulants to determine polymer formulation, binding conditions, and accountability before we addressed actual waste waters. It was found that working at pH 4 gave the best accountability as it minimized adsorption of actinides to the surface of the apparatus. The need for working at lower pH values required the development of new polymers that functioned well in this range (*3,4,5*).

After substantial polymer and methods development (*5*) we were able to concentrate by PF one liter of actual waste water from the RLWTF and compare that with 0.1 liter of the same waste water preconcentrated by evaporation (there were so many solids that we could only evaporate 100 mL to 1 mL). The results indicated 318 (duplicate 314) cpm/L gamma activity in the waste water treated by PF and 310 cpm/L in the normalized evaporated solution. Although we used direct gamma counting techniques for this test, it was possible to use other measurement techniques such as ICP-MS, alpha scintillation spectroscopy, or alpha plate counting. Sample preparation for other modes of analysis was accomplished by quantitatively dissolving the cellulose acetate UF membrane with its polymer filter cake in sulfuric acid (*5*).

Bench-Scale/Glovebox Studies for Actinide Removal from Plutonium Facility Distillate Waters. The results from the analytical application were so encouraging that it was decided to determine if this method could be developed into a process application. There are a number of facilities at LANL and other DOE sites where the technology could be applied; for instance, in the Los Alamos Plutonium Facility (TA-55) before the waste water is discharged to the RLWTF. Alternatively, it could be used for waste stream polishing on neutral waters after they reach the RLWTF at TA-50. The discharge limit for the Plutonium Facility acid waste line to the RLWTF is currently 7×10^7 dpm/L total alpha (30 µCi/L) with typical nitric acid concentrations of 1-6 M. The new discard target in the industrial waste line is 1×10^6 dpm/L total alpha (0.5 µCi/L) in 0.1 mM nitric acid. Part of this goal will be accomplished by nitric acid recycling from distillation. The target alpha activity from the fractionator is 10^2 to 10^5 dpm/L total alpha (50-50,000 pCi/L).

We have been performing bench-scale testing to remove alpha activity (^{238}Pu and ^{241}Am) from distillate waters on two different scales. The first bench-scale testing experiments used a small UF unit with a peristaltic pump similar to that shown in Figure 7. The test results are shown in the flow diagram in Figure 8. In this case we were able to treat 11 liters having 4300 cpm/mL at a flow rate of 80 mL/min to give 10.5 liters of permeate waste water having 97 ± 50 cpm/mL (scintillation counter at 100% efficiency) and 0.5 L of a concentrate with greater than 39,000 cpm/mL. This test met the upper limit of our 10^5 dpm/L goal, but also

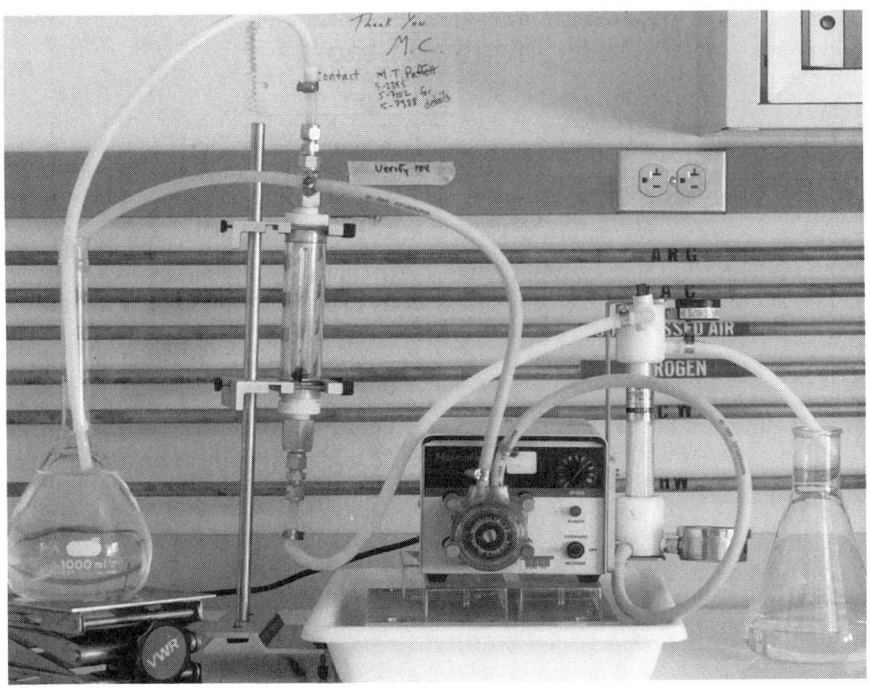

Figure 7. Small UF Unit with a Peristaltic Pump

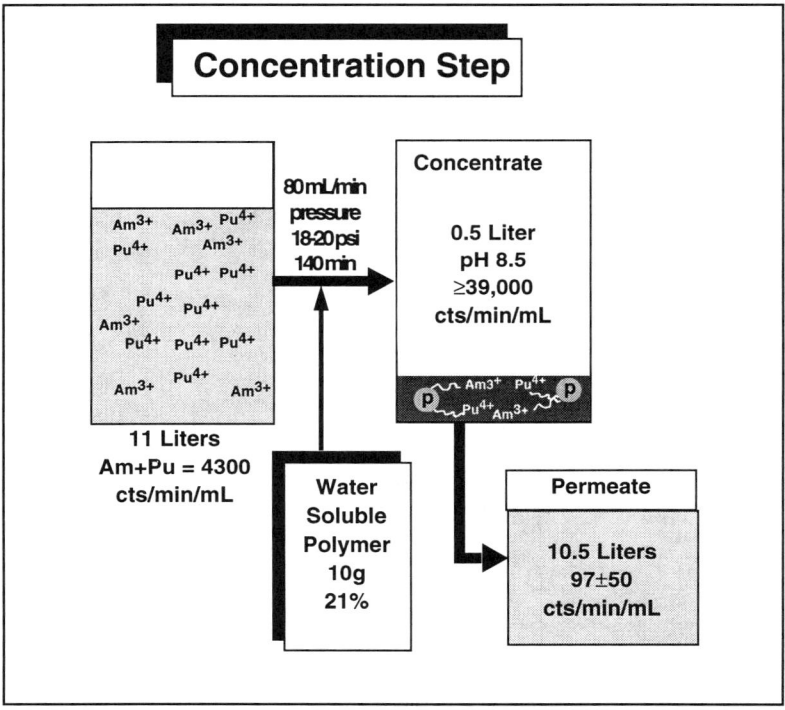

Figure 8. Flow Diagram and Results of a Test on Distillate Waters from TA-55.

illustrated that if we are to reach the lower goal of 10^2 dpm/L (30 pCi/L) that we would need a better polymer formulation and/or two stages of PF.

A larger two stage unit was built as shown in Figure 9, having the flow diagram shown in Figure 10. The reservoir holds 10 L and the flux rates were approximately 1 L/min. The pilot-scale unit was initially tested with neodymium(III) nitrate solutions. These Nd solutions are a reasonable surrogate for americium in the process and waste waters. The two stages that we built into the PF unit worked as expected. One gram of polymer (5) was employed in the first 10 L reservoir and 0.1 gram in the second 2 L reservoir. The initial Nd level of 14 ppm was reduced to less than 10 ppb (detection limit for the ICP-AES analysis) in the process during processing of the first 10 L batch which loaded the polymer to about 20% of "capacity" (assuming a 1:1 mole ratio of the chelating groups to Nd ions). The mole ratio of chelating groups to the Nd(III) ion may be 2:1 or 3:1 in the actual polymer/metal-ion complex and may change with the degree of loading. The loading of the polymer was continued with two additional 10 L batches of Nd solution to observe the expected "breakthrough" of Nd. This work gave us information on the metal-ion capacity of the water-soluble chelating polymer. This unit was placed into a glovebox in the Los Alamos Plutonium Facility and is undergoing further testing with various process and waste solutions, in addition to further optimization of the polymer formulation.

Bench-Scale/Pilot-Scale Studies for Actinide Removal from LANL Radioactive Liquid Waste Treatment Facility(TA-50). The RLWTF at LANL receives all the water from the radioactive acid waste lines from many sites around LANL, including that from over 1000 sinks and drains at approximately 20,000 to 30,000 gal/day. The characteristics of the influent waste water can vary dramatically depending on the status of waste generating activities. During the testing period, the total alpha activity of the waste water ranged from 48,800 to 451,500 pCi/L. The vast majority of the alpha activity can be attributed to three nuclides: ^{241}Am, ^{238}Pu, and ^{239}Pu. Bench-scale testing with ion exchange resins confirmed that these three nuclides exist as both cationic and anionic species. Total suspended solids varied in size with the total quantities measured between 2 to 200 mg/L. It was determined that the split of alpha activity between solid and solution species varied and that the soluble fraction ranged from 1.5 to 5% based on activity levels. Turbidity ranged from 20 to 590 NTU, total organic carbon from 10 to 50 ppm and conductivity from 400 to 1000 µS/cm. The pH ranged from 3.4 to 9.5.

The waste water is presently treated by an iron/lime precipitation method (20). The problems associated with this method are that the removal of actinides by precipitation generates large volumes of waste sludge, and that LANL is facing new actinide discharge limits that are not readily attainable with the existing precipitation method. Storage of large quantities of low-level waste is expensive and requires constant monitoring, plus these wastes will use a substantial quantity of limited landfill space. The best possible solution to both these problems is to demonstrate a

Figure 9. Pilot-Scale Two Stage Polymer Filtration Unit for Installation in a Glovebox.

313

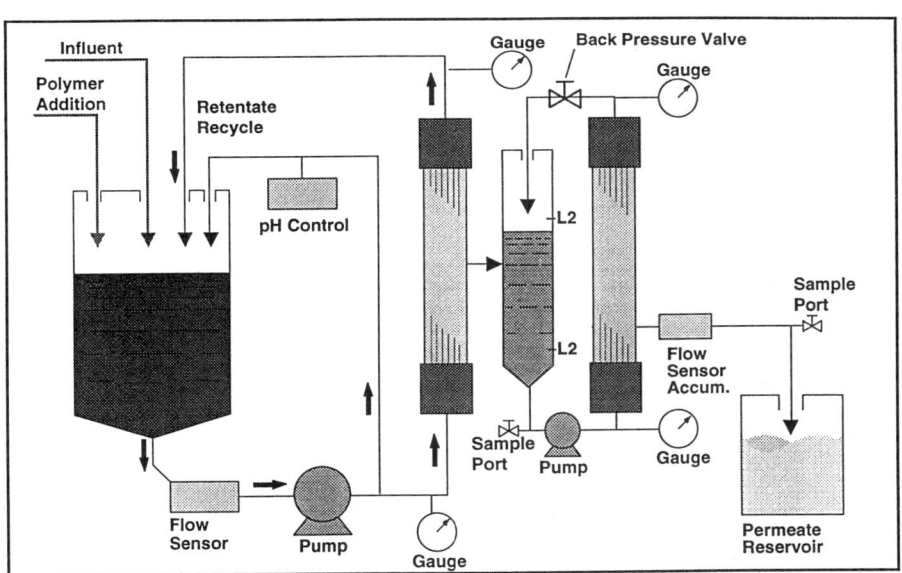

Figure 10. Schematic of Pilot-scale, Two Stage Polymer Filtration Process.

new process for removal of actinides without generating large secondary waste streams.

From previous experiments we have shown that a large percentage of the alpha activity could be removed by simple micro-filtration techniques as shown in Figure 11. Though this technique provides substantial alpha activity removal, it does not consistently meet the new discharge level requirements of 30 pCi/L. To meet this low level, we have proposed prefiltration of the waste water to remove the particulate fraction and then treatment of the filtrate with the PF process to polish to the required levels. Another approach would be to add the water-soluble polymer to the waste stream and remove all alpha activity simultaneously by UF of suspended solids and the water-soluble polymer-bound metals. Both approaches have been tested.

Experiments were performed on a number of different days at the TA-50 RLWTF using equipment similar to that displayed in Figure 7. Table I shows the influent alpha activity followed by the results of several different treatment approaches. The first treatment used a small amount of Betz 1175 flocculating polymer followed by a 5 μm filter. The second treatment was simple UF (10,000 MWCO) which removed more activity than the flocculated system. The final treatment employed PF and gave the best results, <100 pCi/L, which was the limit of detection from facility background (evaporated 10 mL of sample on a planchet and counted). Controlled laboratory measurement of one sample showed ≤ 30 pCi/L alpha per nuclide.

Table I. Test Results on TA-50 Waste Water at pH 6.5

Run Date	Net Alpha Activity (pCi/L)	Rough Bag Filter, 5 μm Pretreat Betz 1175 (pCi/L)	UF (pCi/L)	PF 100 ppm Polymer (pCi/L) In Plant	^{238}Pu (pCi/L)	^{241}Am (pCi/L)
3/26/96	451,552	3,737	3,080	<100		
3/27/96	218,550	17,180	5,000	<100		
3/28/96	118,017	3,690	340	<100		
4/1/96	349,997	1,422	740	<100	29	3.2
4/2/96	116,830	1,657	<100	<100		

Based on these results, a PF unit was designed and assembled for a full-scale demonstration. The unit used open tubular membranes in the first stage so that solids could be removed simultaneously either with or without the water-soluble polymer, eliminating the need for bag filters which helped to minimize waste. This first stage was followed by a second smaller stage having a hollow-fiber UF cartridge. We tested several different process conditions such as pH, polymer formulation and polymer contact time. The unit, shown in Figure 12, treated over 1,000 gallons of waste water. Because of these facility constraints the unit was placed at a second waste water treatment facility at LANL, TA-21. This move

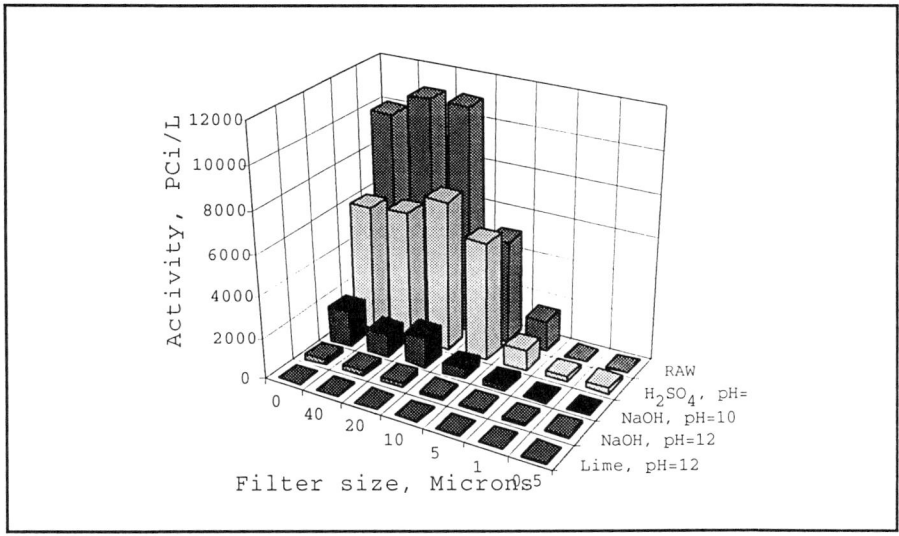

Figure 11. Results of a Filtration Study Using Different Filter Sizes and Different Waste Water Treatments to Determine the Amount of Activity Removed by Simple Filtration.

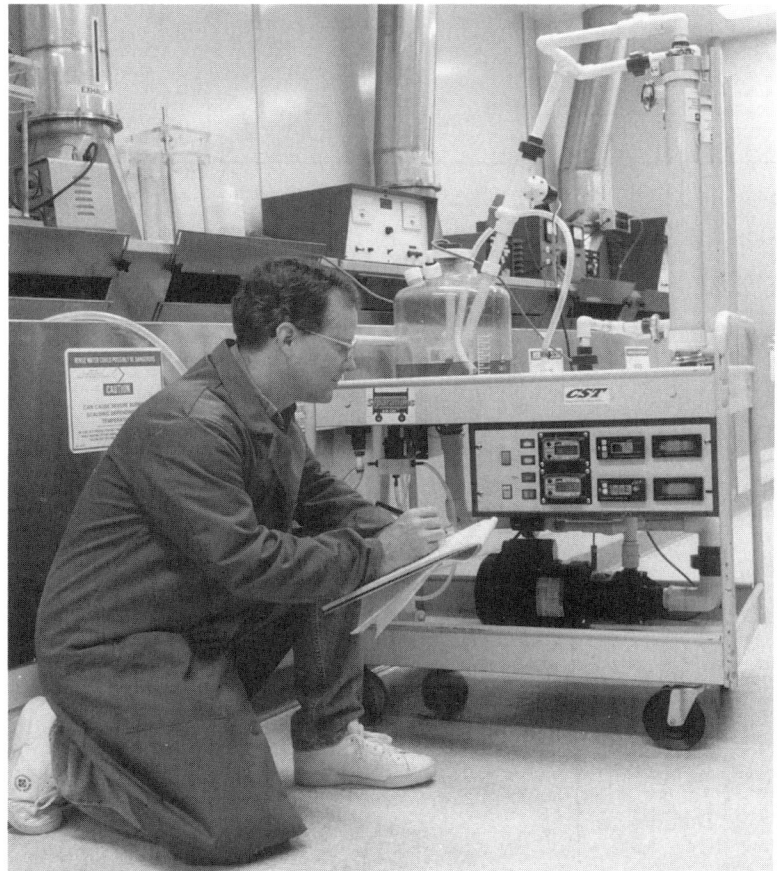

Figure 12. Photograph of PF Unit Assembled for a Demonstration to Remove Actinides at a Waste Water Treatment Facility at LANL.

required further bench-scale testing of the waste water because the influent to this plant was different than at TA-50. A different polymer formulation was developed for this waste stream and we were able to meet our goal of ≤ 30pCi/L after process optimization. For one set of data the raw feed had 1086 pCi/L alpha at pH 7.7. The first tubular UF stage, where no polymer was present in this particular run, gave 173 pCi/L total alpha in the permeate (simple UF), and the second hollow fiber UF stage (with the water-soluble polymer formulation present) gave 23 pCi/L total alpha.

In sum, we have tested on the analytical-scale and moved to the bench- and pilot-scale applications of PF on a variety of different actinide-containing aqueous streams. The strategy has been to use polymers with the highest binding constant and not to recycle the polymer for these applications. We have chosen this method in order to attain our goal of low discharge levels and because the binding constants are so high that it may be difficult to reverse the equilibrium. Generally, there is no need to reprocess the actinides because it is more desirable to stabilize the metal for final waste management. Lastly, although the solution activity may be high, the actual weight of actinide metal is quite small and does not give high loadings on the polymer. If there was a need to recover the metal, redox reactions and competitive chelators would be the likely approaches.

Electroplating Waste Minimization. There are over 10,000 electroplating facilities in the United States that discharge an average 55,000 gal/day of waste water that has to meet the regulations of the EPA clean water act. Presently, about 90% of the technology used to meet EPA and state discharge limits involves hydroxide precipitation. This process requires that sludge be contained, shipped, and buried, steps that are costly and constitute industrial liabilities. The metals are valuable and their replacement represents a considerable energy cost. Consequently, in contrast to the approach taken for actinide removal and concentration, it is preferable for the electroplating industry to concentrate, recover, and recycle the metals in-house.

PF technology was first demonstrated at the pilot-scale at Boeing Defense and Space Group in Seattle, WA for nickel-zinc recovery and recycle from new alloy baths under development (7). In all cases, we were able to obtain a permeate that had < 0.1 ppm of nickel and zinc, well below the state and EPA discharge limit. Figure 13 is a photograph of the unit built for the Boeing demonstration. It has a flow diagram similar to Figure 10, but it is a single stage unit. This application is currently being commercialized for general electroplating applications such as zinc, nickel, and copper rinse baths (8).

The summary of the test results taken from a series of baths at the electroplating facility at LANL is shown below in Table II. The samples were removed from actual electroplating baths and diluted 100 to 1 or 1000 to 1 (based on their original concentration) to represent rinse baths. Metal-Set-Z polymer (8) was added to the pH 7 adjusted solutions to give a 1% w/v solution. The solutions were ultrafiltered (Centricon-10, Amicon) and analyzed by ICP-AES for metal-ion in the permeate. In almost all cases, the permeate was less than 1 ppm metal-ion except where it was apparent we were near the capacity of the polymer (e.g., Cu plate and

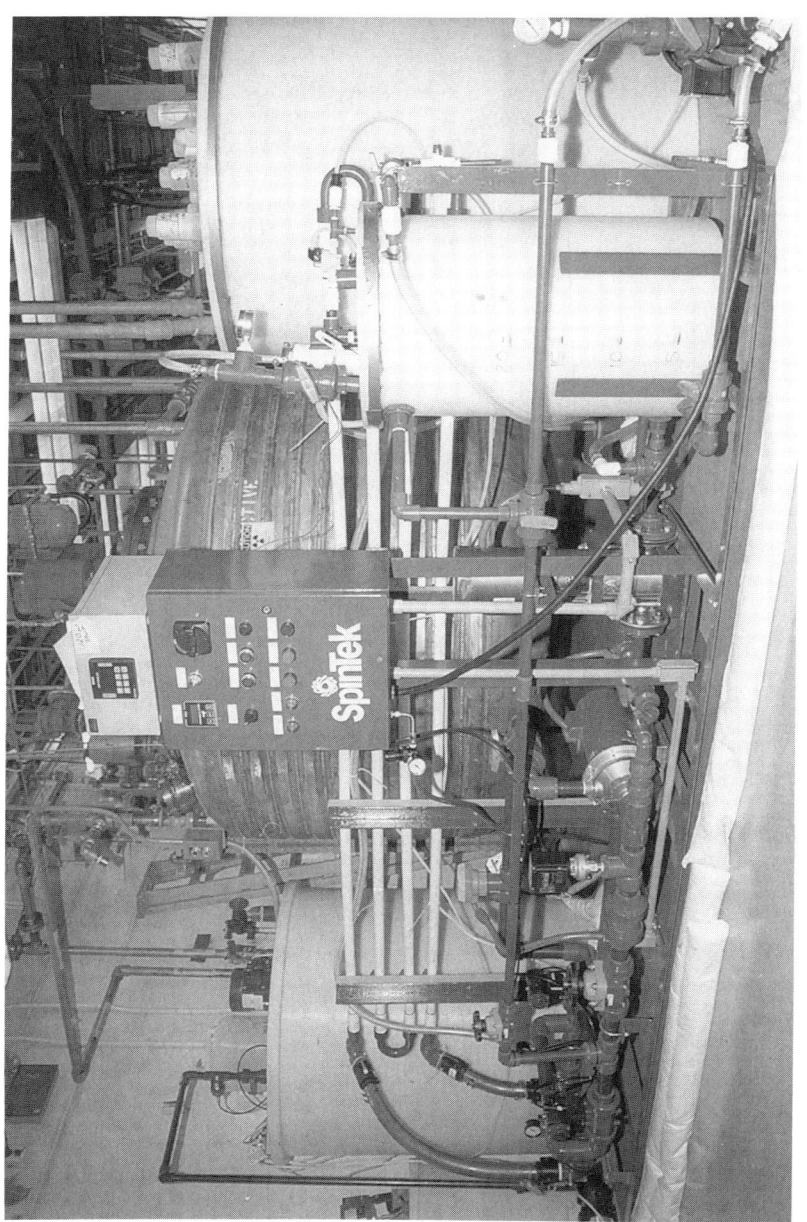

Figure 13. Photograph of PF Unit Built for the Demonstration at Boeing.

Ni plate). Thus, not only can the zinc/nickel alloy rinse baths be readily treated, but a series of other electroplating baths including lead, copper, nickel, and zinc can be processed to remove the metal from a variety of different counterions and additives. Further tests using Metal-Set-Z were performed to determine the recovery of metal-ions that are often found in electroplating baths, either as impurities or as plating metals, from solutions that have chloride, sulfate, or nitrate counterions. Individual solutions containing 0.1 M sulfate, 0.1 M nitrate, and 0.1 M chloride with 0.1% w/v of the polymer were prepared at a pH range of 2 to 7. All solutions contained copper(II), nickel(II), aluminum(III), iron(III), chromium(III), zinc(II), lead(II), and cadmium(II) ions at the 10 to 20 ppm range (low end concentration range expected in electroplating rinse waters). Ten milliliters of the resulting solutions were centrifuged using the Centriprep-10 unit having a MWCO of 10,000 until eight milliliters passed through the membrane. The top (retentate) and bottom (permeate) portions were analyzed by ICP-AES for metal-ion content. The results are summarized below in Tables III and IV.

Table II. Results of Treating a Variety of Diluted Electroplating Baths with 1% w/v Metal-Set-Z Adjusted to pH 7

Bath Name	Product Name	Composition	Original Bath pH	ppm Metal	Dilution Factor	ppm Metal in Permeate
Cu Strike	M&T Harshaw	$CuSO_4$, KOH, Strike complexer	9	11,070 ppm Cu	1/100	0.09
Ni Strike	Made at LANL	$NiCl_2$, HCl	<1	72,960 ppm Ni	1/1000	0.43
Bright Ni	Udylite, OMI Int. Corp.	$NiSO_4$, $NiCl_2$ $B(OH)_3$, org. brightener	5	126,100 ppm Ni	1/1000	0.84
Pb Plate	Made at LANL	$Pb(BF_4)_2$, HBF_4,	<1	284,900 ppm Pb	1/1000	< 1
Cu Plate	Udylite, OMI Int. Corp	$CuSO_4$, H_2SO_4, HCl, UBAC R-1	<1	71,610 ppm Cu	1/100	1.75
Ni Plate	Made at LANL	Ni NSO_3H $B(OH)_3$	3.5	64,780 ppm Ni	1/100	1.49
Zincate	Made at LANL	ZnO $B(OH)_3$	14	2,595 ppm Zn	1/1000	0.25

Table III. Metal Concentrations in the Permeate with 0.1% w/v Metal-Set-Z in 0.1 M Chloride

pH	ppm Cu(II)	ppm Ni(II)	ppm Al(III)	ppm Fe(III)	ppm Cr(III)	ppm Zn(II)	ppm Pb(II)	ppm Cd(II)
2.00	5.22	14.33	13.03	11.50	11.68	12.73	10.58	12.55
2.87	0.10	14.25	12.58	10.93	12.85	12.85	11.48	12.68
4.03	0.02	2.89	11.63	2.32	7.89	10.94	12.11	10.29
4.78	0.02	0.04	2.19	0.44	0.63	0.63	9.60	0.20
5.94	0.67	0.32	0.31	0.31	0.04	0.14	1.34	0.06

Table IV. Metals Concentrations in the Permeate with 0.1%w/v Metal-Set-Z in 0.1 M Nitrate

pH	ppm Cu(II)	ppm Ni(II)	ppm Al(III)	ppm Fe(III)	ppm Cr(III)	ppm Zn(II)	ppm Pb(II)	ppm Cd(II)
2.04	3.96	5.19	4.99	3.75	3.71	4.76	2.95	4.31
3.05	0.15	5.16	4.88	2.83	3.47	5.39	2.97	4.28
4.02	0.05	3.72	4.72	0.44	2.67	5.34	2.80	4.36
4.97	0.02	0.04	1.86	0.05	1.73	2.57	2.54	2.70
6.21	0.03	0.02	0.01	0.01	1.15	0.09	0.33	0.04
6.86	0.05	0.02	0.01	0.01	1.49	0.08	0.03	0.05

This polymer was insoluble at < pH 5 in sulfate solutions, but was completely soluble under all other conditions studied. The solubility data demonstrates the importance of knowing the counterions that exist in the waste streams, along with the anion concentration, to be able to choose the proper polymer for the desired separation.

During a beta test of PF at a large electroplating facility in the Midwest, we encountered an interesting situation where chromate was being splattered (unknown to the facility personnel) into the nickel electroplating rinse bath that we were testing. This problem resulted in a performance reduction for the polymer formulation we were testing. Further evaluation indicated that a different formulation was needed and that some feed adjustment was necessary to convert chromium (VI) to chromium (III). Once that adjustment was made we were able to readily remove both chromium and nickel and selectively recovery the nickel as shown in Figure 14. The two different ratios of polymers tested gave similar results.

A variety of other polymers, polymer formulations, and electroplating systems have been tested that are too numerous to report here. In several cases other polymers perform better than those reported here, but the cited examples illustrate the utility of the process for general electroplating applications.

Evaluation of PF for Applications to Oxyanions. The recovery of oxyanions such as tungstate and chromate is a recurring need in the electroplating industry. Other process or waste streams that contain oxyanions include molybdate from mining wastes, selenate and arsenate from agricultural wastes, and antimonate from

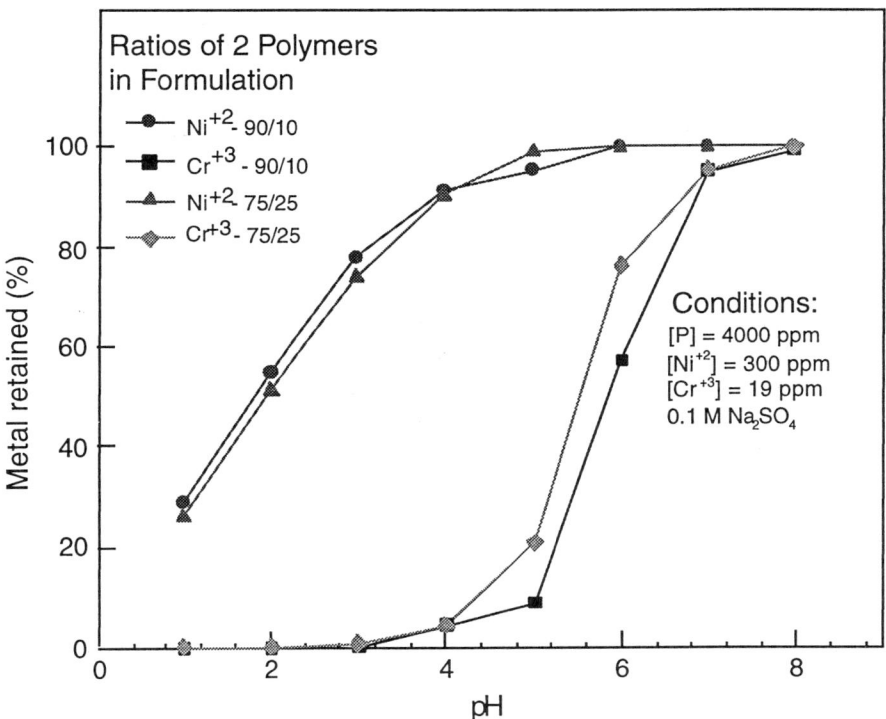

Figure 14. Chromium(III) and Nickel(II) Retention as a Function of pH.

manufacturing. Because of their highly soluble nature, these oxyanions can easily enter surface and groundwaters.

We undertook a survey of oxyanions to evaluate the usefulness of PF for a variety of potential applications. Some previous work on removal of selected oxyanions from aqueous streams using water-soluble anion exchangers has been reported (9). Our study was intended as a baseline to determine if further polymer development was necessary.

The first approach was to evaluate the effect of simple weak and strong base anion exchangers and determine their retention ability under a variety of competing salt conditions and several different pH values. These were performed by preparing 100 ppm solutions of the respective oxyanion at the proper pH value and contacting them with a solution of the water-soluble anion exchange polymer to give a final 1% w/v concentration of the polymer. The solutions were mixed, filtered through a Centricon-10 unit, and the permeate analyzed for the oxyanion using ion chromatography (Dionex). Figures 15 and 16 show the results at pH 7 and 12 in the presence of increasing NaCl concentration. Experiments were not performed at lower pH values as some oxyacids precipitate from solution.

It is apparent that retention decreases for all anions at higher salt concentrations for both the strong and weak anion exchangers. At higher pH, the weak anion exchanger does not bind well compared to the strong anion exchanger as is expected when the ammonium ion exchange sites are deprotonated. Molybdate, tungstate, and selenate showed the highest overall retention with a permeate concentration of < 0.5 ppm under 0.01 M sodium chloride conditions with the strong base anion exchanger and almost ten times that amount for the weak base anion exchanger under the same conditions. Arsenate gave poor results under the experimental conditions with the best removal of 3.4 ppm under 0.01 M NaCl conditions. Arsenate binding is poor because it is not fully deprotonated to the dianionic species until pH 8.5. Though feed adjustment to pH 8.5 may be desirable for process waste waters, for drinking water treatment it would be less desirable to require chemical adjustment to optimize arsenic removal.

From our results, it can be seen that there are many situations where dilute salt solutions could be readily treated to remove oxyanions using simple anion exchange polymers. One of those dilute solution systems would be from groundwater, and chromium removal has been reported using a strong base anion exchanger (19). In those situations where there are high salt concentrations or greater selectivity is needed, other polymers will be required.

Recovery of Metals from Acid Mine Drainage: Treatment of Berkeley Pit Waters (BPW). Water flows into the Berkeley Pit in Butte, Montana, (Figure 17) with a volume of approximately 20 billion gallons from runoff and underground water sources at >3500 gallons per minute. Acidic water produced from bacterial action on the sulfide ores leaches toxic levels of metal-ions from the surrounding mining district in the form of sulfates causing the water to have a pH of about 2.6. It is anticipated that the pit will reach capacity by approximately the year 2015 (21),

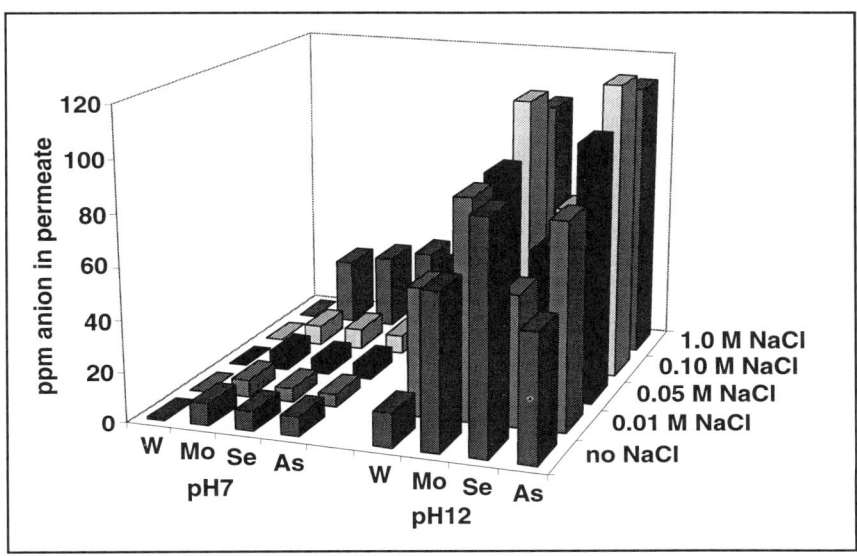

Figure 15. Plots of Permeate versus pH and Salt Concentration for a Variety of Oxyanions Using a 1% w/v Weak Base Anion Exchanger.

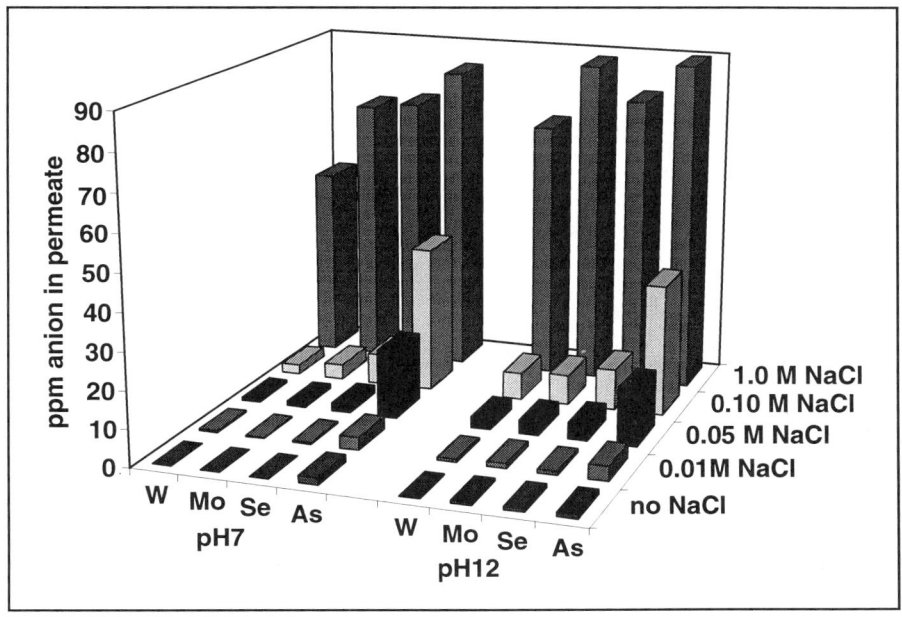

Figure 16. Plots of Permeate versus pH and Salt Concentration for a Variety of Oxyanions Using a 1% w/v Strong Base Anion Exchanger.

Figure 17. Photograph of the Berkeley Pit in Butte, MT.

and treatment is required to prevent water from entering the local rivers and aquifers. The current baseline technology for treatment of the BPW is precipitation of all metals with lime such that the liquid discharge meets criteria for discharge into a POTW or into the local rivers, and that the sludge resulting from precipitation will be buried in the original mine tailings or accumulate on the pit bottom (22). In this way, no metal value is recovered and only the overflow issue is addressed. Precipitation does not address the issue of excess sulfate in the waters, nor does it assure that fresh acid leach water will not redissolve the metals from the hydroxide/carbonate precipitate in the future.

In the estimated 20 billion gallons of contaminated water in the Berkeley Pit there are 61 kilotons of copper and 176 kilotons of zinc. In most of the western states there are thousands of abandoned mine sites of which 10% are thought to have a problem with acid mine drainage. In addition, there are many active mine sites that require management of water and acid drainage.

We have completed a preliminary proof-of-principle evaluation of PF technology for removal of hazardous metal-ions and recovery of valuable metal-ions from BPW. In concert with several other separations technologies, PF can both remove nuisance metals such as aluminum and iron and recover valuable metal such as copper and zinc while removing other hazardous trace metals such as lead and chromium. A water stream suitable to be added to the local streams or used for irrigation could result.

Our strategy for solving the BPW problem is based on the fact that there are large amounts of aluminum and iron present that need to be separated from the more valuable and hazardous metals. Non-hazardous metals, such as calcium, potassium, and sodium, can be discharged with the water. It is uncertain if aluminum and iron have any value at this time, but it might be desirable to remove them separately. The main metals of value in large amounts are copper and zinc. Only small amounts of nickel are present. Table V gives the analytical composition of BPW averaged over three depths.

To increase the efficiency of the PF process we felt it would be advantageous to initially remove as much of the iron and aluminum from the water as possible. If left in the waters, their high concentrations would potentially deplete the polymer's metal-binding capacity, necessitating higher polymer concentration. Additionally, if any electrochemical recovery approaches are used, iron is a major interference because it is quite electroreactive. For optimum metal-ion-binding of the polymers used in our experiments, it is necessary that the pH be raised to approach neutral values (ultimately for discharge to POTW the waters must be nearly neutral).

The addition of hydroxide (i.e., as potassium hydroxide or sodium hydroxide, etc.) presents a number of problems. As the pH is raised, iron and aluminum will slowly precipitate from solution. The precipitation of metal hydroxides can result in the inclusion of large amounts of other metals (e.g., copper, and zinc) within the iron and aluminum hydroxides generating complex mixtures of metals.

Table V. Composite Composition from Three Depths of Acid Mine Drainage Water* from the Berkeley Pit, Butte, Montana

Element (ppm)	Element (ppm)	Element (ppm)
Ca (478)	B (0.40)	V (0.11)
Mg (418)	Cd (1.67)	Zn (528)
Na (69)	Cu (184)	As (0.53)
K (18)	Li (0.26)	Co (1.75)
SiO_2 (97.5)	Mo (0.058)	Cr (0.055)
Fe (875)	Ni (1.06)	Cl (12)
Mn (186)	Sr (1.36)	SO_4 (7643)

* pH = 2.6

We decided to take advantage of one of the water-soluble polymer's chelating abilities, in addition to its basicity, to eliminate the disadvantages of hydroxide precipitation. By adding sufficient polymer to complex all the metal-ions except iron and aluminum, we anticipated that copper, zinc, and other metals in solution would remain bound as polymer-metal complexes, selectively precipitating the iron and aluminum (solubility at pH 4.8; Al(III) = 0.1 ppm, Fe(III) = 0.001 ppm).

A general procedure involved treating the BPW with a dilute solution of basic polymer to adjust the pH of the water to near 4.8. The solid precipitate was separated by centrifugation. The pH of the supernatant was then increased so that most of the valuable and toxic metal-ions became completely polymer bound. The supernatant was ultrafiltered to concentrate the metal-ions and to give a permeate free of hazardous metals for discharge (after sulfate and manganese are removed).

The first test involved treating BPW (20 mL) with dilute polymer solution (2 mL, 50,000 ppm of polymer) to adjust the pH of the water to pH 4.8. It was noted that upon the addition of the polymer solution, the pH of the water increased and immediately stabilized at 4.9 resulting in rapid and nearly complete iron and aluminum precipitation. The supernatant was separated from the solid by centrifugation and allowed to sit overnight. No additional precipitate was observed and the pH changed by no more than 0.1 units. This result is evidence of the rapid kinetics of the precipitation process. The procedure was repeated a number of times and was reproducible. Table VI gives typical metal-ion concentrations found in the supernatant in comparison to the metal concentrations found in 20 mL of raw BPW. The analysis indicated all the iron and 95% of the aluminum was removed from solution with zinc (100%), magnesium (100%), manganese (98%), and copper (72%) remaining in solution. A water wash of the precipitate yielded another 10% of copper bringing the recovery to 82%. We thought that a continuous wash step would recover more of the polymer-copper complex from the iron/aluminum sludge.

The above experiment was repeated on 80 mL of BPW using 10 mL of polymer-solution with similar results. This approach might indicate that the process can be readily scaled.

Previous work (see Tables III and IV) has shown that we can selectively separate the zinc from the copper by simple pH adjustment once the majority of the iron and aluminum is removed from solution. The polymer forms a less stable complex with zinc, allowing it to be selectively stripped from the polymer at a higher pH value. Once the zinc is collected in a concentrated solution, the copper can also be removed from the polymer as a concentrated solution. Although these results were encouraging, we need to emphasize that the process has not been optimized.

To show PF effectiveness in meeting metal-ion discharge levels after the precipitation of iron and aluminum, the BPW (20 mL) was treated in a similar manner as above with a different polymer, Metal-Set-C, resulting in a final pH of 5.6. The supernatant was removed from the iron/aluminum precipitate and ultrafiltered through a 10,000 MWCO membrane. Analysis of the permeate discharge waters by ICP-AES indicated metal concentrations below detectable limits for copper, aluminum, iron, and zinc of <0.01 ppm and nickel, chromium, lead, and

Table VI. Metal Content (ppm) of Berkeley Pit Water After Various Treatments

Element	Untreated Sample pH 2.6[a]	After NaOH pH 3.0[b]	After KOH pH 3.8[b]	After NH4OH pH 3.8[b]	After Metal-Set-Z pH 4.8[b]	After Metal-Set-C pH 5.6[c]
Fe	17.21	4.47 (26%)	2.97 (17%)	0.02 (<1%)	0.0 (<1%)	<0.01
Mn	3.99	---	---	---	4.12 (103%)	---
Al	5.92	4.51 (76%)	4.29 (72%)	3.64 (62%)	0.31 (5%)	<0.01
Cu	3.74	3.33 (89%)	3.25 (87%)	3.04 (81%)	3.04 (81%)	<0.01
Mg	8.82	---	---	---	8.7 (98%)	---
Zn	10.74	10.95 (103%)	10.8 (100%)	9.82 (92%)	11.2 (104%)	<0.01

[a] Values are the amount in mg of each metal in 20 mL of BPW.
[b] Values are reported as mg of metal remaining in solution after precipitation from 20 mL of BPW. Values in parentheses are the percent of metal remaining in solution based on original BPW. Metal-Set-Z, PolyIonix, Dayton, NJ.
[c] Sample from NaOH precipitation treated with 0.12%w/v Metal-Set-C, PolyIonix, Dayton, NJ.

cadmium of <0.3 ppm. The preliminary evaluation of PF as a technology for the recovery of metal-ions from the Berkeley Pit has proven to be very encouraging.

We have successfully shown that the addition of the appropriate amount of a basic water-soluble polymer to the untreated water results in the selective precipitation of iron and aluminum from the copper, zinc, and manganese by a kinetically rapid process. This process has the potential advantage of an in-line treatment as compared with a batch-type process. It has been shown in previous work that the zinc and copper can be separated selectively by pH adjustment. The polymers are recycled in the system for additional metal-ion recovery; therefore, a secondary waste stream is not generated. Analysis of discharge or permeate waters were below detectable levels of metals which demonstrates the effectiveness of PF as a polishing step for low-level concentrations of metals.

Summary

The PF system is a technology for removing, concentrating, and recycling metal-ions from industrial waste water, thereby conserving valuable resources and reducing pollution. In its current commercial application for the electroplating industry, the system can be sized for both large and small operations. PF is a new process that will see broad application in industry. The following applications are under investigation in our laboratory:

(1) Analytical preconcentration,
(2) Nuclear power/nuclear facility waste streams,
(3) Electroplating rinse waters,
(4) Photofinishing waste streams,
(5) Acid Mine drainage/advanced mining techniques,
(6) Treatment of ground water/drinking water,
(7) Precious metals industry,
(8) Catalyst waste streams,
(9) Electronics waste streams,
(10) Cooling tower water,
(11) Textile waste streams,
(12) Municipal waste streams,
(13) Soil remediation/surface decontamination.

Polymer Filtration is a technology worthy of consideration in any situation where dilute waste stream polishing or dilute metal-ion recovery is needed.

Acknowledgments

The authors would like to acknowledge the DOE, Office of Industrial Technology, Industrial Waste Reduction Program for support of the electroplating waste minimization efforts, the DOE, Environmental Management (EM-50), Efficient

Separation and Processing Crosscutting Programs for supporting the actinide separations work, Los Alamos Waste Management Program (DOE, EM 30) for the TA-50 demonstration work, and Los Alamos National Laboratory Directed Research and Development (LDRD) and program development (LDRD-PD) for supporting the rest of the projects including the oxyanion work and the Berkeley Pit work. The work was performed under DOE. Contract #W-7405-ENG-36.

There are a number of people who have contributed to these numerous projects over the years including Nancy Sauer, Norman Schroeder, Peter Stark, Ken Bower, Rich Barrans, Ken Mullen, Rowena Gibson, Yvonne Rogers, Mavis Lin, Kennard Wilson, Doris Ford, Brandy Duran, Trudi Foreman, Noline Clark, Debbie Ehler, Gery Purdy, Johnnie Anderson, Greg Hirons, Man Lu, Jim Jarvinen, Eric Santos, Susan Folkert, Everett Neal, David Soran, Chris Lubeck, Monica Martinez, Anna Mack, Bryan Carlson, Cecilia Olivares, Neal Martin, and Raj Jain.

Literature Cited

(1) Michaels, A. S. In *Advances in Separations and Purifications*; Perry, E. S., Ed.; John Wiley: NY, 1968.

(2) Geckeler, K. E.; Bayer, E.; Shkinev, V. M.; Spivakov, B. Y. *Naturwissenschaften* **1988**, *75*, 198.

(3) Novikov, A. P.; Shiknev, V. M.; Myasoedov, B. F.; Geckeler, K. E.; Bayer, E. *Radiochim. Acta* **1989**, *46*, 35.

(4) Smith, B. F.; Gibson, R. R.; Jarvinen; G. D.; Jones, M. M.; Lu, M. T.; Robison, T. W.; Schroeder, N. C.; Stalnaker, N. In *Evaluation of Synthetic Water-Soluble Metal Binding Polymers with Ultrafiltration for Selective Concentration of Americium and Plutonium*, MARC IV Proceedings, *J. Radioanal. Nucl. Chem.*, **1998**, in press.

(5) Smith, B. F.; Gibson, R. R.; Jarvinen; G. D.; Robison, T. W.; Schroeder, N. C. *Preconcentration of Ultra-Low Levels of Americium and Plutonium from Waste Waters by Water-Soluble Metal-Binding Polymers with Ultrafiltration*, MARC IV Proceedings, *J. Radioanal. Nucl. Chem.*, **1998**, in press.

(6) Sauer, N .N.; Smith, B. F.; *Metal Ion Recycle Technology for Metal Electroplating Waste Waters*; Los Alamos National Laboratory Report LA-12532-MS, UC-367, **1993**.

(7) Smith, B. F.; Robison T. W.; Cournoyer, M. E.; Wilson, K. V.; Sauer, N. N.; Lu, M. T.; Groshart, E. C.; Nelson, M. C. In *Polymer Filtration: A New Technology for Selective Metals Recovery,* International Technical Proceedings from *SURFIN 95*, **1995**, p 607.

(8) PF Technology is commercially available for electroplating applications from PolyIonix Separation Technologies, Inc., a Division of PGI, Dayton, N.J.

(9) Shkinev, V. M.; Vorob'eva, G. A.; Spivakov, B. Y. ; Geckeler, K. E.; Bayer, E. *Sep. Sci. Technol.* No. 1, **1987**, *22*, 2165.

(10) Smith, B. F.; Cournoyer, M. E.; Duran, B.; Ford; T. W.; Gibson, R.; Lin, M.; Meck, A.; Robison, P.; Robison, T. *Chelating Water-Soluble Polymers for Waste Minimization*, Los Alamos National Laboratory Report LA-UR-96-3224, **1996**.

(11) For a general discussion of the ultrafiltration process see: Cheryan, M., *Ultrafiltration Handbook*, Technomic, Lancaster, United Kingdom, 1986; and Winston, W. S.; Sirkar, K. K. *Membrane Handbook*, Van Nostrand Reinhold: N.Y., 1992.

(12) Strathmann, H. *Sep. Sci. Tech.* **1980**, *15*, 1135.

(13) Tuncay, M.; Christain, S. D.; Tucker, E. E.; Taylor, R. W.; Scamehorn, J. F. *Langmir*, **1994**, *10*, 4688.

(14) Nguyen, Q. T.; Aptel, P.; Neel, J. J. *Mem Sci.*, **1980**, Vol. 71.

(15) Mitsubishi Rayon Co. LTD., **1977**, *466,* 150.

(16) Geckeler, K.; Volchek, K. *Env. Sci. Tech.* **1996**, *30*, 727.

(17) PF Systems applied to electroplating waste minimization won a 1995 R&D 100 award, August 1995, LALP-95-184.

(18) Gregor, H. P. *The Specific Binding of Ions By Polyelectrolytes: Correlation with Biological Phenomena*, Proceedings of the New York Academy of Science, **1956**, p. 667.

(19) Tucker, E. E.; Christian, S. D.; Scamehorn, J. F.; Uchiyama, H.; and Guo, W.; In *Transport and Remediation of Subsurface Contaminants*; Sabatina, D. A.; Knox, R. C., Eds.; American Chemical Society: Washington, D. C., 1992; p 84.

(20) Emelity, L. A.; Christenson, C. W.; Kline, W. H.; In *Practices in the Treatment of Low - and Intermediate Level Radioactive Wastes*, IAEA: Vienna, 1966, p 187.

(21) Canonie Environmental Services Corp., *Butte Mine Flooding Operable Unit Remedial Investigation/Feasibility Study, Draft Remedial Investigation Report,* **1994,** ARCO, Vol I.

(22) U.S. Environmental Protection Agency, Superfund Remedy Summary, Mine Flooding Operable Unit, Butte, Montana, Sept. **1994**.

SEPARATIONS USING CHROMATOGRAPHIC AND SUPERCRITICAL FLUID EXTRACTION SYSTEMS

Chapter 21

Different Two-Phase Liquid Systems for Inorganic Separations by Countercurrent Chromatography

Boris Ya. Spivakov, Tatiana A. Maryutina, Petr S. Fedotov, and Svetlana N. Ignatova

Vernadsky Institute of Geochemistry and Analytical Chemistry, Russial Academy of Sciences, 19 Kosygin Street, Moscow 117975, Russia

Within the last several years, some fundamentals and methods of inorganic separations by countercurrent chromatography (CCC) in systems with different solvents and extracting reagents have been developed. A hypothesis on the mechanism of organic phase retention in rotating coil columns and requirements for two-phase liquid systems have been proposed. Reagent concentration, type of organic solvent, and the composition of mobile phase have been shown to affect strongly the stationary phase retention and chromatographic behavior of metal ions. Theoretical and experimental studies have enabled the development of methods for the concentration and separation of different inorganic species. Procedures for the group preseparation of rare earth and some rare (Zr, Hf, Nb, Ta) elements from multicomponent matrices were proposed. It has been demonstrated that CCC can be successfully applied in geochronological studies to the preconcentration and separation of Nd, Sm, Rb, and Sr from geological sample solutions before mass spectrometric determination. The application of CCC to the purification of salt solutions also has been investigated.

Countercurrent chromatography (CCC), a support-free partition chromatography, is currently attracting great interest from investigators working on the separation and preconcentration of organic and inorganic substances. CCC is based on the retention of one phase (stationary) of a two-phase liquid system in a rotating column under the action of centrifugal forces while the other liquid phase (mobile) is being continuously pumped through (1). An important distinguishing feature of CCC as a chromatographic method is the absence of an adsorptive matrix for retaining the stationary phase. This feature determines the main advantages of the method, such as the absence of solute loss due to interaction with the sorbent matrix, a variety of two-phase liquid systems may be used, easy change from one partition system to another,

©1999 American Chemical Society

the possibility to change a volume of the sample solution from 0.1 to 1000 mL or more, and a high preparation capacity is provided by a high ratio (up to 0.9) of the liquid stationary phase volume and the total column volume (this ratio is much higher than that for the stationary solid phase used in HPLC). It should be noted that the problem of column packing is also eliminated in CCC and the stationary phase is relatively inexpensive.

A few devices providing retention of the stationary phase in the field of mass forces in the absence of a solid support have been suggested. Among the various possible designs, the planetary centrifuge retains the liquid stationary phase effectively and enables the fastest and most efficient separation to be achieved. A column (or a column unit) of a certain configuration rotates around its axis and simultaneously revolves around the central axis of the device with the aid of a planetary gear. So far, the technique has been studied and used mainly for preparative and analytical separation of organic and bioorganic substances. The studies of the past several years have shown that the technique can be applied to analytical and radiochemical separation, preconcentration, and purification of inorganic substances in solutions on a laboratory-scale by use of various two-phase liquid systems (2). The chromatographic behavior of inorganic compounds as well as organic compounds is dependent on the properties of the system used, partition coefficients of substances to be separated, and operating parameters of the planetary centrifuge such as rotation and revolution speeds, direction and speed of the mobile phase pumping, internal diameter of the column, and sample volume. However, the systems for inorganic separations are very different from those for organic separations, as in most cases they contain a complexing reagent (extracting ligand) in the organic phase and mineral salts and/or acids in the aqueous phase. Thus, the complexation process, its rate, and the mass transfer rate can play a significant role in the separation process (3).

An important factor that determines the separation efficiency and peak resolution for both organic and inorganic compounds is the S_f value, which is the ratio of the stationary phase volume retained in a column (V_s) to the total column volume (V_c). The value of S_f is dependent on the parameters of the planetary centrifuge (rotation and revolution radii, tube diameter), on the operating conditions (rotation and revolution speeds, flow rate, and direction of pumping of the mobile phase), and on the physico-chemical properties of the two-phase system. The influence of the planetary centrifuge parameters and operating conditions on the stationary phase retention have been well studied for some simple two-phase liquid systems consisting of water and one or two organic solvents (1, 4-13). However, the addition of extracting reagents and mineral salts can strongly affect the physico-chemical properties of the liquid systems and, consequently, their hydrodynamic behavior and S_f value.

The theoretical part of the present work is an attempt to correlate S_f values with the operating conditions of the planetary centrifuge and the composition and physico-chemical properties (interfacial tension, densities, and viscosities of the liquid phases) of two-phase systems used for inorganic separations. Complicated liquid systems containing an organic solvent (n-decane, chloroform, carbon tetrachloride, or methyl isobutyl ketone), an extracting reagent (di-2-ethylhexylphosphoric acid (HDEHP)), water, and a mineral salt (ammonium sulfate) are considered. The investigation of the hydrodynamic behavior of such systems is practically important and is illustrated by several examples. The practical section comprises various applications of CCC to the preconcentration and separation of rare, rare earth, and some other elements in inorganic analyses as well as to the purification of chemical reagents.

Apparatus and Chromatographic Procedure

The chromatographic investigations were performed on a self-designed planetary centrifuge with a vertical one-layer coil column drum (type J according to Ito's classification of planetary motion (*1*)). The rotation and revolution speeds (ω) were 350-500 rpm. The planetary centrifuge has a revolution radius $R = 140$ mm and a rotation radius $r = 50$ mm. The column was made of a Teflon tube with an inner diameter of 1.5 mm and the total inner capacity of the column was 20 mL.

We also have developed a new design of CCC apparatus used in our studies of inorganic separations. The centrifuge consists of six spiral columns wound onto drums of different diameters (7, 10, 15 mm) and mounted in two assemblies bearing three columns each. The tubing diameter ranges from 0.5 to 1.5 mm. The β values ($\beta = r/R$) range from 0.35 to 0.75. The columns can be connected in series or they can be operated independently, which is important for the performance of several simultaneous experiments. In addition, this design enables the rotation axis to be changed from vertical to horizontal. The apparatus can be operated at speeds up to 1000 rpm.

For the separation of inorganic substances, we used the following chromatographic procedure. Before commencing the CCC separation experiment, the components of the two-phase liquid system were stirred and brought into equilibrium to assure mutual saturation of the phases, after which the aqueous phase was used as the mobile phase and the organic as the stationary phase. The spiral column in the stationary mode was filled with the organic phase and, while the column was rotated, the aqueous phase was fed to its inlet. The mass force field which arose during rotation made it possible to retain the stationary phase in the column (V_s) while the mobile phase was continuously pumped through. The amount of the stationary phase in the column depends on the retention factor S_f (ratio of the stationary phase volume to the total column volume). After volumetric equilibrium between the mobile and stationary phase had been established, the sample was introduced into the column. Beginning from the moment of sample introduction, fractions (1-3 mL) of the mobile phase (eluate) were collected to determine the elements of interest by different methods (ICP-AES, AAS, radiometric measurements). The pumping rate (F) was changed from 0.5 to 2.5 mL/min.

Stationary Phase Retention

For about 25 years, CCC has been successfully applied to the separation of different substances. However, regularities and the mechanism of the stationary phase retention in a rotating coil have not been adequately studied. The present study is an attempt to correlate the composition, some physico-chemical parameters of liquid systems, and planetary centrifuge operating conditions with stationary phase retention.

On the Mechanism of Stationary Phase Retention in a Rotating Coil Column.
We are considering a coil rotating with an angular velocity ω (planetary motion,

type J). The coil is filled with two immiscible liquids, their densities are ρ_s and ρ_m, where the subscripts s and m correspond to the stationary and mobile phase, respectively. The interfacial tension is designated γ. The following forces act on an "element" of the stationary phase, which has coordinates of its mass center P(x;y), length L (along a coil), and cross sectional area S (14):

F_A - Archimedean (buoyancy) force due to the difference between ρ_s and ρ_m,
$\Delta\rho = |\rho_s - \rho_m|$;
F_i - inertial force caused by coil motion;
F_η - viscosity force;
F_γ - interfacial tension force;
F_w - adhesion force;
F_h - hydraulic resistance force caused by moving two immiscible phases relative to each other.

The following balance of these forces of different nature has been considered:

$$F_i = F_A + F_\eta + F_\gamma + F_w + F_h \qquad (1)$$

From this, the basic equation of the stationary phase retention process has been derived. The length L of an "element" of the stationary phase in a rotating coil column has been estimated. For the hydrophobic systems, which are characterized by high values of the interfacial tension γ (more than 8 dyn/cm), low values of the viscosity η, and low (a few seconds) hydrodynamic equilibrium settling times, it can be written:

$$L \approx \gamma / [\Delta\rho S^{1/2} \omega^2 R] \qquad (2)$$

The following equation allows us to estimate the cross-sectional area of a stationary phase element for hydrophobic liquid systems:

$$\left(\sqrt{\frac{S_c}{S}} - 1\right)(S_c - S) \approx \frac{v_m^{1/2} r^{1/2} \eta_s}{\rho_m \Delta\rho R \omega^{3/2}} \qquad (3)$$

where v_m is the linear speed of the mobile phase flow, η_s is the viscosity of the stationary phase, and S and S_c are the cross-sectional areas of the stationary phase element and the separation coil column, respectively. It can be rewritten as (with sufficient accuracy for our model):

$$\frac{S}{S_c} \approx 1 - k_3 \frac{\beta^{1/4}}{\omega^{3/4} R^{1/4}} \approx 1 - k_4 \frac{1}{\omega^{3/4}} \qquad (4)$$

where k_3 is a proportional coefficient characterizing peculiarities of the liquid system (it is dependent on the interfacial tension, viscosity of the stationary phase, and density difference between two phases); $k_4 = k_3 \, (r^{1/4}/R^{1/2})$.

The ratio of the cross-sectional area of the stationary phase element to that of the coil tube (S/S_c) governs the volume of the stationary phase retained in the column. The theoretical dependence of S/S_c on the rotation speed ω and the experimental dependence of the S_f value on ω for n-decane - water and chloroform - water liquid systems are in good agreement (*14*). Further work is needed to consider in detail the hydrophilic and intermediate liquid systems.

Influence of the Composition and Physico-Chemical Parameters of Two-Phase Liquid Systems on the Stationary Phase Retention. As mentioned above, the liquid systems for inorganic separations are very different from those of organic separations because, in most cases, they contain an extracting reagent in the organic phase and mineral components (e.g., acids and/or salts) in the aqueous phase. The aqueous phase may also contain some masking agents to enhance metal ion separations. This too can lead to changes in the physico-chemical properties of the system.

We have studied two-phase liquid systems containing an organic solvent (n-decane, n-hexane, chloroform, carbon tetrachloride and methyl isobutyl ketone), an extracting reagent, water, a surfactant (the sodium salt of dodecylbenzenesulphonic acid), and a mineral salt (ammonium sulfate). Varying concentrations of the system constituents allows alteration of certain physico-chemical parameters (interfacial tension γ, density difference between two liquid phases $\Delta\rho$, or viscosity of the organic stationary phase η_{org}). The type of solvent may often have a great effect on the physico-chemical parameters of a liquid system and, consequently, on the stationary phase retention. The composition and physico-chemical properties of the organic phase were modified by adding extracting reagents (di-2-ethylhexylphosphoric acid, tri-n-butylphosphate, trioctylamine). The density and viscosity of the organic phases were varied by changing the amount of reagents in the stationary phase. For example, a small addition (5%) of HDEHP in an organic solvent (n-decane, n-hexane, chloroform, and carbon tetrachloride) leads to a considerable increase in the S_f factor in the organic solvent - $(NH_4)_2SO_4$ - water systems (from 0 to 0.73 in the case of carbon tetrachloride at $\omega = 450$ r/min, $F = 1$ mL/min). Moreover, for the systems containing HDEHP, the retention becomes practically independent of the salt concentration at a reagent concentration of 5% and higher. Therefore, it is possible to attain a constant retention of the stationary phase with different density or viscosity changes between two phases, which can be very useful in practice. It should be also noted that high values of the interfacial tension (20 dyn/cm or more) are apparently not favorable for stationary phase retention.

It has been shown that the higher the rotation speed the lower the influence of the two-phase system composition on the S_f value. When using a stationary phase of higher density ($\rho \geq 1.4$ g/mL), a high rotation speed (500 rpm or more) may be required.

The correlations between the physico-chemical parameters of the complex liquid systems under investigation and their behavior in coiled columns are described in detail (*15, 16*). Further studies are needed to extend the observed trends to other two-phase liquid systems.

It should be noted that elution depends on the operating conditions of the planetary centrifuge which influence the quantity of the stationary phase in the column. The elution band appears earlier and the peak narrows if the volume of the stationary phase is decreased (all other factors being the same). Reagent concentration in the organic solvent also affects the elution curve shape and, therefore, the dynamic partition coefficient values. An increase in the reagent concentration in the organic phase leads to higher partition coefficients and, in general, a better separation is achieved. However, a rather large volume of the mobile phase may be required to elute the elements from the column. The effect of the reagent concentration in the stationary organic phase can be explained in terms of extraction equilibrium by the following example. HDEHP is a cation-exchange (acidic) metal extractant. Metal cation extraction from a diluted aqueous solution can be described by the equation:

$$M^{z+}{}_{(aq)} + zHB_{(org)} = MB_z{}_{(org)} + zH^+{}_{(aq)}$$

where HB is the acidic extractant, MB_z is the adduct species whose formation is responsible for retention, and M^{z+} is the cation involved. If the extracting reagent concentration increases, the equilibrium shifts to the right and, consequently, the partition coefficient of the metal cation increases.

The composition of the mobile phase also has an influence on the partition coefficients of inorganic substances and the separation efficiency. Concentrations of the mobile phase constituents should provide partition coefficients suitable for the enrichment or separation of components under investigation. If a step elution mode is used, partition coefficients higher than 10 and less than 0.1 are favorable for the enrichment of components in the stationary phase and their recovery into the mobile phase, respectively.

Chemical kinetics may also play an important role in the separation of inorganic species by CCC. It has been shown that the values of mass transfer coefficients determine the type of elution (isocratic or step), which is necessary for the desired separation. The data on batch extraction (mass transfer coefficients and partition coefficients) and chromatographic peaks (half-widths) can be interrelated by some empirical expressions (*3*).

Application of Various Two-Phase Liquid Systems to Inorganic Analysis

Group Preseparation and Determination of Rare, Rare Earth, and Some Other Elements in Geological Samples. Due to the high loading capacity and the ability to use many well-known liquid-liquid extraction systems, CCC can be applied as a preconcentration and/or preseparation technique for various trace elements before their instrumental determination (*2, 17-20*). The technique was utilized for preconcentration and separation of rare earth elements (REE) from major constituents of various geological samples and after their decomposition for subsequent determination by inductively coupled plasma atomic emission spectrometry (ICP-AES) (*17, 18*).

Three extraction systems containing HDEHP, trioctylphosphine oxide (TOPO), or diphenyl(dibutylcarbamoylmethylphosphine) oxide (Ph_2Bu_2) were shown to be applicable to the group separation of REE from dissolved samples of rocks, ores, and minerals (basalts, granites, dolomite, fluorite-barite-hydropatite ore, syenite, etc.):

System *1* - 0.5 M HDEHP - n-decane - HCl;
System *2* - 0.3 M HDEHP + 0.02 M TOPO - n-decane + MIBK (v/v=3:1) - 1.0 M NH_4NO_3 / 6 M HCl;
System *3* - 1.0 M Ph_2Bu_2 - chloroform - 3.0 M HNO_3

Some practical capabilities of CCC for the group separation of REE from matrix components of geological samples are illustrated by Figure 1. An extraction system employing HDEHP (System *1*) was used in this case. The separation is achieved by step elution. Alkali, alkaline earth, and other elements with distribution coefficients less than 0.5 are separated from REE in the preconcentration stage from 0.1 M HCl. By eluting with 3 M HCl, REE are selectively eluted from the stationary phase. To remove Fe(III), U(VI), Th(IV), Mo(IV), and Ti from the stationary phase, the column is washed with 5 M HCl. The total time of the separation cycle is about 40 min. The concentrates obtained are aqueous solutions of REE ready for instrumental analysis. Table I illustrates the separation of some matrix components of dolomite and fluorite-barite-hydropatite ore by use of System *1* and subsequent atomic emission analysis. The high capacity of the separation column enables one-stage separation of REE from most of the matrix components to be achieved. ICP-AES determination of REE concentrates were obtained from these two samples as well as from a basalt sample (Table II, System *1*), and has shown that only iron, present in the concentrates in natural quantities, may interfere with the REE determination. However, the residual amounts of Fe as well as Ti and Al do not interfere if the interelement corrections described in reference (*17*) are used. A good agreement with the certified values for light REE and yttrium was obtained with System *1* (Table II). Tm, Yb, and Lu are partially retained in the stationary phase during elution with 3 M HCl and can be removed only by 5 M HCl during column regeneration. Other two-phase systems are required for quantitative and selective simultaneous separation of light and heavy REE.

System *2* allows us to determine 13 rare earth elements and yttrium. Alkali, alkaline earth, Cu, Pb, Cr, and many other elements remain in the aqueous phase (1 M NH_4NO_3) at the REE concentration stage (sample solution pH 2.0-2.3, ascorbic acid as masking agent). The change of eluent to 6.0 M HCl enables the elution of REE to be achieved. This system was used in the analysis of granite (*18*) and basalt (Table II, System *2*). The concentrates obtained contain some Ti, Al, and Fe, but their interferences can be avoided (*17*). One separation run takes one hour.

System *3* allows the separation of all REE by the following procedure. A geological sample solution in 3.0 M HNO_3 is introduced into the column, then 20 mL of 3.0 M HNO_3 is pumped through to remove matrix elements from the organic phase. The next 5 mL portion of the same acid solution elutes all the REE present. The column is then washed with doubly distilled water. System *3* can be used only

Table I. Separation of Some Matrix Components by CCC Recovery of REE from Aqueous Solutions of Geological Samples in 0.5 M HDEHP - n-decane - 0.1 M HCl System[a]

Element	Dolomite		Fluorite-barite-hydropatite ore	
	Content in the sample solution (mg)	Content in the REE concentrate (mg)	Content in the sample solution (mg)	Content in the REE concentrate (mg)
Cr	5.5	≤0.05	5.2	≤0.05
Mn	17.6	≤0.01	3.8	≤0.05
Fe	1700	330	624	72
Co	1.4	≤0.03	0.66	≤0.03
Ni	2.5	≤0.1	0.80	≤0.1
Cu	1.9	≤0.02	0.80	≤0.02
Al	3800	2.5	54.2	≤0.2
Sr	4.3	≤0.5	38.0	≤0.5
Mo	5.0	≤0.05	1.1	≤0.05
Cd	0.5	≤0.02	0.26	≤0.02
Sn	30.0	≤0.2	1.8	≤0.2
Sb	20.0	≤1.0	3.3	≤1.0
Ba	32.0	≤0.01	40.0	≤0.01
Pb	12.0	≤0.2	1.8	≤0.2
Ti	2.0	≤0.1	2.7	≤0.2
Zr	6.0	≤0.03	1.1	≤0.03
B	32.0	≤0.05	1060	2.0
V	5.0	≤0.02	0.47	≤0.02

[a]The volumes of the sample solution and REE concentrate were 12 and 6 mL, respectively.

for the analysis of samples containing relatively high quantities of REE as preconcentration is not achieved in this case. The REE concentrates contain some quantity of matrix components which do not interfere with the ICP-AES determination (Table II, System 3).

It should be stressed that all three systems provide the recovery of REE into a small eluate volume using one chromatographic run. The proposed methods of group separation of REE by CCC have the advantages of simplicity, versatility, relatively short separation times, and they can compete with precipitation (21) and other chromatographic methods (21-26).

Table II. ICP-AES Determination of REE (ppm) in a Reference Sample of Basalt BM-1 After Their Group Separation in Three Systems.
0.5 M HDEHP - n-decane - HCl (1); 0.3 M HDEHP - 0.02 M TOPO - n-decane - MIBK (v/v=3:1) - 1.0 M NH$_4$NO$_3$ / 6 M HCl (2); 1.0 M Ph$_2$Bu$_2$ - Chloroform - 3.0 M HNO$_3$ (3)

Element	Systems			C.V.[a] (21, 26)
	1	2	3	
La	10.7±0.3	10.0±1.0	9.1±0.3	9.0±1.2
Ce	18.0±3.0	19.0±2.0	17.0±3.0	22.0±2.0
Pr	5.0±1.0	4.2±0.3	-	-
Nd	15.0±2.0	14.0±1.0	16.0±3.0	16.0±3.0
Sm	3.6±0.1	3.0±0.5	3.2±0.2	3.6±0.3
Eu	1.0±0.1	1.0±0.1	1.1±0.1	1.12±0.07
Gd	1.8±0.1	1.9±0.3	2.3±0.3	-
Tb	1.3±0.3	1.5±0.2	1.4±0.2	0.9±0.3
Ho	-	1.4±0.3	1.1±0.2	-
Tm	< 0.02	0.50±0.05	0.4±0.1	-
Yb	< 0.03	1.7±0.2	2.0±0.5	3.0±0.5
Lu	< 0.01	0.34±0.03	0.32±0.03	0.41±0.07
Y	22.0±1.0	19.0±3.0	24.0±2.0	27.0±3.0

[a]C.V. - certified values. A digested reference sample (0.2 g) was introduced into the column in 12 mL of an aqueous solution. The volume of the REE concentrate was 6 mL.

Group preseparation of Zr, Hf, Ta, and Nb is desirable in the analysis of geological samples of different composition containing low concentrations of these elements. An extraction system based on a 0.1 M solution of tetraoctylethylenediamine (TOEDA) in chloroform was used for the chromatographic preconcentration of Zr, Hf, Nb, and Ta and allows their elution into a small effluent volume (Figure 2). In the first stage, the four elements are concentrated in the stationary phase, whereas Cu, Al, alkali, alkaline-earth, rare earth, and other elements present in geological samples are eluted shortly after the volume of the mobile phase V_m has passed through. Iron is partially eluted, but its complete elution is achieved using 0.1 M HCl solution containing 5% ascorbic acid. Then, using 2.0 M HCl solution, Zr, Hf, Nb, and Ta are eluted into an eluate volume as small as 7 mL. At the column regeneration stage, Zn, Cd, and other elements are removed. The proposed method for the group separation and preconcentration of Zr, Hf, Nb, and Ta was tested in the analysis of a reference sample "Granit-BM" (27). The results using ICP-AES determination agree well with the certified values.

CCC in Geochronological Studies. Separation of Nd, Sm, Rb, and Sr is needed for geochemical studies of isotopic characteristics of these elements in different geological

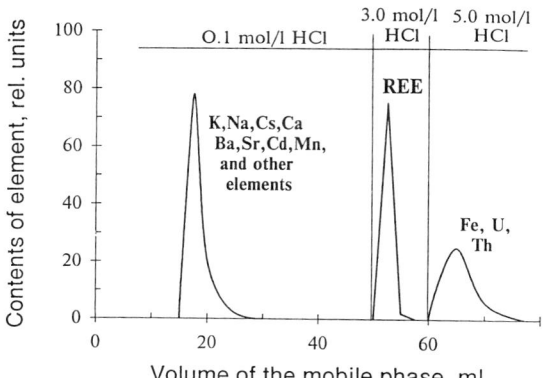

Figure 1. CCC separation of REE from some macrocomponents present in geological samples. Stationary phase: 0.5 M HDEHP in n-decane; $S_f = 0.5$.

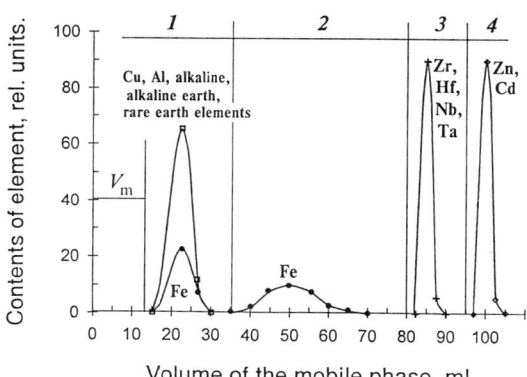

Figure 2. Preconcentration and separation of Zr, Hf, Nb, and Ta from matrix components. Stationary phase: 0.1 M TOEDA - chloroform. Mobile phase: *1* - 0.1 M HCl + 0.01 M $H_2C_2O_4$; *2* - 0.1 M HCl + 5% $H_2C_4H_4O_6$; *3* - 2.0 M HCl; *4* - 1.0 M HNO_3. $S_f = 0.50$; $F = 1.0$ mL/min.

samples and very pure fractions are needed for mass spectrometric (MS) measurements. Complete separation of the four elements is attained by subsequent use of two extraction systems.

The 1.0 M HDEHP - *n*-decane - HCl system allowed us to separate Nd and Sm as well as to separate both elements from the majority of the principal constituents of geological materials. At the first stage (sample introduction, mobile phase - 0.1 M HCl), Nd and Sm are concentrated in the stationary phase, whereas Rb, Sr, and most of matrix elements (Al, Ca, Mg, Na, K, Mn, Ba, etc.) are mainly eluted. Then, 5 mL more of 0.1 M HCl are pumped through to remove all of the matrix components. Afterwards, Nd and Sm are isolated in two small fractions by use of 0.5 and 3.0 M HCl, respectively. Other REE are distributed between the two fractions, but they do not interfere with the MS determination of Nd and Sm. At the column regeneration stage (5.0 M HCl), Fe, Th, and U are eluted. Depending on the purpose, a single fraction containing Nd and Sm can be taken (if quantitative isolation is necessary), or the fractions can be used separately (if very pure Nd and Sm are required for estimating the isotopic ratios).

Strontium can then be selectively separated by using the 0.1 M dicyclohexano-18-crown-6 (DCH18C6) - chloroform - 5.0 M HNO_3 system. A sample is introduced in 5.0 M HNO_3, and at the concentration step, Rb, Ca, and Ba are removed with the acid flow. Rb and Ca are eluted together with Al, Mg, Na, K, and Mn. The change of eluent to 0.5 M CH_3COOH solution results in the elution of strontium in a small volume.

Selective separation of Rb also can be achieved in a DCH18C6-based system, but with the use of picrate (Pi^-) instead of nitrate as the former is a better counter-ion for extraction of metal - crown ether cationic complexes. A sample is injected to 0.005 M picric acid solution with pH=4.0-6.0. Rubidium is concentrated in the stationary phase at the sample introduction stage, whereas calcium and other elements are removed. Twenty mL more of 0.005 M picric acid are passed through to complete their elution. Rubidium is then eluted with a small volume of 2.0 M HCl.

All the elements of interest are separated quantitatively and can be determined by MS. The total blank levels are less than 0.2 ng for Nd and Sm and less than 2 ng for Rb and Sr, if commercially available reagents are used.

Separation of Transplutonium and Rare Earth Elements. Two extraction systems based on bidentate neutral organophosphorus reagents were shown to be applicable to the group separation of trivalent transplutonium elements (TPE) from trace and macroamounts of rare earth elements (*28, 29*):

System *1* - 0.005 M tetraphenylmethylenediphosphine dioxide (TPMDPD) - chloroform - 0.5 M NH_4SCN - 1 M HCl / 0.025 M hydroxyethanediphosphonic acid (HEDPA);
System *2* - 0.02 M 2,4,6-tris(ditolylphosphoryl)-1,3,5-triazine (Tol-triazine) - chloroform - 0.5 M NH_4SCN - 1 M HCl / 0.025 M HEDPA

The main advantage of these systems is the higher partition coefficients for all TPE than for all REE.

To develop a method for the group separation of TPE from REE, preliminary studies of the batch extraction of Am, Cm, Bk, Cf, and all REE in both systems were made. Am and Cm were shown to have the lowest (amongst TPE) and very similar partition coefficients, whereas La and Ce were found to possess the highest (amongst REE) and quite similar partition coefficients. Therefore, the Am/Ce separation factor characterizes the group separation of TPE and REE.

The effect of the total REE concentration on the separation of TPE (Am) from REE (Ce) in both systems was investigated (*29*). Under the chosen conditions, when the REE are eluted with the first portion of the eluent, the Am peak position and the Am/Ce separation factor depend on the REE concentration. Increasing the REE concentration results in better Am retention and a higher separation factor, possibly due to a salting-out effect of the REE. Further increasing the REE concentration leads to a reverse effect; at the higher REE concentration, Am retention decreases along with the separation factor. The decreasing retention and separation can be explained by a decrease in the free extractant concentration when the extraction of the REE is appreciable. The group separation of TPE from trace and macroamounts of REE was also achieved in a system containing 0.1 M α-pyridylmethylene-bis(diphenylphosphine)dioxide (DPS) - chloroform - 3 M HNO_3 (*30*).

Separation of Cesium and Strontium Radionuclides. CCC can be used to solve other radioanalytical problems, for example, in the separation of Cs and Sr radionuclides (*2, 19, 20*). Systems of 0.01 M cobalt dicarbollide (CD) - nitrobenzene - HNO_3 solution and 0.05 M DCH18C6 - 3% HDEHP - chloroform - HNO_3 solution were used for this purpose. The use of a stationary phase consisting of CD enables the concentration of Cs and Sr to be achieved. Elution of Cs and Sr from the stationary phase containing DCH18C6 and HDEHP into a mobile phase containing barium nitrate, polyethylene glycol (PEG), and nitric acid, was studied. Different degrees of cesium and strontium separation can be realized by changing the composition of the mobile phase (at the same extractant concentration in the stationary phase). The quantitative elution of Cs and Sr by one eluent, or their complete separation by step elution is feasible. The latter procedure is more rapid and enables stripping of the elements into smaller mobile phase volumes to be achieved.

Purification of Salt Solutions. The applicability of CCC to the purification of salt solutions (to obtain high-purity reagents, which, after evaporation, can be used for fusion decomposition procedures in trace analysis of high-tech ceramic and other materials) has been investigated (*31*). There is a difference in the objectives of analytical preconcentration and purification procedures: in the latter, a purified constituent is the goal, while the trace elements are impurities to be separated and discarded. In this case, the maximum number of possible trace elements should be separated in one chromatographic run. Other requirements to be met in the use of any purification method are connected with the purity of all the chemicals employed and with the necessary purification of the solutions used and decontamination of the device materials. In the case of a planetary centrifuge, the solutions are only in contact with Teflon, which is a quite inert material.

Application of CCC to the purification of aqueous solutions of inorganic salts, such as $(NH_4)_2SO_4$, NH_4F, or NH_4Cl, from a number of common metal impurities was shown by use of *N, N*-hexamethylenedithiocarbamic acid (HMDTCA), diethylammonium diethyldithiocarbamate, 8-hydroxyquinoline, dibenzo-18-crown-6 (DB18C6), and DCH18C6 as extracting reagents. The results of the purification of the salt solution from trace elements by 1% HMDTCA are presented in Table III. The treated solution is free of Fe, Cu, Zn, Co, Cd, Ni, and Mn and shows the virtually complete removal of the metallic impurities to below the detection limits for electrothermal atomic absorption spectroscopy (ETAAS) measurements. However, some Al and Cr remain in the solution. Contamination of the solutions by the column material, reagents, and organic solvents was minimized. Concentrations of impurities in water and 2 M HCl passed through the column were less than the detection limits.

A mixture of HMDTCA, DCH18C6, and DB18C6 in chloroform allows for the purification of 1% $(NH_4)_2SO_4$ solution (pH=5.5). K, Fe, Cu, Zn, Co, Cd, Ni, Al, and Mn were removed to below the detection limits for ETAAS. 5% Ca, 8% Mg, and 55% Cr remained in the purified salt. This example shows that the purification of salt solutions can be attained by varying the organic phase composition.

Table III. Purification of 1% $(NH_4)_2SO_4$ Solution by 1% HMDTCA in Chloroform ($S_f = 0.5$)

Element	Concentration in salt solution, ng/mL	
	before purification	after purification
Fe	15.0	< 0.1
Zn	2.5	< 0.1
Cd	1.6	< 0.1
Co	15.0	< 0.1
Cu	15.0	< 0.2
Ni	15.0	< 0.1
Mn	15.0	< 0.2
Al	10.0	1.5
Cr	5.0	2.5

Acknowledgments

The authors are grateful to the Russian Foundation of Basic Research, grant N 97-03-33399, for the support of this work.

Literature Cited

(1) *Countercurrent Chromatography. Theory and Practice;* Mandava, N. B.; Ito, Y., Eds.; Marcel Dekker: New York, 1988.
(2) Zolotov; Yu. A.; Spivakov, B. Ya.; Maryutina, T. A.; Bashlov, V. L.; Pavlenko, I. V. *Fresenius Z. Anal. Chem.* **1989**, *35*, 938.
(3) Fedotov, P. S.; Maryutina, T. A.; Pukhovskaya, V. M.; Spivakov, B. Ya. *J. Liq. Chromatogr.* **1994**, *17*, 3491.
(4) Berthod, A.; Schmitt, N. *Talanta* **1993**, *40*, 1489.
(5) Menet, J.-M.; Thiébaut, D.; Rosset, R.; Wesfreid, J. E.; Martin, M. *Anal. Chem.* **1994**, *66*, 168-176.

(6) Conway, W. D. *Countercurrent Chromatography. Apparatus, Theory, and Applications*; VCH: New York, 1990.
(7) Menet, J.-M.; Rolet, M.-C.: Thiébaut, D.; Rosset, R.; Ito, Y. *J. Liq. Chromatogr.* **1992**, *15*, 2883.
(8) Berthod, A. *J. Chromatogr.* **1991**, *550*, 677.
(9) Foucault, A. P.; Le Goffic, F. *Analysis* **1991**, *19*, 227.
(10) Bousquet, O.; Foucault, A. P.; Le Goffic, F. *J. Liquid Chromatogr.* **1991**, *14*, 3343.
(11) Foucault, A. P.; Bousquet, O.; Le Goffic, F. *J. Liquid Chromatogr.* **1992**, *15*, 2691.
(12) Foucault, A. P.; Bousquet, O.; Le Goffic, F. *J. Liquid Chromatogr.* **1992**, *15*, 2721.
(13) Drogue, S.; Rolet, M.-C.; Thiébaut, D. *J. Chromatogr.* **1992**, *593*, 363.
(14) Fedotov, P. S.; Kronrod, V. A.; Maryutina, T. A.; Spivakov, B. Ya. *J. Liq. Chrom. & Rel. Technol.* **1996**, *19*, 3237.
(15) Maryutina, T. A.; Ignatova, S. N.; Fedotov, P. S.; Spivakov, B. Ya; Thiébaut, D. *J. Liq. Chrom. & Rel. Technol.* **1998**, *21*, 19.
(16) Fedotov, P. S.; Thiébaut, D. *J. Liq. Chrom. & Rel. Technol.* **1998**, *21*, 39.
(17) Pukhovskaya, V. M.; Maryutina, T. A.; Grebneva, O. N.; Kuz'min, N. M.; Spivakov, B. Ya. *Spectrochim. Acta* **1993**, *48B*, 1365.
(18) Pukhovskaya, V. M.; Grebneva, O. N.; Maryutina, T. A.; Kuz'min, N. M.; Spivakov, B. Ya. *Spectrochim. Acta* **1995**, *50B*, 5.
(19) Spivakov, B. Ya.; Maryutina, T. A.; Bashlov, V. L.; Pukhovskaya, V. M.; Zolotov, Yu. A. *Proceedings of 5th Japan - USSR Symposium on Analytical Chemistry (ITAS'90);* Sendai and Kiryu, Japan. 1990, pp 241-250.
(20) Spivakov, B. Ya.; Maryutina, T. A.; Zolotov, Yu. A. *Proceedings of the International Solvent Extraction Conference (ISEC'90);* Kyoto, Japan. Elsevier: 1992; *A*, pp 451-456.
(21) Abbey, S.; Gladney, F. S. *Geostand. Newslett.* **1986**, *10,* 1.
(22) Kuz'min, N. M.; Pukhovskaya, V. M.; Varshal, G. M.; Spivakov, B. Ya.; Maryutina, T. A.; Volynets, M. P.; Ryabukhin, V. A.; Chkhetiya, N. N.; Grebneva, O. N.; Pavlutskaya, V. M. *Zh. Anal. Khim.* **1993**, *48,* 898.
(23) Bauer-Wolf, E.; Wegscheider, W.; Posch, S.; Knapp, G. *Talanta* **1993**, *40,* 9.
(24) Croudace, I. W.; Marshall, S. *Geostand. Newslett.* **1991**, *15*, 139.
(25) Watkins, P. S.; Novan, S. *Chemical Geology* **1992**, *95,* 131.
(26) Govindaraja, K. *Geostand. Newslett.* **1989**, *13,* 1.
(27) Fedotov, P. S.; Maryutina, T. A.; Grebneva, O. N.; Kuz'min, N. M.; Spivakov, B. Ya. *J. Anal. Chem.* **1997**, *52,* 1034.
(28) Chmutova, M. K.; Maryutina, T. A.; Spivakov, B. Ya.; Myasoedov, B. F. *Radiokhimiya* **1992**, *6,* 56.
(29) Chmutova, M. K. *Proceedings of 3rd Finnish-Russian Symposium on Radiochemistry;* Helsinki, 1994; p 44.
(30) Chmutova, M. K.; Ivanova, L. A.; Bodrin, G. V.; Myasoedov, B. F. *Radiokhimiya* **1996***, 38,* 520.
(31) Maryutina, T. A.; Spivakov, B. Ya.; Tschöpel, P. *Fresenius Z. Anal. Chem.* **1996**, *356*, 430.

Chapter 22

Fundamental Aspects of Metal-Ion Separations by Centrifugal Partition Chromatography

Subramaniam Muralidharan and Henry Freiser

Strategic Metals Recovery Research Facility, Department of Chemistry, University of Arizona, Tucson, AZ 85721

Centrifugal partition chromatography (CPC), a multistage countercurrent liquid-liquid distribution technique employing discrete stages and two immiscible bulk liquid phases, is ideally suited for the detailed examination, through evaluation of separation factors and efficiencies, of the influence of bulk aqueous and liquid-liquid interfacial equilibria and kinetics on the separations of metal ions. This has been demonstrated by studies of the separation of transition metals, platinum group metals, and trivalent lanthanides. The results indicate that separation efficiencies in CPC are mainly limited by back-extraction kinetics that occur in the bulk aqueous phase and at the organic-aqueous interface as indicated by a direct linear correlation between the half-lives ($t_{1/2}$'s) of the dissociation reactions and the reduced plate height. In addition, the interfacial areas calculated through this correlation are much larger in many cases than those generated in highly stirred two phase mixtures. Finally, addition of surfactants and interfacial catalysis of the formation and dissociation of the complexes dramatically improve efficiencies.

Solvent extraction is a particularly appropriate technique for difficult metals separations problems since it incorporates selectivity, versatility, and convenience. It is a powerful separation technique applicable both to trace analytical and macro- or process- scales (1). Because of their great selectivity, solvent extraction techniques can be used to recover metals from multicomponent solutions, which makes them ideal for environmental remediation. Extractants that incorporate chelating functionalities are employed to achieve the highest possible selectivity. Much research has been devoted to the elucidation of the many factors that affect metal chelate stability and extractability. Conversely, equilibrium and kinetic studies of metal extraction systems are key to the more complete understanding of separations involving chelation. Similarly, the path to full understanding of multistage separations such as countercurrent distribution and liquid partition chromatography must be, of necessity, based on thoroughly characterized single-stage solvent extraction processes.

The separation of metal ions, especially closely related ones such as the trivalent lanthanides, by single stage methods poses daunting challenges even to the most selective of extractants (2). Therefore, the use of multistage methods is necessary for their separation. Multistage methods consist of solid-liquid and liquid-liquid partition, with the former involving solid supports such as silica and cross-linked organic polymers, derivatized, coated, or impregnated with a ligand which serves to separate metal ions by complexation in a conventional liquid chromatographic mode. The latter involves two bulk liquid phases with the extractant dissolved in the organic phase (3). Centrifugal partition chromatography (CPC), which is a liquid-liquid separation technique, is the only multistage technique with discrete stages among the various solid-liquid and liquid-liquid techniques (4). Because of the use of discrete stages and two bulk liquid phases with well-defined volumes, CPC is ideally suited for studying the fundamental factors that influence metal ion separations. CPC, being a multistage technique, is far more sensitive to equilibrium and kinetic phenomena in the bulk aqueous phase and at the organic-aqueous interface than single stage techniques and thus can provide more accurate and detailed information than can batch experiments. In addition, CPC is an excellent model for understanding separations with solid-liquid techniques, since it does not suffer from many of the difficulties associated with solid supports such as diffusion into pores, irreversible adsorption, and so on.

We were one of the first to adapt CPC, which was originally developed for the separation of organic compounds and biochemicals, to metals separations (5). During the past several years, we have examined the CPC separation of several families of metals (the transition metals, platinum group metals, and the trivalent lanthanides) using a variety of chelating extractants (acylpyrazolones, organophosphorous acids, arylhydroxyoximes, etc.) to discern the influence of bulk and interfacial kinetics, interfacial activities of the ligands and their metal chelates, and interfacial areas generated on the separation efficiencies and resolution of closely related metal ions.

A correlation between the half-lives of the slow chemical kinetic steps, and the interfacial areas generated and the inefficiencies of separations have been drawn from these studies. Moreover, the significance and influence of the liquid-liquid interface in multistage separations of metal ions has been illustrated.

Centrifugal Partition Chromatography

The CPC apparatus (manufactured by Sanki Engineering Company, Japan (4)) consists of a series of cartridges, each of which contains 40 - 400 channels, depending upon the internal volume. These channels serve as stages in the separation experiment. The total number of channels is 400 - 4800, depending upon the number of cartridges employed. These cartridges are arranged in a rotor that is rotated at 700 - 1200 rpm. The centrifugal force generated keeps one of the two phases (usually the organic phase) stationary while the other phase (usually the aqueous phase) is moved through it at a constant flow rate. The injected analyte mixture is carried by the aqueous mobile phase into the cartridges where the mixture components are extracted into the organic stationary phase by simple distribution if they are organic, or by complexation with a suitable ligand if they are metals. When the mobile phase is depleted of the analytes, further flow of the mobile phase of the same (isocratic elution) or different (gradient elution) composition causes the back-extraction of the analytes, which can be detected by a suitable method. If the analytes are completely separated, they appear as discrete peaks such as those observed in conventional chromatographic methods like HPLC (hence the name centrifugal partition chromatography). CPC has a number of unique features, among them a large number of discrete stages (400 - 4800 depending upon the operational volume chosen), high loading capacity for extractants and analytes, negligible loss of stationary phase due to bleeding, flexible organic-aqueous phase volume ratios, a high stationary phase to mobile phase ratio, and ready adaptability to process-scale.

Four basic parameters are employed in the analysis of CPC chromatograms: the retention volume, V_r, which is related to the stationary phase and mobile phase volumes (V_s and V_m, respectively) and the distribution ratio of the analyte, D, through equation 1; the chromatographic efficiency, as measured by the number of theoretical plates, N, which is calculated from the retention volume (V_r) and the width of the chromatogram (w), as per equation 2; the chromatographic inefficiency, represented by the channel equivalent of a theoretical plate (CETP), which is analogous to reduced plate height and is the ratio of the number of channels, CH (2400 in our experiments), to N, according to equation 3; and the selectivity, α, achieved in the separation of two analytes (designated 1 and 2), which is the ratio of their distribution ratios, D_1 and D_2, and is given by equation 4 (6).

$$V_r = V_m + D V_s \qquad (1)$$

$$N = 16 \left(\frac{V_r}{w}\right)^2 \qquad (2)$$

$$CETP_{obs} = \frac{CH}{N} \qquad (3)$$

$$\alpha = \frac{D_2}{D_1} \quad (D_2 > D_1) \qquad (4)$$

Optimization of the CPC Operational Parameters

A major difference between CPC and conventional LC is that in the former, even at the smallest attainable ratio of the volume of the stationary phase to the volume of the mobile phase, there are still two bulk phases. Thus, optimizing conditions such as the phase volume ratio, rotational speed, flow rate of the mobile phase, and nature of the stationary and mobile phases, (i.e., organic or aqueous) are important in obtaining the best efficiencies. This is especially true in metals separations, where the CPC bandwidth is determined by the rates of mass transfer and diffusion and by slow chemical kinetics. It is necessary to minimize the contribution to CETP from mass transfer and diffusion in order to properly elucidate the kinetic factors.

The influence of flow rate of the mobile phase, the ratio of the volume of the stationary phase to the volume of the mobile phase (V_s/V_m), and the nature of the stationary phase (organic or aqueous), on the CPC efficiencies have been determined using three organic-aqueous phase pairs (1,2-dichloroethane, toluene, and heptane) with three analytes of different types: 3-picoline (organic), (tetraheptylammonium)$_2$IrCl$_6$ (ion pair), and PdCl$_2$(TOPO)$_2$ (coordination complex; TOPO = trioctylphosphine oxide). The results are summarized below and those for the 1,2-dichloroethane-H$_2$O system are displayed in Figure 1.

The efficiencies and hence, CETP values of 3-picoline and Q$_2$IrCl$_6$ (Q = tetraheptylammonium) were very similar at different flow rates and phase volume ratios for the various solvent pairs examined. This strongly suggests that such CETP values represent the contribution of diffusion and mass transfer effects and, therefore, can be termed CETP$_{dif}$ for the metal complexes and ion pairs. The CETP values of PdCl$_2$(TOPO)$_2$ were significantly higher than the values for 3-picoline and Q$_2$IrCl$_6$.

The efficiencies decreased with increasing flow rate for all the analytes, with PdCl$_2$(TOPO)$_2$ exhibiting a much larger decrease compared to the other analytes. This suggests an interfacial contribution to the back-extraction kinetics, as the mobile phase droplet size increases and hence interfacial area decreases with increasing flow rate. Support for this interpretation comes from the CPC behavior of the Ni(II)-HPMBP (HPMBP = 1-phenyl-3-methyl-4-benzoyl-5-pyrazolone) and the Ni(II)-LIX 860 (LIX 860 = dodecylsalicylaldoxime) systems, which exhibit bulk aqueous and interfacial kinetics, and the trivalent lanthanide-HPMBP and HPMCP (HPMCP = 1-phenyl-3-methyl-4-capryloyl-5-pyrazolone) systems, which will be discussed in the following sections.

The efficiency improved dramatically below V_s/V_m of one for all analytes and solvent pairs studied. The best efficiencies were obtained below a V_s/V_m of 0.5, with V_s/V_m of 0.3 or less being optimum. This is shown in the upper plot of Figure 1 for the three analytes with 1,2-dichloroethane as the stationary phase. The efficiencies when the aqueous phase rather than the organic phase was held stationary were generally much lower (lower plot of Figure 1), by as much as a factor of three. At V_s/V_m ratios below 0.5, the efficiencies for the three analytes at a given flow rate of

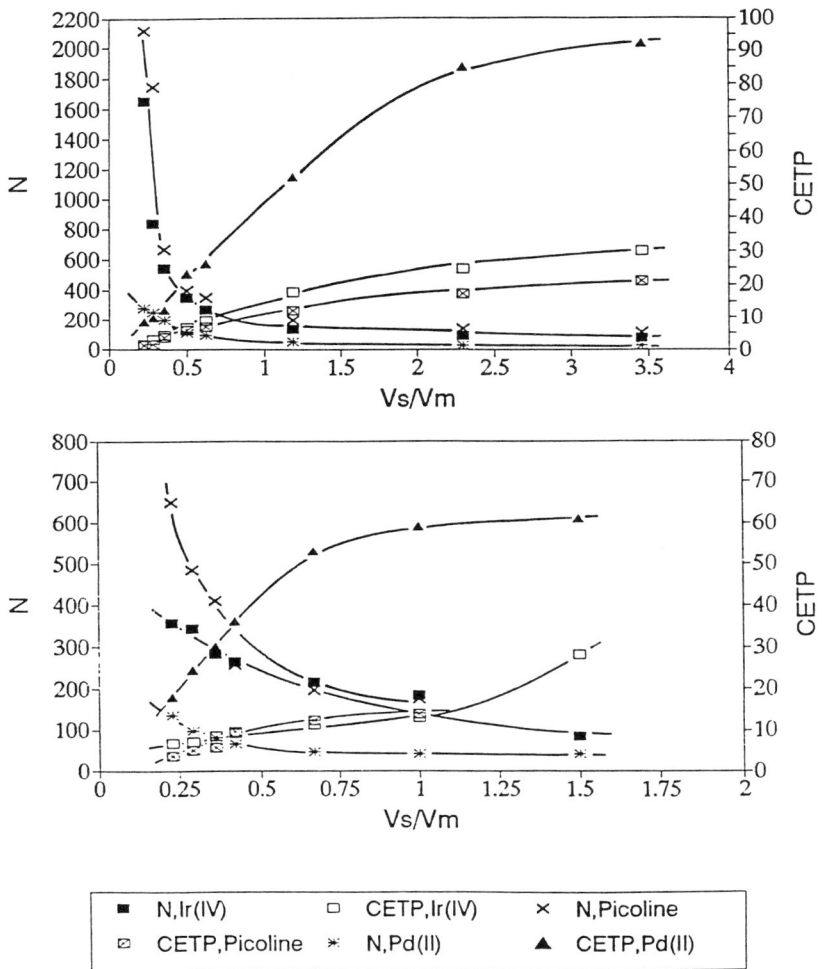

Figure 1. Efficiency and CETP as a function of V_s/V_m for the analytes 3-picoline, Q_2IrCl_6 (Q = tetraheptylammonium), and $PdCl_2(TOPO)_2$ in the 1,2-dichloroethane-H_2O phase pair at a flow rate of 2 mL/min, upper: 1,2-dichloroethane as stationary phase, lower: H_2O as the stationary phase.

aqueous mobile phase decreased in the order 1,2-dichloroethane > heptane > toluene.

Separation of Platinum Group Metals

Separation, extraction, and purification of the platinum group metals (PGM) Pt, Pd, Ir, and Rh in their various oxidation states continue to be challenging and interesting areas of research *(1-4)*. Several techniques have been investigated to recover these precious metals, among them solvent extraction *(1-4)*, ion exchange *(5,6)*, membrane absorption *(7- 9)*, and biomagnetic separation *(10,11)*. The separation of PGM from chloride media by solvent extraction can be achieved either by complexation with a suitable ligand or through ion pair formation with a large cation. Complexation with a ligand is more selective but generally suffers from slow complex formation and dissociation kinetics. By contrast, ion pair formation is diffusion-controlled and not very selective, but is necessary to separate kinetically inert species such as $PtCl_6^{2-}$ and $IrCl_6^{2-}$.

Trioctylphosphine oxide is an organophosphorus compound and is a stable and inexpensive extractant. TOPO, as we have shown *(12-14)*, is unique in that it can function as a monodentate ligand and as a cation for ion pair extraction when protonated. Our initial work focused on the separation by CPC of Pd(II) and Pt(II) from Rh(III) and Ir(III) with TOPO as ligand in the heptane-water solvent pair. We elucidated the extraction equilibria involved and showed $PdCl_2(TOPO)_2$ to be the extracted species at [HCl] ≤ 10^{-3} M. A very important practical aspect of our work is that these separations were performed under relatively mild conditions in contrast to traditional methods which often involve harsh conditions such as high acidity.

The CPC chromatogram for the separation of Pd(II) and Pt(II) with TOPO in a heptane stationary phase, with $V_s/V_m = 0.2$ and at a mobile phase pH of 3 and at 0.08 M chloride concentration, is shown in Figure 2. The separation was performed as a function of the concentrations of TOPO, Cl⁻, and pH. Only the neutral complex

$$MCl_n^{(n-2)-} + 2\,TOPO_o \underset{}{\overset{K_{ex}}{\rightleftharpoons}} MCl_2(TOPO)_{2,o} + (n-2)\,Cl^- \tag{5}$$

$MCl_2(TOPO)_2$ (M = Pd(II), Pt(II)), was extracted (equation 5 where n = 2 - 4). The K_{ex} values for $PdCl_2$, $PdCl_3^-$, and $PdCl_4^{2-}$ are 794.3 M^{-2}, 2.75 M^{-1}, and 0.14 respectively. A single peak was observed in the CPC chromatogram of Pd(II) at any concentration of Cl⁻ as its hydrolytic equilibria are rapid. The corresponding values for the three Pt(II) chloro species are 48 M^{-2}, 0.047 M^{-1} and 0.018, clearly indicating the better extractability of Pd(II) over Pt(II), also evident from Figure 2. The difference in the K_{ex4} values for the MCl_4^{2-} species can be exploited to obtain an efficient separation of Pt(II) and Pd(II) from Rh(III) and Ir(III). This was achieved using stepwise Cl⁻

gradients of 0.001, 0.08, and 0.5 M in the mobile phase, Figure 3.

Formation of $HTOPO^+$, at HCl concentrations ≥ 0.1 M, resulted in the extraction of $(HTOPO)_2MCl_4$ (M = Pt or Pd):

$$MCl_4^{2-} + 2H^+ + 2TOPO_o \underset{}{\overset{K_{ex}}{\rightleftharpoons}} (HTOPO)_2MCl_{4,o} \qquad (6)$$

The chromatogram of the separation of $RhCl_6^{3-}$, $PdCl_4^{2-}$, and $PtCl_4^{2-}$ by $HTOPO^+$ is shown in Figure 4. The K_{ex} values of Pd(II) and Pt(II) are 93.3 M^{-4} and 1961 M^{-4}, respectively, indicating that Pd(II) elutes ahead of Pt(II) in the ion pair separation while the opposite is true in the separation by complexation. While the chromatogram of Pt(II) involves only the formation of $(HTOPO)_2PtCl_4$, the chromatogram of Pd(II) also involves the formation of $(HTOPO)PdCl_3$. In fact, under the experimental conditions employed in these separations, this is the major Pd ion pair that is extracted. The extraction equilibrium constant for $(HTOPO)PdCl_3$ is 18.25 M^{-1}. Similarly, Pt(IV) and Ir(IV) could be separated by $HTOPO^+$ by ion pair formation with their MCl_6^{2-} species. The K_{ex} values for the Pt(IV) and Ir(IV) species are 1576 M^{-4} and 8035 M^{-4}, respectively.

The concentrations of Cl^- and H^+ can be simultaneously varied to separate $PtCl_4^{2-}$, $PdCl_4^{2-}$, and $PtCl_6^{2-}$ by gradient elution, and such a separation is shown in Figure 5. Here the extraction of Pt(II) occurs by formation of $PtCl_2(TOPO)_2$, that of Pt(IV) by formation of $(HTOPO)_2PtCl_6$, and that of Pd(II) by the formation of $PdCl_2(TOPO)_2$, $(HTOPO)PdCl_3$, and $(HTOPO)_2PdCl_4$.

The ion pair extraction of the chloro anions of PGM by QpTS (pTS = p-toluenesulfonate) in a 1,2-dichloroethane stationary phase also afforded the separation of these species under relatively mild conditions. The overall extraction equilibrium for the MCl_6^{2-} species is given in equation 7.

$$MCl_6^{2-} + 2QpTS_o \underset{}{\overset{K_{ex}}{\rightleftharpoons}} Q_2MCl_{6,o} + 2pTS^- \qquad (7)$$

The K_{ex} values for Pt(IV) and Ir(IV) are 3890 and 7760, respectively, and despite the similarity of these values, baseline separation of these ions can be obtained.

The CPC behavior of both Pt(II) and Pd(II) are dependent on [Cl^-] in the mobile phase. In the absence of any added NaCl (C_{HCl} = 0.0032 M) in the mobile phase, Pt(II) exhibited two peaks and Pd(II) exhibited a broad peak. Addition of 0.1 M NaCl to the mobile phase resulted in three peaks for Pt(II) (Figure 6), and a narrower single peak for Pd(II). The difference in the behavior of Pt(II) and Pd(II) stem from the more sluggish interconversion of the various $MCl_{2+i}^{(i-2)-}$ species when M = Pt. The interconversion of the chloro species is rapid (relative to the time scale of

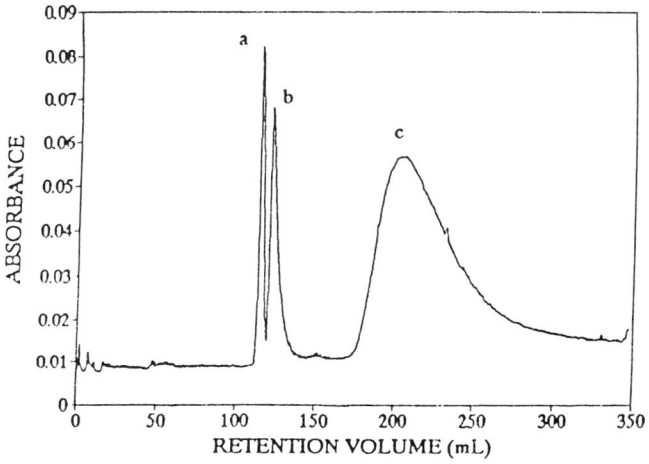

Figure 2. Separation of Rh(III), Ir(III), Pt(II), and Pd(II) using 0.5 M TOPO in heptane, 0.08 M Cl$^-$, $V_s/V_m = 0.2$, and flow rate 0.87 mL/min. (a) 5 x 10^{-4} M Rh(III) and Ir(III); (b) 10^{-4} M Pt(II); (c) 10^{-3} M Pd(II).

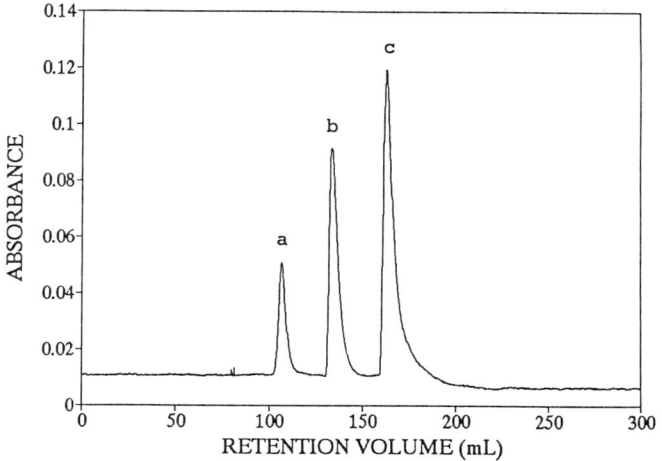

Figure 3. Separation of PdCl$_4^{2-}$ and PtCl$_4^{2-}$ from IrCl$_6^{2-}$ and RhCl$_6^{2-}$ by stepwise Cl$^-$ gradient in the aqueous mobile phase with 0.5 M TOPO, 10^{-3} M HCl, and 2.0 mL/min flow rate. Eluting species are (a) 10^{-3} M IrCl$_6^{2-}$ and RhCl$_6^{2-}$ (no added Cl$^-$), (b) 10^{-4} M PtCl$_4^{2-}$ (0.08 M NaCl), and (c) 10^{-4} M PdCl$_4^{2-}$ (0.5 M NaCl).

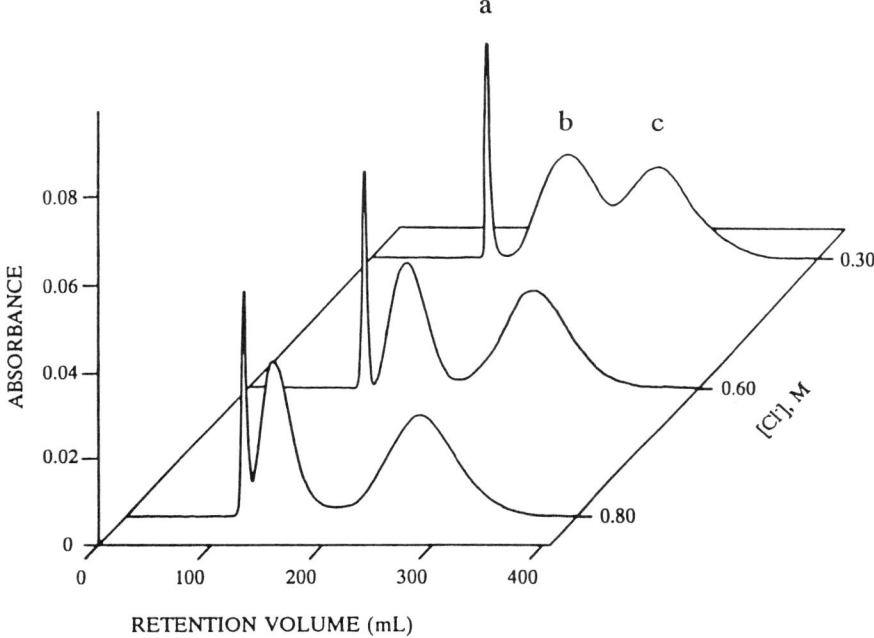

Figure 4. Separation of 10^{-4} M $IrCl_6^{2-}$, $PtCl_4^{2-}$, and 10^{-3} M $PdCl_4^{2-}$ as their ion pairs with HTOPO$^+$ as a function of [Cl$^-$] with 0.5 M TOPO at 0.1 M HCl and 4.0 mL/min flow rate. Eluting species are (a) $IrCl_6^{2-}$, (b) $PdCl_4^{2-}$, and (c) $PtCl_4^{2-}$.

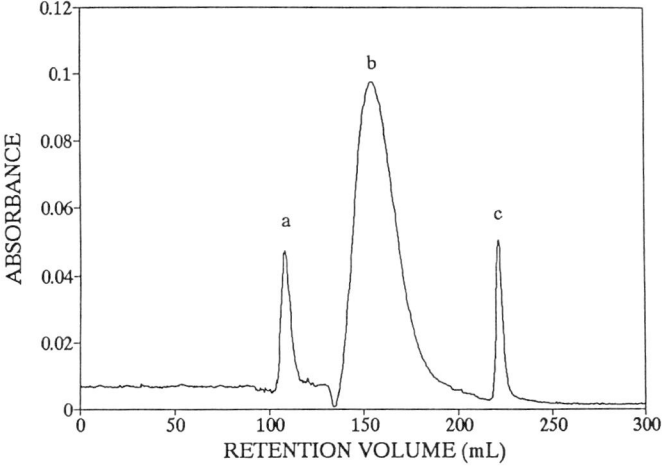

Figure 5. Separation of 10^{-3} M $PtCl_4^{2-}$, $PdCl_4^{2-}$, and $PtCl_6^{2-}$ using H$^+$ and Cl$^-$ gradients in the mobile phase with 0.5 M TOPO and mobile phase flow rate of 4.0 mL/min. The eluting peaks and gradient conditions are: (a) $PtCl_4^{2-}$, 0.01 M HCl, 0.5 M NaCl; (b) $PdCl_4^{2-}$, 0.1 M HCl, 0.5 M NaCl; (c) $PtCl_6^{2-}$, 0.1 M HCl, 0.8 M NaCl.

the experiment) for Pd(II), resulting in a broad peak at low [Cl⁻] where all the species tend to be present, and a narrower peak at high [Cl⁻] where $PdCl_4^{2-}$ is the predominant species. These processes are slow in the case of Pt(II), resulting in the multiple peaks.

Separation Efficiencies of Platinum Group Metals. We observed early on that CETP values are significantly larger for metal ion separations than those for simple organic analytes under the same conditions (Figure 1). They are far larger than could be explained in terms of mass transfer and diffusion factors. Moreover, they increase more rapidly with increasing flow rate than those of organic analytes, indicating a chemical kinetic component affecting the CETP. The CETP values observed with metal ions, after correction for mass transport and diffusion (achieved using an organic analyte with similar distribution characteristics), reflect the half lives of chemical reactions causing the added inefficiencies. Metal complex formation and dissociation reactions with half-lives of milliseconds, that is, rapid enough that in batch experiments they reach equilibrium "instantaneously", will lower the efficiencies of CPC chromatograms. Conversely, CETP values can be used to study rapid reaction kinetics if this relationship is found to be generally valid. Thus CPC is a useful tool not only for uncovering kinetics of metals separations but also for obtaining detailed mechanisms of those reactions responsible for inefficiencies in multistage metals separations. This demonstrates the utility of CPC for examining the kinetics of metal complex formation and dissociation reactions in two-phase systems that are too rapid for the automated membrane extraction system (AMES).

It was evident from the separations of PGM that their experimental CETP values were much larger compared to that for an organic analyte at identical distribution ratios. This is illustrated in Figure 7 where the CPC chromatograms of 3-picoline and $PdCl_2(TOPO)_2$ at the same D values are shown. These results indicated that factors other than mass transfer and diffusion were responsible for the additional bandwidths in the case of the metal ions. The most likely factor is the slow kinetics of back-extraction of the metal ions, as the forward extraction reactions are usually rapid. To test this hypothesis, 3-picoline was used as the model compound for the determination of the CPC bandwidth due to mass transfer and diffusion ($CETP_{dif}$), and the CETP value due to slow chemical kinetics ($CETP_{ck}$) was derived by expressing the experimental CETP ($CETP_{obs}$) as a sum of $CETP_{dif}$ and $CETP_{ck}$, equation 8.

$$CETP_{obs} = CETP_{dif} + CETP_{ck} \qquad (8)$$

The $CETP_{ck}$ values determined by varying the concentrations of the species in the aqueous and organic phases clearly showed that the slow back-extraction kinetics of the metal complexes were indeed responsible for the broad bands in the CPC chromatograms. On the basis of these results, a mechanism of the dissociation step could be deduced. For example, the mechanism of dissociation of $PdCl_2(TOPO)_2$ and

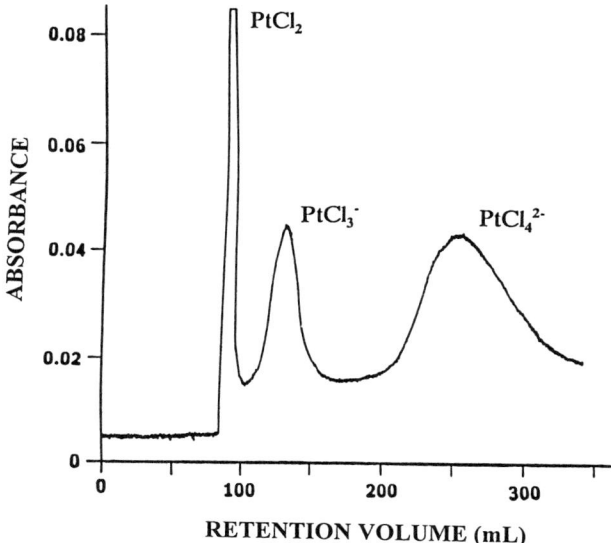

Figure 6. The separation of different chloro species of Pt(II) at 0.1 M NaCl with 0.002 M QpTS (QpTS = tetraheptylammonium p-toluenesulfonate) in 1,2 - dichloroethane-H_2O phase pair, $V_s/V_m = 1$, and aqueous mobile phase flow rate = 2 mL/min.

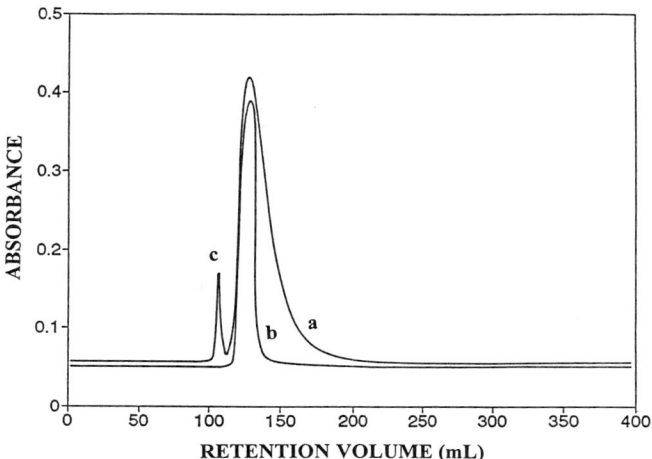

Figure 7. Chromatograms of 3-picoline and $PdCl_2(TOPO)_2$ at identical distribution ratios in the heptane-H_2O phase pair; $V_s/V_m = 0.2$, flow rate = 2 mL/min. (a) 10^{-3} M Pd(II), 0.3 M TOPO, 0.1 M Cl$^-$, pH = 3, (b) 3-picoline, pH =6.1, (c) dead volume peak.

PtCl$_2$(TOPO)$_2$ can be shown by equations 9 - 11 (M = Pd^{2+} or Pt^{2+}), where equation 10 is the rate-limiting step (5).

$$MCl_2(TOPO)_2 \underset{fast}{\overset{K'}{\rightleftharpoons}} MCl_2(TOPO) + TOPO \quad (9)$$

$$MCl_2(TOPO) + Cl^- \underset{slow}{\overset{k_{-2}}{\rightarrow}} MCl_3^- + TOPO \quad (10)$$

$$MCl_3^- + Cl^- \underset{fast}{\rightarrow} MCl_4^{2-} \quad (11)$$

$$k_b^{obs} = \frac{k_{-2} K' [Cl^-]}{K' + [TOPO]} \quad (12)$$

$$k_b^{obs} = k_{-2} K' \frac{[Cl^-]}{[TOPO]} \quad (13)$$

The pseudo-first-order dissociation rate constant is given by equation 12, which has two limiting cases: K' >> [TOPO] and K' << [TOPO], and the latter condition leads to equation 13. This was independently verified by studying the dissociation of MCl$_2$(TOPO)$_2$ in Triton X-100 micelles (as a model for the two phase system) using the stopped flow technique, as this reaction is too fast for conventional spectrophotometric kinetic measurements. As evident from equation 12, the k$_{-2}$K', k$_{-2}$ and K' values can be obtained from the k$_b^{obs}$ values and for Pt^{2+} and Pd^{2+} these are: k$_{-2}$K' = 50.24 ± 2.2 s^{-1} (Pt^{2+}), 0.67 ± 0.02 s^{-1} (Pd^{2+}); k$_{-2}$ = 91.52 ± 4.8 M^{-1}s^{-1} (Pt^{2+}), 168 ± 8 M^{-1}s^{-1} (Pd^{2+}); and K' = 0.55 ± 0.004 M (Pt^{2+}), 0.004 ± 0.0002 M (Pd^{2+}).

According to equation 13, the t$_{1/2}$ (=0.693/k$_b^{obs}$) for the dissociation reaction, equations 9-11, should be proportional to [TOPO] and 1/[Cl$^-$] which was determined to be the case for CETP$_{ck}$ from CPC experiments and t$_{1/2}$ from the stopped flow experiments. The major difference between the PdCl$_2$(TOPO)$_2$ and PtCl$_2$(TOPO)$_2$ dissociation kinetics lies in the preequilibrium step, which involves the dissociation of a TOPO molecule. This equilibrium constant is about two orders of magnitude larger for PtCl$_2$(TOPO)$_2$ compared to PdCl$_2$(TOPO)$_2$, resulting in higher dissociation rate constants and CPC efficiencies for the former. A plot of the $CETP_{ck}$ values for $PtCl_2(TOPO)_2$ against the $t_{1/2}$ values yields a straight line. Further, these points and those for the Pd(II) system fall on a single line indicating a general correlation for the separation of these two metals using TOPO in the heptane-H_2O phase pair.

While the experimental conditions employed in CPC provide the mechanistic and kinetic information on the dissociation of $MCl_2(TOPO)_2$ (M = Pt(II), Pd(II)) complexes, we can also derive information on the formation reaction using the principle of microscopic reversibility (8). This leads us to the mechanism for the formation of $MCl_2(TOPO)_2$ from MCl_4^{2-} and TOPO, equations 14-16.

$$MCl_4^{2-} \underset{}{\overset{K_{-4}}{\rightleftharpoons}} MCl_3^- + Cl^- \qquad (14)$$

$$MCl_3^- + TOPO \underset{slow}{\overset{k_2}{\rightarrow}} MCl_2(TOPO) + Cl^- \qquad (15)$$

$$MCl_2(TOPO) + TOPO \overset{1/K'}{\rightleftharpoons} MCl_2(TOPO)_2 \qquad (16)$$
$$\text{fast}$$

The observed pseudo-first-order rate constant for formation, k_f^{obs}, is given by equation 17 which, like k_b^{obs}, has two limiting cases. The case where $K_{-4} \gg [Cl^-]$ is not encountered in the CPC and stopped-flow experiments due to the small values of K_{-4}, and as a result $[Cl^-] \gg K_{-4}$, equation 18.

$$k_f^{obs} = \frac{k_2 K_{-4}[TOPO]}{K_{-4} + [Cl^-]} \qquad (17)$$

$$k_f^{obs} = \frac{k_2 K_{-4}[TOPO]}{[Cl^-]} \qquad (18)$$

When extraction equilibrium is attained, the rates of formation and dissociation of $MCl_2(TOPO)_2$ are the same and are given by the product of the observed pseudo-first-order rate constants k_f^{obs} and k_b^{obs} and the concentrations of the respective limiting reagents, MCl_4^{2-} and $MCl_2(TOPO)_2$, equation 19.

$$k_f^{obs}[MCl_4^{2-}] = k_b^{obs}[MCl_2(TOPO)_2] \tag{19}$$

Substituting equations 13 and 18 for k_b^{obs} and k_f^{obs}, respectively, we get equation 20:

$$\frac{k_2 K_{-4}}{k_{-2} K'} = \frac{[MCl_2(TOPO)_2][Cl^-]^2}{[MCl_4^{2-}][TOPO]^2} = \beta = \frac{K_{ex} K_L^2}{K_{DC}} \tag{20}$$

where β is the stability constant of the $MCl_2(TOPO)_2$ complex in H_2O, K_L and K_{DC} are the distribution constants for TOPO and $MCl_2(TOPO)_2$, and K_{ex} is the extraction equilibrium constant in the micellar or the two phase heptane-H_2O system. It is reasonable to assume that the values for K_L and K_{DC} are about the same, which yields k_2 as the only unknown quantity in equation 20. These k_2 values are 3.75×10^6 and 5.04×10^6 $M^{-1}s^{-1}$ for $PdCl_2(TOPO)_2$ and $PtCl_2(TOPO)_2$, respectively, using the K_{ex} values from the equilibrium studies and $K_L = 10^5$ in the heptane-H_2O phase pair (15-19), and K_{-4} values for $PdCl_4^{2-}$ (0.0025) and $PtCl_4^{2-}$ (0.018) in H_2O from the literature (17). These values are in excellent agreement with the values determined by stopped-flow formation kinetic experiments (MCl_4^{2-} + TOPO) in Brij 35 micelles and these values are $(3.11 \pm 0.23) \times 10^6$ $M^{-1}s^{-1}$ for $PdCl_4^{2-}$ and $(5.93 \pm 0.36) \times 10^6$ $M^{-1}s^{-1}$. Surprisingly, TOPO complexes with $PdCl_3^-$ and $PtCl_3^-$ (equation 15) with similar rate constants. This could be rationalized on the basis of an associative mechanism for the reaction of TOPO with the coordinatively unsaturated complex, MCl_3^-, and the trans-directing ability of Cl^- (20).

Dissociation reactions with half-lives ranging from milliseconds to seconds can adversely affect the CPC efficiencies. It is important to realize the consequence of these findings: Extraction and back-extraction reactions that appear to be rapid in single-stage equilibrations may still be slow enough to reduce the efficiencies of multistage separations. A further significant finding of this work is that a direct linear correlation exists between $CETP_{ck}$ and $t_{1/2}$, as shown in Figure 8 for the Pd(II)-TOPO and several other systems. Since the $CETP_{ck}$ values are a measure of the half-lives of the slow dissociation steps in metal complex dissociation reactions, CPC is a useful tool for examining the kinetic and the equilibrium aspects of such reactions.

CPC Behavior of Transition Metals

In order to further understand the influence of chemical kinetics on CPC efficiencies, we examined the CPC behavior of Ni^{2+} with the ligands 1-phenyl-2-methyl-4-benzoyl-5-pyrazolone (HPMBP) and dodecylsalicylaldoxime (HDSO) (see structures below) (15,16). In particular, we wanted to understand the reasons for the nongenerality of the $CETP_{ck}$-$t_{1/2}$ correlation; that is, the reason for different slopes for the different systems in Figure 8. The distribution model in Figure 9 is useful in understanding the

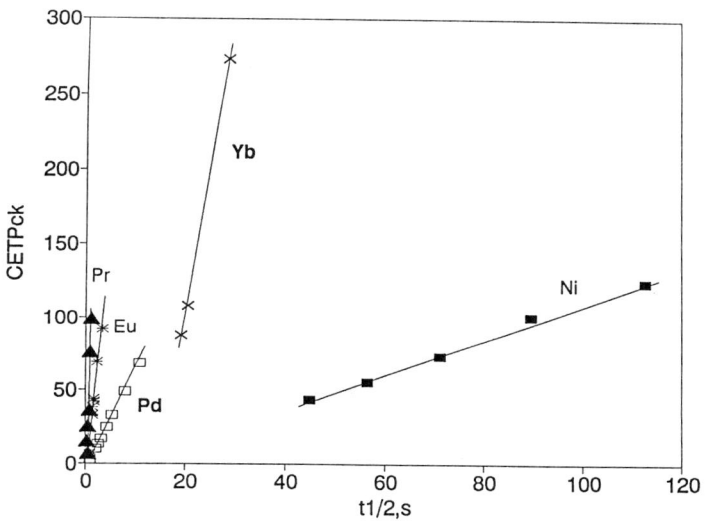

Figure 8. $CETP_{ck}$ vs. $t_{1/2}$ determined in Triton X-100 or Brij 35 micelles. Pr, Eu, and Yb in heptane-H_2O phase pair, 0.1 M Cyanex 272, $V_s/V_m = 0.18$, 1 mL/min flow rate. Pd in heptane-H_2O phase pair, $V_s/V_m = 0.2$, flow rate 2 mL/min, [TOPO] = 0.1 - 0.5 M, [Cl^-] = 0.1 - 0.3 M, pH = 3. Ni in $CHCl_3$-H_2O phase pair, $V_s/V_m = 0.4$, flow rate 0.4 mL/min, pH = 5.8 - 6.2, [HPMBP] = 0.0045 - 0.01 M.

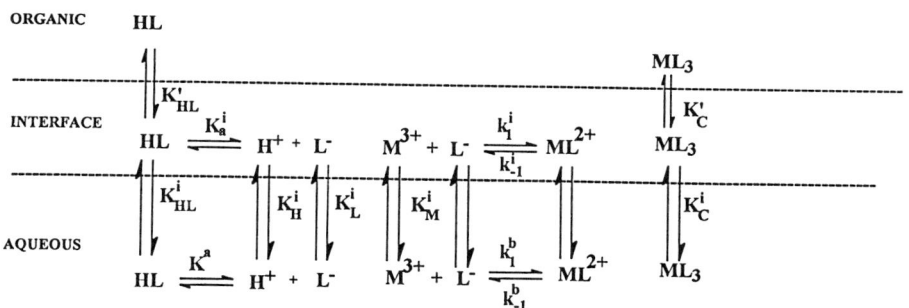

Figure 9. A distribution model for the extraction of metals using trivalent lanthanides as an example.

reasons for this nongenerality. Nickel(II) was chosen because it has fairly slow complex formation and dissociation kinetics, which is useful in discerning the contribution of bulk aqueous and interfacial kinetics to the CPC efficiency. The ligands HPMBP and HDSO were chosen because they have very different bulk properties, such as pK_a values, stability constants with Ni^{2+}, keto-enol equilibrium

HPMBP, $pK_a = 4.0$ HDSO, $pK_a = 8.85$

(HPMBP), and dimer formation (HDSO). The interfacial activities of the neutral and deprotonated ligands at the organic-aqueous interface also differ as indicated by their interfacial excess values Γ (moles/cm^2), which are: HPMBP at the $CHCl_3$-H_2O interface, neutral = 0.14 x 10^{-10}, anion = 0.60 x 10^{-10}, $Ni(PMBP)_2$ = 0.94 x 10^{-10}; HDSO at the hexane-H_2O interface, neutral = 1.87 x 10^{-10}, anion = 2.4 x 10^{-10}, $Ni(DSO)_2$ = 1.26 x 10^{-10}.

The equilibrium for the extraction of Ni^{2+} by HL (HDSO, HPMBP) is given in equation 21:

$$Ni^{2+} + 2HL_o \underset{}{\overset{K_{ex}}{\rightleftharpoons}} NiL_{2,o} + 2H^+ \tag{21}$$

and the rate-limiting steps in the formation of the $Ni(PMBP)_2$ and $Ni(DSO)_2$ complexes by reaction with the protonated and deprotonated ligands are shown in equations 22 and 23.

$$Ni^{2+} + HL \underset{slow}{\overset{k_{HL}}{\rightarrow}} NiL^+ + H^+ \tag{22}$$

The observed pseudo-first-order rate constant, k_f^{obs}, in the presence of excess ligand

(CPC experimental condition) is given by equation 24:

$$Ni^{2+} + L^- \xrightarrow[slow]{k_L} NiL^+ \quad (23)$$

$$k_f^{obs} = [k_{HL}^b + k_L^b \frac{K_a}{[H^+]} + (k_{HL}^i K_{HL}^i K_M^i + k_L^i K_L^i K_M^i \frac{K_a}{[H^+]})ad] \frac{[HL]_o}{K_{DR}} \quad (24)$$

Here k_{HL}^b and k_L^b are the second-order dissociation rate constants in the bulk aqueous phase for HDSO and DSO$^-$, respectively, k_{HL}^i and k_L^i are the corresponding interfacial rate constants, **a** is the specific interfacial area (area per unit volume), d is the thickness of the interface, K_M^i, K_{HL}^i, and K_L^i are the distribution constants of the metal ion, protonated ligand, and free ligand between the interface and the bulk aqueous phase, K_a is the acid dissociation constant, and K_{DR} is the distribution constant of the neutral ligand. The value of K_M^i may be taken to be unity as the interface more closely resembles the bulk aqueous phase. A plot of k_{obs} vs. **a** should yield the bulk and interfacial components of the overall observed rate constants as the intercept and slope, respectively. Such a plot in the pH range 6.2 - 7.5 is shown in Figure 10 for Ni(DSO)$_2$. A plot of the intercept and slopes of Figure 10 as a function of $K_a/[H^+]$ yields the rate constants for the complexation of Ni^{2+} by HL and L$^-$ in the bulk aqueous phase and interface. We have assumed that K_a and [H$^+$] have the same values at the interface as in the bulk aqueous phase. The bulk and interfacial rate constants for HL and L$^-$ are listed in Table I, where the rate constants for 8-mercaptoquinoline (QSH) have been included for comparison.

Table I. Bulk Aqueous and Interfacial Second Order Rate Constants for the Formation of Ni(II) Complexes

Ligand	Solvent Pair	Bulk		Interfacial	
		log k_{HL}^b	log k_L^b	log k_{HL}^i	log k_L^i
QSH	Toluene-H$_2$O	3.18 ± 0.03	4.70 ± 0.11	2.92 ± 0.04	5.11 ± 0.13
HPMBP	CHCl$_3$-H$_2$O	n.d.	4.01 ± 0.02	n.d.	4.61 ± 0.05
HDSO	Hexane-H$_2$O	2.77 ± 0.04	5.34 ± 0.09	2.17 ± 0.05	4.89 ± 0.12

n.d. = not detectable

The dissociation of the NiL_2 complexes under excess $[H^+]$ and $[HL]$ was also investigated. While the dissociation reaction may be expected to proceed by a proton dependent pathway (reverse of equation 22) and a proton independent pathway (reverse of equation 23), it is evident from the formation rate constants in Table I that the proton dependent pathway will be predominant. This assertion is supported by the experimental data which yields a slope of -1.03 ± 0.1 in the plot of log k_d^{obs} (observed dissociation rate constant) vs. pH. Such a plot also indicates that the formation reaction is negligible under the conditions of the dissociation reaction. The mechanisms for the dissociation of NiL_2 are shown in equations 25 and 26, with equation 26 being the rate limiting step.

$$NiL_2 + H^+ \underset{fast}{\overset{K'_{-2}}{\rightleftharpoons}} NiL^+ + HL \qquad (25)$$

$$NiL^+ + H^+ \underset{slow}{\overset{k'_{-1}}{\rightarrow}} Ni^{2+} + HL \qquad (26)$$

The expression for k_d^{obs} is given in equation 27 (analogous to equation 24 for k^f_{obs}) where k_{-1}^b and k_{-1}^i are the bulk and interfacial rate constants which can be obtained by plotting k_d^{obs} vs. **a**.

$$k_d^{obs} = (k_{-1}^b + k_{-1}^i K_{DC}^i \mathbf{a}\, d) \frac{[H^+]}{K_{DC}} \qquad (27)$$

The log K_{DC} values for the $Ni(PMBP)_2$ and $Ni(DSO)_2$ in the $CHCl_3$-H_2O and hexane-H_2O phase pairs are 3.35 and 4.52, respectively. The bulk and interfacial rate constants are given in Table II.

Table II. Bulk and Interfacial Second Order Rate Constants for the Dissociation of Ni(II) Complexes

Complex	Solvent Pair	log k	
		Bulk	Interfacial
$Ni(PMBP)_2$	$CHCl_3$-H_2O	3.92 ± 0.02	3.93 ± 0.06
$Ni(DSO)_2$	Hexane-H_2O	3.49 ± 0.03	4.49 ± 0.05

Unlike Ni(PMBP)$_2$, which has the same bulk and interfacial dissociation rate constants, the interfacial dissociation rate constant for Ni-HDSO is an order of magnitude larger than the bulk value. This is in contrast to the formation reaction for which the interfacial rate constants are smaller than the bulk values. The interfacial formation and dissociation rate constants indicate that the Ni-HDSO complex is less stable at the hexane-aqueous interface compared to the bulk aqueous phase.

The plot of log CETP$_{ck}$ vs. pH for Ni(PMBP)$_2$ and Ni(DSO)$_2$ has a slope close to +1 indicating that the dissociation of NiL$_2$ is the predominant factor in determining the Ni^{2+}-CPC bandwidths. Under the CPC experimental conditions the intercept of this plot yields the composite rate constant, k$_c^{obs}$, which is the sum of the forward and reverse rate constants. This leads to equation 28, from which the interfacial area **a** can be calculated at each pH value of the CPC experiment as all the rate constants and equilibrium constants have been independently determined.

$$(k_{HL}^i K_{HL}^i d \frac{[HDSO]_o}{K_{DR}} + \frac{k_1^i}{K_{DC}} K_{DC}^i d[H^+])\mathbf{a} = k_c^{obs} - k_{HL}^b \frac{[HDSO]_o}{K_{DR}} - k_{-1}^b \frac{[H^+]}{K_{DC}}$$
(28)

In the case of Ni(DSO)$_2$, the complex formation reaction contributes 10% at pH = 4.1 and 30% at pH = 4.7 to k$_c^{obs}$, with the complex dissociation reaction contributing the rest. Equation 28 yields an average specific interfacial area of 207.7 cm^{-1} for the Ni-HPMBP system, which is similar to the specific interfacial area generated in the CHCl$_3$-H$_2$O phase pair in the AMES apparatus. A much larger specific interfacial area of 1370.5 cm^{-1} is generated in the Ni-HDSO system, which is approximately seven times the maximum specific interfacial area generated (208.4 cm^{-1}) in the AMES apparatus. These specific interfacial areas correspond to average mobile phase droplet sizes of 144 μm in the Ni-HPMBP system and 24.0 μm in the Ni-HDSO system. These droplet sizes can be compared with the radius of the capillary ducts, 450 μm, through which the mobile phase droplets enter the channels, and 404 μm, the droplet size calculated by equating the surface tension strength of the drop to the interfacial tension at the [HPMBP], [HDSO], and pH of the mobile phase employed. The radii of the mobile phase calculated from the CETP$_{ck}$ -t$_{1/2}$ correlation are much larger than the radii calculated using Stoke's law, accounting for the centrifugal force (104.5 g) and the linear velocity of the mobile phase in the capillary ducts (1.05 cm s^{-1}), which are 7.4 μm for the Ni-HPMBP system and 6.3 μm for the Ni-HDSO system.

This difference in the CPC behavior of Ni-HDSO and Ni-HPMBP systems may stem from a combination of factors such as the motion of the aqueous phase through an organic phase under a centrifugal force, the motion of the aqueous mobile phase through the lighter hexane (for Ni-HDSO) stationary phase vs. the heavier CHCl$_3$ (for Ni-HPMBP) stationary phase, the interfacial activities of the ligand and the complex (both are higher for Ni-HDSO compared to Ni-HPMBP), and the different rate constants and equilibrium constants. More systems clearly need to be studied to elucidate the role of these factors in determining the efficiencies of multistage separations like CPC.

The $t_{1/2}$ for Ni(DSO)$_2$ and Ni(PMBP)$_2$ under the CPC experimental conditions can be calculated from the k_c^{obs} values obtained using the specific interfacial areas and the expressions for the formation and dissociation reactions, equations 24 and 27. The plot of CETP$_{ck}$ vs. the $t_{1/2}$ calculated in this way is linear with a slope of 1.02 ± 0.03 for Ni-HDSO and a slope of 1.15 ± 0.05 for Ni-HPMBP. As seen from Figure 11, a common straight line can be drawn through the plot of CETP$_{ck}$ vs. $t_{1/2}$ for the Ni-HDSO and Ni-HPMBP systems. This analysis indicates that a given CPC bandwidth should represent a single $t_{1/2}$ irrespective of the system investigated, provided all the distribution constants, rate constants, and interfacial areas have been taken into consideration. These factors are then responsible for the various slopes obtained in the CETP$_{ck}$ vs. $t_{1/2}$ plots for the PGM and lanthanides (Figure 8) where the correlations are made with the $t_{1/2}$ values measured in the micellar pseudo phase. We have conducted CPC separations of the trivalent lanthanides to further understand the various factors affecting the efficiencies of separations and these studies are described in the following section.

Separation of Trivalent Lanthanides

The trivalent lanthanides have been separated using acidic organophosphorous ligands and the acylpyrazolones (17-19). Bis(2,4,4-trimethylpentyl)phosphinic acid (Cyanex 272) in the heptane-water phase pair is dimeric and provides excellent separations of the adjacent light lanthanides at a fixed pH (Figure 12) and a mixture of light and heavy lanthanides using a pH gradient (Figure 13). Cyanex 272 is a

$$M^{3+} + 3\,(HL)_{2(o)} \stackrel{K_{ex}}{\rightleftharpoons} M(HL_2)_{3(o)} + 3\,H^+ \qquad (29)$$

$$M^{3+} + 3\,HL_o \stackrel{K_{ex}}{\rightleftharpoons} ML_{3(o)} + 3\,H^+ \qquad (30)$$

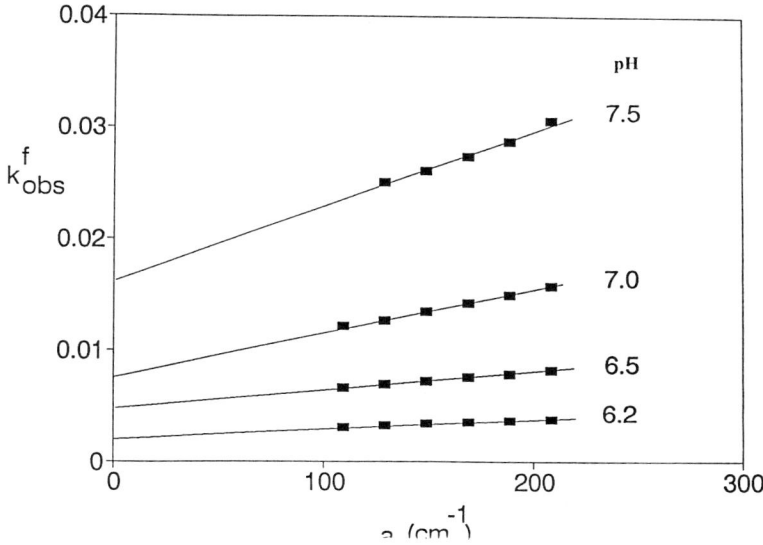

Figure 10. Observed rate constant k^f_{obs} for the formation of $Ni(DSO)_2$ in the hexane-H_2O phase pair as a function of specific interfacial area **a** at different pH values. [Ni(II)] = 2 x 10^{-4} M, [HDSO] = 0.02 M, λ = 394 nm.

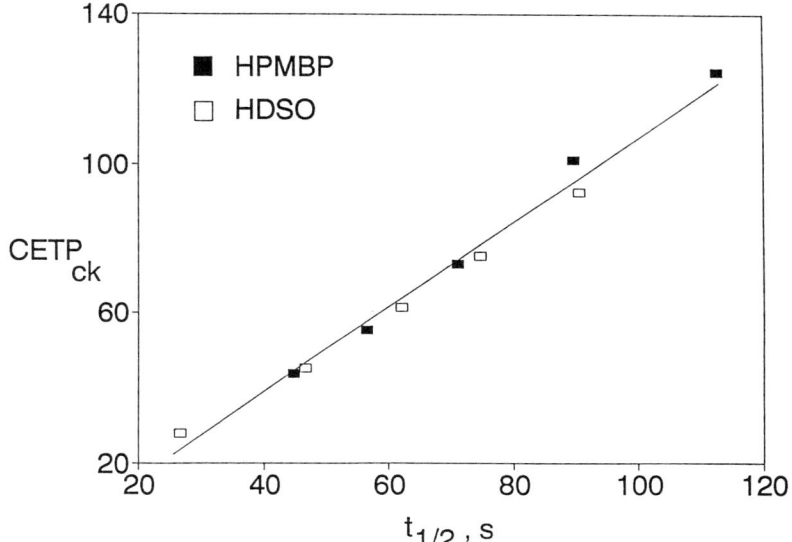

Figure 11. $CETP_{ck}$ vs. $t_{1/2}$ for Ni-HDSO in hexane-H_2O phase pair and Ni-HPMBP in $CHCl_3$-H_2O phase pair.

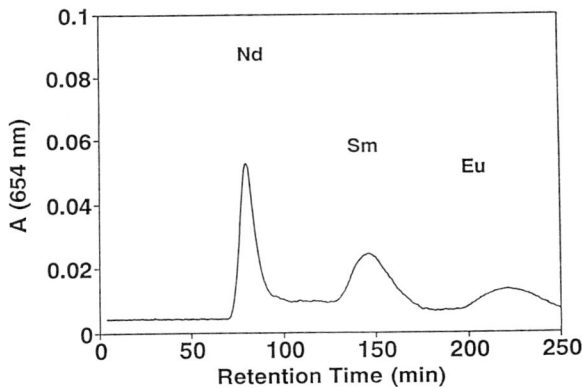

Figure 12. Separation of 2 ppm each of Nd, Sm, and Eu with 0.1 M Cyanex 272 at pH = 2.1 in heptane at $V_s/V_m = 0.18$ and flow rate 1.5 mL/min.

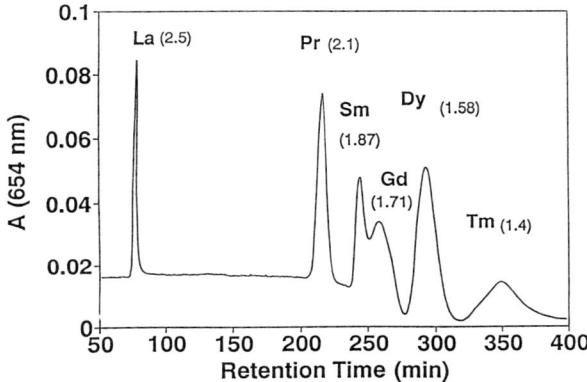

Figure 13. Separation of lanthanides by use of a pH gradient with 0.1 M Cyanex 272 at $V_s/V_m = 0.18$ and flow rate of 1 mL/min. The concentrations and pH of elution are La (2 ppm; 2.5), Pr (6 ppm; 2.1), Sm (4 ppm; 1.87), Gd (4 ppm; 1.71), Dy (10 ppm; 1.58) and Tm (8 ppm; 1.4).

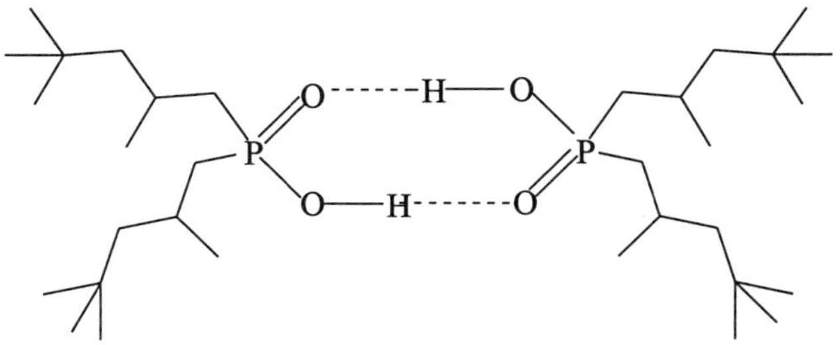

Cyanex 272 (pKa = 3.18)

chelating ligand that extracts the trivalent lanthanides by chelating them in its dimeric form (equation 29). The acylpyrazolones, 1-phenyl-3-methyl-4-benzoyl-5-pyrazolone (HPMBP, see structure above), and 1-phenyl-3-methyl-4-capryloyl-5-pyrazolone (HPMCP, see structure below) have also been used in the toluene-H_2O phase pair for the extraction and separation of the trivalent lanthanides. The extraction equilibrium constants for HPMBP and HPMCP are given in equation 30 and the log K_{ex} values for Cyanex 272 and the acylpyrazolones, along with the log β values for the trivalent lanthanide complexes, are given in Table III.

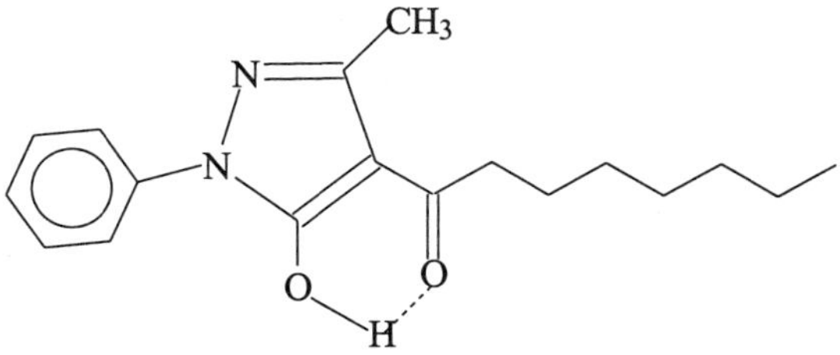

HPMCP

Significant differences are seen between Cyanex 272 and the acylpyrazolones. The extractability of the trivalent lanthanides, as indicated by the log K_{ex} values, is higher with Cyanex 272 than with the acylpyrazolones, where HPMBP shows better extraction than HPMCP. The stability constants of the lanthanides increase from the light to heavy, and the values for the Cyanex 272 and HPMCP complexes are larger than those of HPMBP. The separation factor (or selectivity) for a pair of lanthanides is much better with Cyanex 272 than with HPMBP or HPMCP, which have similar separation factors.

Table III. K_{ex} and β values for Lanthanide Extraction with Cyanex 272 and Acylpyrazolones

Metal Ion	Cyanex 272[a]		HPMBP[b]		HPMCP[b]	
	log K_{ex}	log β	log K_{ex}	log β	log K_{ex}	log β
Pr	-3.83	14.77	-4.08	13.52	-6.98	16.02
Eu	-2.12	16.38	-3.42	14.18	-5.77	17.23
Tb	-1.22	17.38	-2.86	14.74	-5.44	17.56
Ho	-0.62	18.00				
Yb	0.27	18.33	-1.83	15.77	-4.40	18.60

a. Heptane-H_2O phase pair
b. Toluene-H_2O phase pair.

Metallochromic Indicator Method. The kinetics of the dissociation of the Cyanex 272 and acylpyrazolone complexes of the trivalent lanthanides need to be independently studied to obtain the $CETP_{ck}$ - $t_{1/2}$ correlation. These complexes do not have distinct UV-vis absorption spectra and their dissociation reactions are too fast to be monitored by spectrophotometry. We designed the "metallochromic indicator method" to study the dissociation reactions of trivalent lanthanides in micelles formed by neutral surfactants like Triton X-100 (20). The principle of the method is the rapid complexation by Arsenazo III (AZ) of the free trivalent lanthanide after dissociation of the lanthanide phosphonate or acylpyrazolone complex. The formation of the lanthanide-arsenazo III (MAZ) complex is limited by the slow step in the dissociation reaction. As a result, the kinetics of the formation of the MAZ complex directly yields the rate constant for the dissociation of the metal complex, provided the reaction of the MAZ complex with the excess ligand to reform the lanthanide-ligand complex is not significant. The basis of the metallochromic method is given in equations 31 - 33 where ML_3 is the lanthanide complex and the subscript m represents the species in the micellar pseudophase.

$$ML_{3(m)} + 3H^+ \rightleftharpoons M^{3+} + 3HL \tag{31}$$

$$M^{3+} + AZ \rightleftharpoons MAZ \tag{32}$$

$$MAZ + 3HL_m \rightleftharpoons ML_{3(m)} + 3H^+ + AZ \tag{33}$$

Typical absorbance vs. time changes for the Eu-Cyanex 272 complex are shown in Figure 14 at three different initial pH values and a final pH value of 2.35 in all the cases.

The final absorbance in all the cases is the same, but the initial absorbances are different. This is due to the different amounts of free metal present in the aqueous phase at the different pH values as dictated by the proton dependence of the extraction equilibrium constant. The initial absorbance jump, which is instantaneous, yields the free concentration of the metal at the respective initial pH values and hence, the D value of the metal ion. This D value, defined in equation 34, can be used to calculate the extraction equilibrium constant.

$$D = \frac{([M(III)] - [M^{3+}]_{init})}{V_m [M^{3+}]_{init}} \tag{34}$$

Here [M(III)] and $[M^{3+}]_{init}$ are the total free and initial concentrations of the lanthanide, and V_m is the volume fraction of the micellar pseudophase (= ϕ = c - cmc; ϕ = molar volume of the micellar pseudophase; c = concentration of the surfactant; cmc = critical micelle concentration of the surfactant).

The log K_{ex} values in Triton X-100 micelles for Cyanex 272 and the acylpyrazolones are given in Table IV. The log K_{ex} values in the micelles are larger than in the two phase systems (Table III) but the selectivity in the micelles is lower.

These observations can be rationalized by the formation of a 1:1 adduct between the metal complexes and Triton X-100, and this adduct formation constant decreases from the light to the heavy lanthanides for all the ligands, analogous to the adduct formation constants between neutral molecules such as TOPO and the lanthanide complexes in organic-aqueous phase pairs. Adduct formation in micelles is important in understanding the $CETP_{ck}$-$t_{1/2}$ correlations.

Table IV. log K_{ex} in Triton X-100 Micelles

Metal Ion	Cyanex 272	HPMBP	HPMCP
Pr	-1.72	-1.58	-1.42
Eu	-0.19	-1.31	-0.90
Tb	0.35	-1.13	-0.66
Ho	0.79	-0.86	-0.55
Yb	1.52	-0.77	-0.39

Mechanism of Dissociation of Lanthanide-Cyanex 272 Complexes. Two kinds of lanthanide-Cyanex 272 complexes $M(HL_2)_3$ and $M(HL_2)_2L$ are extracted into micelles and dissociate according to equations 35 and 36, respectively. The equilibrium between the two types of complexes is described by equation 37, and the pseudo-first-order rate constant is given by equation 38.

$$M(HL_2)_{3,m} + H^+ \xrightarrow[k_{3m}]{\text{slow}} M(HL_2)_2^+ + (HL)_{2,m} \quad (35)$$

$$M(HL_2)_2L_m + H^+ \xrightarrow[k_{2.5m}]{\text{slow}} M(HL_2)_{2,m}^+ + HL_m \quad (36)$$

$$M(HL_2)_2L_m + 0.5(HL)_{2m} \underset{\text{fast}}{\overset{K_{add}}{\rightleftharpoons}} M(HL_2)_{3m} \quad (37)$$

$$k_{obs} = \left\{ \frac{k_{3m}K_{add}[(HL)_2]_m^{0.5} + k_{2.5m}}{1 + K_{add}[(HL)_2]_m^{0.5}} \right\}[H^+] \quad (38)$$

The $M(HL_2)_2L$ complexes dissociate much more rapidly than the $M(HL_2)_3$ complexes, as indicated by the $k_{2.5m}$ values. For example, the $k_{2.5m}$ values for Eu^{3+} and Yb^{3+} complexes are 400 and 2.5 $M^{-1}s^{-1}$, respectively, which are 4 - 6 times larger than the k_{3m} values for these metals. This detailed study of the kinetics of dissociation of the M^{3+}-Cyanex 272 complexes is essential to understanding their behavior in CPC separations.

The CPC experiments using Cyanex 272, unlike the kinetic studies in the Triton X-100 micelles, were conducted with high concentrations of extractant (0.075 - 0.15 M) and low concentrations of metal ions (~ 10^{-5} M). It is reasonable to expect with such a large excess of ligand that the $M(HL_2)_3$ complex will be exclusively formed in the heptane phase. This is supported by the dependence of the log D values from batch and CPC experiments on pH and the concentration of Cyanex 272. There is no indication of the formation of the $M(HL_2)_2L$ complex in these experiments. As such, we would expect the $CETP_{ck}$ of the lanthanides to exhibit a linear relationship with the concentration of H^+ ($t_{1/2} = 0.693/[H^+]$), with the log $CETP_{ck}$ vs. pH plot having a slope of one. Further, the $CETP_{ck}$ values should be independent of the concentration of the ligand. The CPC experiments, however, reveal quite different dependencies; the log $CETP_{ck}$ vs. pH plots have a slope +2 and the log $CETP_{ck}$ vs. log $[(HL)_2]$ plots have a slope +0.5. The log $CETP_{ck}$ vs. pH plot is shown in Figure 15 for Pr, Eu, and Yb and all exhibit a slope of two. It is also clear from this figure that the $CETP_{ck}$ values are

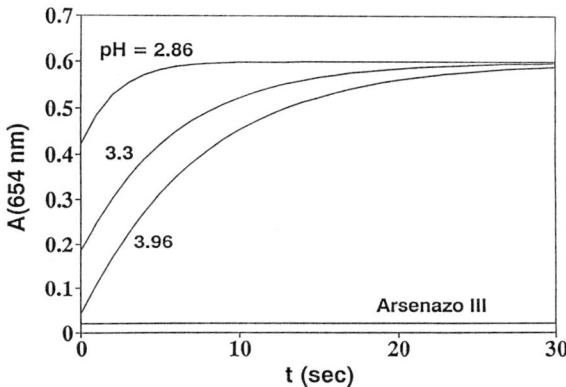

Figure 14. Formation of Eu-Arsenazo III as a function of time for the dissociation of Eu-Cyanex 272 complexes in 1% Triton X-100 following pH jump. $[Eu]_t = 2.2 \times 10^{-5}$ M, [Cyanex 272] = 7.5×10^{-5} M, final pH = 2.35.

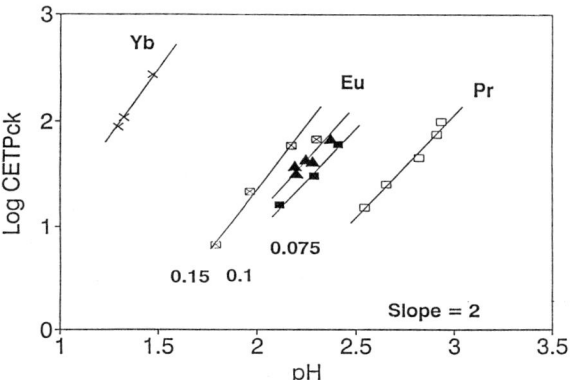

Figure 15. log $CETP_{ck}$ vs. pH plot for Pr, Eu, and Yb. The concentration of Cyanex 272 for Pr and Yb is 0.1 M and for Eu 0.075, 0.1, and 0.15 M.

not independent of the concentration of Cyanex 272 as illustrated by the $CETP_{ck}$ values of Eu^{3+} at various concentrations of Cyanex 272.

A clue to understanding the disagreement between the predicted and observed dependencies of $CETP_{ck}$ on the concentrations of H^+ and Cyanex 272 lies in the kinetic studies conducted in Triton X-100 micelles. These studies revealed that $M(HL_2)_2L$ and $M(HL_2)_3$ complexes are extracted into the micellar pseudophase and that the former has a higher rate constant for dissociation than the latter.

The observed CPC behavior can be rationalized if the dissociation in the heptane-H_2O system proceeds through the $M(HL_2)_2L$ complex formed rapidly from $M(HL_2)_3$, as the former complex dissociates much more rapidly than the latter. The proton induced dissociation of $M(HL_2)_2L$ occurs through a rapid preequilibrium step where $M(HL_2)_2L$ is protonated and this protonated complex reacts with a second proton in a rate limiting step. These reactions are shown in equations 39 - 41, where K_c in equation 39 is the reciprocal of K_{add} defined in equation 37 for the conversion of $M(HL_2)_2L$ to $M(HL_2)_3$.

$$M(HL_2)_3 \underset{fast}{\overset{K_c}{\rightleftharpoons}} M(HL_2)_2L + 0.5(HL)_2 \qquad (39)$$

$$M(HL_2)_2L + H^+ \underset{fast}{\overset{K_p}{\rightleftharpoons}} M(HL_2)_2(HL)^+ \qquad (40)$$

$$M(HL_2)_2HL^+ + H^+ \underset{slow}{\overset{k_{-1}}{\rightleftharpoons}} M(HL_2)HL^{2+} + (HL)_2 \qquad (41)$$

The rate of dissociation of the M^{3+}-Cyanex 272 complex based on this mechanism is as written in equation 42, where the quantity within the braces is the pseudo first order rate constant, k_{obs}. The $t_{1/2}$ and hence, $CETP_{ck}$ clearly have the dependencies on the concentrations of H^+ and Cyanex 272 indicated in equation 43.

$$Rate = \left\{ k_{-1} K_c K_p \frac{[H^+]^2}{[(HL)_2]^{0.5}} \right\} [M(HL_2)_3] \qquad (42)$$

$$t_{1/2} \propto \frac{[(HL)_2]^{0.5}}{[H^+]^2} \qquad (43)$$

This mechanism explains the observed dependencies of log $CETP_{ck}$ on pH (slope 2) and log $[(HL)_2]$ (slope 0.5). The difference between the micelles and the heptane-H_2O systems lies in the preequilibrium step (equation 40), where the complex $M(HL_2)_2L$ is protonated. The product $K_p[H^+]$ is much smaller than one in the two phase system while in the Triton X-100 micelles it is much larger than one.

The dissociation reaction in the micelles occurs almost exclusively at the aqueous-micellar interface, but in the two phase system there could be interfacial and bulk aqueous components. The micellar pseudophase facilitates the protonation of $M(HL_2)_2L$ due to adduct formation between the surfactant molecules and the complex, and in particular, due to the oxyethylene chains of the surfactant molecules. If the dissociation reaction in the two phase system occurs predominantly in the bulk aqueous phase, then we may expect $K_p[H^+]$ to be much larger than one (the case for the micellar pseudophase), as the protonation of the $M(HL_2)_2L$ should proceed readily in H_2O. The kinetic results, however, indicate that $K_p[H^+]$ is much smaller than one in the two phase system, leading to the conclusion that the dissociation reactions are predominantly interfacial in this medium as well. The foregoing analysis not only reveals the subtle difference in the dissociation mechanism between micelles and two phase systems but also helps to determine the location of the reaction in the two phase system.

It is evident from Figure 16 that the $CETP_{ck}$ values bear a linear correlation to the $t_{1/2}$ values determined in micelles. A striking feature of this correlation is that all the points do not fall on a single straight line. In other words the same CPC bandwidth represents a different half-life for each lanthanide. This is partly due to the differences in the K_{DC} values in the micellar and two phase systems, and this point will be further clarified in the separations of lanthanides with the acylpyrazolones. A more significant factor contributing to the different $t_{1/2}$ values for lanthanides at the same $CETP_{ck}$ value is the subtle change in the mechanism of dissociation from the micellar to the two phase system.

Mechanism of Formation and Dissociation of Lanthanide-Acylpyrazolone Complexes. The CPC separation of the lanthanides by HPMBP in the toluene-H_2O phase pair is shown in Figure 17 and the separation by HPMCP in Figure 18. The CPC efficiencies have been correlated with the kinetics of the formation and dissociation of the lanthanide-acylpyrazolone complexes. These kinetics were studied in Triton X-100 micelles as well as in the toluene-H_2O phase pair using the metallochromic indicator method (21). An efficient aqueous phase separator was designed to study the kinetics in two phase systems and used to characterize the equilibrium and kinetics at the organic-aqueous interface. The kinetics of formation and dissociation of the ML_3 complexes were monitored with the aqueous phase separator by following the dissociation and formation of the MAZ complexes with characteristic stability constants, β_{MAZ}, given by equation 44. Both the formation and

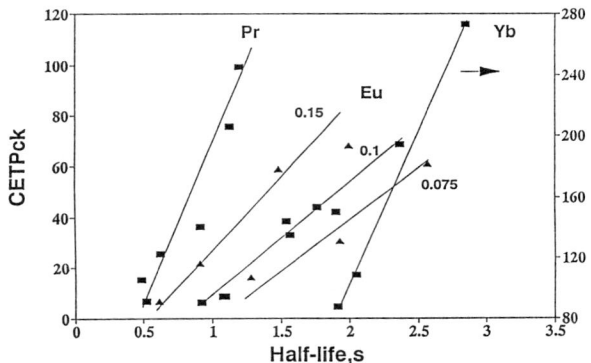

Figure 16. $CETP_{ck}$ vs. $t_{1/2}$ for the lanthanide-Cyanex 272 complexes. The concentration of Cyanex 272 in the case of Pr and Yb is 0.1 M and for Eu 0.075, 0.1, and 0.15 M. Note that the data for Yb is plotted against the right hand axis.

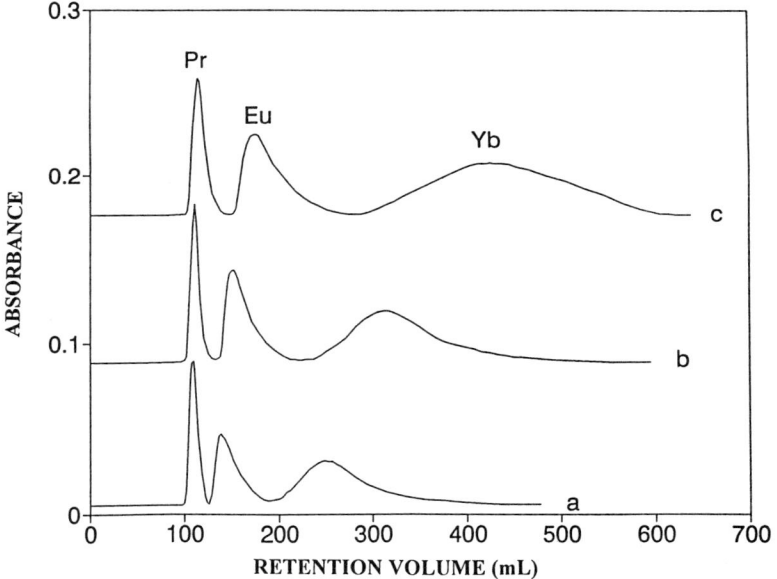

Figure 17. Separation of Pr^{3+}, Eu^{3+}, and Yb^{3+} with [HPMBP] = 0.08 M, at I = 0.1 ($NaClO_4$), and pH (a) 2.36, (b) 2.41, and (c) 2.46 (displaced by arbitrary A values).

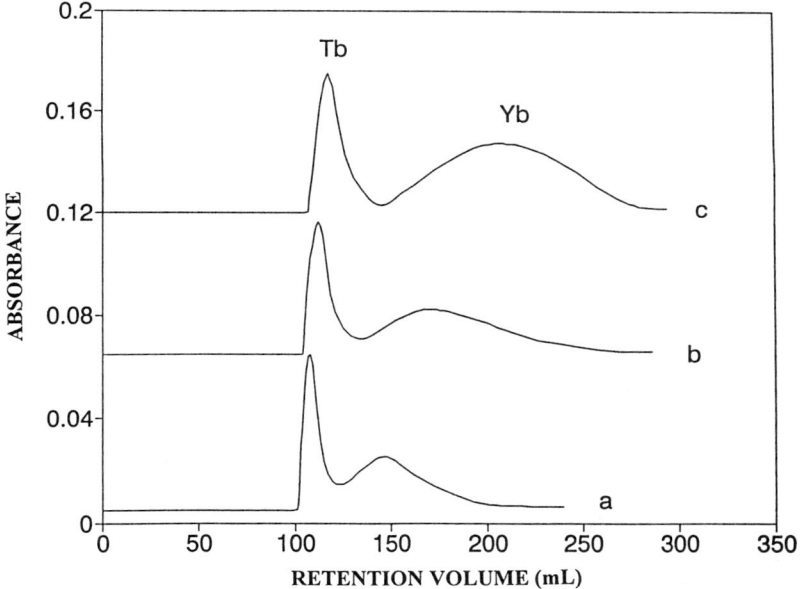

Figure 18. Separation of Tb^{3+} and Yb^{3+} with [HPMCP] = 0.06 M, at I = 0.1 (NaClO$_4$) and pH (a) 2.85, (b) 2.90, and (c) 2.95 (displaced by arbitrary A values).

dissociation reactions were catalyzed by AZ as indicated by the mechanisms of these reactions. The rate limiting steps for the formation reactions are given in equations 45 and 46, and the corresponding observed pseudo first- order rate constant in equation 47. Here, H_3AZ^{5-} represents the predominant Arsenazo III species under the conditions of the experiments.

$$M^{3+} + H_3AZ^{5-} \overset{\beta_{MAZ}}{\rightleftharpoons} M(H_3AZ)^{2-} \tag{44}$$

$$MAZ + L^- \overset{k_{AZ}}{\underset{slow}{\rightarrow}} ML^{2+} + H_3AZ^{5-} \tag{45}$$

$$M^{3+} + L^- \overset{k_1}{\underset{slow}{\rightarrow}} ML^{2+} \tag{46}$$

$$k_{fi}^{obs} = (k_1^i + k_{AZ}^i \beta_{MAZ} K_{MAZ}^i [H_3AZ^{5-}]) \left\{ \frac{[HL]_o K_a}{K_{DR}[H^+]} \right\} (K_L^i d) \tag{47}$$

The k_1 and k_{AZ} values for the lanthanide ions Pr^{3+}, Eu^{3+}, Tb^{3+}, and Yb^{3+} with HPMBP and HPMCP are listed in Table V. The k_1 values for PMBP$^-$ and PMCP$^-$ increase from the light (Pr^{3+}) to the heavy lanthanides (Yb^{3+}) and are 1 - 3 orders of magnitude smaller than their reactions with polyazapolycarboxylic acids (where the formation of the 1:1 metal:ligand complex is also the rate limiting step). The complexation rate constants of M^{3+} with the acylpyrazolones and the polyazapolycarboxylic acids are significantly lower than the water exchange rates of the M^{3+} ions. The structures and the mode of complexation of the two families of ligands are also different; that is, the acylpyrazolones are bidentate ligands requiring three ligands to form a neutral lanthanide complex, while the polyazapolycarboxylic acids possess several ionizable carboxylate groups and generally form 1:1 metal:ligand complexes. The better reactivity of PMCP$^-$ over PMBP$^-$ (Table V) is not due to differences in their interfacial activities and specific interfacial areas generated, as these ligands have similar Γ values (interfacial excess at the toluene-H_2O interface) and yield similar **a** values. The reactivity difference may stem from different steric constraints imposed by the alkyl and phenyl substituents in the four-position of HPMCP, and HPMBP, respectively (see structure), and should be reflected in their entropies of activation. Determination of the activation parameters is in progress.

Table V. Complexation Rate Constants for Trivalent Lanthanide-Acylpyrazolone Complexes in the Toluene-H_2O Phase Pair Catalyzed by Arsenazo III

Metal	HPMBP		HPMCP	
	$\log k_1^i$	$\log k_{AZ}^i$	$\log k_1^i$	$\log k_{AZ}^i$
Pr	4.15	3.79	5.77	4.03
Eu	4.70	3.67	6.07	4.49
Tb	4.80	3.47	6.13	4.52
Yb	5.05	2.84	6.35	3.89

k_1^i and k_{AZ}^i in $M^{-1}s^{-1}$

It can be seen from Table V that there is an increase in $\log k_1$ and decrease in $\log k_{AZ}$ from the light to the heavy lanthanides. It is interesting that the MAZ complexes, like the free metal ions, react more rapidly with PMCP⁻ than with PMBP⁻. The difference in the rate constants is close to an order of magnitude for EuAZ, TbAZ, and YbAZ. These k_{AZ} values are much larger than the values reported for the reactions of the AZ complexes of Lu^{3+} and Y^{3+} with polyazapolycarboxylic acids (25), where the direct reaction of the free metal ions with the ligands was not determined. The catalysis of the formation of the ML_3 complexes by AZ can be understood using equation 47 and the different equilibrium constants and rate constants. For example, in the formation of $Pr(PMBP)_3$, the k_1 and $k_{AZ}K_{MAZ}^i\beta_{MAZ}[H_3AZ^{5-}]$ values corresponding to the reactions of Pr^{3+} and PrAZ with L^- at $[AZ] = 10^{-5}$ M are 6.16×10^3 and 4.26×10^4 $M^{-1}s^{-1}$, respectively, indicating that the contribution to the observed pseudo-first-order rate constant k_{fi}^{obs} from the catalyzed pathway is an order of magnitude larger at this concentration of AZ than by the uncatalyzed pathway. The relative contributions from the uncatalyzed and catalyzed pathways are a function of the rate constants and equilibrium constants, and $[H_3AZ^{5-}]$, and the experimental conditions were chosen such that these can be accurately determined.

The rate-limiting steps in the dissociation of the ML_3 complexes catalyzed by AZ are given in equations 48 and 49, with the corresponding pseudo-first-order rate constant given in equation 50. The dissociation rate constants, k_{-1}^i, were calculated using equation 50 from a plot of k_{di}^{obs} vs. $[H_3AZ^{5-}]$, as all the other quantities are known. Equation 50 yields the products $k_{-1}^i K_{-2}$ and $k_{-AZ}^i K_{-2}$ from which the k_{-1}^i and k_{-AZ}^i values can be obtained using the K_{-2} values calculated from the stability constants of the ML_3 complexes (β, listed in Table III). The various dissociation rate constants in the toluene-H_2O phase pair and Triton X-100 micelles are given in Table VI.

$$ML^{2+} \xrightarrow[\text{slow}]{k_{-1}} M^{3+} + L^- \qquad (48)$$

$$ML^{2+} + H_3AZ^{5-} \xrightarrow{k_{-AZ}} M(H_3AZ)^{2-} + L^- \tag{49}$$

$$k_{obs}^{di} = (k_{-1}^i + k_{-AZ}^i K_{AZ}^i [H_3AZ^{5-}]) \frac{K_{-2}}{K_a} \frac{K_{DR}}{K_{DC}} \frac{[H^+]}{[HL]_o} (K_{DC}^i d) \tag{50}$$

Table VI. Dissociation Rate Constants for Lanthanide-Acylpyrazolone Complexes in the Toluene-H$_2$O Phase Pair and Triton X-100 Micelles

Metal	HPMBP					HPMCP				
	Toluene-H$_2$O			Triton X-100		Toluene-H$_2$O			Triton X-100	
	log k_{-1}^i	log k_{-AZ}^i	log K_{-2}	log k_{-1}	log $k_{-1}K_{-2}$	log k_{-1}^i	log k_{-AZ}^i	log K_{-2}	log k_{-1}	log $k_{-1}K_{-2}$
Pr	0.05	7.38	-4.51	2.91	-1.60	0.43	6.79	-5.34	3.41	-1.93
Eu	-0.03	7.34	-4.73	2.46	-2.27	0.33	7.15	-5.74	3.11	-2.63
Tb	-0.11	7.36	-4.91	2.49	-2.42	0.28	7.47	-5.85	2.62	-3.23
Ho					-2.71					-3.72
Yb	-0.21	6.88	-5.26	2.42	-2.84	0.15	6.99	-6.20	1.86	-4.34

k_{-1}^i and k_{-AZ}^i in M^{-1}s^{-1}; K_{-2} in M

It can be seen from Table VI that the equilibrium constant K_{-2} and the dissociation rate constant k_{-1}^i decrease from Pr^{3+} to Yb^{3+} for both HPMBP and HPMCP as observed in the case of the polyazapolycarboxylic acids. The decrease in K_{-2} reflects the variation of β values. The dissociation rate constant for the HPMCP complex for a given lanthanide is higher than that for the HPMBP complex, paralleling the trend in the formation rate constants. Like the formation reaction, the dissociation rate constants of ML$^+$ catalyzed by AZ do not exhibit much variation among the various lanthanides. The k_{-AZ}^i values (equation 49) are larger by almost three orders of magnitude than the k_{AZ}^i values for every lanthanide ion studied, and this difference has a significant influence on the CPC efficiencies of separations of these metal ions using HPMBP and HPMCP in the presence of AZ. The relative contributions of k_{-1}^i and $k_{-AZ}^i K_{AZ}^i[H_3AZ^{5-}]$ to k_{di}^{obs}, will depend on the pH and [AZ]. For example, in the case of Pr(PMBP)$_3$ at pH = 3 and [AZ] = 2 x 10^{-5} M ([H$_3$AZ^{5-}] = 4 x 10^{-9} M) the values are 1.12 and 0.2, respectively. This indicates that the major dissociation pathway is the uncatalyzed reaction, and that the catalyzed pathway becomes more significant at higher pH values and AZ concentrations.

The kinetics of dissociation of the ML$_3$ complexes were also measured in Triton X-100 micelles, and these rate constants are included in Table VI. The mechanism of dissociation of ML_3 is not altered by the formation of adducts with Triton X-100, as indicated by the independence of k_{di}^{obs} on the concentration of Triton X-100. Equation 50 applies to the dissociation reactions in micelles also, and in contrast to the toluene-H_2O system, the observed pseudo-first-order rate constant k_{di}^{obs} does not exhibit a dependence on [H_3AZ^{5-}]. This indicates that dissociation of the ML^{2+} complex by reaction with free AZ is an insignificant pathway in micelles; that is, AZ does not catalyze the dissociation in this medium. The log k_{-1} values in micelles can be obtained from the log $k_{-1}K_{-2}$ in micelles; these k_{-1} values and the K_{-2} values in H_2O are also given in Table VI. A comparison of the k_{-1} value (uncatalyzed component of k_{di}^{obs}) with the k^i_{-AZ}[H_3AZ^{5-}] (catalyzed component of k^{di}_{obs}) indicates the reason for the catalyzed dissociation pathway being insignificant in micelles. For example, in the dissociation of Pr(PMBP)$_3$ in micelles at pH = 3, the catalyzed and uncatalyzed components (k_{-1} and $k_{-1}K_{-2}$, respectively) are 812 and 0.024, clearly indicating the uncatalyzed pathway to be the dominant one. The log $k_{-1}K_{-2}$ values in micelles for HPMBP and HPMCP decrease from Pr^{3+} to Yb^{3+} as in the toluene-H_2O phase pair, the decrease being about an order of magnitude for HPMBP (similar to in toluene-H_2O) and 2.5 orders of magnitude for HPMCP (larger than in toluene-H_2O). The k_{-1} values for HPMBP in micelles do not vary much from Pr^{3+} to Yb^{3+} (as is the case for the toluene-H_2O phase pair), but for HPMCP, decrease by more than an order of magnitude from Pr^{3+} to Yb^{3+} (in contrast to the toluene-H_2O system). The k_{-1} and K^i_{-1} can be used to estimate the k_1 values in the micelles for the formation of ML^{2+}, and these values for both HPMBP and HPMCP will be 2 - 3 orders of magnitude larger than the values in the toluene-H_2O phase pair. This would also render the AZ catalyzed pathway for the formation reaction insignificant in the micellar system, which is an important difference between the kinetic behavior in the pseudophase and the toluene-H_2O systems.

The plot of log CETP$_{ck}$ as a function of pH and log [HL] for HPMBP and HPMCP have slopes of one for all the lanthanides. The plot of CETP$_{ck}$ vs. $t_{1/2}$, for $t_{1/2}$ values calculated using the dissociation rate constants in Triton X-100 micelles, yields a separate straight line for each trivalent lanthanide ion and ligand (Figures 19a and 19b). When the rate constants and equilibrium constants determined in the toluene-H_2O phase pair are used to calculate the $t_{1/2}$ values, a single CETP$_{1/2}$-$t_{1/2}$ correlation is obtained for all the lanthanide ions and both HPMBP and HPMCP, as shown in Figure 20. A comparison of the equilibrium and kinetic results in the toluene-H_2O and Triton X-100 micellar systems indicates that three factors contribute to the difference in the CETP$_{ck}$-$t_{1/2}$ correlations when the $t_{1/2}$ values in micelles and two phase systems are used. They are: (1) differences in the distribution constants of the metal complexes (K_{DC}) and the ligands (K_{DR}) between the two phase system and the micellar system, (2) differences in the rate constants between the two phase system and the micellar system and, (3) the fact that dissociation reactions in the two phase system can occur both in the bulk aqueous phase and the aqueous-organic interface, while in the micellar systems, they occur exclusively at the aqueous-micelle interface. In the present

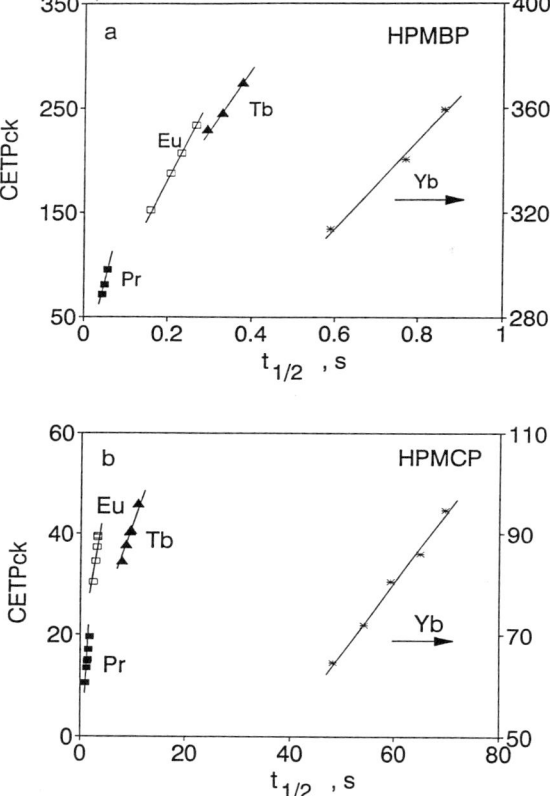

Figure 19. CETP$_{ck}$ vs t$_{1/2}$ for CPC separations of lanthanides with HPMBP and HPMCP for t$_{1/2}$ values calculated with dissociation rate constants in Triton X-100 micelles, (a) HPMBP and (b) HPMCP. Note that the data for Yb is plotted against the right hand axis.

studies, only factors (1) and (2) are important, as the dissociation reactions both in the toluene-H_2O phase pair and the micelles exclusively occur in the interfacial region. The CPC separation of lanthanides with the toluene-H_2O phase pair and the ligands HPMBP and HPMCP is the first demonstration of a countercurrent separation with two bulk phases where the efficiency is determined exclusively by interfacial kinetics, analogous to conventional LC separations.

Many dissociation reactions in two phase systems are too fast for spectrophotometric measurements using the automated membrane extraction system (AMES), necessitating the use of homogeneous media like micelles (which are good models for the two phase system) for the measurement of dissociation rate constants by rapid kinetic techniques like stopped-flow. Even though the dissociation rate constants ($k_{-1}K_{-2}$) for the HPMCP complexes in the toluene-H_2O phase pair are smaller than those for the HPMBP complexes, the efficiencies for the former ligand are significantly larger than the latter. This is because $CETP_{ck}$ is not only a function of the metal complex dissociation rate constant but also of the various equilibrium constants and the interfacial area generated.

The specific interfacial areas, **a,** generated in the CPC experiments can be calculated from the intercepts of the plots of log $CETP_{ck}$ vs. pH and log $CETP_{ck}$ vs. log [HL] as all the other quantities are known. The **a** values for HPMBP and HPMCP determined from such plots are 183.43 cm^{-1} and 130.17 cm^{-1}, respectively, which are similar to the **a** values generated at a stirring speed of 5000 rpm in the AMES experiments (160.74 cm^{-1} for HPMBP and 114.96 cm^{-1} for HPMCP. The areas generated in CPC correspond to aqueous mobile phase droplet sizes of 163.6 μm for HPMBP and 230.5 μm for HPMCP which are similar to the droplet sizes generated at 5000 rpm in the AMES experiments (186.6 μm for HPMBP and 261.0 μm for HPMCP).

Effect of Triton X-100 on the CPC Efficiencies and Separations. To further understand the influence of the liquid-liquid interface in CPC separations, the neutral surfactant Triton X-100 (TX) was added to the toluene phase and its effect on the separation efficiency and resolution were examined. The separation of Pr^{3+} and Eu^{3+} using 0.08 M HPMBP with 0.001 M Triton X-100 added is shown in Figure 21. It is evident from a comparison of this figure to Figure 17 that the addition of Triton X-100 not only improves the chromatographic efficiencies, but also changes the distribution ratios of the metal ions. This indicates that surfactant addition affects both interfacial kinetics and extraction equilibria. To understand better the role of the added Triton X-100 in the CPC separations, experiments were conducted as a function of the concentrations of HL, H^+, and TX. The extraction equilibrium in equation 51 was deduced from the dependencies of log D of the metal ion as a function of the log of the concentrations of HL, H^+, and TX.

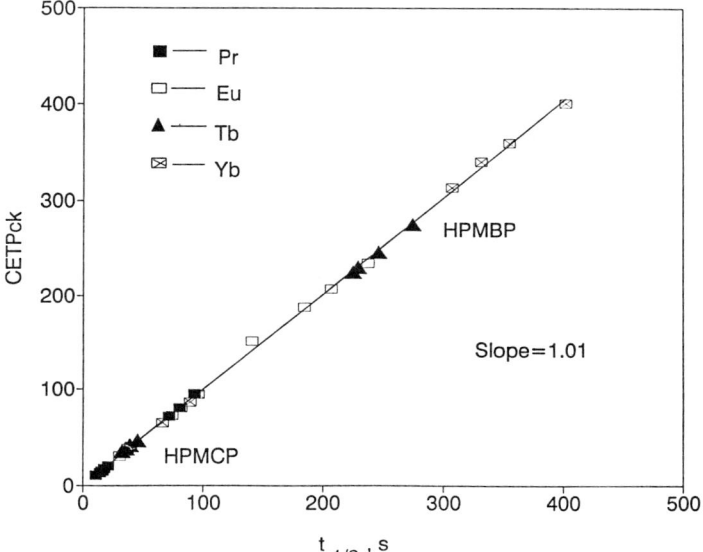

Figure 20. $CETP_{ck}$ vs. $t_{1/2}$ for CPC separations of lanthanides with HPMBP and HPMCP for $t_{1/2}$ values calculated with dissociation rate constants in the toluene-H_2O phase pair.

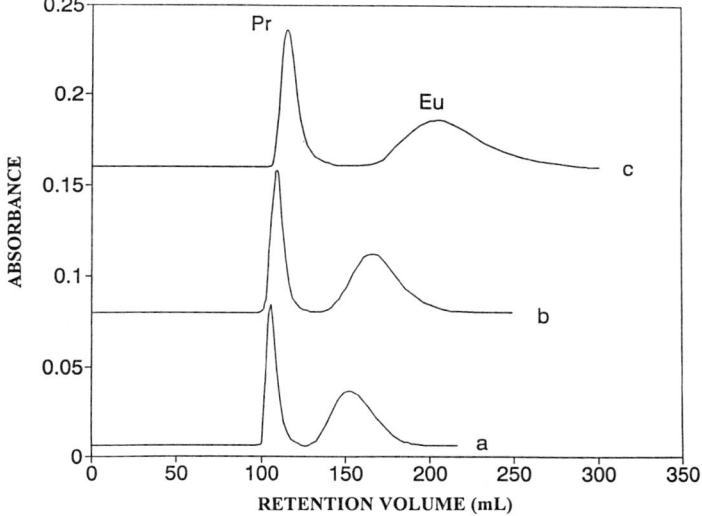

Figure 21. Separation of Pr^{3+} and Eu^{3+} with [HPMBP] = 0.08 M and [Triton X-100] = 0.001 M at I = 0.1 ($NaClO_4$) and pH (a) 2.37, (b) 2.41, and (c) 2.46 (displaced by arbitrary A values).

$$M^{3+} + 3HL_o + TX_o \underset{}{\overset{K_{ex}^T}{\rightleftharpoons}} ML_3 \cdot TX_o + 3H^+ \qquad (51)$$

The equilibrium studies indicate that the complex ML_3 forms a 1:1 adduct with Triton X-100 in the toluene phase.

The addition of Triton X-100 has opposite effects in lanthanide separations with HPMBP and HPMCP, increasing the efficiency and resolution for HPMBP and decreasing them for HPMCP. Since adduct formation does not alter the dissociation mechanism of ML_3 complexes in Triton X-100, we can expect this to be the case in the toluene-H_2O phase when Triton X-100 is added to toluene to form the adduct. The changes in the $CETP_{ck}$ values in the presence of Triton X-100 stem from changes in **a**, $K^i_{DC}d$ and K_{DC}. The specific interfacial area **a** and $K^i_{DC}d$ in the toluene-H_2O interface for both HPMBP and HPMCP using the AMES apparatus were determined to increase by a factor of two in the presence of 0.001 M Triton X-100 (**a** at 5000 rpm, PMBP$^-$ = 335.23 cm^{-1}, PMCP$^-$ = 241.45 cm^{-1}, $K^i_{DC}d$: Eu(PMBP)$_3$ = 0.00969 L/cm^2, Eu(PMCP)$_3$ = 0.01176 L/cm^2) and these changes would also occur in the CPC experiments. The lower efficiencies for HPMCP in the presence of Triton X-100 can be attributed to the K_{DC} values of M(PMCP)$_3$·TX complexes being larger (factor of 5 - 10 depending on the lanthanide ion) than the values for M(PMCP)$_3$. The experiments in the presence of Triton X-100 again emphasize the critical role played by the liquid-liquid interface in CPC separations and the interplay of the kinetic and equilibrium parameters in determining the efficiencies of separations.

CPC Separations in the Presence of Arsenazo III. We have shown that the interfacial formation and dissociation reactions of ML_3 are catalyzed by the reaction of the MAZ complex adsorbed at the toluene-H_2O interface. We conducted CPC separations of the trivalent lanthanides with HPMBP and HPMCP in the presence of AZ in the aqueous mobile phase to determine if interfacial catalysis resulted in significant improvement in the efficiencies. In the presence of AZ, the M^{3+} is present as MAZ. Its extraction equilibrium was established by examining the dependence of its log D on pH, log [HL], and log [AZ], which yielded slopes of +1, +3, and -1, respectively. These observations can be rationalized on the basis of the extraction equilibrium given in equation 52.

$$M(H_3AZ)^{2-} + 3HL_o \underset{}{\overset{K_{ex}^{AZ}}{\rightleftharpoons}} ML_{3,o} + H^+ + H_5AZ^{3-} \qquad (52)$$

An examination of the extraction equilibrium constants with HPMBP and HPMCP indicates that for a lanthanide metal ion, $K^T_{ex} > K^{AZ}_{ex} > K_{ex}$; also the separation factor for a given pair of lanthanides in the presence of AZ is lower with or without Triton X-100. This poor selectivity in the presence of AZ allows only the partial separation of Pr^{3+} and Yb^{3+}. This can be seen from Figure 22, in which their separation with HPMBP at various concentrations of AZ is displayed. Unlike the selectivities, the efficiencies are high in the presence of AZ as expected from the kinetic studies. Unfortunately the higher efficiencies in the presence of AZ do not lead to better separations, as the resolution of a pair of analytes is a direct function of efficiency, selectivity, and retention. While the efficiency is higher in the presence of AZ, both selectivity and retention are lower, and a combination of these factors leads to poor resolution.

We can understand further the reason for the poor selectivity in the presence of AZ using equation 53, which relates the stability constant, β_{MAZ}, of the MAZ complex to the extraction equilibrium constants, K_{ex} and K^{AZ}_{ex}, and the pK_4 (3.40) and pK_5 (6.27) of AZ (21).

$$\log \beta_{MAZ} = \log K_{ex} - \log K^{AZ}_{ex} + pK_4 + pK_5 \quad (53)$$

The $\log \beta_{MAZ}$ values calculated using equation 53 are in excellent agreement with the values independently determined in our studies using Job's method, and found to be 8.1, 8.4, 8.8, and 9.3 for the lanthanides Pr^{3+}, Eu^{3+}, Tb^{3+}, and Yb^{3+}, respectively. This analysis indicates that the smaller separation factors in the presence of AZ are due to the closeness in the β_{MAZ} values. A ligand such as AZ can simultaneously improve efficiency and selectivity if its stability constants for the different lanthanides are sufficiently different.

The **a** values calculated from the $CETP_{ck}$ values and equation 50 are 2625.8 ± 963.0 cm^{-1} (mobile phase drop size = 12.9 ± 0.4 µm) for HPMBP and 225.5 ± 81.9 cm^{-1} (mobile phase drop size = 148.7 ± 4.4 µm) for HPMCP. Comparing these values with the values of **a** generated in the absence of AZ, we find that **a** increases by a factor of 14 for HPMBP and by a factor of 2 for HPMCP. Thus, the improvement in efficiency in the presence of AZ for HPMBP is entirely due to an increase in the specific interfacial area, while in the case of HPMCP it is due to the catalysis of the dissociation reaction by AZ, as well as an increase in the specific interfacial area.

The reasons for the large increase in **a** in the case of HPMBP are not clear, but are similar to the observation we had previously made in the case of the Ni-HDSO system. In CPC, where the mobile phase is moved through capillary ducts, the coadsorptions of the ligands and their lanthanide complexes along with AZ and MAZ could lead to much different **a** values for HPMBP and HPMCP. This requires further investigation. The specific interfacial areas generated and the sizes of the aqueous

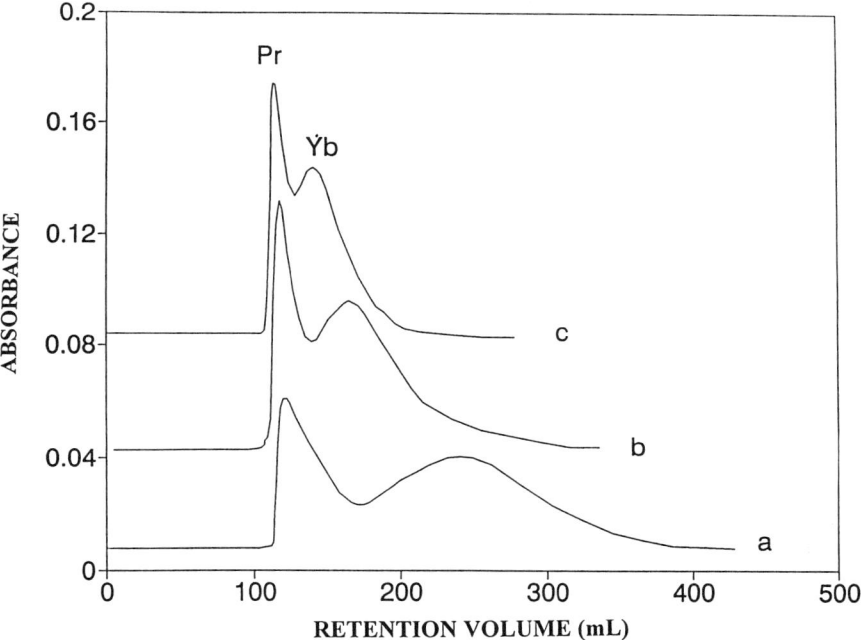

Figure 22. Separation of Pr^{3+} and Yb^{3+} with [HPMBP] = 0.06 M, at pH = 2.82, I = 0.1 ($NaClO_4$), and [Arsenazo III] of (a) 2.0×10^{-5} M, (b) 4.0×10^{-5} M, and (c) 6.0×10^{-5} M (displaced by arbitrary A values).

mobile phase droplets in the CPC separations of the lanthanides and transition metals are summarized in Table VII.

Table VII. Specific Interfacial Areas and Mobile Phase Droplet Sizes Generated in CPC

System	Specific Area (cm^{-1})	Drop Size (μm)
M^{3+} - HPMBP[a]	183.4	163.6
M^{3+} - HPMCP[a]	115.0	230.5
M^{3+} - HPMBP + Triton X-100	335.2	89.5
M^{3+} - HPMCP + Triton X-100	241.5	124.2
M^{3+} - HPMBP + AZ	2625.8	12.9
M^{3+} - HPMCP + AZ	225.5	148.7
Ni^{2+} - HPMBP[b]	207.7	144.0
Ni^{2+} - HDSO[c]	1880.5	16.1

a. Toluene-H_2O.
b. $CHCl_3$-H_2O.
c. Hexane-H_2O.

Conclusions

A number of significant results have emerged from studies of the CPC separations of various families of metal ions: (1) separation efficiencies are a function of bulk and interfacial equilibria and kinetics of metal complex formation and dissociation, with the dissociation reaction being the major factor limiting efficiencies in many cases; (2) the chromatographic inefficiency, as indicated by the parameter CETP, exhibits a direct linear correlation with the $t_{1/2}$ values for the slowest step in the formation and dissociation reactions; (3) this correlation is not general when the $t_{1/2}$ values are determined in micelles, but is general when they are determined in two phase systems; (4) CPC separations can be entirely interfacially driven, analogous to conventional liquid chromatography; (5) CPC efficiencies can be improved by an increase in the interfacial area generated in the CPC experiments and interfacial catalysis of the metal complex formation and dissociation reactions; and (6) very large interfacial areas can be generated in CPC experiments and can provide much higher efficiencies than those expected from the rate constants for the complex formation and dissociation reactions. CPC will continue to grow as a useful tool for analytical and macroscale separations and for obtaining a fundamental understanding of such separations. It may also prove to be an invaluable tool for the development of highly selective ligands based on organized molecular assemblies such as chelating micelles and dendrimers, and for gaining an understanding of their metal ion recognition mechanisms.

Acknowledgments

This research was supported by a grant from the Chemistry Division of the National Science Foundation.

Literature Cited

(1) *Value Adding Through Solvent Extraction*, Shallcross, D. C.; Paimin, R.; Prvcic, L. M., Eds.; University of Melbourne: Melbourne, 1996, Vol. I-II.
(2) *Separation of f Elements*; Nash, K. L.; Choppin, G. R., Eds.; Plenum Press: New York, 1995.
(3) *Modern Countercurrent Chromatography*; Conway, W. D.; Petroski, R. J., Eds.; ACS Symposium Series 593, American Chemical Society: Washington, DC, 1995.
(4) *Centrifugal Partition Chromatography*; Foucault, A. P., Ed.; Marcel Dekker: New York, 1994.
(5) Muralidharan, S.; Freiser, H. In *Recent Progress in Actinides Separation Chemistry*; Yoshida, Z.; Kimura, T.; Meguro, Y., Eds.; World Scientific: Singapore, 1997, pp 191-208.
(6) Berthod, A.; Armstrong, D. W. *J. Liq. Chromatogr.* **1988**, *11*, 567.
(7) Ma, E.; Freiser, H.; Muralidharan, S. In, *Value Adding Through Solvent Extraction*, Shallcross, D. C.; Paimin, R.; Prvcic, L. M., Eds.; University of Melbourne: Melbourne, 1996, pp 457-462.
(8) Cote, B.; Demopoulos, G. *Solvent Extr. Ion. Exch.* **1994**, *12*, 393.
(9) Zhang, B.; Zhang, Y.; Grote, M.; Kettrup, S. *React. Polym.* **1994**, *22*, 115.
(10) Pyrzynska, K. *Anal. Chim. Acta* **1991**, *225*, 169.
(11) Watson , J.; Ellwood, D. *Miner. Eng.* **1994**, *7*, 1017.
(12) Surakitbanharn, Y.; Muralidharan, S.; Freiser, H. *Solvent Extr. Ion Exch.* **1991**, *9*, 45.
(13) Surakitbanharn, Y.; Muralidharan, S.; Freiser, H. *Anal. Chem.* **1991**, *63*, 2642.
(14) Surakitbanharn, Y.; Freiser, H.; Muralidharan, S. *Anal. Chem.* **1996**, *68*, 3934.
(15) Chen, F.; Freiser, H.; Muralidharan, S. *Langmuir* **1994**, *10*, 2139.
(16) Chen, F.; Ma, H.; Freiser, H.; Muralidharan, S. *Langmuir* **1995**, *11*, 3235.
(17) Cai, R.; Muralidharan, S.; Freiser, H. *J. Liq. Chromatogr.* **1990**, *13*, 3651.
(18) Inaba, K.; Freiser, H.; Muralidharan, S. *Solv. Extr. Res. Develop. Japan* **1994**, *1*, 13.
(19) Ma, G.; Freiser, H.; Muralidharan, S. *Anal. Chem.* **1997**, *69*, 2835.
(20) Inaba, K.; Muralidharan, S.; Freiser, H. *Anal. Chem.* **1993**, *65*, 1510.
(21) Ma, G.; Freiser, H.; Muralidharan, S. *Anal. Chem.* **1997**, *69*, 2827.

Chapter 23

Extraction and Separation of Uranium and Lanthanides with Supercritical Fluids

Chien M. Wai[1], Yuehe Lin[1], Min Ji[1], Karen L. Toews[1], and Neil G. Smart[2]

[1]Department of Chemistry, University of Idaho, Moscow, ID 83844
[2]Research and Technology, BNFL, Sellafield, Cumbria CA20 1PG, United Kingdom

Uranyl ions in nitric acid solutions can be effectively extracted by supercritical CO_2 containing tributyl phosphate (TBP) or organophosphine oxides. The form of the extracted uranyl nitrate-TBP complex and the kinetics of the supercritical extraction are similar to those reported for the conventional solvent extraction of uranyl nitrate with TBP. On-line back-extraction of uranium in supercritical CO_2 with an aqueous solution has also been demonstrated. The results suggest that supercritical CO_2 could be as effective as the organic solvents used in the PUREX process. Supercritical CO_2 containing organophosphinic acids such as Cyanex 301 and Cyanex 302 has been shown to extract heavy lanthanides selectively from the light lanthanides in aqueous solutions. This *in situ* chelation-SFE technique is also capable of removing leachable uranium from solid samples such as mine tailings as indicated by the EPA Toxicity Characteristics Leaching Procedure.

There has been considerable interest in the past two decades to utilize supercritical fluids as solvents for chemical separations (*1*). The reasons for developing supercritical fluid extraction (SFE) technologies are mostly due to the environmental regulations and waste disposal costs for conventional solvents. Supercritical fluids have both gas-like and liquid-like properties. The solvation power of a supercritical fluid depends on pressure and temperature; thus, one can achieve the optimum conditions for a particular separation process by manipulating the temperature and pressure of the fluid phase. The high diffusivity and low viscosity of supercritical fluids enable them to penetrate and transport solutes from solid matrices. Carbon dioxide is the most widely used gas for SFE because of its moderate critical constants ($T_c = 31.3\ °C$, $P_c = 72.9$ atm), nontoxic nature, and availability in pure form. In SFE processes, solutes dissolved in supercritical carbon dioxide are separated by reducing the pressure of the fluid phase causing precipitation of the solutes. The fluid phase is usually expanded into a collection vessel to remove the solutes and the gas is recycled for repeated use. Typical examples of large-scale industrial applications of the SFE technology using supercritical CO_2 include the preparation of decaffeinated coffee and hop extracts (*1*).

Until recently, little information was available in the literature regarding SFE of metal species. Direct extraction of metal ions is highly inefficient because of the charge neutralization requirement and the weak solute-solvent interactions. However, when metal ions are chelated with organic ligands, they may become quite soluble in supercritical CO_2 (*2*). This *in situ* chelation-SFE technique appears to have a wide range of applications including the treatment of metal contaminated or radioactive waste materials and mineral processing.

Background

Quantitative measurements of metal chelate solubilities in supercritical CO_2 were first made by Wai and co-workers in 1991 using a high pressure view cell and UV/Vis spectroscopy (*3*). In this report, the authors showed that the solubility of metal dithiocarbamates depends on metal coordination and that fluorine substitution can greatly enhance their solubility in supercritical CO_2. The demonstration of copper extraction from solid and liquid materials using supercritical CO_2 containing the fluorinated ligand *bis*(trifluoroethyl)dithiocarbamate (FDDC) was reported in 1992 (*4*). Since then over 50 papers regarding SFE of metal species from different sample matrices have appeared in the literature. A variety of chelating agents, including dithiocarbamates, β-diketones, organophosphorus reagents, macrocyclic compounds, fluorinated surfactants, etc. have been tested for SFE of metals (*4-9*). The feasibilies of extracting organometallic compounds, heavy metals, lanthanides, and actinides from solid and liquid materials using the *in situ* chelation-SFE method have been evaluated by a number of research groups (*7-9*). According to the literature, the important parameters controlling SFE of metal species appear to be: (i) solubility and stability of chelating agents, (ii) solubility and stability of metal chelates, (iii) water and pH, (iv) temperature and pressure, (v) chemical form of metal species, and (vi) matrix.

Chelation and extraction of lanthanides and actinides in supercritical fluids are of interest to separation scientists because of their potential applications in nuclear waste management and nuclear fuels reprocessing. Several recent reports demonstrated that uranyl (UO_2^{2+}), thorium (Th^{4+}), and trivalent lanthanide ions (Ln^{3+}) in liquid and in solid matrices can be effectively extracted by supercritical CO_2 containing organophosphorus reagents and β-diketones (*6-8*). The structures of some organophosphorus reagents tested for SFE of metals are shown in Figure 1. The possibility of dissolution of uranium oxides directly in supercritical CO_2 containing organic ligands has also been reported recently (*10*). The SFE technology may offer a method of separating uranium and transuranic elements without utilization of conventional acid dissolution and organic solvent extraction. This could lead to a significant waste reduction utilizing the SFE technology in contrast to the current acid leaching and solvent extraction processes. This paper reports some recent developments concerning the separation of uranium and lanthanides using supercritical

CO_2 as an extraction medium to further illustrate the utility of this novel extraction technology.

Extraction of Uranyl Ions from Nitric Acid Solutions with Supercritical CO_2

In a previous communication from our group, the feasibility of extracting uranyl and thorium ions from nitric acid solutions using TBP dissolved in supercritical CO_2 was reported (7). TBP is stable and very soluble in supercritical CO_2 (11). The solubility of TBP in supercritical CO_2 at 60 °C and 120 atm is about 11 mole percent (11). The distribution coefficient and the rate of extraction of uranyl ions from 6 M nitric acid with TBP in supercritical CO_2 have been determined recently (12). The extraction of uranyl ions in the supercritical fluid system can be expressed by the following equation:

$$UO_2^{2+}{}_{(aq)} + 2\,NO_3^-{}_{(aq)} + n\,TBP_{(sf)} \leftrightarrow [UO_2(NO_3)_2 \bullet n\,TBP]_{(sf)} \quad (1)$$

where the subscript sf denotes the supercritical CO_2 phase. The equilibrium constant for the reaction can be expressed as:

$$K = [UO_2(NO_3)_2 \bullet n\,TBP]_{sf} / [UO_2^{2+}]_{aq} \bullet [NO_3^-]_{aq}^2 \bullet [TBP]_{sf}^n \quad (2)$$

Defining D_U, the distribution coefficient of uranium, as the ratio of the concentration of U(VI) in the supercritical CO_2 phase to its concentration in the aqueous phase at equilibrium and rearranging equation 2, we obtain the following equation in logarithmic form:

$$\log D_U = \log K + 2 \log [NO_3^-]_{aq} + n \log [TBP]_{sf} \quad (3)$$

If the TBP concentration in the supercritical CO_2 phase is varied and all other conditions that may affect the extraction are maintained constant, the slope of a logarithmic plot of D_U versus $[TBP]_{sf}$ should yield a line with slope n, the TBP solvation number.

The D_U values were determined using a high pressure liquid extraction cell described elsewhere (12). Figure 2 shows the data collected at 60 °C and 200 atm for the extraction of uranyl ions (4.2×10^{-4} M) in 6 M HNO_3 with supercritical CO_2 containing varying concentrations of TBP. At 0.3 to 10^{-4} M TBP, the slope approaches a value of 1.8±0.1, while below 10^{-4} M TBP the extraction is nonlinear. The results indicate that with high TBP concentrations, the extracted uranyl nitrate species is in the form $UO_2(NO_3)_2 \bullet 2TBP$, which is similar to the stoichiometry observed in the conventional solvent extraction of uranyl ions from nitric acid solutions with TBP.

It was observed in this study that the nitric acid concentration in the sample solution was reduced after the extraction. Separate experiments with a nitric acid solution in contact with supercritical CO_2 containing TBP showed that HNO_3 can be extracted by the supercritical fluid phase. Coextraction of nitric acid with uranyl nitrate by TBP in supercritical CO_2 is likely the cause of deviation of the plot of log D_U versus log [TBP] shown in Figure 2. The observation, nevertheless, suggests the possibility of utilizing TBP modified supercritical CO_2 for removing nitric acid from aqueous

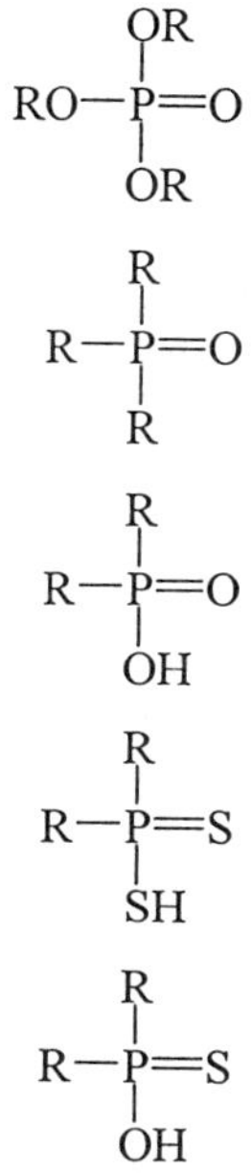

OR
|
RO—P=O   TBP   (R = n-C$_4$H$_9$)   Tributylphosphate
|
OR

R
|
R—P=O   TBPO (R = n-C$_4$H$_9$)   Tributylphosphine oxide
| TOPO (R = n-C$_8$H$_{17}$) Trioctylphosphine oxide
R

R
|
R—P=O   Cyanex 272 (R = (CH$_3$)$_3$CCH$_2$-CH(CH$_3$)-CH$_2$-)
| Bis(2,4,4-trimethylpentyl)phosphinic acid
OH

R
|
R—P=S   Cyanex 301 (R = (CH$_3$)$_3$CCH$_2$-CH(CH$_3$)-CH$_2$-)
| Bis(2,4,4-trimethylpentyl)dithiophosphinic acid
SH

R
|
R—P=S   Cyanex 302 (R = (CH$_3$)$_3$CCH$_2$-CH(CH$_3$)-CH$_2$-)
| Bis(2,4,4-trimethylpentyl)monothiophosphinic acid
OH

Figure 1. Structures of the organophosphorus reagents tested for SFE of uranium and lanthanides.

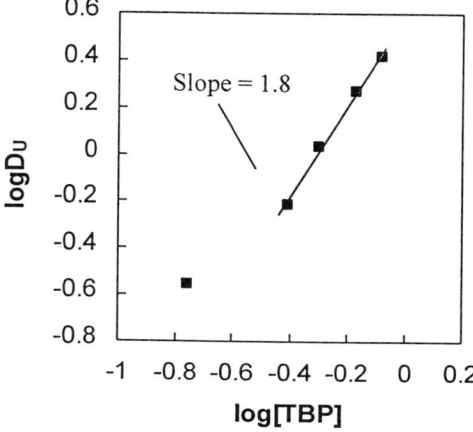

Figure 2. Effect of TBP concentration on the distribution coefficient of U(VI) between supercritical CO_2 and 6 M HNO_3.

solutions. It should be noted that coextraction of HNO_3 by TBP is also observed in the conventional solvent extraction process.

The efficiency of extracting uranyl ions by TBP modified CO_2 was found to decrease with reducing nitric acid concentration. In 0.1 M HNO_3, only about 10% of the uranyl ions in the aqueous phase can be extracted by supercritical CO_2 containing 6% TBP at 60 °C and 200 atm. This observation suggests a method of recovering uranium by on-line back-extraction with an aqueous solution in supercritical CO_2. Experimentally, we have proven that the uranyl nitrate-TBP complex extracted from 6 M HNO_3 by supercritical CO_2 can be back-extracted by bubbling the supercritical fluid phase on-line through an aqueous solution at pH > 1. Because TBP is very soluble in supercritical CO_2, the ligand is carried away by the supercritical fluid phase leaving uranyl ions in the aqueous phase.

Organophosphine oxides such as TBPO and TOPO are also effective ligands for SFE of uranyl ions from nitric acid solutions. The solubility of TBPO, however, is about an order of magnitude lower than that of TBP. TBPO is a stronger Lewis base than TBP and thus is expected to form a stronger complex with the uranyl cation. This is reflected in the high SFE efficiency of TBPO for uranyl ions from aqueous solutions. The SFE efficiency of TBPO for uranyl ions in 0.1 M HNO_3 is about 97% and ~99% in 6 M HNO_3 at 60 °C and 200 atm under the same conditions described above for the TBP experiments. Back-extraction of uranium from the TBPO complex may be problematic if this phosphine oxide is used for SFE of uranyl ions from acid solutions.

The rate of extraction of uranyl ions (0.01 M) from 6 M HNO_3 with 6% (v/v) TBP modified supercritical CO_2 is shown in Figure 3. In this study, supercritical CO_2 containing TBP was bubbled through the nitric acid solution in a high pressure stainless steel cell at a flow rate of 1.5 mL/min at 60 °C and 200 atm. The extraction of uranyl ions in this system approaches completion after about 40 minutes. A plot of ln $[UO_2^{2+}]_{aq}$ versus time yields a straight line, indicating the reaction is first order with respect to the uranyl ion. The slope of the line is 1.0×10^{-3} sec^{-1}.

The rate of extraction of uranyl ions may be expressed by the following equation:

$$-(dC/dt) = (A/V) k_f C_{sf} \qquad (4)$$

where C and C_{sf} are the concentrations of uranyl in aqueous phase and in supercritical fluid phase, respectively, A is the interfacial area, V is the volume of the aqueous phase, and k_f is the observed pseudo-first order rate constant. In our SFE system, the i.d. of the extraction vessel is 1.0 cm (A = 0.78 cm^2) and V is 6 mL. The k_f value calculated from the slope shown in Figure 3 is 7.7×10^{-3} cm sec^{-1}. The k_f value obtained from this study is similar to those observed in the solvent extraction experiments reported in the literature (*13,14*). The first order k_f value obtained by Horner et al. in their solvent extraction studies using different uranium concentrations and different methods varied from 5.3×10^{-3} to 8.5×10^{-3} cm sec^{-1} with an average value of 7.3×10^{-3} cm sec^{-1} (*13*). Based on these results, it seems that the chemical behavior of uranyl ions in these supercritical CO_2 systems is similar to that in conventional solvent extraction systems. This may provide a guideline for predicting the behavior of extraction and separation of uranium from other metal ions in supercritical fluids based on the known behavior of uranium in solvent extraction systems.

Supercritical CO_2 Separation of Trivalent Lanthanides with Organophosphinic Acids

Recently, Smart et al. reported the use of organophosphinic acids for extraction of heavy metals in supercritical CO_2 (15). The structures of two commercially available organophosphinic acids (from Cytec Industries Inc. under the trade names Cyanex 301 and 302) used by Smart et al. are shown in Figure 1. Both of the organophosphinic acids are stable and soluble in supercritical CO_2. The solubilities of Cyanex 301 and 302 in CO_2 at 200 atm and 40 °C are 22 g/L and 24 g/L, respectively. Organophosphates and organophosphoric acid such as TBP and di(2-ethylhexyl)phosphoric acid (D2EHPA) are widely used for intra-lanthanide separations in solvent extraction processes. To date, no systematic study has been reported that evaluates the feasibility of separating the trivalent lanthanide ions within the lanthanide series using organophosphinic acids as extractants in supercritical fluids.

Our SFE experiments with the organophosphinic acids were performed with 5 mL of an aqueous solution at pH 3 containing 20 ppm each of La and the following lanthanides Nd, Eu, Tb, Dy, Ho, Er, Tm, Yb, and Lu. It is known that when water is in contact with supercritical CO_2 under normal SFE conditions, the pH of the water is about 2.9 because of the formation and dissociation of carbonic acid (16). For this reason, we used an acetate buffer (0.01 M sodium acetate) to control the pH of the aqueous system at 3.0±0.1 for this study. The solution was placed in a high pressure liquid cell with 0.2 mmol of a Cyanex reagent and extracted with supercritical CO_2 at 40 °C and 200 atm statically for 30 minutes. This was followed by dynamic flushing of the system with a flow rate of 1.5 mL/min to collect the lanthanides dissolved in the supercritical fluid phase. Analysis of the lanthanides was done using ICP-AES (IRIS Model, Thermo Jarrell Ash).

As shown in Figure 4, both Cyanex 301 and 302 are effective for the extraction of lanthanides in supercritical CO_2 under the specified experimental conditions. They also show good separation factors for the heavy lanthanides. For example, the ratio of the distribution coefficients for Lu/Yb and Yb/Tm are 2.3 and 2.4, respectively, using Cyanex 302 as the extractant in supercritical CO_2. Cyanex 302 also shows a slightly higher extraction efficiency and separation factor than Cyanex 301 for the SFE of the trivalent lanthanide ions. The SFE results of Cyanex 272 are also given in Figure 4 for comparison. Cyanex 272 shows higher extraction efficiencies for the lanthanides, but its selectivity for heavy lanthanides is poor compared with Cyanex 301 and 302. The lanthanide ions are considered hard Lewis acids, hence oxygen-containing ligands such as phosphoric acids (e.g., Cyanex 272) are effective extractants for them. Substitution of soft sulfur donors, as in the case of organophosphinic acids, tends to decrease a ligand's affinity for the lanthanides. The efficiency of extracting the trivalent lanthanide ions at pH < 1 with Cyanex 302 in supercritical CO_2 is negligible (< 2%). Recovery of the extracted lanthanides from the supercritical fluid phase can be achieved by passing the fluid on-line through a dilute nitric acid solution in the SFE system.

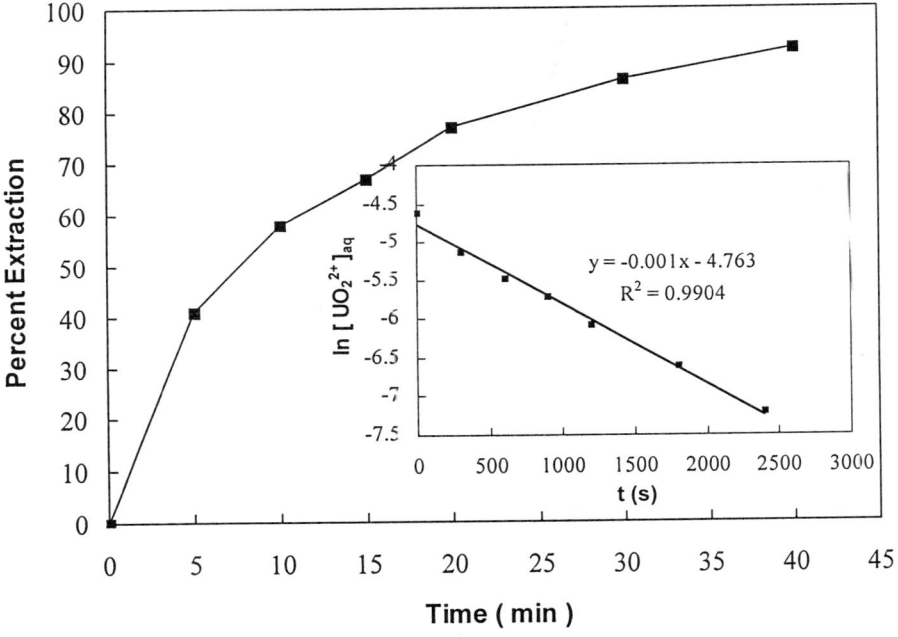

Figure 3. Rate of extraction of U(VI) from 6 M HNO_3 with supercritical CO_2 containing 6% TBP.

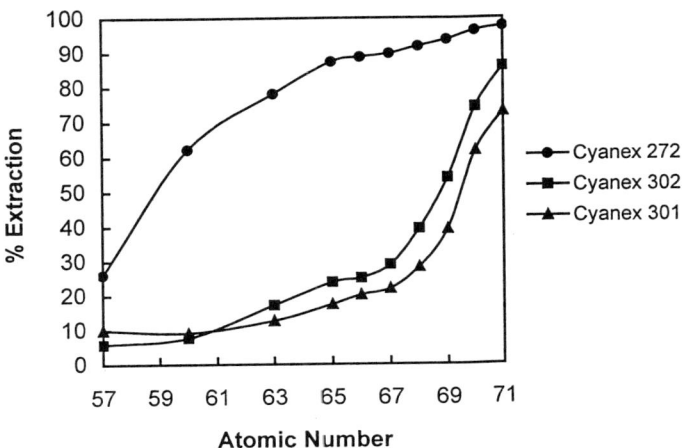

Figure 4. Percent extraction of lanthanides (from La Z=57 to Lu Z=71) with Cyanex reagents in supercritical CO_2

Supercritical Fluid Extraction of Mobile Uranium from Environmental Samples

It was reported by Lin et al. that uranyl ions spiked on solid materials such as filter papers and sand can be effectively extracted by supercritical CO_2 containing fluorinated β-diketones (6). β-diketones react with metal ions to form neutral chelates through the enolate anions formed by the following equilibria:

$$R_1\text{-}\overset{O}{\overset{\|}{C}}\text{-}CH_2\text{-}\overset{O}{\overset{\|}{C}}\text{-}R_2 \leftrightarrow R_1\text{-}\overset{O}{\overset{\|}{C}}\text{-}CH\text{=}\overset{OH}{\overset{|}{C}}\text{-}R_2 \leftrightarrow R_1\text{-}\overset{O}{\overset{\|}{C}}\text{-}CH\text{=}\overset{O^-}{\overset{|}{C}}\text{-}R_2 + H^+$$

Preliminary data obtained from our laboratory show that the solubility of acetylacetone (AA, $R_1=R_2=CH_3$) in CO_2 at 60 °C and 130 atm is about 4×10^{-4} mole fraction. At the same temperature and pressure, the solubility of the fluorinated β-diketone, thenoyltrifluoroacetylacetone (TTA, $R_1 = CF_3$ and $R_2 =$ thienyl) in CO_2 is about 2.3×10^{-2} mole fraction. TTA is a solid at room temperature and is easy to handle compared to the other liquid fluorinated β-diketones such as hexafluoroacetyl acetone (HFA) and trifluoroacetyl acetone (TFA). Recent NMR studies show that the fluorinated β-diketones HFA and TTA are almost exclusively in the enol form under the temperature and pressure conditions relevant to SFE systems (17). The non-fluorinated ligand AA is found to be only partly in the enolate form in supercritical CO_2.

The extraction of spiked analytes by the *in situ* chelation-SFE method, though a valuable indicator of the method's potential, is ineffective in simulating real world conditions. Spiked metal ions are usually in a highly extractable form and the use of inert matrices such as sand or cellulose based filter papers also neglects the possible difficulties involved in the extraction of metals from inherently heterogeneous samples such as soils, tailings, sludges, etc. Recently, we have evaluated the extraction of uranium from a tailings sample obtained from CANMET (Canada Centre for Mineral and Energy Technology, Ottawa, Canada) using supercritical CO_2 containing the fluorinated β-diketone TTA as an extractant. The tailings sample (2 g) was extracted with supercritical CO_2 at 60 °C and 150 atm for 20 minutes followed by 20 minutes of dynamic flushing. The TTA (1 g) was loaded in a stainless steel vessel placed upstream from the sample cell in the SFE system. The extraction was repeated several times under the same conditions and the analytes collected each time were analyzed by an ICP-MS (Varian Ultra Mass). After 4 or 5 repeated extractions, the residual sediment was removed for EPA's Toxicity Characteristics Leaching Procedure (TCLP) test. The TCLP test is the recommended method for determining environmental mobility of metals (18). TCLP tests of a solid environmental sample before and after SFE should reveal the effectiveness of the SFE procedure in removing mobile metals, and function as an indicator of SFE's potential in remediation schemes.

The original uranium containing tailings contained 1,010 µg of uranium/g of tailings. After one extraction of the dry tailings, about 46% of the uranium was removed from the sample by supercritical CO_2 under these conditions (Figure 5). Successive extractions did not dramatically increase uranium removal from the tailings.

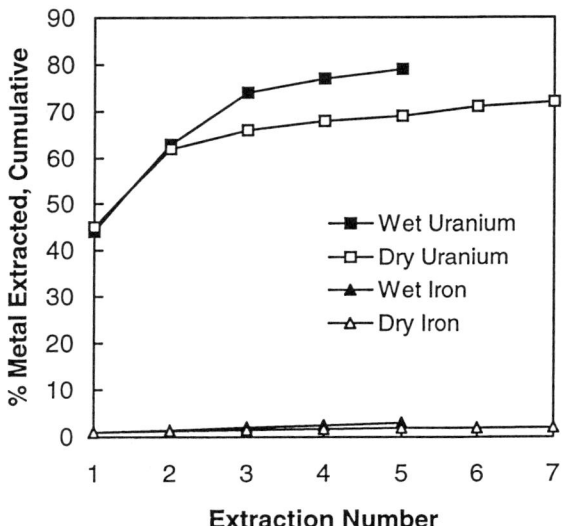

Figure 5. Extraction of uranium from tailings with supercritical CO_2 containing TTA.

After 3 extractions, only about 2 percent more of the uranium in the tailings could be removed from the sediment. About 70% of the uranium was removed from the tailings after 5 successive extractions. Addition of water (100 µL H_2O to 2 g sample) resulted in a slightly enhanced extraction efficiency. A fraction of the uranium in the tailings apparently cannot be removed by TTA in supercritical CO_2 even after 5 repeated extractions. The amount of iron removed from the tailings in the same SFE experiments was also measured. As shown in Figure 5, only a small fraction (< 2%) of the total iron originally present in the tailings (2.6%) was removed after 5 successive extractions. TCLP tests indicated that > 97% of the leachable uranium was removed from the tailings after repeated SFE with TTA at 60 °C and 200 atm. The SFE technique is effective in removing leachable uranium from solid mill tailings. Further studies on the nature of the uranium species removed from environmental samples by the SFE technique are currently in progress.

Acknowledgments

This work was supported by the NSF-Idaho EPSCoR program and by BNFL (British Nuclear Fuels plc.)

Literature Cited

1. Phelps, C. L.; Smart, N. G.; Wai, C. M. *J. Chem. Ed.* **1996**, *73*, 1163.
2. Smart, N. G.; Carleson, T. E.; Kast, T.; Clifford, A. A.; Burford, M. D.; Wai, C. M. *Talanta* **1997**, *44*, 137.
3. Laintz, K. E.; Wai, C. M.; Yonker, C. R.; Smith, R. D. *J. Supercrit. Fluids* **1991**, *4*, 194.
4. Laintz, K. E.; Wai, C. M.; Yonker, C. R.; Smith, R. D. *Anal. Chem.* **1992**, *64*, 2875.
5. Wang, S; Wai, C. M. *Anal. Chem.* **1995**, *67*, 919.
6. Lin, Y.; Smart, N. G.; Wai, C. M. *Environ. Sci. Technol.* **1995**, *29*, 2706.
7. Lin, Y.; Brauer, R. D.; Laintz, K. E.; Wai, C. M. *Anal. Chem.* **1993**, *65*, 2549.
8. Meguro, Y.; Iso, S.; Takeishi, H.; Yoshida, Z. *Radiochim. Acta*, **1996**, *75*, 185.
9. Yazdi, A. V.; Beckman, E. J. *Ind. Eng. Chem. Res.* **1996**, *35*, 3644.
10. Wai, C. M.; Smart, N. G.; Phelps, C. L. U.S. Patent No. 5,606,724, **1997**.
11. Page, S. H.; Sumpter, S. R.;Goats, S. R.; Lee, M. L. *J. Supercrit. Fluids* **1993**, *6*, 95.
12. Lin, Y.; *Supercritical Fluid Extraction and Chromatography of Metal Chelates and Organometallic Compounds*, Ph.D. Dissertation, Department of Chemistry, University of Idaho, Moscow, ID, May, 1997.
13. Horner, D.; Mailen, J.; Thiel, S.; Scott, T.; Yates, R. *Ind. Eng. Chem. Fund.* **1980**, *19*, 103.
14. Orth, D. A.; Wallace, R. M.; Karraker, D. G. *Science and Technology of Tributyl Phosphate*, Schulz, W. W.; Navratil, J. D.; Kertes, A. S.; Eds., CRC Press, Boca Raton, FL, **1984**;Vol. 1, p 161.
15. Smart, N. G.; Carleson, T. E.; Elshani, S.; Wang, S.; Wai, C. M. *Ind. Eng. Chem. Res.* **1997**, *36*, 1819.

16. Toews, K. L.; Shroll, R. M.; Wai, C .M.; Smart, N. G. *Anal. Chem.* **1995**, *67*, 4040.
17. Wallen S. L.; Yonker, C. R.; Phelps, C. L.; Wai, C. M. *J. Chem. Soc., Faraday Trans.* **1997**, *93*, 2391.
18. U.S. Environmental Protection Agency, *Toxicity Characteristics Leaching Procedure (TCLP)*, Vol. 1C, Ch. 8, Sec. 8.4, 7/92, Method 1311.

INDEXES

Author Index

Alexandratos, Spiro D., 183, 206
Ashley, Kenneth R., 219
Ball, Jason R., 219
Bartsch, Richard A., 146, 183
Behrens, Elizabeth A., 168
Bond, Andrew H., 2, 234
Bordunov, Andrei V., 133
Bortun, Anatoly I., 168
Bortun, Lyudmila N., 168
Bradshaw, Jerald S., 133
Bruening, Ronald L., 251
Cahill, Roy A., 168
Chiarizia, Renato, 206
Choppin, Gregory R., 13
Christian, Sherril D., 280
Clearfield, Abraham, 168
Dietz, Mark L., 2, 234
Drake, Lawrence R., 260
Fedotov, Petr S., 333
Freiser, Henry, 347
Goken, Garold L., 251
Griffin, Scott T., 79
Gula, Michael, 206
Hay, Benjamin P., 102
Hayashita, Takashi, 183
Horwitz, E. Philip, 20, 206, 234
Huddleston, Jonathan G., 79
Hussain, Latiff A., 183
Ignatova, Svetlana N., 333
Izatt, Reed M., 133, 251
Jackson, Paul J., 260
Jarvinen, Gordon D., 294
Ji, Min, 390
Krakowiak, Krzysztof E., 251
Lin, Shan, 260
Lin, Yuehe, 390
Maryutina, Titiana A., 333
Moyer, Bruce A., 114
Muralidharan, Subramaniam, 347
Nash, Kenneth L., 52
Pastushok, Victor N., 133
Poojary, Damodara M., 168
Radzinski, Susan D., 219
Rayson, Gary D., 260
Robison, Thomas W., 294
Rogers, Robin D., 2, 79
Sachleben, Richard A., 114
Scamehorn, John F., 280
Schovanec, Annette L., 280
Schroeder, Norman C., 219
Schulz, Wallace W., 20
Shadizadeh, Susan B., 280
Smart, Neil G., 390
Smith, Barbara F., 294
Spivakov, Boris Ya., 333
Strauss, Steven H., 156
Taylor, Richard W., 280
Toews, Karen L., 390
Wai, Chien M., 390
Whitener, Glenn D., 219
Willauer, Heather D., 79
Xia, Hongying, 260
Zhang, Jinhua, 79
Zhang, Xian Xin, 133

Subject Index

A

ABEC. *See* Aqueous biphasic extraction chromatography
ABS. *See* Aqueous biphasic systems
Acid group identity, proton-ionizable lariat ethers, 153
Acid mine drainage, Berkeley Pit water treatment, 322, 324f-328
Acidic extractant processes
 bis(2-ethylhexyl) phosphoric acid (HDEHP), 21-23, 66-68, 235
 bis(hexoxyethyl) phosphoric acid (HDHoEP), 22-24
 diisodecyl phosphoric acid (DIDPA), 22-23
Acrylpyrazolones
 Arsenazo III effect, 385-388
 neutral surfactant effect, 383-385
 1-phenyl-3-methyl-4-benzoyl-5-pyrazolone (HPMBP)
  lanthanide complex formation mechanism, 375-376, 378-383
  structure, 362f
 1-phenyl-3-methyl-4-capryloyl-5-pyrazolone (HPMCP)
  lanthanide complex formation mechanism, 375, 377-383
  structure, 369f
 trivalent lanthanide separations, 369-371
Actinide-lanthanide systems, aqueous complexation in, 64-73
Actinide removal
 by polymer filtration, distribution coefficients, 302, 304-305f
 from aqueous streams by polymer filtration, 307
 from mixed-waste, 207, 213-214
 from tailings, fluorinated diketones used in supercritical fluid extraction, 397-399
 studies at Los Alamos National Laboratory (LANL) radioactive liquid waste treatment facility, 311, 314
Actinides
 aqueous complexation, role in metal separation, 15-16, 64-73
 coordination numbers, ligand influence, 15
 extraction chromatography, 240-242
 ion exchange separations, 16-17, 61-64
 medium effects and solvation, 60-63
 nonaqueous processes, role in metal separation, 17-18
 oxidation states, 14, 59
 precipitation, early isolation of plutonium, 58-59
 solution chemistry, weak complexing agents, 59-60
 solvent extraction, 15-16, 20-50
 See also individual elements; Transuranium elements
Acyclic dibenzo polyether carboxylic acid resins, alkaline earth metal cation sorption by, 188-189
Affinity chromatography measurements, binding isotherms, 272-275
^{27}Al NMR in metal binding studies in biomaterials, 271-273t
Alkali metal cation sorption by
 lariat ether carboxylic resins, 186-189
 lariat ether phosphonic acid monoethyl ester resins, 189-190
 lariat ether sulfonic acid ester resins, 189-190
Alkali metal salts, extraction by crown ethers in 1-octanol, 118-120

Alkaline earth metal cation sorption by acyclic dibenzo polyether carboxylic acid resins, 188-189
Alkaline earth metals in extraction chromatography, 238-240
Alkaline extractant processes, 74
Alkyl-substituted 14-crown-4 ethers, modeling extraction behavior, 120-122
Aluminum removal in acid mine drainage treatment, Berkeley Pit water, 325-328
Amberlite XAD-2. *See* Macroporous styrene copolymers
Americium separation
by exchange chromatography, 240-241f
from europium, 62
See also Actinides; Extraction chromatography, process-scale application; Transuranium elements
Apparatus, centrifugal partition chromatography, 348-349
Aqueous biphasic extraction chromatography, compared to ABS, 97-98
Aqueous biphasic systems
metal ion separations, 79-100
polyethylene glycol-based, Gibbs free energy of hydration, 81-83
prediction, metal ion partitioning, 91-96
quantitative recovery, pertechnetate anion, 85-90
Aqueous complexation in lanthanide actinide systems, 64-73
Aqueous phase role, 52-78
See also Solvent matrix role
Arm length, proton-ionizable lariat ethers, 151
Argonne National Laboratory
substituted methane diphosphonic acids development, 240-244
TRUEX process experience, 29-31
Argonne National Laboratory—East, actinide removal from mixed-waste, 207, 213t

Assay techniques, pertechnetate anion, 220-221
Azacrown ethers
N-substituted-phenol-containing, synthesis, 133-144
with hydroxyaromatic substituents, structure, 136f
Azamacroheterocyclic compounds containing phenol units, synthesis, 137-139

B

Benzocrown ethers, condensation polymers, 184
Berkeley Pit water treatment, 322, 324f-328
Binding isotherms from affinity chromatographic measurements, 272-275
Biological application, aqueous biphasic systems, 85, 91
Biomass, silica-based immobilization, 263
Biomaterials
heavy ions selective removal, 261
metal ion binding 260-276
Bis(2,4,4-trimethylpentyl) dithiophosphinic acid (Cyanex 301), structure, 393f
Bis(2,4,4-trimethylpentyl)-monothio-phosphinic acid (Cyanex 302), structure, 393f
Bis(2,4,4-trimethylpentyl)phosphinic acid (Cyanex 272)
lanthanide complex dissociation kinetics from metallochromic indicator method, 370-371
mechanism steps, 371-375
structure as dimer, 369f, 393f
trivalent lanthanide separation, 366, 368-370
Blue-green algae, biologically-generated material for metal-ion bonding, 262
Boeing Defense and Space Group,

polymer filtration minimizing electroplating waste, 317-318f
Bryopsis (algae), biologically-generated material for metal-ion bonding, 268-269f

C

Cadmium
  ^{113}Cd NMR, in metal binding studies in biomaterials, 267-271, 273t
  extraction using MoS_2, 164
  sorption by HDEHP-loaded Amberlite XAD-2, 235-236
  See also Heavy metals
Calixarenes in extraction chromatography, 237
Carbon dioxide, uranyl extraction by supercritical, 391-395
Carboxylic acid resins, lariat ethers, 186-189
Centrifugal partition chromatography
  chemical kinetics influence on efficiencies, 360-366
  chromatogram analysis parameters, 349
  dissociation mechanism for metal chloride trioctylphosphine oxide complexes, 356-359
  experimental apparatus, 348-349
  nickel dissociation rate constants, 362-366
  nickel extraction equilibria, 360-365
  optimizing conditions, 350-352
  platinum group metals extraction equilibria, 352-356
  platinum group metals separation efficiencies, 356-360
  rate-limiting step mechanism, 358, 362
  stability constant for metal chloride trioctylphosphine oxide complexes, 359
  trivalent lanthanide separation, 366-388
Centrifugal solvent extraction, 347-389
Cerium. *See* Lanthanides
Cesium
  extraction of radionuclide by cobalt dicarbollide process, 39, 41-45
  radionuclide separation by counter-current chromatography, 344
  removal from high-level waste, 214-215
Chelating resins, functional groups, 196-198
Chelex-100. *See* Iminodiacetic acid resin
Chemical kinetics influence on efficiencies, centrifugal partition chromatography, 360-366
Chromatogram analysis parameters, centrifugal partition chromatography, 349
Chromatography, extraction. *See* Extraction chromatography
Chlorella pyrenoidosa (algae), biologically-generated material metal-ion bonding, 268-269f
Cladophora (algae), biologically-generated material for metal-ion bonding, 268-269f
Cobalt dicarbollide, in extraction of cesium and strontium, 39-42, 45
Cobalt removal from waste at Millstone nuclear power plant, 214
Column efficiency, extraction chromatography, 236-237
Column performance experiments, technicium elution, 227-232
Commercial processes, polymer filtration
  actinide removal from aqueous streams, 307
  analytical procedure for precon-centration of actinides, 307-308
  bench-scale/glovebox studies for actinide removal from plutonium facility distillate waters, 308-311
Computational methods, ligand binding site analysis, 102-113
Copper
  removal by hexadecyloxybenzyl-

iminodiacetic acid, 283-292
removal by micellar-enhanced ultrafiltration, 280-292
sorption by HDEHP-loaded Amberlite XAD-2, 235-236
uptake by macroporous styrene-divinylbenzene copolymers, 242, 245-246
Countercurrent chromatography
absent adsorptive matrix for retaining stationary phase, 333-334
cesium and strontium radionuclides separation, 344
rare earth group preseparation, 339-341
salt solution purification, 344-345
stationary phase retention in a rotating coil column, 335-338
transplutonium and rare earths separation, 343-344
See also Extraction chromatography, process-scale application
Countercurrent solvent extraction, 333-346
Crown ethers
alkali metal salts extraction in 1-octanol, 118-120
cesium extraction, 39-44
dibenzo-14-crown-4, derivatives, 115*f*
extraction chromatography, 237-238
hexadentate, potassium formation constants, 104-105
lithium with 14-crown-4 derivatives, 105-108, 114-132
nonamethyl-14-crown-4, diluent effect on extraction behavior, 122-125
polymeric structures for formaldehyde condensation products, 184
ring rigidity in proton-ionizable lariat ethers, 153
ring size in proton-ionizable lariat ethers, 151
strontium extraction, 39-44
strontium with dicyclohexano-18-crown-6 derivatives, 108-110
substituted 14-crown-4, equilibrium modeling, 126-129
synthesis, 116-117
Cryptands in extraction chromatography, 237
Crystal structure, tunnel type inorganic ion exchangers, 168-182
Curium. *See* Actinides; Transuranium elements
Cyanex 272. *See* Bis(2,4,4-trimethylpentyl)phosphinic acid
Cyanex 301. *See* Bis (2,4,4-trimethylpentyl)dithiophosphinic acid
Cyanex 302. *See* Bis(2,4,4-trimethylpentyl)monothiophosphinic acid

D

Datura innoxia, biologically-generated material for metal-ion bonding, 261-269*f*, 271-275
DIAMEX process. *See* Neutral extractant processes, bifunctional amides
Dibenzocrown ethers, condensation polymers, 184
Diluent effect on extraction behavior, crown ethers, 122-125
Diphonix-A. *See* Modified Diphonix resins
Diphonix-CS. *See* Modified Diphonix resins
Diphonix resins, 206-218
See also Gem-diphosphonic acid ligand
Diphosil resins. *See* modified Diphonix resins
Diphosphonic acids, substituted methane, 240-244
Dissociation mechanism for metal chloride trioctylphosphine oxide complexes, centrifugal partition chromatography, 356-359

Distribution coefficient
  actinide removal by polymer filtration, 302, 304-305f
  pertechnetate ion, 87-90
  strontium with dicyclohexano-18-crown-6 derivatives, 108-110
  uranyl extraction by supercritical CO_2, 392-394
  See also Distribution ratio; Equilibrium constant; Stability constant
Distribution ratio, 55-57, 64-65, 68f, 70-71f
  See also Distribution coefficient; Equilibrium constant; Stability constant
Dodecylsalicylaldoxime, ligand for centrifugal partition chromatography
  nickel extraction studies, 360-366
  structure, 362f
Donnan equilibrium effect on uncomplexed cations, 281
Double Shell Slurry. See Tank waste simulants, pertechnetate ion removal
Double Shell Slurry Feed. See Tank waste simulants, pertechnetate ion removal
Dowex A-1. See iminodiacetic acid resin
Dual mechanism bifunctional polymers, phosphorus-based metal ion complexing, 199-200
Dysprosium. See Lanthanides

E

Electroplating waste minimization, polymer filtration, 317-321f
Empore membranes. See Membranes in separation technology
Enhanced selectivity in separations process, future goal, 3

Entiomorpha (algae), biologically-generated material for metal-ion bonding, 268-269f
Environmental Protection Agency (EPA)
  discharge limit, 317
  Resource Conservation and Recovery Act (RCRA) metals, 301
  toxicity characteristic leaching procedure, 300, 397-399
Environmental remediation as social goal, 3-4, 79-80
Environmental sample, fluorinated diketones in uranyl supercritical fluid extraction, 397-399
EPA. See Environmental Protection Agency
Equilibrium constant
  azacrown ethers with hydroxyaromatic sidearms with metal ions, 141t
  lithium extraction constants with 14-Crown-4 derivatives, 105-107
  See also Distribution coefficient; Distribution ratio; Stability constant
Equilibrium modeling, 126-129
Erbium. See Lanthanides
Europium ion
  luminescence measurements, 264-267
  redox chemistry in water, 59
  separation from Americium, 62
  See also Lanthanides
Extraction chromatography (EXC)
  acidic organophosphorus extractants, 235-237
  actinide separations, 240-242
  alkaline earth metals, 238-240
  column efficiency expressed as plate height, 236-237
  crown ethers as extractants, 238
  material preparation, inert substrate with extractant, 234-235
  process-scale application, 246-247t
  retention of extractant on the support, 235
  sulfonic acid derivatives as extractants, 245

supporting materials development, 242-245
tri-*n*-butylphosphate as extractant, 235
Extraction studies
   alkali metal salts by crown ethers in 1-octanol, 118-120
   redox-recyclable ions from aqueous media, 156-165
   thermodynamic calculations, 52-57
   See also Free energy of formation, lanthanide complexes

F

f-Elements. *See* Actinides; Lanthanides; Transuranium elements
Fluorinated diketones in uranyl supercritical fluid extraction, 397-399
Formation constants
   lithium with 14-crown-4 derivatives, 107-108
   sodium with hexadentate polyether ligands, 110-111
   See also Stability constants
Free energy of formation, lanthanide complexes, 69, 72-74
   See also Extraction studies, thermodynamic calculations
Functional group classification, ion exchange resins, 196
Future research in separations technology, 3, 10-11, 14

G

Gadolinium. *See* Lanthanides
Gem-diphosphonic acid ligand, 200, 202*f*
   See also Diphonix resins
Geochronological studies, separations by countercurrent chromatography, 341, 343

Geological samples, two-phase liquid systems for metal determination, 338-341
Gold. *See* Extraction chromatography, process-scale application

H

Hafnium preseparation by countercurrent chromatography, 341-342
Hanford site (Washington State)
   complex high level waste, 194
   pertechnetate anion separation, 219
   strontium and cesium removal from high-level waste, 214-215
   TRUEX experience, 30*f*, 32-33
   waste storage issues, 14, 21
Heavy metal divalent cations, sorption by resins, 191-192
Heavy metal ions, redox-recyclable from aqueous media, 156-165
   See also individual metals
Hexadecyloxybenzyliminodiacetic acid
   compound structure, 283
   ligand in copper removal, 283-292
   ligand solubility in water and partitioning, 284-285
   metal ion analysis in solutions by atomic absorption spectrometry, 284
   metal-ligand stoichiometry in surfactant micelles, 285-288
   semi-equilibrium dialysis studies, 290-292
   test solutions preparation, 283-284
   ultrafiltration studies, stirred cell method, 285, 288-292
Hexadecylpyridinium chloride as cationic surfactant, 283, 289
Hexadentate polyether ligands with sodium, formation constants, 110-111
High-level liquid waste (HLW), solvent extraction in treatment, 20-50

Historical review, radioactive element separations, 13-14
Hofmeister separation, natural bias based on free energies of hydration, 159
Holmium. *See* Lanthanides
Hydrated cations, movement across phases, 52-55

I

Idaho National Engineering Laboratory, 21
See also Lockheed Martin Idaho Technologies
Iminodiacetic acid resin, 196-198
Immobilized phosphorus ligands, synthetic routes, 197, 199
Inorganic ion exchangers, crystal structure 168-182
Ion-exchange resins
 amphoteric resins, 196
 anion exchange resins, 196
 cation exchange resins, 196
 chelating resins, 196
 coordination resins, 200-201
 functional groups, 194-196
 intra-ligand cooperation, 200, 202
 lariat ethers, 183-193
 precipitation resins, 200-201
 redox resins, 199-201
 Reilex-HPQ resin, 219-233
 See also individual resins
Ion selective extractants, design and synthesis, 102-144
Iron complexes as extractants, 162
Iron control by Diphonix resins, 215
Iron removal in acid mine drainage treatment, Berkeley Pit water, 325-328

K

Kraft temperature determination in ultrafiltration, 290

L

Lanthanide actinide systems, aqueous complexation in, 64-73
Lanthanides
 aqueous complexation, 64-73
 extraction with supercritical fluids, 395-399
 ion exchange techniques, 61-64
 oxidation state in aqueous solutions, 59-60
 precipitation, early separations by fractional crystallization, 58
 separation by centrifugal partition chromatography
  Arsenazo III effect, 385-388
  neutral surfactant effect, 383-385
  1-phenyl-3-methyl-4-benzoyl-5-pyrazolone (HPMBP), 362f, 375-376, 378-383
  1-phenyl-3-methyl-4-capryloyl-5-pyrazolone (HPMCP), 369f, 375, 377-383
 separation by countercurrent chromatography, 343-344
 separation with supercritical CO_2 with organophosphinic acids, 395-396
 solution chemistry, 59-63
Lanthanum. *See* Lanthanides
Lariat ethers
 carboxylic acid resins, 186-189
 in metal ion separation, 146-155
 ion-exchange resins, 183-193
 metal ion receptors, structure, 136f
 phosphonic acid monoethyl ester resin, 189-190
 proton-ionizable, 184-185
 sulfonic acid ester resins, 189-190
Lead extraction using MoS_2, 164
Lead test kit, 254-255t
Levextrel resins. *See* Macroporous styrene-divinylbenzene copolymers
Ligand binding site organization analysis, 102-113

Ligand reorganization energy, 103
Ligand strain, 128-129
Lipophilic group(s) attachment site, proton-ionizable lariat ethers, 153
Liquid-liquid chromatography. *See* Extraction chromatography
Liquid-liquid extraction. *See* Solvent extraction
Liquid-liquid separation modes, 146-165
Lithium-crown ether extraction constants, 105-107
Lithium-intercalated metal chalcogenides, redox-recyclable extractants, 163-164
Lithium with 14-crown-4 derivatives, 114-132
Lockheed Martin Idaho Technologies, TRUEX experience, 30-32
See also Idaho National Engineering Laboratory
Los Alamos National Laboratory (LANL) radioactive liquid waste treatment facility, actinide removal studies, 311, 314
Los Alamos Plutonium Facility, 308-309
Lutetium. *See* Lanthanides

M

Macrocyclic ligands, insertion of proton-ionizable groups, 147, 150*f*
Macroporous styrene-divinylbenzene copolymers, 235, 242-243, 245-246
Manhattan project, nuclear weapons development, 13
Mannich modified aminomethylation reaction, 133-144
Medicago sativa (alfalfa sprouts), biologically-generated material metal-ion bonding, 262
Membrane-based processes for metal ion extraction, 3
Membrane filtration, 280-330
Membranes in separation technology, 251-254
Mercuric ion, reduction by dual mechanism bifunctional polymers, 199
Mercury recovery by heavy-metal ion-exchange, 163-164
Metal chelate solubilities in supercritical carbon dioxide, 391
Metal ion binding affinity, predicting ligand structure influence, 102-113
Metal ion binding by biomaterials, 260-276
Metal ion complexes with phenol- and CHQ-substituted azacrown ethers, 139, 141-143
Metal ion extraction
  actinide separations, aqueous complexes in f-element separation science, 52-78
  aqueous biphasic extraction chromatography, compared to ABS, 97-98
  aqueous biphasic systems, 79-100
  column chromatography, 219-233
  countercurrent chromatography, 333-346
  extraction chromatography, 234-250
  future research needs, 3, 10-11, 14
  ion exchange resins, Diphonix class, 206-218
  lanthanide separations, aqueous complexes in f-element separation science, 52-78
  membrane-based processes, 3, 251-258, 280-330
  polymer-supported reagent design, 194-205
  supercritical fluid processes using CO_2, 3, 390-400
  with lariat ether ion-exchange resins, 183-193
  with proton-ionizable lariat ethers, 146-155
  See also Solvent extraction techniques

Metal ion partitioning prediction, 91-97
Metal ion polymer-bound release methods, 298-300
Metal ion removal from aqueous streams, 294-330
Metal ion separation. *See* metal ion extraction
Micelluar-enhanced ultrafiltration (MEUF)
 definition as a surfactant-based separation technique, 280
 selective copper removal, 280-292
Millstone nuclear power plant, radioactive cobalt and zinc removal, 214
Modified Diphonix resins
 kinetic properties, metal ion uptake, 207-212f
 radioactive waste treatment, analytical results, 207, 213-215
Molecular mechanics method used to predict metal ion binding affinity, 102-113
Molecular recognition technology, 251-258

N

Neodymium, separation in geochronological studies, 341, 343
See also Lanthanides
Neptunium, redox chemistry use in actinide processing, 59
See also Actinides; Transuranium elements
Neutral extractant processes
 advantages over acidic extractants, 38-39
 bifunctional amides, 28f, 36-37
 dihexyl-N,N-diethylcarbamoylmethyl-phosphonate as ligand, 24-28
 diphenyl-N,N-di-n-butylcarbamoyl-methyl-phosphine oxide as ligand, 28f, 35-36

octyl (phenyl)-N,N-diisobutylcarba-moylmethylphosphine oxide as ligand, 26-35, 74-75
trialkylphosphine oxide mixture, 28f, 37-38
Neutral lariat ether compared to proton-ionizable lariat ether, solvent extraction, 149f
Nickel dissociation rate constants in centrifugal partition chromatography, 362-366
Nickel extraction equilibria in centrifugal partition chromatography, 360-366
Niobium preseparation by countercurrent chromatography, 341-342
Nonamethyl-14-crown-4, extraction alkali metal nitrates, solvent effect study, 122-125
Nonaqueous separation processes, 17
Nuclear defense, solvent extraction, 2, 20-50
Nuclear energy programs, solvent extraction, treatment acidic HLW, 2, 20-50
Nuclear medicine, yttrium extraction, 189
Nuclear radioactive materials, small scale separation by extraction chromatography, 246
Nuclear waste treatment, solvent extraction, treatment acidic HLW, 2, 20-50

O

Oak Ridge National Laboratory, TRUEX experience, 30t, 32
Optimizing conditions for centrifugal partition chromatography, 350-352
Organophosphorus reagents, bifunctional, in extraction chromatography 237

Oxyanions, polymer filtration applications, 320, 322-324f

P

Palladium. *See* Platinum group metals extraction
Peat moss, biologically-generated material for metal-ion bonding, 268-269f
Pecan shells, biologically-generated material for metal-ion bonding, 268-269f
Perchlorate effect on water structure, 62
Perrhenate anion
 ion-exchange chromatography, 161-162
 liquid-liquid extraction, 159-160
Pertechnetate anion
 assay techniques, 220-221
 extraction in aqueous biphasic systems and using ABEC, 84-98
 ion-exchange chromatography, 161-162
 liquid-liquid extraction, 159-161
 reduction by stannous ion to form cationic species, 227
 separation from neutralized Hanford waste studies, 219-233
Pharmacosiderites, tunnel structure, 176-180
1-Phenyl-2-methyl-4-benzoyl-5-pyrazolone, ligand in centrifugal partition chromatography, 360-366
Phosphorus acid ligands, polymer-supported, 197, 199
Phosphorus acid resins, synthesis by phosphorylation of polystyrene-DVB copolymers, 197, 199
Phosphorylation of polystyrene-DVB copolymers, 197, 199
Platinum. *See* Platinum group metals extraction
Platinum group metals extraction
 centrifugal partition chromatography rate-limiting step, 358
 equilibria for centrifugal partition chromatography, 352-356
 separation efficiencies for centrifugal partition chromatography, 356-360
 triotylphosphine oxide extractant in centrifugal partition chromatography, 352-353, 355-360
Plutonium
 isolation by precipitation, early separation, 58
 isolation by solvent extraction using methyl isobutyl ketone (REDOX process), 58
 isolation by solvent extraction using tributyl phosphate (PUREX Process), 58
 redox chemistry used in actinide processing, 59-60
 See also Actinides; Extraction chromatography, process-scale application; Transuranium elements
Plutonium Uranium Recovery and Extraction (PUREX) process, 2, 13-16, 20-21
Polyether carboxylic acid resins, divalent heavy and transition metal cation sorption, 191-192
Polyether phosphonic acid monoethyl ester resins, divalent heavy and transition metal cation sorption, 191-192
Polyether sulfonic acid resins, divalent heavy and transition metal cation sorption, 191-192
Polyethylene glycol salt-based aqueous biphasic systems
 biological application, 85, 91
 cloud point, 91
 Gibbs free energy of hydration, 81, 83t, 87, 89f
 metal ion partitioning prediction, 91-97
 pertechnetate distribution, 84-98
 phase diagram, 81-82
 salt selection, 81, 83t

Polymer-bound metal ion release methods, 298-300
Polymer filtration
 acid mine drainage treatment, 322, 324f-328
 actinide removal from aqueous streams, 307
 analytical procedure for preconcentration of actinides, 307-308
 application, oxyanions, 320, 322-324f
 bench-scale/glovebox studies for actinide removal from plutonium facility distillate waters, 308-311
 comparison to other separation technologies, 304, 306-307
 concentration factor calculation, 301-302
 concentration mode, 296-298
 diafiltration mode, 298-299f
 electroplating waste minimization, 317-321f
Polymer filtration unit at Los Alamos National Laboratory (LANL) radioactive liquid waste treatment facility, 314, 316f-317
Polymer leakage through ultrafiltration membrane, 300-301
Polymer-supported reagent design, 194-205
Polystyrene-divinylbenzene-based cation-exchange resin, 245
Polystyrene-divinylbenzene-based copolymers, 197, 199
Polystyrene, support in ion exchange, 195-196
Power Reactor and Nuclear Fuel Development Corp. (Japan), TRUEX, 30f, 33-34
Praseodymium. See Lanthanides
Principle of complementarity, 103
Process-scale application, extraction chromatography, 246-247t
Proton-ionizable groups, insertion into macrocyclic ligands, 147, 150f
Proton-ionizable lariat ether compared to neutral lariat ether, solvent extraction, 149f

Proton-ionizable lariat ethers
 acid group identity, 153
 arm length, structural variation influence, 151
 crown ether ring rigidity in preorganizing the binding site, 153
 crown ether ring size, lithium ion best fit within ring, 151
 lipophilic group(s) attachment site, 153
 structural variations in ring size, 151-153
 structures, 184-185
 with metal ion separation, 146-155
PUREX process. See Plutonium-Uranium Recovery and Extraction
Pyrochemical separation processes in treatment of spent nuclear fuel, 17-18

R

Radioactive element separations, historical review, 13-14
Radioactive metal-ion discharge, 294
Radioactive wastes, storage processing, 14
Radionuclides, redox-recyclable from aqueous media, 156-165
Radium rad disk, 256-257t
RAP membranes. See Membranes in separation technology
Rapid analysis products (RAP)
 lead test kit, 254-255t
 mercury test kit, 254
 radium rad disk, 256-257t
 strontium rad disk, 255-256t
Rare earth group preseparation, 339-341
Rare earths. See Lanthanides
Reactive ion exchange, method to enhance ion exchange separations, 199
Recovery, redox-recyclable ions from aqueous media, 156-165
Redox-recyclable extractants
 design criteria, 157-158

lithium-intercalated metal
chalcogenides, 163-164
organometallics, 162
substituted bis(hydridotris(pyrazollyl-
borate) iron complexes, 162
1,1´,3,3´-tetrakis(2-methyl-2-
hexyl)ferrocene, 159-162
Regulatory limits, radioactive metal-ion
discharge, 294
Reillex-HPQ anion exchange column
chromatography, 219-233
batch distribution coefficients, 221-224
breakthrough curves, 224-226
column performance, 225-227
column set-up, 222-223
elution reagents, 222
pertechnetate anion preparation, 220-
221
stripping, 223
waste simulants, 221
Resource Conservation and Recovery
Act (RCRA) metals, 301
Rhenium. *See* Platinum group metals
extraction
Rubidium, separation in geochronological studies, 341, 343

S

Salsola spp.(tumbleweeds), biologically-
generated material metal-ion
bonding, 262, 268-269f
Samarium, separation in geochronological studies, 341, 343
See also Lanthanides
Savannah River site (South Carolina)
PUREX process, 20
strontium and cesium removal from
high-level waste, 214- 215
Scandium. *See* Extraction
chromatography, process-scale
application
Selectivity, structural basis in tunnel type
inorganic ion exchangers, 168-
182
Separation efficiency in countercurrent
chromatography, 334-335

Separation factor for extractions, 65-67
Separation process, definition as intermixed components forced into
different spatial locations, 4-5
Siderophores, microbially produced
chelating agents, 18
Silver ion, reduction by dual mechanism
bifunctional polymers, 199
Sodium titanium silicate, tunnel
structure, 169-173
Sodium with hexadentate polyether
ligands, formation constants,
110-111
Solid-liquid separation modes, 168-278
Solid phase extraction system, 251-258
Solvation effects, thermodynamic
energies, 60-64, 128-129
Solvent effect study, extraction alkali
metal nitrates, 122-125
Solvent extraction in nuclear defense and
energy programs, 2, 13-50
Solvent extraction techniques
centrifugal partition chromatography,
347-389
countercurrent chromatography, 333-
346
equilibria, distribution ratio
calculations, 55-58
membrane filtration, 280-330
supercritical fluids using CO_2, 390-400
See also Metal ion extraction
Solvent extraction used in nuclear
reprocessing and waste
treatment, 2, 14-50
Solvent matrix role, 4
See also Aqueous phase role
Stability constants
actinide lanthanide complexes, 69, 72-
74
metal chloride trioctylphosphine oxide
complexes, 359
See also Distribution coefficients;
Distribution ratios; Equilibrium
constants; Formation constants
Stannous ion, used in pertechnetate
reduction, 227-228f
Stationary phase retention in a rotating
coil column, 335-338

Strontium rad disk, 255-256t
Strontium radionuclides
  removal from high-level waste, 214-215
  separation by countercurrent chromatography, 344
  separation by extraction chromatography, 238-239f
  separation in geochronological studies, 341, 343
  separation from yttrium, 189
  solvent extraction from acidic high-level waste, 39-45
  with dicyclohexano-18-crown-6 derivatives, distribution coefficients, 108-110
Structural modeling, extraction behavior, alkyl-substituted 14-crown-4 ethers, 120-122
Sulfonic acid derivatives as extractants, 245
Supercritical fluid extraction technology
  carbon dioxide as widely used gas, 390-399
  environmentally friendly process, 390-392
  See also Supercritical fluid processes for metal ion extraction
Supercritical fluid processes for metal ion extraction, 3, 390-400
  See also Supercritical fluid extraction technology
Supporting materials in columns
  extraction chromatography, 242-245
  organic polymers in ion exchange resins, 195
Synthesis, organic compounds used as ligands
  azamacroheterocyclic compounds containing phenol units, 137-139
  benzoazacrown ethers, 139-140
  benzoazamacrocycles, 140f
  bis-CHQ-containing diaza-18-crown-6, 136f
  crown ethers, 116-117
  cryptohemispherand, 140f
  N-substituted-phenol-containing azacrown ethers, 133-144
  phenol-containing macrobicycle, 138f
  phenol-containing macrotricycle, 138f

T

Tank waste simulants, pertechnetate ion removal, 219-233
Tantalum preseparation by countercurrent chromatography, 341-342
Technetium complex, reduced, elution from anion exchange resin, 227-232
Technetium recovery, volume reduction, 159-161
Terbium. *See* Lanthanides
1,1´,3,3´-Tetrakis(2-methyl-2-hexyl)ferrocene
  ion-exchange chromatography, 161-162
  liquid-liquid extraction, 159-161
Thenoyltrifluoroacetylacetone, supercritical fluid extraction of uranium, 397-399
Thermochemical cycle, role of ligand strain, 128f
Thulium. *See* Lanthanides
Thorium uptake by tri-*n*-butyl phosphate-loaded Amberlite XAD-4, 235-236, 240-241
  See also Actinides
Titanosilicate, acid form, titration curves, 173-176
Transition metal divalent cations, sorption by resins, 191-192
Transplutonium elements, separation by countercurrent chromatography, 343-344
Transuranium elements
  early identification and separation methods, 69
  solvent extraction, 20-50
  See also Actinides
Tributylphosphate, 391-395

Trioctylphosphine oxide 338-341, 352-353, 355f-360, 393f
TRPO process. *See* Neutral extractant processes, trialkylphosphine oxide mixture
TRU process. *See* Neutral extractant processes, diphenyl-N,N-di-n-butylcarbamoylmethylphosphine oxide
TRUEX process. *See* Neutral extractant processes, octyl (phenyl)-N,N-diisobutylcarbamoylmethylphosphine oxide
Tunnel type inorganic ion exchangers, 168-182
Typha latifolia (cattails), biologically-generated material for metal-ion bonding, 262, 268-269f

U

Ultrafiltration centrifugal units, 302-304
Ultrafiltration membrane efficiency, 296
Ultrafiltration membrane with polymer leakage, 300-301
Ultrafiltration studies
 applied pressure effects, 288
 electrolyte effect, 289-290
 separation behavior at equilibrium 290-292
Ultrafiltration, water-soluble metal-binding polymers, 294-330
Uranium
 extraction from nitric acid solutions with supercritical CO_2, 392-395
 on-line back extraction with supercritical CO_2, 394
 purification by gaseous diffusion process, 17
 removal from tailings, fluorinated diketones used in supercritical fluid extraction, 397-399
 uptake by tri-n-butyl phosphate-loaded Amberlite XAD-4, 235-236, 240-241
 See also Extraction chromatography, process-scale application
Uranyl extraction by supercritical CO_2 containing tributyl phosphate, 391-395

V

Volatility separation processes, 17

W

Waste from electroplating minimization, polymer filtration, 317-321f
Waste, nuclear. *See* Nuclear waste treatment, solvent extraction
Water hydrogen bonded structure, 60-62, 64
Water soluble metal-binding polymers, 294-330

X

XAD resins. *See* Macroporous styrene-divinylbenzene polymers

Y

Yttrium, chromatographic separation from strontium, 189
Ytterbium. *See* Lanthanides

Z

Zirconium preseparation by counter-current chromatography, 341-342
Zinc in process-scale application, extraction chromatography, 246-247t
Zinc removal from waste at Millstone nuclear power plant, 214
Zinc sorption by HDEHP-loaded Amberlite XAD-2, 235-236

Bestsellers from ACS Books

The ACS Style Guide: A Manual for Authors and Editors (2nd Edition)
Edited by Janet S. Dodd
470 pp; clothbound ISBN 0–8412–3461–2; paperback ISBN 0–8412–3462–0

Writing the Laboratory Notebook
By Howard M. Kanare
145 pp; clothbound ISBN 0–8412–0906–5; paperback ISBN 0–8412–0933–2

Career Transitions for Chemists
By Dorothy P. Rodmann, Donald D. Bly, Frederick H. Owens, and Anne-Claire Anderson
240 pp; clothbound ISBN 0–8412–3052–8; paperback ISBN 0–8412–3038–2

Chemical Activities (student and teacher editions)
By Christie L. Borgford and Lee R. Summerlin
330 pp; spiralbound ISBN 0–8412–1417–4; teacher edition, ISBN 0–8412–1416–6

Chemical Demonstrations: A Sourcebook for Teachers, Volumes 1 and 2, Second Edition
Volume 1 by Lee R. Summerlin and James L. Ealy, Jr.
198 pp; spiralbound ISBN 0–8412–1481–6
Volume 2 by Lee R. Summerlin, Christie L. Borgford, and Julie B. Ealy
234 pp; spiralbound ISBN 0–8412–1535–9

The Internet: A Guide for Chemists
Edited by Steven M. Bachrach
360 pp; clothbound ISBN 0–8412–3223–7; paperback ISBN 0–8412–3224–5

Laboratory Waste Management: A Guidebook
ACS Task Force on Laboratory Waste Management
250 pp; clothbound ISBN 0–8412–2735–7; paperback ISBN 0–8412–2849–3

Reagent Chemicals, Eighth Edition
700 pp; clothbound ISBN 0–8412–2502–8

Good Laboratory Practice Standards: Applications for Field and Laboratory Studies
Edited by Willa Y. Garner, Maureen S. Barge, and James P. Ussary
571 pp; clothbound ISBN 0–8412–2192–8

For further information contact:
Order Department
Oxford University Press
2001 Evans Road
Cary, NC 27513
Phone: 1-800-445-9714 or 919-677-0977

Highlights from ACS Books

Desk Reference of Functional Polymers: Syntheses and Applications
Reza Arshady, Editor
832 pages, clothbound, ISBN 0–8412–3469–8

Chemical Engineering for Chemists
Richard G. Griskey
352 pages, clothbound, ISBN 0–8412–2215–0

Controlled Drug Delivery: Challenges and Strategies
Kinam Park, Editor
720 pages, clothbound, ISBN 0–8412–3470–1

Chemistry Today and Tomorrow: The Central, Useful, and Creative Science
Ronald Breslow
144 pages, paperbound, ISBN 0–8412–3460–4

Eilhard Mitscherlich: Prince of Prussian Chemistry
Hans-Werner Schutt
Co-published with the Chemical Heritage Foundation
256 pages, clothbound, ISBN 0–8412–3345–4

Chiral Separations: Applications and Technology
Satinder Ahuja, Editor
368 pages, clothbound, ISBN 0–8412–3407–8

Molecular Diversity and Combinatorial Chemistry: Libraries and Drug Discovery
Irwin M. Chaiken and Kim D. Janda, Editors
336 pages, clothbound, ISBN 0–8412–3450–7

A Lifetime of Synergy with Theory and Experiment
Andrew Streitwieser, Jr.
320 pages, clothbound, ISBN 0–8412–1836–6

Chemical Research Faculties, An International Directory
1,300 pages, clothbound, ISBN 0–8412–3301–2

For further information contact:
Order Department
Oxford University Press
2001 Evans Road
Cary, NC 27513
Phone: 1-800-445-9714 or 919-677-0977
Fax: 919-677-1303